P. Gómez-Romero, C. Sanchez
Functional Hybrid Materials

Related Titles from WILEY-VCH

Ajayan, P., Schadler, L. S., Braun, P. V.

Nanocomposite Science and Technology

2003
ISBN 3-527-30359-6

Caruso, F.

Colloids and Colloid Assemblies

2003
ISBN 3-527-30660-9

Decher, G., Schlenoff, J. B.

Multilayer Thin Films
Sequential Assembly of Nanocomposite Materials

2003
ISBN 3-527-30440-1

Komiyama, M., Takeuchi, T., Mukawa, T., Asanuma, H.

Molecular Imprinting
From Fundamentals to Applications

2003
ISBN 3-527-30569-6

Krenkel, W.

High Temperature Ceramic Matrix Composites

2001
ISBN 3-527-30320-0

Manners, I.

Synthetic Metal-containing Polymers

2004
ISBN 3-527-29463-5

Functional Hybrid Materials

edited by
Pedro Gómez-Romero, Clément Sanchez

WILEY-VCH

WILEY-VCH Verlag GmbH & Co. KGaA

The Editors of this Volume

Pedro Gómez-Romero
Materials Science Institute of Barcelona
(ICMAB)
Campus della UAB, 08193, Bellaterra
Barcelona
Spain

Clément Sanchez
Laboratoire de Chimie de la
Matière Condensée (CMC)
4, place Jussieu
75252 Paris cedex 05
France

Library of Congress Card No.: applied for

British Library Cataloguing-in-Publication Data: A catalogue record for this book is available from the British Library.

Bibliographic information published by Die Deutsche Bibliothek
Die Deutsche Bibliothek lists this publication in the Deutsche Nationalbibliografie; detailed bibliographic data is available in the Internet at <http://dnb.ddb.de>

Printed on acid-free paper.

Composition, Printing and Bookbinding
Druckhaus »Thomas Müntzer«,
Bad Langensalza
Cover Design Schulz Grafik-Design,
Fußgönheim

Printed in the Federal Republic of Germany.

ISBN 3-527-30484-3

Table of Contents

Preface

The book you have in your hands is the result of a thrilling struggle. A struggle to depict, in a bit more than a handful of chapters, the blooming and multifaceted world of hybrid materials with functional properties and applications.

Hybrid organic-inorganic materials constitute indeed a remarkable and growing category within the world of Materials Science. A realm where engineering the combination of dissimilar components at the nanometric and molecular level leads both to new challenges and opportunities for the development of novel and improved materials. This is a field where the boundaries between molecular and extended materials blur out, a field where ceramics and polymers meet at the chemical dimension to yield new materials that go well beyond conventional composites, a domain in which nanocomposites push forward the frontier of discovery. In this exciting field, remarkable structural materials, halfway between glass and polymers have been developed. Yet, the hybrid approach also offers great opportunities for the development of functional materials, a fertile ground to harness the chemical, physical, electrochemical or biological activity of a myriad organic and inorganic components and put them to work in the materials of tomorrow.

Collecting a thorough taxonomic list of contents that could fairly represent this fascinating family of materials would be impossible. Instead, we have strived to select a few topics that would criss-cross the field revealing in some detail both a variety of materials and a variety of functional properties and applications. Thus, beginning with some historical perspective – if that is possible at all in a field that has developed in the last two or three decades-the book goes from mineral intercalates, sol-gel hybrids and polysiloxanes, to other radically different types of hybrids and approaches, such as hybrids based on conducting polymers. Also very varied are the functional properties and multifunctional combinations and applications you will find in these chapters, ranging from optical or magnetic properties, to energy storage and conversion or from the wealth of electroactive materials used in sensors, batteries or solar cells, to the fascinating bioactive materials discussed in the final chapter.

We hope this impressionistic portrait of a very dynamic field will contribute to give the reader a feeling of the great potential, the multiple possibilities and the many promising trends behind the development of functional hybrid materials.

August 2003 *Pedro Gómez-Romero Clément Sanchez*

List of Contributors

Pierre Audebert
Laboratoire de Photophysique et
Photochimie Supramoléculaires et Macromoléculaires
Ecole Normale Supérieure de Cachan
61 Avenue du Président Wilson
94230 Cachan
France
audebert@ppsm.ens-cachan.fr

Jean-Pierre Boilot
Chimie du Solide
Physique de la Matière Condensée
Ecole Polytechnique
91128 Palaiseau
France
jean-pierre.boilot@polytechnique.fr

Bruno Boury
Laboratoire de Chimie Moléculaire et Organisation du Solide
UMR 5637 CNRS
Université de Montpellier II
Place Eugène Bataillon
34095 Montpellier cedex 5
France
boury@crit.univ-montp2.fr

Frédéric Chaput
Chimie du Solide
Physique de la Matière Condensée
Ecole Polytechnique
91128 Palaiseau
France
frederic.chaput@polytechnique.fr

Thibaud Coradin
Chimie de la Matière Condensée
CNRS/UPMC UMR 7574
4 place Jussieu
75252 Paris Cedex 05
France
coradin@ccr.jussieu.fr

Eugenio Coronado
Instituto de Ciencia Molecular
Departamento de Quimica Inorganica
Universidad de Valencia
Dr. Moliner 50
46100 Burjassot
Valencia
Spain
eugenio.coronado@uv.es

Robert Jean-Pierre Corriu
Laboratoire de Chimie Moléculaire et Organisation du Solide
UMR 5637 CNRS
Université de Montpellier II
Place Eugène Bataillon
34095 Montpellier cedex 5
France
corriu@crit.univ-montp2.fr

Marc Drillon
IPCMS-GMI
23, rue du Loess
67037 Strasbourg Cedex
France
drillon@ipcms.u-strasbg.fr

José Ramon Galán-Mascarós
Instituto de Ciencia Molecular
Departamento de Quimica Inorganica
Universidad de Valencia
Dr. Moliner 50
46100 Burjassot
Valencia
Spain
jose.r.galan@uv.es

Pedro Gómez-Romero
Materials Science Institute of Barcelona (ICMAB) (CSIC)
Campus de la UAB
08193 Bellaterra
Barcelona
Spain
pedro.gomez@icmab.es

Nicola Hüsing
Technische Universität Wien
Institut für Anorganische Chemie
Getreidemarkt 9/165
A-1060 Wien
Austria
nhuesing@mail.zserv.tuwien.ac.at

Monica Lira-Cantu
ExxonMobil Research & Engineering
1545 Route 22 East (LA244)
Annandale, NJ 08801
USA
monica_m_lira-cantu@email.mobil.com

Bénédicte Lebeau
Laboratoire de Matériaux Minéraux
CNRS UPRES-A 7016 - ENSCMu
3 rue Alfred Werner
68093 Mulhouse Cedex
France
b.lebeau@uha.fr

Jacques Livage
Chimie de la Matière Condensée
CNRS/UPMC UMR 7574
4 place Jussieu
75252 Paris Cedex 05
France
livage@ccr.jussieu.fr

Douglas A. Loy
Hybrid Organic-Inorganic Materials Group
Sandia National Laboratories
Albuquerque, NM 87185-0888
USA
daloy@sandia.gov

David B. Mitzi
IBM Research Division,
Thomas J. Watson Research Center
P.O. Box 218
Yorktown Heights, NY 10598
USA
dmitzi@us.ibm.com

Joël Moreau
ENC Montpellier
Hétérochimie Moléculaire et Macromoléculaire
8 rue de L'Ecole Normale
34296 Montpellier cedex 5
France
jmoreau@cit.enscm.fr

Pierre Rabu
IPCMS-GMI
23, rue du Loess
67037 Strasbourg Cedex
France
rabu@ipcms.u-strasbg.fr

Eduardo Ruiz-Hitzky
Instituto de Ciencia de Materiales de Madrid (CSIC)
Cantoblanco
28049 Madrid
Spain
eduardo@icmm.csic.es

Francisco Romero
Instituto de Ciencia Molecular
Departamento de Quimica Inorganica
Universidad de Valencia
Dr. Moliner 50
46100 Burjassot
Valencia
Spain
fmrm@uv.es

Cécile Roux
Chimie de la Matière Condensée
CNRS/UPMC UMR 7574
4 place Jussieu
75252 Paris Cedex 05
France
roux@ccr.jussieu.fr

Clément Sanchez
Laboratoire de Chimie de la Matière Condensée
CNRS/UPMC UMR 7574
4, place Jussieu
75252 Paris cedex 05
France
clems@ccr.jussieu.fr

Ulrich Schubert
Technische Universität Wien
Institut für Anorganische Chemie
Getreidemarkt 9/153
A-1060 Wien
Austria
uschuber@mail.zserv.tuwien.ac.at

Kenneth J. Shea
Department of Chemistry
University of California
Irvine, CA 92697-2025
USA
kjshea@uci.edu

Alain Walcarius
LCPME-CNRS
405 rue de Vandoeuvre
54600 Villers les Nancy
France
walcariu@lcpe.cnrs-nancy.fr

1
Hybrid Materials, Functional Applications. An Introduction

Pedro Gómez-Romero and Clément Sanchez

1.1
From Ancient Tradition to 21st Century Materials

In 1946, at a site in eastern Chiapas (Mexico) known as Bonampak (painted walls) a startling archaeological discovery was made. This ancient Maya site contained an impressive collection of fresco paintings characterized by bright blue and ochre colors that had been miraculously preserved (Figure 1.1). A specially striking feature of these wall paintings was precisely their vivid blue hues, characteristic of what turned out to be an hitherto unknown pigment which came to be known as Maya blue [1].

In addition to its beautiful tones, that seemed to span all the shades of the Caribbean Sea, the most remarkable feature of Maya blue was its durability. Despite the unavoidable deterioration of the Bonampak painted scenes, that particular blue pigment had withstood more than twelve centuries of a harsh jungle environment looking almost as fresh as when it was used in the 8th century. Maya blue is indeed a robust pigment, not only resisting biodegradation, but showing also unprecedented stability when exposed to acids, alkalis and organic solvents.

Only after half a century from its archaeological discovery and not without scientific controversy [1] could sophisticated analytical techniques uncover the secret of Maya blue. The pigment is not a copper mineral, nor is it related to natural ultramarine, ground Lapis Lazuli or Lazurite as originally thought. Maya blue is a hybrid organic-inorganic material with molecules of the natural blue dye known as indigo encapsulated within the channels of a clay mineral known as palygorskite [1]. It is a man-made material that combines the color of the organic pigment and the resistance of the inorganic host, a synergic material, with properties and performance well beyond those of a simple mixture of its components.

Maya blue is a beautiful example of a remarkable hybrid material and a very old one to be sure, but its conception was most likely the fruit of a fortunate accident, an ancient serendipitous discovery.

More than twelve centuries later, when a deep knowledge of atomic and molecular structure is replacing trial and error tradition in the design of novel useful

Gómez-Romero: Organic-Inorganic Materials. Pedro Gómez-Romero and Clément Sanchez

ISBN: 3-527-30484-3

Fig. 1.1 Mayan wall paintings like this at Bonampak are twelve centuries old. Yet one of their most remarkable characteristics is the good preservation of tones of blue. This so called Maya blue pigment is indeed a hybrid organic-inorganic material formed by indigo dye molecules entrapped in a palygorskite clay mineral, a synergic nanocomposite material that has passed with excellent marks the test of centuries (courtesy Professor Constantino Reyes-Valerio)

materials, when our refined analytical techniques have allowed us to understand the true nature and structure of this pigment, Maya blue stands as an inspiration and a challenge for the designers of novel hybrid materials, for the explorers of a technological territory with quickly expanding frontiers of which this book wants to provide an account.

Aside from serendipity – the fruits of which should never be undervalued – the deliberate effort to combine properties of organic and inorganic components in a single composite material is an old challenge starting with the beginning of the industrial era. Some of the earliest and best known organic-inorganic admixtures are certainly derived from the paint and polymer industries, where inorganic pigments or fillers are dispersed in organic components (solvents, surfactants, polymers, etc.) to yield or improve optical and mechanical properties. However, the concept of "hybrid organic-inorganic" materials has more to do with chemistry than with physical mixtures. Thus, as the size of interacting particles gets reduced in going from mixtures to composite materials, the importance of the interface in determining final properties grows, and as we move towards nanocomposite materials, where components interact at a molecular level, the concept of organic-inorganic composites gets a new dimension, a chemical dimension.

The development of hybrid organic-inorganic materials stemmed from several different areas of chemistry, including intercalation chemistry (see Chapter 2), but exploded only very recently with the birth of soft inorganic chemistry processes ("Chimie Douce"), where mild synthetic conditions open a versatile access to chemically designed hybrid organic-inorganic materials [2, 3]. Later on, research shifted towards more sophisticated nanocomposites with higher added values [4, 5]. Nowadays the field of organic-inorganic materials has grown to include a large variety of types, extending to other fields as diverse as molecular and supramolecular materials or polymer chemistry [6]. Furthermore, a very significant trend has been the growing interest in functional hybrids, which broadens the field even further. Thus, in addition to structural hybrid materials bringing the best of glass and plastics together, there is a quickly expanding area of research on functional materials in which mechanical properties are secondary – though certainly not unimportant – and the emphasis is on chemical, electrochemical, or biochemical activity, as well as on magnetic, electronic, optical or other physical properties, or a combination of them [6a].

Numerous new applications in the field of advanced materials science are related to functional hybrids. Thus, the combination at the nanosize level of active inorganic and organic or even bioactive components in a single material has made accessible an immense new area of materials science that has extraordinary implications in the development of multi-functional materials [2–6]. The chemical nature of this emerging class of hybrids varies wildly, from molecular and supramolecular adducts [7] to extended solids, mineral or biomineral phases [8]. These functional hybrids are considered as innovative advanced materials, and promising applications are expected in many fields: optics, electronics, ionics, energy storage and conversion, mechanics, membranes, protective coatings, catalysis, sensors, biology, etc. [2–6]. Many interesting new materials have already been prepared with mechanical properties tunable between those of glasses and those of polymers, with improved optical properties, or with improved catalytic or membrane based properties [4, 5]. For example, hybrid materials having excellent laser efficiencies and good photostability [9a], very fast photochromic response [9b], very high and stable second order non-linear optical response [9c], or being original pH sensors [9d] and electroluminescent diodes [9e] have been reported in the past five years. And some hybrid products have already entered the applied field and the market. Examples include organically doped sol-gel glassware sold by Spiegelau [10a], sol-gel entrapped enzymes sold by Fluka [10b], or the one million TV sets sold annually by Toshiba, the screens of which are coated with hybrids made of indigo dyes embedded in a silica/zirconia matrix [10c]; interestingly, a 21st century material which brings us echoes of ancient Maya Blue.

1.2
Hybrid Materials. Types and Classifications

When it comes to formal classifications hybrid materials tend to resist rigid categorizing. Their variety is too large – and growing – to allow for a systematic grouping criterion. Figure 1.2 tries to convey this variety by showing examples of general types of hybrids spreading on a field of organic and inorganic dimensions. In this scheme material types are arranged according to the approximate dimensions of their organic and inorganic components. The limited space prevents an exhaustive list of materials and only several representative types are shown. Yet, this visual arrangement provides a first general overview of the area, spanning from molecular to extended organic-inorganic combinations. In this respect the graph also shows the greater richness of the field in the twilight region of supramolecular and nanostructured materials, forming a broad continuum between molecular and solid state chemistry.

The bidimensionality of this graph could suggest some type of classification according to the nature of the predominant phase in the hybrid, i.e. organic-inorganic vs. inorganic-organic materials depending on whether the extended, host or matrix phase were organic or inorganic respectively. Such classification has been conveniently used to categorize a particular type of polymer-based hybrid (see Chapter 7) although it could be difficult to generalize due to the abundance of intermediate cases and to the indistinct use of both terms in the literature, where the label organic-inorganic is most commonly used in a generic way.

On the other hand, a classification most widely used for all sorts of hybrid materials relies on the nature of interaction between organic and inorganic components.

The particular nanostructure, the degree of organization and the properties that can be obtained for hybrid materials certainly depend on the chemical nature of their components, but they are also heavily influenced by the interaction between these components. Thus, a key point for the design of new hybrids is the tuning of the nature, the extent and the accessibility of the inner interfaces. As a consequence, the nature of the interface or the nature of the links and interactions

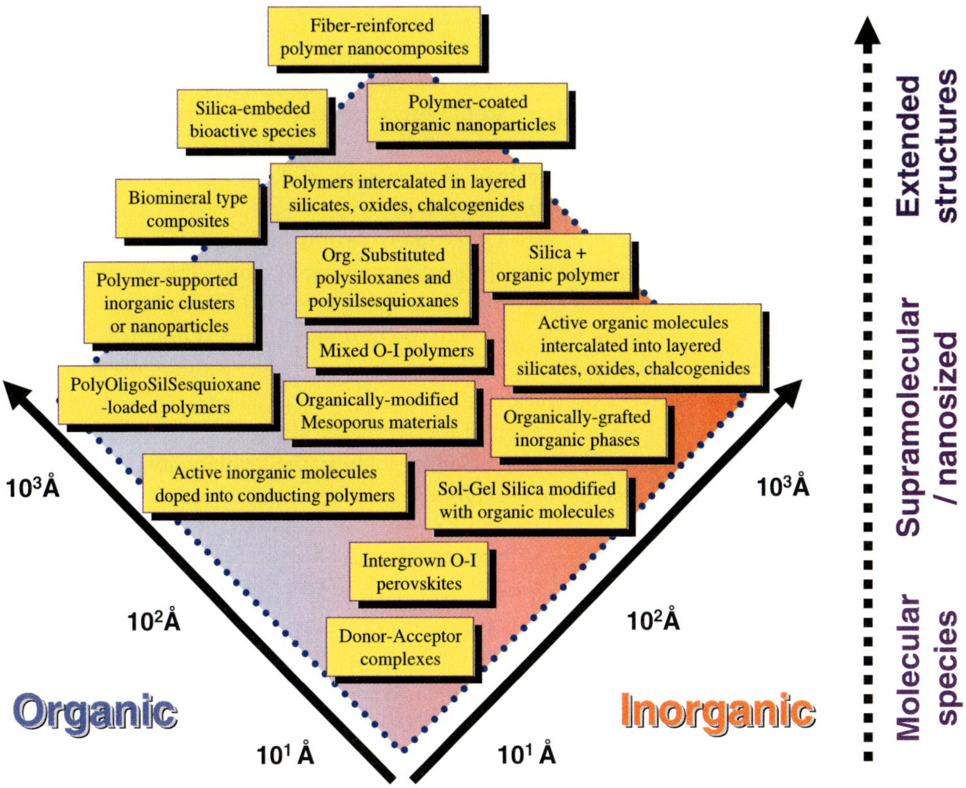

Fig. 1.2 The field of hybrid organic-inorganic materials has bloomed at the interface of many conventional disciplines, and is producing an amazing variety of materials and applications, ranging from molecular and supramolecular structures, to cluster-polymer adducts, sol-gel hybrids, or to nanocomposite materials based on extended phases

exchanged by the organic and inorganic components has been used to categorize these hybrids into two main different classes [3e, h, i]. Class I corresponds to all the systems where no covalent or iono-covalent bonds are present between the organic and inorganic components. In such materials, the various components only exchange weak interactions (at least in terms of orbital overlap) such as hydrogen bonding, van der Waals contacts, π–π interactions or electrostatic forces. On the contrary, in class II materials, at least a fraction of the organic and inorganic components are linked through strong chemical bonds (covalent, iono-covalent or Lewis acid-base bonds).

The chemical strategy followed for the construction of class II hybrid networks depends of course on the relative stability of the chemical links that associate the different components. Thus, under hydrolytic conditions Sn–C_{sp3} and Si–C_{sp3} are usually stable bonds that can be used for organic functionalization whereas for transition metal cations complexing organic ligands (such as carboxylic acids, phosphonates, hydroxyacids, polyols or betadiketones etc...) could be used to anchor organic components [3e, 2e, 19].

Finally, the obvious classification of materials according to their properties and applications and, in particular, into the broad groups of structural and functional materials will help to put in perspective the scope of this book, which, notwithstanding the importance of mechanical properties, will put the emphasis on functionality and on functional hybrid materials.

The book includes a solid series of chapters dealing both with the chemistry and design of hybrids as well as with properties and applications. The emphasis goes from the former to the latter as we go from the first to the last chapters of the book but the reader will find a systematic attempt to bridge chemical design with physical properties and final applications in each single chapter, spanning overall a wide range of different types of hybrids and their applications. Thus, the book includes chapters devoted to the description of the synthesis, structure and chemical nature of several major kinds of hybrids, including intercalation compounds, sol-gel nanocomposite hybrids, polymer-based hybrids, or donor acceptor molecular materials as well as chapters dealing with the design of mesoporous hybrid materials and derivatives. Sol-gel chemistry of hybrids [3, 11] and organized matter sol-gel chemistry [12–16] have been reviewed extensively very recently [3s, 16] and consequently will not be reported with special detail in the present book. On the other hand, properties and applications are well covered by several chapters dealing with mechanical, optical, electrochemical, magnetic and multifunctional properties, as well as specific applications such as energy storage and solar energy conversion, electroanalytical, magnetic or microelectronics applications, to finish with a chapter on the novel and fascinating bioactive hybrid materials.

1.3 General Strategies for the Design of Functional Hybrids

Independently of the types or applications, and in addition to the nature of the interface between organic and inorganic components, a second important feature in the tailoring of hybrid networks concerns the chemical pathways that are used to design a given hybrid material. But, as has been hinted in the previous section, the design and synthesis of hybrid materials depend markedly on the type of hybrid sought. Class I and II hybrids differ radically in the type of synthetic approaches adequate to their successful preparation. In the same way, the widely varied types of hybrids shown in Figure 1.2 will require equally varied strategies for their synthesis.

Yet, general strategies can be considered within each subfield. For instance, we could point out several general approaches for the synthesis of sol-gel derived hybrid materials. These main chemical routes are schematically represented in Figure 1.3.

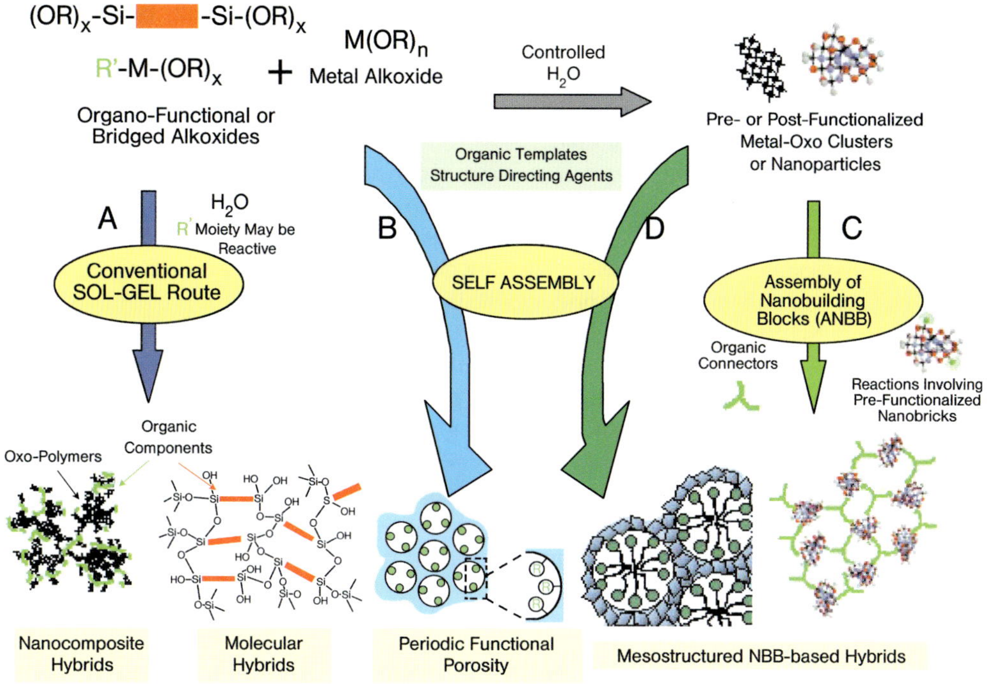

Fig. 1.3 a) One key factor in the development of hybrid materials is the understanding and control of synthetic mechanisms and approaches, which allows the design of tailor-made materials with predictable properties for specific applications. This figure summarizes several general approaches for the design of sol-gel derived hybrid materials (see text)

Path A corresponds to conventional sol-gel chemistry. Hybrid networks are obtained through hydrolysis of organically modified metal alkoxides or metal halides condensed with or without simple metallic alkoxides. Examples of such compounds are $R'_nSi(OR)_{4-n}$ (n = 1, 2) $OR_3Si{-}R'{-}SiOR_3$, or $R'_nSn(OR)_{4-n}$, with R′ being a simple non-hydrolyzable group. R' will have a network modifying effect if it contains for example a phenyl, an alkyl group or an organic dye. R′ will act as a network former if it bears any reactive group which can, for example, polymerize or copolymerize, (e. g. pyrrol, methacryl, epoxy or styryl groups) or $M(OR)_{m-n}(LZ)_n$, (where LZ is a functional complexing organic ligand with L an anchoring function and Z a general organic group)[19].

The solvent may or may not contain a specific organic molecule, a biocomponent or polyfunctional polymers that can be crosslinkable or that can interact or be trapped within the inorganic components through a large set of fuzzy interactions (H-bonds, π–π interactions, Van der Waals). These strategies are simple, low cost and yield amorphous nanocomposite hybrid materials. These materials that exhibit an infinity of microstructures can be transparent and easily shaped as films or bulks. However, they are generally polydisperse in size and locally heterogeneous in chemical composition.

Better understanding and control of the local and semi-local structure of these materials and their degree of organization are important issues, especially if tailored properties are sought for.

Five main approaches may be conceived to achieve such a control of the materials structure; they are schematized in Figure 1.3:

1. The use of **bridged precursors of silsesquioxanes** $X_3Si–R'–SiX_3$ (R′ is an organic spacer, X=Cl, Br, –OR), following the route A, allows the making of homogeneous molecular hybrid organic-inorganic materials [17].
2. **Self assembling procedures** (route B) [12–16]:
   In the last ten years, a new field has been explored, which corresponds to the organization or the texturation of growing inorganic or hybrid networks, templated by organic structure-directing agents (Figure 1.3, routes B, D). The success of this strategy is also clearly related to the ability that materials scientists have to control and tune hybrid interfaces. In this field, hybrid organic-inorganic phases are very interesting, due to the versatility they demonstrate in the building of a whole continuous range of nanocomposites, from ordered disper-

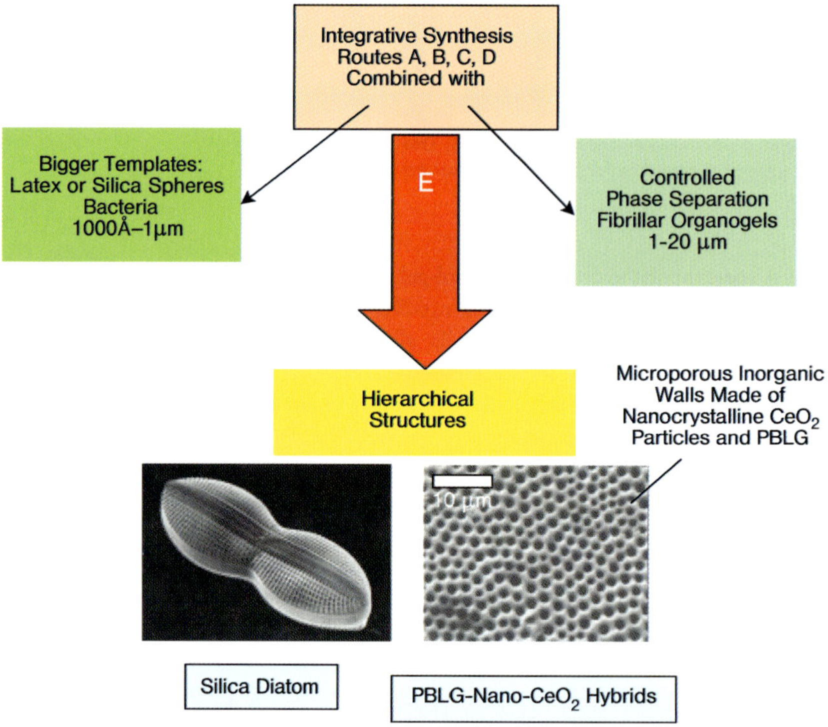

Fig. 1.3 b)

sions of inorganic bricks in a hybrid matrix to highly controlled nanosegregation of organic polymers within inorganic matrices. In the latter case, one of the most striking examples is the synthesis of mesostructured hybrid networks (routes B and D).

3. **The assembling of well-defined nanobuilding blocks** (NBB, route C) [3s]:
A suitable method to reach a better definition of the inorganic component consists in the use of perfectly calibrated preformed objects that keep their integrity in the final material. These NBB can be clusters, organically pre- or post- functionalized nanoparticles (metallic oxides, metals, chalcogenides, etc ...), nano-core-shells [3s, 11] or layered compounds able to intercalate organic components [18]. NBB can be capped with polymerizable ligands or connected through organic spacers, like telechelic molecules or polymers, or functional dendrimers (Figure 1.3, route C) [2m]. The use of highly pre-condensed species presents several advantages:
They exhibit a lower reactivity towards hydrolysis or attack of nucleophilic moieties than metal alkoxides.
The nanobuilding components are nanometric and monodispersed, and with perfectly defined structures, which facilitate the characterization of the final materials.
The variety found in the nanobuilding blocks (nature, structure, and functionality) and links allows one to build an amazing range of different architectures and organic-inorganic interfaces, associated with different assembling strategies. Moreover, the step-by-step preparation of the materials usually allows for a high degree of control over their semi-local structure.
4. **The combination of self-assembly and NBB approaches** (route D)[3s]:
Strategies combining the nanobuilding blocks approach with the use of organic templates that self-assemble and allow one to control the assembling step are also appearing (Figure 1.3, route D). This combination between the "nanobuilding block approach" and "templated assembling" will have a paramount importance in exploring the theme of "synthesis with construction". Indeed, such materials exhibit a large variety of interfaces between the organic and the inorganic components (covalent bonding, complexation, electrostatic interactions, etc.). These NBB with tunable functionalities can, through molecular recognition processes, permit the development of a new vectorial chemistry.
5. **Integrative synthesis** (route E) [13–16]:
The strategies reported above mainly allow control of the design and the assembly of hybrid materials in the 1 Å to 500 Å range. Recently, micro-molding methods have been developed, in which the use of controlled phase separation phenomena, emulsion droplets, latex beads, bacterial threads, colloidal templates or organogelators leads to controlling the shapes of complex objects on the micron scale [16]. The combination between these strategies and those described along routes A, B, C and D allows the construction of hierarchically organized materials, in terms of structure and functions [16, 8c]. Such synthesis procedures are inspired by those observed to take place in natural systems for some hundreds of million years. Learning the "*savoir faire*" of hybrid living systems

and organisms from understanding their rules and transcription modes could enable us to design and build ever more challenging and sophisticated novel hybrid materials.

1.4 The Road Ahead

Looking forward to the 21st century, nano-sciences will be, as well as biology, one of the fields that will contribute to a high level of scientific and technological development. Hybrid (organic-bio-inorganic) materials must play a major role in the development of advanced functional materials.

Nowadays the molecular approaches of solid state chemistry and nanochemistry have reached a high level of sophistication. Today the synthesis of many or any organic ligands or molecules, coordination metal complexes, functional organo or functional metalo-organic precursors, functional nanobuilding units carrying magnetic, electrical, optical or catalytic properties, is very close to being mastered.

On the other hand, a large amount of research has been carried out to obtain organic templates (surfactants, dendrimers, organogelators, polymers, block copolymers, multifunctional organic connectors, biopolymers etc ...) and to understand and rationalize their physicochemical properties. Indeed, many research programs or research actions have been devoted to OMS (Organized Molecular Systems) or OPS (Organized Polymeric Systems).

As a consequence, nowadays, chemists can practically tailor-make any molecular species from molecules to clusters or even to nanosized particles, nanolamellar compounds, nanotubes etc. Clusters are mainly used as model compounds while nanoobjects can directly enter into more applied research fields.

In the near future, original materials will be designed through the synthesis of new hybrid nanosynthons (hybridons), selectively tagged with complementary connectivities, allowing for the coding of hybrid assemblies presenting a spatial ordering at different length scales. Hybridons carrying chirality or/and dissymmetry, and multiple or complementary functionalities will open new pathways for the synthesis of these materials.

Numerous scientific breakthroughs can be expected in this field through a closer involvement of skilled chemists in original pathways of materials processing. The synergy between chemistry and chemical engineering will permit access to materials having complex structures allowing a high degree of integration.

In particular, the synthesis and construction of materials through the simultaneous use of self-assembly processes and morphosynthesis (exploiting chemical transformation in spatially restricted reaction fields) together with external factors like gravity, electrical or magnetic fields, mechanical stress, or even through the use of strong compositional flux variations of the reagents during the synthesis (open systems) are particularly interesting to explore. The chemical strategies offered by such coupled processes allow, through an intelligent and tuned coding, to develop a new vectorial chemistry, able to direct the assembling of a large variety of structurally well-defined nanoobjects into complex architectures. These bio-inspired strategies trying, in some naive way, to mimic the growth processes occurring in biomineralization will open a field of opportunities for designing innovative multiscale structured hybrid materials (from nanometer to millimeter scale), hierarchically organized in terms of structure and functions.

The strong interest of the field of functional hybrid organic-inorganic materials will be amplified in the future by the growing interest of biologists, chemists, physicists and materials scientists to fully exploit this opportunity for creating materials and devices benefiting from the best of the three realms: inorganic, organic and biologic. In addition to their high versatility, which offers a wide range of possibilities in terms of chemical and physical properties and shaping, hybrid nanocomposites present the paramount advantage of facilitating both integration and miniaturization of the devices, thereby offering a prospect of promising applications in many fields: optics, electronics, ionics, mechanics, membranes, functional and protective coatings, catalysis, sensors, biology, medicine, etc.

Finally the explosion of new strategies that we are presently witnessing for synthesis of innovative hybrid materials allows us to dream of further challenging steps in the design of intelligent materials. Thus, we can envision the possibility of building in the future advanced materials which will respond to external stimuli, adapt to their environment, self replicate, self repair or self destroy at the end of their useful life cycle. The possibilities are only limited by our imagination.

References

1 H. Van Olphen, *Science* 1966, 154, 645; b) M. Miller, *National Geographic* 1995 February), 50; c) M. J. Yacamán, L. Rendon, J. Arenas, M. C. Serra, Science 1996, 273, 223; d) L. A. Polette, N. Ugarte, M. José Yacamán, R. Chianelli, *Scientific American Discovering Archaeology*, 2000 (July–August), 46; e) G. Chiari, R. Giustetto, C. Reyes-Valerio, G. Richiardi, Poster presented at the XXX Congresso Associazione Italiana di Cristallografia, Martina Franca, October 2000; f) L. Polette, N. Ugarte, R. Chianelli, Presentation at the Workshop on Synchrotron Radiation in Art and Archaeology, SSRL, 18 October 2000.

2 H. Schmidt, A. Kaiser, H. Patzelt, H. Sholze, *Journal de Physique* 1982, 12, C9–275; b) G. L. Wilkes, B. Orler, H. H. Huang, Polymer Prep. 1985, 26, 300; c) G.-S. Sur, J. E. Mark, Eur. Polym. J. 1985, 21, 1051; d) D. Avnir, D. Levy, R. Reisfeld, J. Phys. Chem. 1984, 88, 5956; e) H. Schmidt, B. Seiferling, *Mater. Res. Soc. Symp. Proc.* 1986, *73*, 739.

3 A. Morikawa, Y. Iyoku, M. Kakimoto, Y. Imai, *J. Mater. Chem.* 1992, 2, 679; b) Y. Chujo, T. Saegusa, *Advances in Polymer Science* 1992, *100*, 11; c) B. M. Novak, *Adv. Mater.* 1993, 5, 422; d) *Proceedings of the First European Workshop on Hybrid Organic-Inorganic Materials*, eds C. Sanchez, F. Ribot, *New J. Chem.* 1994, 18; e) C. Sanchez, F. Ribot, *New J. Chem.* 1994, *18*, 1007; f) U. Schubert, N. Hüsing, A. Lorenz, Chem. Mater. 1995, 7, 2010; g) D. A. Loy, K. J. Shea, Chem. Rev. 1995, 95, 1431; h) P. Judeinstein, C. Sanchez, *J. Mater. Chem.* 1996, 6, 511; i) *Matériaux Hybrides*, Série Arago 17, Masson, Paris, 1996; j) U. Schubert, *J. Chem. Soc., Dalton Trans.* 1996, 3343; k) R. J. P. Corriu, D. Leclercq, *Angew. Chem. Int. Ed.* 1996, 35, 1420; l) R. J. P. Corriu, C. R. *Acad. Sci. Paris. Ser. II* 1998, 1, 83; m) F. Ribot, C. Sanchez, *Comments in Inorganic Chemi-*

stry 1999, 20, 327; n) C. Sanchez, F. Ribot, B. Lebeau, *J. Mater. Chem.* 1999, 9, 35; o) S. I. Stupp, P. V. Braun, *Science* 1999, 277, 1242; p) P. V. Braun, P. Osenar, V. Tohver, S. B. Kennedy, S. I. Stupp, *J. Am. Chem. Soc.* 1999, *121*, 7302; q) *Chem. Mater.* 2001, *13*, Special Issue on Nanostructured and Functional Materials, eds H. Eckert and M. Ward, and all articles therein, r) *MRS Bulletin* 2001, 26, Special Issue on Hybrid Materials, ed D.A. Loy, and all articles therein; s) C. Sanchez, G. J. A. A. Soler-Illia, F. Ribot, C. Mayer, V. Cabuil, T. Lalot, *Chem. Mater.* 2001, *13*, 3061; t) P. Gomez-Romero, M. Lira-Cantu, Hybrid Materials, in *Kirk-Othmer Encyclopedia of Chemical Technology*, Wiley Interscience, New York, 2002.

4 Better Ceramics Through Chemistry VI, in *Mater. Res. Soc. Symp. Proc.*, eds A. Cheetham, C. J. Brinker, M. MacCartney, C. Sanchez, 1994; b) Better Ceramics Through Chemistry VII: Organic/Inorganic Hybrid Materials, in *Mater. Res. Soc. Symp. Proc.*, eds B. K. Coltrain, C. Sanchez, D. W. Schaefer, G. L. Wilkes, 1996; c) Hybrid Materials, in *Mater. Res. Soc. Symp. Proc.*, eds R. Laine, C. Sanchez, C. J. Brinker, E. Gianellis, 1998; d) Organic/Inorganic Hybrid Materials *II*, in Mater. *Res. Soc. Symp. Proc.*, eds L. C. Klein, L. F. Francis, M. R. De Guire, J. E. Mark, 1999; e) Hybrid Materials, in *Mater. Res. Soc. Symp. Proc.*, eds R. M. Laine, C. Sanchez, E. Gianellis, C. J. Brinker, 2000.

5 Sol-Gel Optics I, eds J. D. Mackenzie, D. R. Ulrich, Proc. SPIE 1990, 1328; b) Sol-Gel Optics II, ed J. D. Mackenzie, *Proc. SPIE* 1992, 1758; c) Sol-Gel Optics III, ed J. D. Mackenzie, Proc. *SPIE* 1994, *2288*; d) Sol-Gel Optics IV, ed J. D. Mackenzie, Proc. *SPIE* 1997, *3136*; e) Organic-Inorganic hybrids for Photonics, eds L. G. Hubert-Pfalzgraf, S. I. Najafi, Proc. SPIE 1998, 3469; 192; f) Sol-Gel Optics V, eds B. S. Dunn, E. J. A. Pope, H. K. Schmidt, M. Yamane, *Proc. SPIE* 2000, *3943*; g) *Hybrid Organic-Inorganic Composites*, eds J. E. Mark, C. Y. C. Lee, P. A. Bianconi, American Chemical Society, Washington, DC, 1995; i) B. S. Dunn, J. I. Zink, *J. Mater. Chem.* 1991, *1*, 903; j) B. Lebeau, C. Sanchez, *Current Opinion in Solid State and Materials Science*, 1999, *4*, 11; k) J. P. Boilot, F. Chaput, T. Gacoin, L. Malier, M. Canva, A. Brun, Y. Lévy, J. P. Galaup, *C. R. Acad. Sci. Paris*, 1996, *322*, 27; l) D. Avnir, S. Braun, O. Lev, D. Levy, M. Ottolenghi, M., in *Sol-Gel Optics, Processing and Applications*, ed L. Klein, Kluwer Academic Publishers, Dordrecht, 1994; m) J. Livage, *C.R. Acad. Sci. Paris* 1996, *322*, 417; n) D. Avnir, *Acc. Chem. Res.* 1995, *28*, 328; o) K. Moller, T. Bein, *Chem. Mater.* 1998, *10*, 2950.

6 P. Gomez-Romero, *Adv. Mater.* 2001, *13*, 163; b) J. Portier, J.-H. Choy, M. A. Subramanian, *Int. J. Inorg. Mat.* 2001, *3*, 581; c) R. Gangopadhyay, A. De, *Chem. Mater.* 2000, *12*, 608; d) E. Ruiz-Hitzky, P. Aranda, *An. Quim. Int. Ed.* 1997, *93*, 197; e) M. Lira-Cantu, P. Gomez-Romero in *Recent Res. Dev. Phys. Chem.* ed S.G. Pandalai, 1997, *1*, 379; f) T. Bein, *Stud. Surf. Sci. Catal.* 1996, *102*, 295; g) L. Nazar, T. Kerr, et al., Ionic and Electronic Transport Properties of Layered Transition Metal Oxide/Conductive Polymer Nanocomposites, in *Access Nanoporous Mater.* eds T. J. Pinnavaia, M. F. Thorpe, Plenum, New York 1995, 405–427; h) H. R. Allcock, *Adv. Mater.* 1994, *62*, 106; i) E. Ruiz-Hitzky, *Adv. Mater* 1993, *55*, 334; j) P. Day, R. D. Ledsham, *Mol. Cryst. Liq. Cryst.* 1982, *86*, 163; k) P. Gomez-Romero, M. Lira-Cantu, *Adv. Mater.* 1997, *9*, 144; l) P. Gomez-Romero, N. Casan-Pastor, M. Lira-Cantu, *Solid State Ionics* 1997, *101–103*, 875; m) M. Lira-Cantu, P. Gomez-Romero, *Ionics* 1997, *3*, 194; n) M. Lira-Cantu, P. Gomez-Romero, *Chem. Mater.* 1998, *10*, 698; o) M. Lira-Cantu, K. Cuentas-Gallegos, G. Torres-Gomez, P. Gomez-Romero, *Bol. Soc. Esp. Ceram. Vidrio* 2000, *39*, 386; p) G. Torres-Gomez, P. Gomez-Romero, *Synth. Met.* 1998, *98*, 95; q) G. Torres-Gomez, M. Lira-Cantu, P. Gomez-Romero, *J. New Mater. Electrochem. Syst.* 1999, *2*, 145; r) G. Torres-Gomez, K. West, S. Skaarup, P. Gomez-Romero, *J. Electrochem. Soc.* 2000, *147*, 2513; s) M. Lira-Cantu, P. Gomez-Romero, *J. Solid State Chem.* 1999, *147*, 601; t) M. Lira-Cantu,

P. Gomez-Romero, *J. New Matls. for Electrochem. Syst.* 1999, *2*, 141; u) M. Lira-Cantu, P. Gomez-Romero, *J. Electrochem. Soc.* 1999, *146*, 2029.

7 C. R. Kagan, D. B. Mitzi, C. Dimitrakopoulos, *Science* 1999, *286*, 945; b) D. B. Mitzi, Synthesis, Structure, and Properties of Organic-Inorganic Perovskites and Related Materials, in *Progress in Inorganic Chemistry*, John Wiley & Sons, Inc., New York, 1999, 1, Vol. 48; c) D. B. Mitzi, S. Wang, C. A. Feild, C. A. Chess, A. M. Guloy, *Science* 1995, *267*, 1473; d) S. Wang, D. B. Mitzi, C. A. Feild, A. Guloy, *J. Amer. Chem. Soc.* 1995, *117*, 5297; e) M. Era, S. Morimoto, T. Tsutsui, S. Saito, *Appl. Phys. Lett.* 1994, *65*, 676; f) D. B. Mitzi, C. A. Feild, W. T. A. Harrison, A. M. Guloy, *Nature* 1994, *369*, 467; g) D. B. Mitzi, *Bull. Amer. Phys. Soc.* 1993, *38*, 116; h) E. Coronado, J. R. Galán-Mascarós, C. J. Gómez-García, J. Ensling, P.Gütlich, *Chemistry Eur. J.* 2000, *6*, 552; i) E. Coronado, J. R. Galán-Mascarós, C. Giménez-Saiz, C. J. Gómez-García, S. Triki, *J. Am. Chem. Soc.* 1998, *120*, 4671; j) M. Clemente-León, E. Coronado, P. Delhaes, J.R. Galán Mascarós, C. J. Gómez-García, C. Mingotaud, in *Supramolecular Engineering of Synthetic Metallic Materials. Conductors and Magnets*, NATO ASI Series, eds J. Veciana, C. Rovira, D. B. Amabilino, Kluwer Academic Publishers, Dordrecht 1998, 291–312, Vol. C518; k) E. Coronado, J.R. Galán Mascarós, C. Giménez-Saiz, C. J. Gómez-García, C. Ruiz-Pérez, S. Triki, *Adv. Mater.* 1996, *8*, 737; l) C. J. Gómez-García, C. Giménez-Saiz, S. Triki, E. Coronado, P. Le Magueres, L. Ouahab, L. Ducasse, C. Sourisseau, P.Delhaes, Inorg. Chem. 1995, *34*, 4139; m) J. R. Galán-Mascarós, C. Giménez-Saiz, S. Triki, C. J. Gómez-García, E. Coronado, L. Ouahab, *Angew. Chem. Int. ed. Engl.* 1995, *34*, 1460.

8 S. Mann, *Biomineralization. Principles and Concepts in Bioinorganic Materials Chemistry*, Oxford University Press, Oxford, 2001; b*) Biomineralization: Chemical and Biological Perspectives*, eds S. Mann, J. Webb, R. J. P. Williams, VCH, New York, 1989; c) *Biomimétisme et Matériaux*, ed C. Sanchez, OFTA Edition Trans. Tech., Arago, Paris, 2001; d) A. Okada, A. Usuki, *Materials Science and Engineering* 1995, *C32*, 109; e) R. Cingolani, R. Rinaldi, et al., *Physica E: Low-dimensional Systems and Nanostructures* 2002 (in press, uncorrected proof); f) Z. Ahmad, J. E. Mark, *Mat. Sc. Eng.* 1998, *C6*, 183.

9 M. Faloss, M. Canva, P. Georges, A. Brun, F. Chaput, J. P. Boilot, *Appl. Opt.* 1997, *36*, 6760; b) B. Schaudel, C. Guermeur, C. Sanchez, K. Nakatani, J. Delaire, *J. Mater. Chem.* 1997, *7*, 61; c) B. Lebeau, C. Sanchez, S. Brasselet, J. Zyss, *Chem. Mater.* 1997, *9*, 1012; d) C. Rottman, G. Grader, Y. De Hazan, S. Melchior, D. Avnir, *J. Am. Chem. Soc.* 1999, *121*, 8533; e) T. Dantas de Morais, F. Chaput, K. Lahlil, J. P. Boilot, *Adv. Mater.* 1997, *11*, 107.

10 G. Schöttner, J. Kron, K. J. Deichmann, *Sol-Gel Sci. Technol.* 1998, *13*, 183; b) Aldrich, Fluka, Sigma and Supelco Chiral Products Catalog 1997, p. 250; c) T. Itou, H. Matsuda, *Key Eng. Mater.* 1998, *67*, 150; d) M. Reetz, A. Zonta, J. Simpelkamp, *Angew. Chem. Int. Ed.*, 1995, *34*, 301.

11 E. Bourgeat-Lami, *Nanomaterials*, 2002 .? volume & page numbers? ?

12 C. T. Kresge, M. E. Leonowicz, W. J. Roth, J. C. Vartuli, J. S. Beck, *Nature*, 1992, *359*, 710; b) J. S. Beck, J. C. Vartuli, *Curr. Opin. in Solid State & Mater. Sci.* 1996, *1*, 76, and references therein.

13 S. Mann, S. L. Burkett, S. A. Davis, C. E. Fowler, N. H. Mendelson, S. D. Sims, D. Walsh, N. T. Whilton, *Chem. Mater.* 1997, *9*, 2300.

14 S. Mann, G. A. Ozin, *Nature*, 1996, *382*, 313.

15 J. Y. Ying, C. Mehnert, M. S. Wong, *Angew. Chem. Int. Engl. Ed.* 1999, *38*, 57; b) T. J. Barton, L. M. Bull, W. G. Klemperer, D. A. Loy, B. McEnaney, M. Misono, P. A. Monson, G. Pez, G. W. Scherer, J. C. Vartuli, O. M. Yaghi, *Chem. Mater.* 1999, *11*, 2633; c) D. Zhao, P. Yang, Q. Huo, B. F. Chmelka, G. D. Stucky, *Curr. Opin. in Solid State & Mater. Sci.* 1998, *3*, 111; d) C. G. Göltner, M. Antonietti, *Adv. Mater.* 1997, *9*, 431; e) C. J. Brinker, Y. Lu, A. Sellinger, H. Fan, *Adv. Mater.* 1999, *11*, 579; f) G. A. Ozin, *Chem. Commun.* 2000, 419.

16 G. J. A. A. Soler-Illia, C. Sanchez, B. Lebeau, J. Patarin, *Chem. Review* 2002, (in press).
17 K. J. Shea, D. A. Loy, O. J. Webster, *J. Am. Chem. Soc.* 1992, *114*, 6700; b) G. M. Jamison, D. A. Loy, K. J. Shea, *Chem. Mater.* 1993, *5*, 1193; c) K. M. Choi, K. Shea, *Chem. Mater.* 1993, *5*, 1067; d) R. J. P. Corriu, J. J. E. Moreau, P. Thepot, M. Wong Chi Man, *Chem. Mater.* 1996, *8*, 100; e) R. J. P. Corriu, J. J. E. Moreau, P. Thepot, M. Wong Chi Man, *Chem. Mater.* 1992, *4*, 1217; f) P. M. Chevalier, R. J. P. Corriu, J. J. E. Moreau, M. Wong Chi Man, *J. Sol-Gel Sci. Tech.* 1997, *8*, 603.
18 V. Laget, C. Hornick, P. Rabu, M. Drillon, P. Turek, R. Ziessel, *Adv. Mater.* 1998, *10*, 1024; b) D. O'Hare, in *Inorganic Materials*, eds D. W. Bruce and D. O'Hare, John Wiley and Sons, Chichester 1992, 165; c) E. Ruiz-Hitzky, *Adv. Mater.* 1993, *5*, 334.
19 C. Sanchez, J. Livage, M. Henry, F. J. Babonneau, *Non-Cryst. Solids* 1988, *100*, 650; b) J. Livage, M. Henry, C. Sanchez, *Progr. Solid State Chem.* 1988, *18*, 259.

2 Organic-Inorganic Materials: From Intercalation Chemistry to Devices

Eduardo Ruiz-Hitzky

2.1 Introduction

Nowadays the development of new materials has to be focused on the preparation of systems with properties that can be predicted and controlled, and aiming at predetermined technological applications. This means that research carried out within the field of Materials Science and Technology should be based on criteria of multidisciplinarity which allow the design and preparation of specific materials. This perspective is particularly necessary in the case of *functional materials,* in which the final goal is to make a material with specific properties that are related in general to the presence of a particular atomic or molecular group within the system. This confers a suitable functionality providing a basis for building all sort of devices.

To reach this goal a methodology must be planned in order to answer the following basic questions: which material? and, why and how to make it? In practice, appropriate methods to answer these questions could be found by considering the procedures and techniques of *molecular engineering,* that can be applied to *nanostructured solids.* The manipulation of atoms and molecules in a solid material corresponds to a scale as small as the nanometer (1 nm = 10^{-9} m) and could be included as a subdiscipline of Solid State Chemistry, named as *Nanochemistry* [1]. In this way, a spectacular challenge to modern Science and Technology of Materials consists in the synthesis of nanostructured materials taking into account their final application and expected performance as a response to technological demands. Many such necessities are associated with a predetermined *function,* and the designed materials have to be prepared by a combination of physical and chemical methods (Fig. 2.1) following the *modus operandi* characteristic of the procedures of molecular engineering [2] (Fig. 2.2).

Among nanostructured materials, certain *nanocomposites* and other organic-inorganic materials could be considered as *functional hybrid materials.* The simplest preparation routes for this type of compound are based on the controlled modification of inorganic solids by using different organic and organometallic reagents.

Gómez-Romero: Organic-Inorganic Materials. Eduardo Ruiz-Hitzky

ISBN: 3-527-30484-3

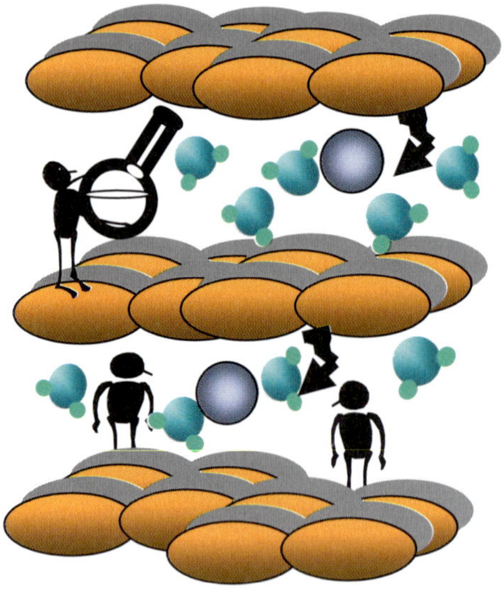

Fig. 2.1 Idealized representation of a *nanochemist* applying methods for deliberate modifications of the functional properties of a solid material. The picture includes some laborers of theoretical formation studying physical processes in the solid

The aim of this chapter is to present and discuss different strategies to synthesize hybrid materials furnished with specific functions. In this context, we intend to select some illustrative examples, derived mainly from our own experience, rather than make an exhaustive report of methods, materials and devices. The methodology considered here is related to diverse preparative aspects, ranging from intercalation processes of a large variety of organic compounds into layered solids, to the anchoring or grafting of organic functions on inorganic solid surfaces in order to obtain organic derivatives of these solids, and to the development of organic-inorganic networks by sol-gel procedures. Moreover, inclusion within the pores of inorganic matrices such as zeolites and other zeotypes, mesoporous solids such as the

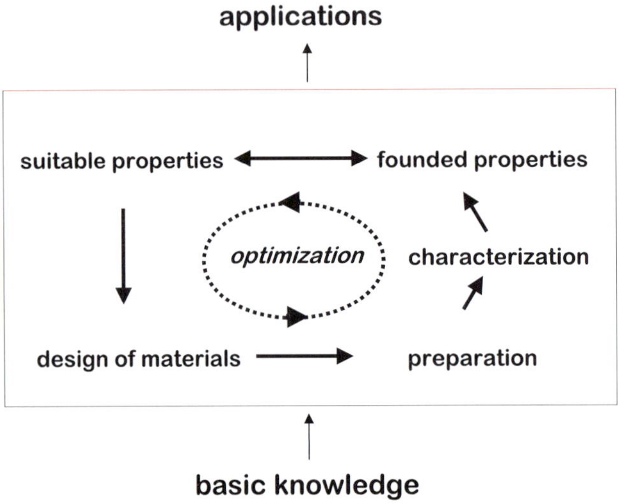

Fig. 2.2 *Modus operandi* in molecular engineering [2]

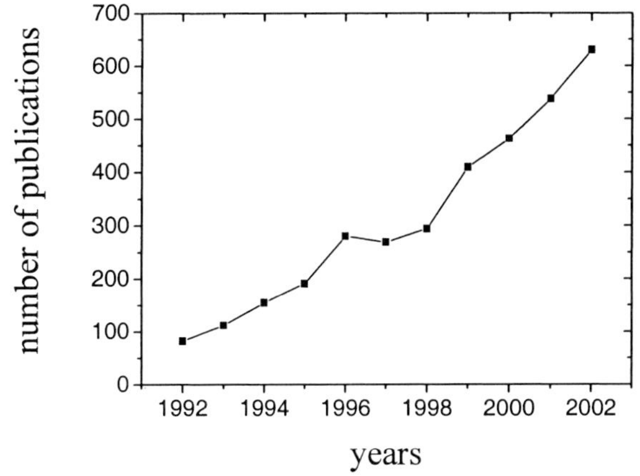

Fig. 2.3 Variation in the number of publications on organic-inorganic hybrids during the last decade (data from Materials Science Citation Index, ISI)

MCM family, or porous membranes, as well as the self-assembling of organic molecules on the surface of small metal-oxide particles, are also considered here as alternative methods for preparing hybrid systems. As a consequence of the variety of preparation methods, the number of organic-inorganic materials is continuously increasing (Fig. 2.3) and this circumstance opens in turn new ways for the development of devices with useful functions as selective sensors, adsorbents and catalysts or as improved optical, electronic or electrochemical systems.

The definition of organic-inorganic hybrid materials is not always clear, since materials containing simultaneously organic and inorganic moieties are not necessarily included within the group. This is the case, for example, for physical mixtures or combinations at the micro- and macroscale level of both types of matter, organic and inorganic. The requirement for considering a solid as an organic-inorganic hybrid material is that the combination between the two integrating parts should occur at the molecular level, i.e. at the nanometer scale. Thus, conventional hybrid systems, as for instance typical commercial polymers containing mineral charges, or pure molecular systems such as the metal salts of organic anions, or macromolecules containing small inorganic counter-ions of polymers such as $(Polyaniline)^{+}/ClO_4^{-}$, should not be included. Exceptions, related to these last materials, are the systems based on much bigger counter-ions, as for instance the polyoxometalates based on Keggin or Wells-Dawson structures which could be regarded as metal oxide clusters and which present their own chemical and electrochemical activity [3].

On the other hand, it should be clearly explained that the concept of hybrid materials is usually referred to in the literature using either of the terms *organic-inorganic* or *inorganic-organic*. As has been pointed out in Chapter 1, the two different terms have sometimes been used to indicate the nature of the predominant component among constituting materials, as for instance the predominance of the inorganic matrix over the organic part or *vice versa*. In our opinion, either one of the two is correct, but taking into account that *organic-inorganic* is the more widely used and better accepted expression we propose to use it in the future as the preferred term for the sake of consistency.

2.2 Types of Hybrid Organic-Inorganic Materials

In the preceding section several different methods for the preparation of organic-inorganic materials have been mentioned. All of these new materials have the common characteristic that they are prepared at moderate temperatures (<200 °C), typically at room temperature. Such processes give materials where the organic matter remains associated with an inorganic skeleton, being homogeneously distributed among the whole of the material. However, there are significant peculiarities from one type of hybrid to another, peculiarities that depend both on the different association energy between the organic and the inorganic moieties, and on the different microstructural arrangement of the two parts. Consequently, the associated physical and chemical properties of different types of materials can also be quite different.

A basic classification of the most common types of organic-inorganic materials normally includes the following three main groups: *intercalation compounds, organic derivatives of inorganic solids, and sol-gel hybrid materials.* An additional classification, introduced for the latter class of solids but applicable to other types of hybrids, separates hybrids into two groups, *class I* and *class II*, depending on the presence (class II) or absence (class I) of covalent strong bonds between organic and inorganic components (see Chapter 1). Moreover, there are other organic-inorganic materials not included in this classification, as for instance: (i) compounds based on amines or organo-ammonium species associated with silica or metal-oxide substrates, that are intermediates in the synthesis of zeolites, ALPOs, MCMs and other micro- and meso-porous materials, and (ii) organic polymer-based hybrids such as those formed between conducting polymers and polyoxometalate oxide-clusters. The former are normally prepared with a view to the eventual elimination of the organic matter (e.g. by combustion) to create structural intracrystalline cavities – channels, windows and nanotubes – conferring nanosized porosity to the solids that can act then as specific adsorbents and selective heterogeneous catalysts [4]. As was pointed out in Sect. 2.1, polymer-polyoxometalate adducts could be considered as a special type of hybrid, in which atomic clusters based on oxygen and transition elements such as W, Mo, V, Nb and Ta, are heteropolyanions that act as counterions of conducting polymers such as polyaniline (PANI), polypyrrole (PPy) and polythiophenes [3].

2.2.1 Intercalation Compounds

Chronologically (Table 2.1) the first organic-inorganic materials were the intercalation compounds resulting from the intracrystalline insertion of organic compounds inside the layers of certain lamellar solids. Although the term intercalation refers to reversible and topotactic insertion in 2D and 3D solids, in this contribution we will exclusively discuss the inclusion into layered inorganic hosts. As the number of potential 2D host candidates is very high and the number of possible

Tab. 2.1 Chronology of some relevant milestones showing progress in the development of hybrid organic-inorganic materials

Year	Organic-inorganic hybrids	Examples	Authors & References
1939	Clay minerals intercalated by organic cations	Montmorillonite/quaternary ammonium species	Gieseking [5]
1948	Clay minerals intercalated by neutral species	Montmorillonite/neutral polar molecules	MacEwan [7]
1953	Organic derivatives of silica	Esterification of precipitated silica with alcohols	Iller [170]
1959	Graphite oxide (GO) intercalation compounds	GO/alkylamines	Aragón et al. [171]
1961	Polymer-clay intercalation compounds	Montmorillonite/ polyacrylonitrile	Blumstein [32]
1965	Layered phosphates intercalation compounds	α-$[Zr(PO_4)2H_2 \cdot H_2O]$/ alkylamines	Michel et al. [172]
1967	Transition metal oxychlorides intercalation compounds	FeOCl/amines	Hagenmuller et al. [173]
1969	Transition metal dichalcogenides intercalation compounds	TiS_2/amides	Weiss & Ruthard [174]
1973	LDH intercalated by organic anions	$[Zn_3Al(OH)_8]$/dicarboxylates	Miyata & Kumora [18]
1964–68	Organic derivatives of silicates	Organosilanes grafted on olivine, chrysotile,...	Lentz [60], Fripiat & Mendelovici [61]
1976–78	Organic derivatives of layered phosphates	Grafting of epoxides & direct synthesis of organophosphonates in α-Zr phosphate	Yamanaka [175], Alberti [176]
1979	Intercalation compounds based on layered silicic acids	Magadiite/alkylamines	Lagaly [177]
1980	Organic derivatives of layered silicic acids	Interlamellar grafting of trimethylsilyl groups on H-magadiite	Ruiz-Hitzky & Rojo [56]
1980	Sol-gel heteropolysiloxane materials	Organically modified silicates (ORMOSILS)	Schmidt [178]
1986	Conducting polymers- 2D intercalated materials	FeOCl/polypyrrole	Kanatzidis [30]
1990	Ion-conducting polymers – 2D intercalated materials	Layered silicates/PEO	Ruiz-Hitzky & Aranda [35]
1992	Silsesquioxanes molecular composites	Bridged polysesquioxanes	Corriu et al. [83], Shea et al. [82]
1991	Metal oxides-surfactant mesophases	Silica-alkylammonium intermediates towards MCM mesoporous solids	Beck [77]
1997	Polyoxometalates doping conducting polymers	$[PMo_{12}O_{40}]^{3-}$/polyaniline	Gómez-Romero & Lira-Cantú [179]
2002	Silica grafted surfactants (SGS)	Self-assembled organosilica	Ruiz-Hitzky et al. [81]

organic molecular or ionic guest species is vast, the combination of inorganic with organic moieties is a basis on which to obtain a great variety of hybrid materials. Therefore the possibilities of preparation of this class of solids are enormous, being limited merely by the imagination. The only prerequisite is the host-guest affinity based on thermodynamic behavior and geometrical impediments (topochemistry).

2.2.1.1 **Intercalation of Ionic Species**

Among the typical 2D materials acting as host for intercalation of organic compounds, the phyllosilicates are a group of solids belonging to the clay minerals family. In particular, charged 2:1 phyllosilicates are structurally composed of two silicic tetrahedral sheets sandwiching a central octahedral sheet showing isomorphous substitutions, as for instance Al by Si in the tetrahedral layer, or Mg by Al in the octahedral sheets (Fig. 2.4). These substitutions are characteristic of a large variety of smectite minerals (montmorillonite, beidellite, hectorite, etc.) being responsible for framework charge deficiency, which is compensated by interlamellar cations. Such cations are named *exchangeable cations* as they can easily be exchanged by treatment of the clay with salt solutions. In just this way the ability of those solids to form the first organic-inorganic materials was realized. Deriving from these clay minerals, Gieseking in 1939 [5] showed that the exchangeable metal-ions located in the interlamellar space of montmorillonite may be replaced by different organic cations, such as alkylammonium, methylene blue and anilinium, in water solutions. This author, and a little later Hendricks [6], claimed that this class of guest organic species was strongly held to the negatively charged layers by electrostatic forces, although the additional contribution of Van der Waals forces between the alkyl chains and the oxygen atoms belonging to the host surface was also invoked. The term *interlamellar sorption* was introduced by MacEwan in 1948 [7] to designate the penetration of ions – and also neutral species – between the layers of a 2D solid, resulting in a reversible, one-dimensional swelling. The major structural change observed by X-ray diffraction was the increase of the interlayer distances (d_L, *basal spacing*) whose magnitude is directly related to the molecular size of the guest specie. From these diffraction data, Jordan [8] showed in 1949 the possibility of included alkylammonium species to adopt different dispositions as mono- or bi-layers with the alkyl chains lying in the parallel direction of the silicate (montmorillonite) layer. More recently, Lagaly [9] proposed different arrangements of the alkyl chains with respect to the plane defined by the silicate layer, which could be in mono- or bi-layers (Fig. 2.5) and other geometrical dispositions, as a pseudo-trimolecular arrangement. The orientation of intercalated molecules was initially studied by Serratosa [10] analyzing the dichroic effect of IR bands associated with a pyridinium ring into a 2:1 phyllosilicate such as montmorillonite. The geometry of the intercalated species could be also investigated by application of other techniques such as UV-Vis and solid state NMR spectroscopies [11–12]. Ogawa and Kuroda have published a well-focused review on organo-ammonium intercalation in layered silicates, discussing the most salient properties and potential applications derived from this type of organic-inorganic material [13].

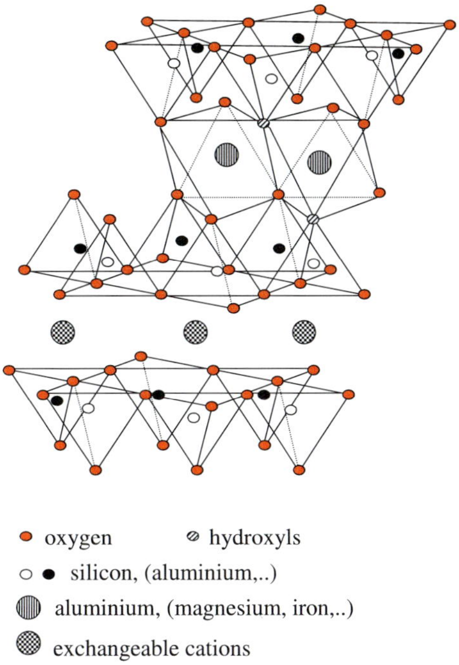

Fig. 2.4 Structural representation of a 2:1 phyllosilicate of the smectite clay minerals family

The so-called *organoclays*, commercialised under the name of Bentones, are prepared by treatment of smectites (bentonites) with aqueous solutions of different quaternary ammonium salts. The most important applications of these hybrid solids are based on the rheological behaviours developed by their dispersions in organic solvents, allowing uses as thickeners and thixotropic agents.

As we report later, more complex cationic species have also been intercalated in 2:1 charged phyllosilicates, many of them having functional behaviors as in the case of some organic dyes and charged polymers.

Similarly, other layered inorganic hosts such as V_2O_5, $VOPO_4$ and MX_2 are also able to insert cationic organic species to balance the negative electrical charge deliberately created in the inorganic skeleton. Thus, redox intercalation of quaternary ammonium species into the intracrystalline region of 2D solids containing transi-

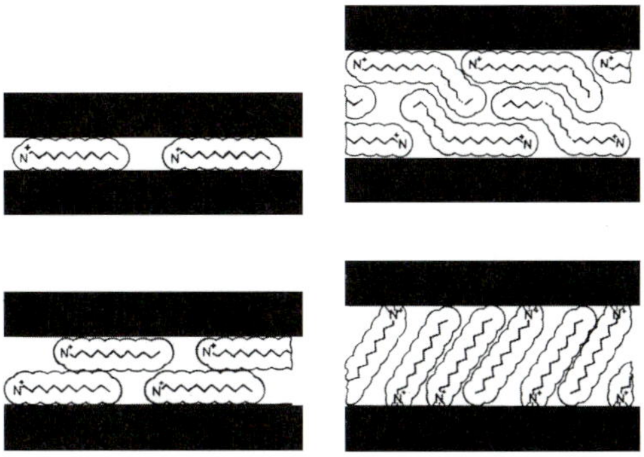

Fig. 2.5 Structural model representing different possible arrangements of alkylammonium species when they are intercalated in layer silicates, after Lagaly [9]

tion elements, as for instance vanadium (V), is an alternative procedure for inserting this class of ions. The use of mild reducing agents, such as iodide in the form of a counter-ion of the ammonium species, reduces the oxidation state of vanadium belonging to the host lattice, with the simultaneous oxidation of the I^- ions to iodine (Fig. 2.6). Following this redox mechanism, layered materials such as vanadyl phosphate or vanadium pentoxide insert alkyl- and aryl-ammonium species compensating the V^{4+} that are formed by reduction of V^{5+} ions belonging to the starting solid [14–15]. Reduction of the hosts could also be assured by using powerful reducing agents such as n-BuLi that promote the Li^+ insertion into the crystal structure of certain MX_2 transition metal dichalcogenides. The Li-containing phases react with water, giving colloidal solutions in which delaminated MX_2 species could associate organic compounds present in the water solution. In this way, species from small ionic to big macromolecular species could remain entrapped after restacking of the MX_2 elemental lamellae, giving hybrid organic-inorganic materials equivalent to conventional intercalation compounds [16].

Another approach is the intercalation of certain cationic species, such as anilinium, into 2D transition metal oxides or chalcogenides, then allowing their further transformation into polymeric species (i.e. PANI) giving hybrid materials with interesting properties. This is for instance the case for V_2O_5/PANI hybrids that show electrical and electrochemical behavior useful for applications in different devices [15]. Layered double hydroxides (LDHs) act in the opposite sense to smectite clay minerals, intercalating anionic species. These host solids are structurally formed by $[M^{3+}_x\ M^{2+}_{1-x}\ (OH)_2]^{n+}$ layers in which M^{3+} (Al, Cr, Fe, …) and M^{2+} (Mg, Zn, Fe, Cu, …) are alternately occupying octahedral sites in a brucite-type sheet. Such layers are separated by X^- anions which are in many cases associated with water molecules. These anions can be exchanged by inorganic or organic negatively charged species in solution. The natural compound of formula $Al_2Mg_6(OH)_{16}CO_3 \cdot 4H_2O$ is the hydrotalcite mineral, which is considered as the LDH of reference [17]. The interlayer X^- species could be replaced by organic sulfates, sulfonates, phosphates, carboxylates and other anions, giving very stable organic-inorganic intercalation compounds [18]. Organic compounds, including polymers, of interest for their photoactive, conductive or biological behavior have been reported as suitable species to form functional LDH based materials [19].

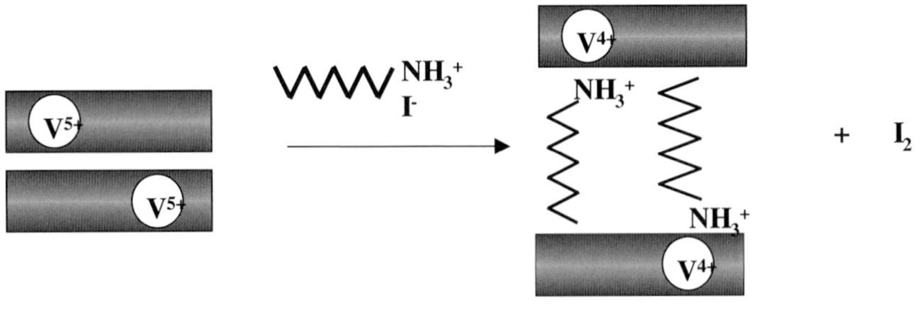

Fig. 2.6 Redox intercalation of alkylammonium species in host solids such as vanadium pentoxide or vanadyl phosphate

2.2.1.2 Intercalation of Neutral Species

The formation of organic-inorganic compounds by intercalation of neutral molecules in 2D solids was also accomplished for the first time using phyllosilicates of the clay minerals family. The first systems of this sort were reported by Bradley [20] and MacEwan [7]. The polarity of the guest molecules plays an important role in their tendency to be inserted between the inorganic layers and in the stability of this new class of resulting complexes. A large variety of compounds of very different organic functionality (alcohols, carbonyl compounds, amines, nitrogenated heterocyclics, aminoacids, etc.) are able to intercalate phyllosilicates like kaolinite, smectites and vermiculites [21–23]. Later, this phenomenon was also observed for other inorganic layered materials such as 2D transition metal halides, oxyhalides, dichalcogenides, graphite and graphite oxide, layered phosphates, phosphorous trichalcogenides, etc. [17] [24].

Different mechanisms have been proposed to explain the host-guest interactions occurring in these intercalation materials derived from neutral species:

- Van der Waals forces
- hydrogen bonding & water bridges
- ion-dipole & coordination
- proton transfer
- electron transfer.

Van der Waals forces are specially relevant in the insertion of long-chain alkylammonium species in the different 2D solids, as for instance in the so-called *organoclays* [23]. The mutual interactions between neighboring hydrocarbon chains are specially significant when they are perpendicularly directed – or tilted – with respect to the plane defined by the inorganic layers. When the alkyl chains lie flat into the interlamellar space, the van der Waals forces are also now established between oxygen atoms of the layer and methyl and methylene groups of the alkylammonium. Typically the interactions between those oxygen atoms and the OH groups of intercalated alcohols or polyhydric molecules are explained by hydrogen bonding and experimentally proved mainly from IR and NMR spectroscopic analysis. Interlayer adsorption of polar molecules of different functionality is frequently governed by direct coordination with respect to certain metal atoms belonging to either the interlayer surface or the exchangeable cations. This is for instance the case for the coordination between nitrogenated molecules, such as amines or pyridine, and transition metal oxides, chalcogenides and oxyhalides, as well as the interaction of such molecules with transition-metal cations introduced in 2 : 1 phyllosilicates by ion-exchange. When these interlayer cations preserve their hydration shell, which occurs frequently in samples exposed to air, then the association of molecules takes place with water molecules acting as hydrogen-bond bridges. By heating these intercalation compounds, the elimination of water produces direct coordination between the guest specie and the involved *Mt* transition metal (Eq. 1):

$$\mathrm{R{-}CH_2X \ldots HOH \ldots Mt} \rightarrow \mathrm{R{-}CH_2X \ldots Mt + H_2O} \qquad (1)$$

Direct coordination is not exclusive of transition-metal cations and organic ligands. Alkaline and alkaline-earth interlayer cations could also be involved in this class of mechanism of association between organic species and inorganic host lattices. This is, for instance, the case for macrocyclic compounds such as crown-ethers and cryptands, which spontaneously penetrate the interlayer space of phyllosilicates and other layered solids such as phosphorotrichalcogenides or disulfides, giving stable intracrystalline complexes [25].

On the other hand, when the interlayer environment of certain layered solids exhibits an acid character, such as for instance some aluminosilicates of the clay minerals group, the interaction with basic species produces a proton transfer between the inorganic host and the organic guest molecules. The organic molecules become protonated, giving rise to organic cations balancing the electrical charge of the silicate. The degree of anomalous dissociation ($\times 10^3$ times) [26] of interlayer water molecules is produced by the electrical field associated with the exchangeable cations, provoking the high dissociation of H_2O molecules belonging to their hydration shell (Eq. 2).

$$[M(H_2O)_x]^{n+} \rightarrow [M(OH)_y(H_2O)_{x-y}]^{(n-y)+} + y\,H^+ \tag{2}$$

The extent of this dissociation is governed by the nature of the cations being dependent on their charge/radius ratio, and so *acidic cations* such as Al^{3+} are very active in proton generation. In this way ammonia, amines and nitrogenated compounds like pyridine are adsorbed in the interlayer space of smectites giving their corresponding cationic form (Eq. 3) as proved by IR and other spectroscopic techniques.

$$R{-}NH_2 + H^+ \rightarrow R{-}NH_3^+ \tag{3}$$

This interlamellar acid character is also responsible for catalytic transformations of adsorbed species in the intracrystalline region of certain 2D solids, which could operate in a different way from those carried out in homogenous media [27].

In addition to the proton transfer mechanisms, redox reactions (electron transfer mechanisms) have also been invoked to explain the intercalation of organic and organometallic species into various 2D solids containing transition metal ions, either as exchangeable cations or belonging to the host structure. Thus, clays containing interlayer cations like Cu(II) are able to interact with aromatic compounds such as benzene, giving intercalation compounds characterized by the existence of σ or π bonds between the host solid and the guest molecule [28]. Interestingly, these clays may also induce intercalative redox reactions such as the polymerization of pyrrole or aniline, forming conducting polymers that result in hybrid compounds of interesting behavior [29]. These reactions occur also for the intercalation of these molecules in FeOCl, as first reported by Kanatzidis [30] . In Sect. 2.2.1.1 above the intercalations of alkyl- and aryl-ammonium species in matrices such as V_2O_5 xerogel, which are also good examples of intercalation mechanisms based on redox processes, were noted.

The driving force determining the intracrystalline insertion of organic species is in some cases intricate and various of the above mechanisms could be simultaneously invoked. This is for instance the case for pyridine intercalation in V_2O_5 xerogel. In fact, these molecules remain bonded to the host solid : (i) by electrostatic forces after protonation, (ii) through water bridges with V(V) species, and (iii) by direct coordination to the transition metal when water is eliminated by heating. Intercalation of pyridine in transition metal chalcogenides such as TaS_2 is also controlled by complex mechanisms, as indicated by the formation of 2,2-bypyridine, among other products in their interlayer space, implying again the existence of redox processes in the host-guest interactions [31].

2.2.1.3 **Polymer Intercalations: Nanocomposites**

Blumstein [32] in 1961 and some years later Friedlander [33] reported the ability of neutral unsaturated monomers such as acrylonitrile, methyl methacrylate and vinyl acetate to intercalate into montmorillonite. The *in situ* polymerization takes place by the action of radical initiators such as azo-*bis*isobutyronitrile (AIBN), the polymer remaining strongly associated to the mineral substrate. These stable organic-inorganic materials are nowadays referred to as polymer-clay nanocomposites. From these pioneering works to the present, the study of intercalation of organic polymers into different 2D host lattices has been very intense. The resulting materials are considered as *nanocomposites* with interactions at atomic level between inorganic hosts and polymer guests.

As for conventional composites, polymer-2D nanocomposites could be classified into two groups of materials: structural and functional nanocomposites. The first ones result from combinations of polymers whose interest is centered on their mechanical or rheological properties, such as nylon-clay nanocomposites [34], whereas the latter group refers to nanostructured materials with specific chemical or physical properties and behavior, as in for instance the case of conducting nanocomposites [29].

Depending on the nature of each component, 2D host and guest polymer, the following procedures of synthesis can be adopted:

- direct intercalation from polymer solutions or by polymer melting
- *in situ* intercalative polymerization
- delamination & entrapping-restacking
- template synthesis.

The synthesis of PEO-smectite materials is a representative example of polymer-2D nanocomposites obtained by direct adsorption from solutions. These are functional nanocomposites with attractive ion-conducting behavior as first reported by Ruiz-Hitzky & Aranda [35]. Two possible arrangements of PEO in the interlayer space of the silicate have been proposed: (i) helical conformation of the polymer in a monolayer (Fig. 2.7), and (ii) planar disposition of polymer chains in zig-zag conformation [36–38]). Giannelis and co-workers report an alternative method for direct intercalation, involving the melt of the polymer [39]. This innovative procedure has been successfully applied to prepare PEO and other polymer-clay nanocomposites

[40–41]. The use of microwave irradiation that simultaneously provokes the water elimination of the silicate and the melting of the polymer appears as an useful tool saving time and energy compared to conventional thermal heating which has also been applied to PEO-montmorillonite preparations [42].

Kanatzidis and co-workers in 1986 reported the preparation of nanocomposite materials by the so-called *in situ intercalative polymerization* procedure applied to polypyrrole formation in the interlayer space of FeOCl [30]. Polypyrrole (PPy) is also spontaneously formed when pyrrole is adsorbed on transition-metal exchanged smectites [43] . This polymer, and others containing conjugated unsaturated bonds, doped by different compounds (e.g. iodine), is able to act as an electrical conductor, and therefore the nanocomposites containing such *conducting polymers* are regarded as functional materials, as discussed in Sect. 2.3.4.

In a similar way to pyrrole, aniline gives intercalated PANI after interaction with different 2D hosts. Thus, transition-metal oxides (e.g. V_2O_5 xerogel), halides (e.g. α-$RuCl_3$) and oxyhalides (e.g. FeOCl), as well as chalcogenides (e.g. MoS_2), phosphates (e.g. $Cu(UO_2PO_4)_2 \cdot 10\,H_2O$), phosphorotrichalcogenides (e.g. $CdPS_3$), layered double hydroxides (LDHs) and graphite oxide, are able to give PANI-nanocomposites [44]. Alternatively, PANI can also be formed in the interlayer space of smectites previously exchanged with anilinium cations (C_6H_5–NH_3^+), followed by oxidation with ammonium peroxodisulfate that gives the emeraldine salt form of the polymer [45].

The third preparative method for polymer intercalations consists in the delamination of the host lattice followed by entrapping-restacking (*encapsulative precipitation*, after Kanatzidis) [46] of polymers in solution. In some cases, previous treatment of the 2D solid is necessary to produce expandable intermediate phases. This is the case for molybdenum disulfide that forms the Li_xMoS_2 phase by reaction with *n*-BuLi, able to insert polymers as PEO [47]. Poor crystalline materials are in general obtained following this last preparative route.

The *templated synthesis* is a procedure recently reported by Carrado and co-workers [48–50] based on the *in situ* hydrothermal crystallization of gels of silica, magnesia and LiF composition around a water-soluble polymer. In this way, PANI, PAN and PVP give hybrid materials containing variable polymer/silicate ratios, that con-

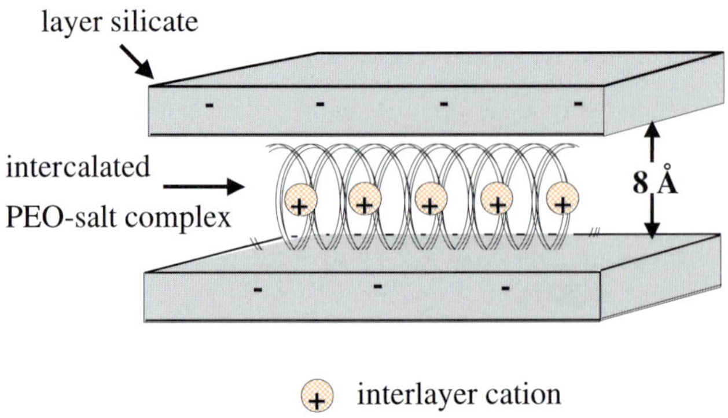

Fig. 2.7 Structural model representing a polymer-clay nanocomposite based on PEO-silicate as an ion-conducting organic-inorganic material, after Ruiz-Hitzky and Aranda [35]

fer suitable behavior on these nanocomposites, such as structural features or electrical conductivity. Similarly, LDHs can be formed in the presence of polymers such as certain PEO derivatives in anionic form [51].

2.2.2 Organic Derivatives of Inorganic Solids

This class of hybrid materials is formed by grafting of organic groups onto inorganic surfaces, the attachment between both moieties being assured through covalent bonds. The first attempts to obtain these materials were based on the so-called "esterification of silica" by reaction (Eq. 4) between alkanols and silanol groups (≡Si–OH) present in precipitated silica (considered as "silicic acid") [52].

$$\text{[silica]Si–OH} + CH_3–(CH_2)_n–OH \rightarrow \text{[silica]Si–O–}(CH_2)_n–CH_3 + H_2O \qquad (4)$$

In these hybrid materials, alkyl chains remain attached to the inorganic surface of the silica particle through ≡Si–O–C-bonds, which show low stability because they are very sensitive to water molecules being easily hydrolyzed.

More stable compounds were obtained by reaction of silica with a large variety of mono-, di- or tri-chloro (or alkoxy)-silanes containing alkyl, alkenyl, aryl, amino, mercapto and other organic groups that may confer different functionality on the inorganic substrate. Thus, organosilanes containing ≡Si–X groups (X=OR, Cl) are able to react with surface silanol groups giving very stable bonds through siloxane bridges between the inorganic substrate and the organic groups linked to the Si atoms of the silane (Eq. 5).

$$\text{[surface]Si–OH} + \text{X–Si} [R_1R_2R_3] \rightarrow \text{[surface]Si–O–Si}[R_1R_2R_3] + XH \qquad (5)$$

Direct reaction of organosilanes with 2:1 layered silicates such as vermiculite mainly gives organosiloxanes in place of grafted silicates, due to the rapid hydrolysis of the X functions of the reagent by the water molecules adsorbed in the interlayer space of the silicate [53]. The first step of these secondary processes takes place through the very reactive [R_3SiOH] intermediate organosilanol species (Eq. 6).

$$R_3Si–X + H_2O \rightarrow [R_3Si–OH] \rightarrow R_3Si–O–SiR_3 \qquad (6)$$

Layered silicic acids, such as the so-called H-magadiite which derives from the sodium silicate magadiite [54], have the particular property that silanol groups are contained inside the interlayer region of these solids [55]. The direct reaction of H-magadiite with organochlorosilanes exclusively affects the external surface of the microcrystalline particles, the coupling between internal silanol groups and the reagents being only achieved using polar molecules such as DMSO or NMF to expand the silica layers. Remarkably, the formation of intermediate H-magadiite/polar molecules allows the penetration of the organosilanes to the interior of the solid, facilitating the corresponding intracrystalline grafting reactions (Fig. 2.8) [56]. In this way, nanocomposites that are formed by the stacking of planar silicones of about 2 nm high by 10^3 nm long can be designed and prepared provided with controlled functionality [57]. Alkylammonium exchanged magadiite appears also as an intermediate intercalation compounds able to give organic derivatives containing long-chain alkylsilyl grafted groups, with potential interest as host for selective molecular adsorption of organophilic species [58]. Following similar procedures, fluoroalkyl groups have also been grafted in the interlayer space of magadiite, giving stable compounds of interest as oil-repellent materials [59].

The silanol groups necessary to react with organosilanes can be produced by acid treatment of silicates. Lentz in 1964 [60] developed an ingenious procedure to obtain organic-inorganic solids involving the simultaneous hydrolysis of a silicate (e.g. olivine) and an organosilane (e.g. hexamethyldisiloxane). Afterwards, these cohydrolysis processes were applied to prepare organic derivatives of various minerals such as chrysotile, vermiculite, wollastonite and sepiolite, having a more intri-

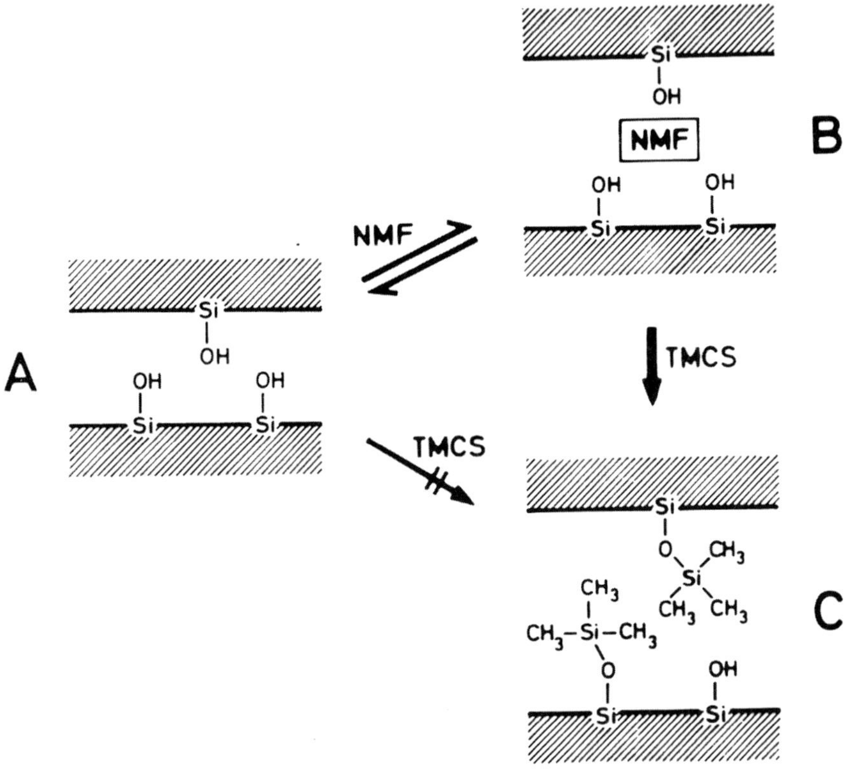

Fig. 2.8 Intracrystalline grafting of trimethylsilyl groups in layer silicic acids such as H-magadiite (A). The method needs a previous expansion of the starting solid by intercalation of a polar molecule as *N*-methylformamide (B) giving in a second step the intracrystalline grafting (C) , after refs. [56] and [57]

cate silicic framework than olivine [61–63]. The extraction of octahedral cations (e.g. Mg^{2+} ions) by the action of acid solutions (e.g. hydrochloric acid) leads to silica formation with *fresh* silanol groups, able to give further reactions with the organosilanols produced from chloro- or alkoxy-organosilanes (Eq. 7).

$$\equiv Si–O–Mg–O–Si\equiv + [H^+/H_2O] \rightarrow [surface]\ Si–OH + Mg^{2+} \quad (7)$$

The case of sepiolite is particularly interesting because this natural magnesium silicate is able to give its organic derivatives by direct reaction with organochlorosilanes such as trimethylchlorosilane and methylvinyldichlorosilane, in vapor phase or in solution in organic solvents [64]. This is due to the presence of available silanol groups on the external surface of the microparticles of the natural silicate. The surface of sepiolite grafted by organosilanes such as trimethylchlorosilane [64] becomes organophilic and hydrophobic, whereas the use of functional silanes, such as those containing sulfoaryl groups, strongly modifies the chemical reactivity of the mineral [65]. This is the basis for the preparation of hybrid organic-inorganic materials with complementary properties given by both, the inorganic skeleton and the organic matter being associated by covalent bonds.

The organic derivatives of silicates obtained using organosilanes, either by cohydrolysis or by direct reactions, are extraordinarily stable towards chemical agents or thermal treatments. The presence of selected organic functions allows further reactions as well as important applications as structural or functional materials, as will be later discussed.

Organic-inorganic materials obtained by grafting of organic groups on silicic substrates have also been prepared using compounds of different functionality, such as isocyanates and epoxides, which react with ≡Si–OH groups of silica (Cabosil or Aerosil) and sepiolite [67–70]. Some reactions of isocyanates were also carried out using vermiculite [66].

Isocyanates are grafted to inorganic substrates through surface silanol groups forming silyl urethane (≡Si–O–CO–NH-R) bridges (Eq. 8) [67].

$$\equiv Si–OH + R–N{=}C{=}O \rightarrow \equiv Si–O–CO–NH–R \quad (8)$$

The formation of N,N′-disubstituted ureas [$(R–NH)_2C{=}O$] by reaction of isocyanates with water molecules adsorbed on the surface of clays is a secondary process, that in certain cases could predominate over the grafting reactions. For instance, phenyl isocyanate reacts with sepiolite giving large amounts of diphenyl urea, whereas butyl isocyanate gives stable grafted compounds deriving from this silicate [67]. Diisocyanates, such as hexamethylene diisocyanate and 2,4-toluene isocyanate, give organic derivatives of silica and sepiolite containing excess –C=N=O functions. Therefore, the resulting organic-inorganic compounds exhibit an improving surface reactivity that could be useful for further coupling molecular materials of predetermined physical behavior. Thus, some of these hybrids are of particular interest with a view to preparing polycondensates from subsequent reaction with polyols and polyesters [68]. Grafting reactions of 1,2-epoxides on silica and sepiolite occur also through the silanol groups present at the mineral surfaces (Eq. 9) [69–70].

$$\equiv Si-OH + \underset{\diagdown O \diagup}{CH_2CH-R} \longrightarrow \equiv Si-O-\underset{\displaystyle CH_2OH}{\underset{|}{CH}}-R \qquad (9)$$

The organic groups (R) remain covalently attached to the surface through ≡Si–O–C bonds, whose stability towards the hydrolysis depends on the nature of the R substituent [69]. The main interest of the organic-inorganic materials obtained from epoxide treatment of silica and silicates could be related to applications as reinforcing fillers of elastomers and thermoplastics. This is particularly interesting when the involved epoxides contain unsaturated groups, such as for instance allylglycidyl ether, allowing copolymerization reactions [71]. As occurs with other grafting agents, the reaction of epoxides with mineral surfaces of clays produces also secondary reactions, in this case consisting in rearrangement processes towards carbonyl compounds. The extent of these reactions is principally significant for the reaction of epoxystyrene with clays, that gives large amounts of phenyl acetaldehyde. To explain the role of the clays in catalysing those secondary reactions, a mechanism has been proposed involving silico-alumina species present on mineral surfaces [72].

The epoxide functions could also be opened and grafted onto intracrystalline inorganic surfaces of certain lamellar phosphates. Thus, ethylene oxide reacts with OH groups of layered zirconium phosphate [73], giving an intracrystalline environment composed of oxyethylene units resembling those corresponding to the crown-ethers or PEO clay complexes, which are interesting for selective complexation of ionic species. These organic-inorganic materials have potential significance as ionic conductors of interest for applications in different electroactive devices [25].

2.2.3 Sol-Gel Hybrid Materials

As this category of materials is largely covered in detail in other chapters of this book, only a short overview of this topic is here presented. Additional information is also available in several reviews and books (see for instance [74–76]). Our aim in this section is exclusively centered on the introduction of some relevant points concerning sol-gel methods for preparation of functional hybrid materials, which are complementary or alternative to those concerning the intercalation and grafting procedures, above reported.

Molecular species such as certain organometallic compounds based on alkoxydes of silicon, titanium, tin, aluminium or zirconium act as precursors of this class of hybrids characterized by the presence of –Mt-O-R metal-organic bonds in a metal oxo-network. Organopolysiloxane materials are the most important group of hybrids prepared by sol-gel methods. They contain the very stable ≡Si–O–Si≡ bonds which can be produced from a big variety of organosilane precursors provided with hydrolyzable alkoxy or halogenide functions. Various steps are usually invoked to explain the sol-gel reactions from such precursors to the polysiloxane network, starting from solutions of molecular species (*sol*) that give the *gel*, after hydrolysis, intermediate organosilanols, and condensation of these species. The whole process comprises also other important steps such as the evaporation of solvents, rearrangement reactions, shrinkage of the formed polysiloxanes, etc. [76]. The process is therefore very complex and the microstructure, stability and other properties of final materials are strongly dependent on many factors regarding both thermodynamic and kinetic aspects. Note for instance that precursor species or their oligomers, as well as molecules of the involved solvents (alcohols, water, ethers, etc.), can remain entrapped within the polysiloxane network, to an extent depending on the nature of the involved species, temperature, concentrations and other additional experimental conditions. This is often the reason why the point signaling a complete stabilization of the organic-inorganic systems is very difficult to determine. The ability to entrap molecular species has been largely made use of to prepare functional materials useful for catalysis, optical, electrochemical and other applications, as will be illustrated in the next paragraph and also treated more extensively throughout this book. One of the most significant characteristics of the sol-gel processes, that is of crucial importance from the point of view of enabling applications, is related to the easy processing of these materials, such as films, monoliths, fibers, or powders with particles of controlled size and shape. The resulting materials obtained by drying the gels in the atmosphere are known as xerogels, whereas the solids obtained by drying at supercritical conditions are known as aerogels. This class of hybrids has received different names, such as ORMOSILS and ORMOCERS®, referring to organically modified silicates or ceramics respectively.

The nature of hybrids obtained by sol-gel can be extremely diverse as the number of potential precursors is very high. Moreover, the mixture of various precursors makes practically unlimited the number of different organic-inorganic materials that can be designed and prepared in a more or less controlled way. The selection of molecular or polymeric species of quite different natures and properties that could also be incorporated into these organic-inorganic systems offers also a great number of alternatives. Finally, the use of precursors containing reactive groups such as alcohol, mercapto, amino, carboxyl, carbonyl phenyl or unsaturated functions, presents the possibility of developing further reactions of the resulting polysiloxane networks with appropriate molecular or polymeric species.

One interesting feature of sol-gel processes carried out in the presence of surfactants is the possibility of preparing self-assembled organic-inorganic materials. The surfactants form micelles that act as templates organizing around them inorganic (or organic-inorganic) materials arranged in lamellar, cubic or hexagonal structures, by polycondensation of the hydrolyzed precursors (e.g. metal alkoxydes). This is the basis of the preparation of MCM type mesoporous solids provided with long range periodical order, that results from the elimination of the organic template by combustion or extraction with solvents [77–78].

A large number of these last type of materials is based on silica or organo-silicic compounds containing reactive functions (amino, mercapto, aryl, etc.) belonging to the starting precursors (e.g. organoalkoxysilanes) [79]. Alternatively, the organic functions could be incorporated by further reaction of the MCM solids with the corresponding organosilanes. Selective grafting of organic functions on the internal tubular surface of MCM silica, following various steps, has been reported [80] . More recently, the preparation of very stable organic-inorganic compounds arranged as tubular mesophases, by the so-called *self-templating* procedure, has been proposed to obtain a novel class of organic derivatives of silica named as SGS (silica grafted surfactants) provided with intratubular swelling ability [81]. In this case, the starting reagent is a cationic surfactant covalently bonded to an alkoxysilane which is able to hydrolyze and polymerize giving spontaneously self-assembled entities with layered (like organoclays) or hexagonal (like silica-micelle intermediates of MCM porous solids) arrangements (Fig. 2.9). These hybrids show the property of further incorporation of lipophylic and anionic compounds, being promising materials to incorporate molecular or ionic species of variable functionality.

Polyhedral oligomeric silsesquioxanes (POSS^{TM}) and polysilsesquiloxane bridges are organic-inorganic compounds also based on the formation of ≡Si–O–Si≡ bonds from organosilane precursors, being prepared by sol-gel techniques [82–86]. Starting reagents of POSS are R′–(SiX)$_3$, organotrialkoxy (X=OR) or trichloro (X=Cl) silanes, which are able to polymerize as $(R–SiO_{3/2})_n$ molecular cages. In the case of the polysilsesquioloxane bridges, the precursors are also organotrialkoxy- (or chloro-) silanes containing here two or more trialkoxysilyl groups separated by R′ variable bridging organic groups [(OR)$_3$–Si–R′–Si–(OR)$_3$]. This type of hybrid can contain

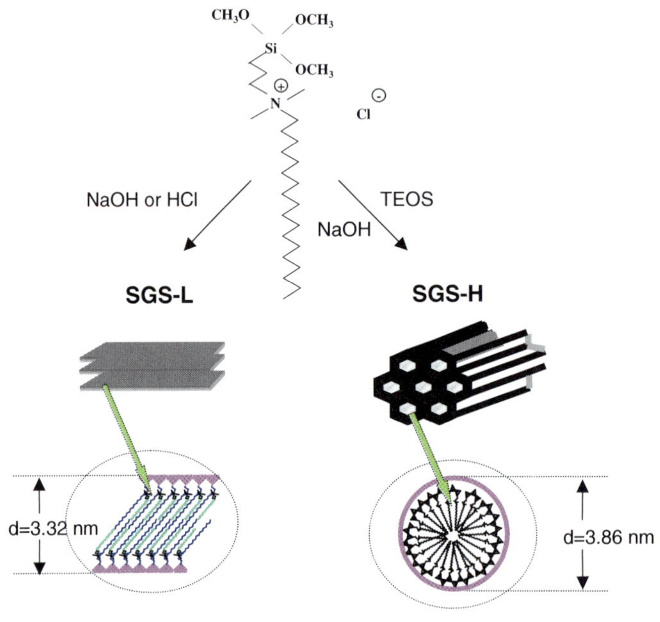

Fig. 2.9 Self-templating synthesis of organic-inorganic SGS (*silica grafted surfactant*) materials with layered and hexagonal arrangement, after Ruiz-Hitzky et al. [81]

one or more covalently bonded reactive functionalities suitable for further polymerization, grafting or surface bonding that provide applications as described in Sect. 2.3.3.

2.3 Functions & Devices Based on Organic-Inorganic Solids

As recounted above, *molecular engineers* have the capacity to design and prepare an unlimited number of organic-inorganic materials by selecting among a large choice of inorganic substrates and a large variety of organic functions. The ability to introduce suitable functionality to these substrates is certainly a challenge that very often reaches reality. Thus, by the rational use of organic or organometallic reagents the nature of solid surfaces can be modified in a controlled manner. Consequently, the chemical reactivity of inorganic solids can be deeply changed, the introduced organic functions offering the faculty to give further reactions which are typically and exclusively presented by organic compounds. They also confer physical characteristics related to optical, magnetic, electronic or electrochemical properties inherent to molecular species. As the combination of organic reagents with inorganic hosts allows the construction of flexible and sophisticated organic-inorganic systems, the resulting synergetic behavior afforded by both moieties makes possible the development of novel physico-chemical devices. So, highly specific molecular sorbents, selective catalysts for enantioselective synthesis, photoactive intercalation compounds, ion-conducting hybrids, electroactive nanocomposites, and other nanostructured organic-inorganic solids are the basis of functional materials operative in singular emerging devices.

2.3.1 Selective Sorbents, Complexing Agents & Membranes

Intercalation compounds based on clays receive great interest for applications based on their capacity for selective adsorption of molecules. Although water and small polar molecules can be adsorbed by organoclays, the organophilic character of the interlayer region of clays and other 2D solids intercalated by alkylammonium species is adequate for the uptake of a large variety of organic compounds [87]. These types of organic-inorganic materials have been used for applications in chromatography separations [88–89], to remove organic pollutants from air and water, and to develop improved formulations of pesticides [23] [90–93]. The design of chemical sensors based on the selective molecular adsorption by alkylammonium-clays (organoclays) intercalated compounds has been reported by Yan and Bein, showing the discrimination of small from large molecules, acting as molecular sieves [94]. The nature of the intercalated cations is determinant in the adsorption behavior. So, thin films of trimethylammonium-hectorite compounds deposited on piezoelectric sensors (quartz crystal microbalances) display a selective uptake of aromatic compounds such as benzene by comparison with alicyclic species such as cyclohexane (Fig. 2.10).

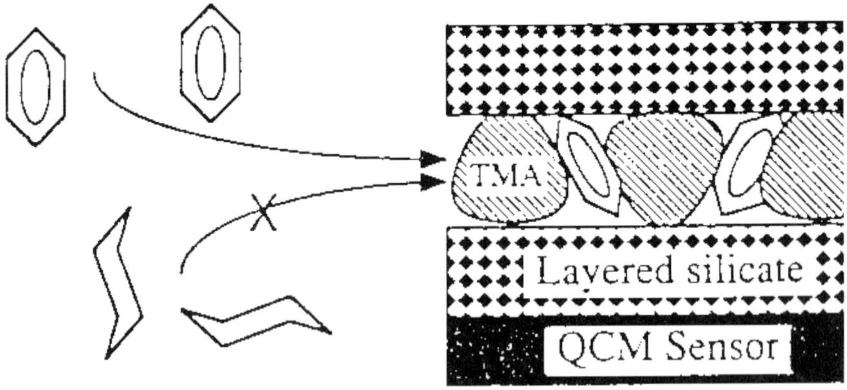

Fig. 2.10 Device for molecular recognition based on trimethylammonium (TMA)-hectorite films deposited on a piezoelectric sensor (QCM, quartz crystal microbalance), after Yan and Bein [94]

It is well know that liquid membranes containing macrocyclic compounds are very efficient in the separation of ions, but they are strongly limited in their practical applications [95]. Membranes inmobilizing crown-ethers and cryptands are an excellent alternative, and several ways to prepare organic-inorganic ion-selective membranes using macrocyclic compounds associated with inorganic substrates have been reported [25] [96].

Intercalation compounds based on 2:1 phyllosilicates and crown-ethers have been used in the preparation of composite membranes which are conformed in a sandwich-like arrangement by encapsulation of the organic-inorganic intercalation compounds with polybutadiene coatings [97]. In this manner, the resulting materials show improved mechanical resistance compared to free-standing films of phyllosilicates/macrocycles, being convenient to be used without damage in aqueous solutions. Metal-ions pass freely through the pores of the polymer alone, but the presence of intercalation compounds modulates the transport properties of the cations that can be discriminated depending on their nature. The mechanism controlling the ion-selectivity of these composite membranes is not clear and can be hypothetically ascribed to various factors such as (i) cation hindrance based on the ion size and/or hydration energy, (ii) migration of the electrolytes through the channels of the membrane following a diffusion mechanism (passive transport), and (iii) the action of a facilitated transport assisted by the macrocyclic compounds [97]. Electrochemical devices based on this class of membranes prepared by coverage of electrodes could be of great interest with a view to preparing sensors provided with selectivity towards cations in solution [96, 97].

An optional route to immobilizing crown-ethers giving materials operative as membranes consists in the entrapping of those macrocycles into polyorganosiloxane matrices generated via sol-gel processes [98]. Synthesis of the active phase of these composite membranes has been achieved by controlled hydrolysis and polycondensation of different organosilanes (*precursors*) as for instance tetramethoxyorthosilicate (TMOS), tetraethoxyorthosilicate (TEOS), ethyltriethoxysilane (ETEOS) or γ-methacryloxypropyltrimethoxysilane (MAPTMS), in the presence of variable amounts of a selected crown-ether. The crown-ethers are incorporated into the sol-phase giving, after drying and ageing, xerogels that evolve to colorless and transparent glasses. To improve mechanical behaviors of final composite membranes, supports of borosilicate or PAN have been employed [98]. Using mixtures of precursors such as MAPTMS and TMOS, organopolysiloxane matrices with good physical properties are obtained, being useful for processing as continuous films and also to produce uniform coating of electrodes [99].

Grafting of macrocycles to polysiloxane networks was introduced as a way to avoid the possible leaching of entrapped macrocycles from the sol-gel materials. Thus, alkoxysilanes containing benzo15-crown-5 give polysiloxane chains containing macrocycles that are operative in the transport mechanism of ions as proposed by Lacan et al. [100–101]. Membranes derived from these organic-inorganic materials have also been tested for applications as pH sensors and processes based on gas separation and nanofiltration technologies [100].

Immobilization by grafting reactions of macrocycles on the surface of silica gel were carried out by (i) anchoring of trialkoxysilanes and further reaction with functionalised macrocycles [102], and (ii) synthesis of reactive silanes containing macrocyclic rings, followed by direct reaction with surface silanol groups [103]. The grafting of benzo-crowns on silica, reported by Nakajima *et al.* [104] and Iwachido *et al.* [105], shows the dependence between the separation of the macrocycle and the silica surface and the ion-binding ability of the organic-inorganic materials. The first applications of such compounds were focused towards chromatography separation of both cations (*e.g.*, alkaline and alkaline earth, heavy and noble metals) and anions (*e.g.*, halides, heteropolyacids) [106–107]. A good example of this is the use of benzo-18C6 grafted on silica as a stationary phase of chromatography columns, showing an excellent discrimination between cations in the sequence:

$$Li^+ < Na^+ < Rb^+ < Cs^+ < K^+$$

This is directly correlated with the stability sequence of the corresponding crown-ether/cation complexes [105]. It should be remarked that racemic mixtures have been successfully resolved by using chiral crowns grafted on silica [108] with the advantage of heterogeneous systems (easy handling, recovering, etc.) compared to homogeneous conditions. On the other hand, membranes based on crown-ethers grafted on silica showing facilitated transport properties have been applied in other fields such as the development of ion-selective electrodes [100–101] [109–110], similarly to membranes based on crown-ether intercalation compounds and sol-gel matrices [96–97] [99] [111].

2.3.2
Heterogeneous Catalysts & Supported Reagents

Certain organoclays containing protonated species in the interlayer space of the silicates could be operative not only as selective sorbents of organic molecules but also as catalysts for such adsorbates. For instance, the diprotonated form of the 1,4-diazabicyclo [2,2,2] octane (triethylene diamine, TED) intercalated in montmorillonite, acts as a set of pillars between consecutive layers allowing the uptake of molecules by the organoclay and also their catalytic transformation [112]. This is the case for acetonitrile that gives selectively the corresponding amide after adsorption by the TED^{2+}-smectite. This reaction was previously detected from direct adsorption of certain nitriles, such as trichloromethylacetonitrile in smectites, but acetonitrile remained unchanged when it was adsorbed on unmodified clays.

Other reactions catalyzed by intercalation compounds are related to metal-complexes formed or directly introduced into the interlayer space of clays [113]. An illustrative example is the intercalation of the rhodium-triphenylphosphine species in its cationic form $[Rh(PPh_3)_3]^+$ by ion-exchange with smectites such as montmorillonite and hectorite. The resulting material acts efficiently in the hydrogenation of terminal olefins without the isomerization-competitive reactions that usually take place in similar reactions carried out in homogeneous media [17].

Alternatively, Rh and other active transition metal complexes can be immobilized on inorganic solids by grafting reactions. So, Rh-complexes with nitrogen-donor ligands, with potential ability to induce hydrogenation and hydroformilation reactions, can be fixed either on silica surfaces *via* functionalized silanes or in a polysiloxane network generated by sol-gel processes [114]. Similarly, phosphine ligands can be immobilized using $(EtO)_3Si(CH_2)_3PPh_2$ in the presence of TEOS obtaining polysiloxane-phosphine materials [115], or by direct grafting on different inorganic substrates such as SiO_2, MgO, Al_2O_3 and TiO_2 [116]. These functionalized polysiloxanes are able to take up transition metals such as Co(II) and Ni(II), and also to form stable complexes with metal ligands based on ruthenium, rhodium and iridium. The formation of these last complexes gives organic-inorganic catalysts operative for selective terminal olefin hydrogenations [115].

Heterogenized catalysts based on the immobilization of osmium species were obtained by direct reaction of OsO_4 with vinyl groups previously grafted by the reaction of methylvinyldichlorosilane with silica or sepiolite (Eq. 10).

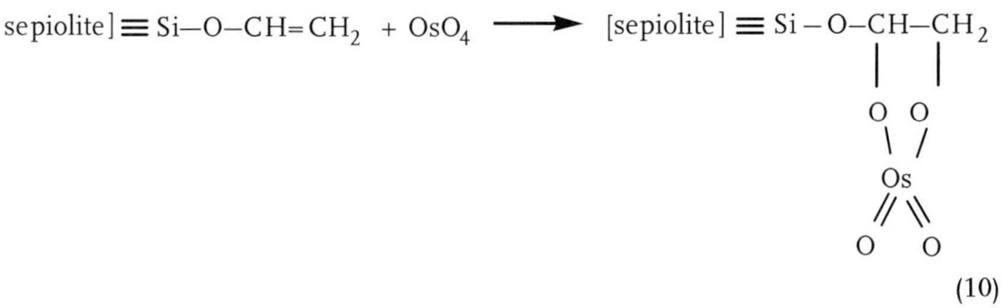

(10)

After controlled hydrogenation, these stable Os-derivatives act as effective oxidative catalysts, [117] giving free hydrogen from photodecomposition of water [118]. Interestingly, the opaque effect of a heavy metal such as Os towards the electron beam of the electron microscope was used to reveal the localization and distribution of grafted groups on the mineral surface as they are observed as dark dots of about 1 nm diameter (Fig. 2.11) [119].

The immobilization of chiral catalysts can be attained following the general procedures based on direct grafting, adsorption or entrapment (*ship in bottle*) [120]. Grafting of enantioselective catalysts has been carried out on common inorganic supports of high specific surface area: oxides, such as silica and, in minor extent, in alumina or zirconia, clays and pillared clays, zeolites and mesoporous solids like MCM-41. Hydrogenations, oxidations, dehydration, hydrolysis, epoxidation, Diels-Alder, and other processes have been developed using these systems. The most simple way to graft active chiral species is here also the direct reaction using previously conditioned organosilanes. For instance, the grafting of chiral Rh-complexes as represented in Eq. 11, on inorganic supports, gives catalysts for hydrogenation of α-acylaminocynnamate derivatives that are much more effective using zeolites than silica substrates [121–122].

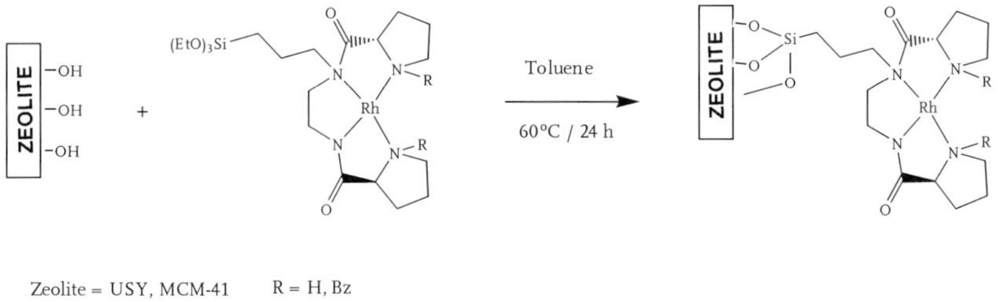

(11)

On the other hand, organic-inorganic materials offer an interesting possibility as they could be used as support for organic synthesis, in the so-called *dry media conditions,* i.e. replacing organic solvents which are toxic, dangerous and pollutant, by fine powders dispersing the mixture of reagents. For instance, the immobilization

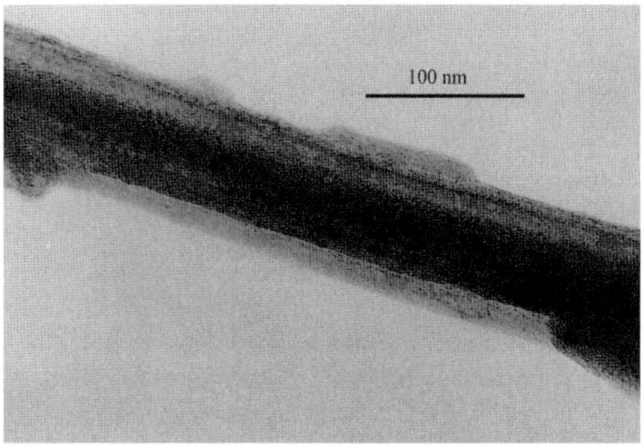

Fig. 2.11 Photomicrograph of a vinyl-sepiolite treated with OsO_4. Dark dots indicate the presence of clusters of Os atoms, after Barrios-Neira et al. [119]

of crown-ethers allows the easy alkylation of potassium acetate under mild conditions [107] (Eq. 12).

$$n\text{–}C_8H_{17}\text{–}Br + CH_3COOK \rightarrow KBr + n\text{–}C_8H_{17}CH_3COO \qquad (12)$$

The affinity of the organic moieties of hybrid materials towards the reagents might be useful for selective organic synthesis in the absence of solvents.

2.3.3
Photoactive, Optical and Opto-Electronic Materials & Devices

The unique optical qualities of sol-gel materials prepared from alkoxysilanes has been probably the most important reason why the applications of those class of hybrids has been directed to fields such as conventional optical materials, colored glasses, photoprotection, photo-imaging, laser, information-recording and other devices. As this topic is extensively treated in other chapters of this book, we refer here only to organic-inorganic materials based on intercalation compounds and grafting of photoactive organic species, avoiding duplication of information on sol-gel methods used to entrap and to immobilize such species.

Intercalation of cationic dyes in layered solids, such as phyllosilicates belonging to the smectite group of clay minerals, has been a subject broadly investigated [23]. In particular, cationic dyes such as methylene blue, methyl green, crystal violet, safranin T, thioflavine T, acrydine orange, rhodamine 6G and a large group of heterocyclic polyaromatic compounds, have been intercalated from aqueous solutions at room temperature in clay minerals following ion-exchange processes. UV-Vis spectroscopy is considered a powerful technique for obtaining information about the resulting colored organic-inorganic materials [11]. The most interesting feature in the visible spectra of cationic dyes such as methylene blue intercalated in clays such as montmorillonite, is the metachromasy effect. This effect consists in a decrease in the absorption intensity of bands assigned to isolated species (*monomers*), accompanied by an increase in intensity at higher energies assigned to formation of aggregated species [123]. This effect, related to the self-aggregation of dyes to give dimmers, trimmers and higher aggregates, was spectroscopically deduced from dye adsorptions on hectorite, laponite [124] and sepiolite [125] clay minerals. Dimerization of methylene blue in the interlayer space of V_2O_5 xerogel has also been deduced from visible spectra [126]. In this case, the driving force for intercalation is more complex than ion-exchange processes invoked in clays, as a redox reaction between V^{5+} ions of the host and methylene blue species could also be implicated. Usually adsorbed cationic dyes are disposed in planar conformation with the aromatic rings lying flat on 2D-solid layers, but the use of clay samples of high charge density as the synthetic clay named fluor-taenolite determines a nearly perpendicular orientation in the interlayer space of dye species [127]. Cationic dyes are adsorbed on the external surface of sepiolite, and also to some extent inserted into the structural channels of this silicate [125] [128–129]. The resulting organo-clays are of interest for applications in the photostabilization of co-adsorbed labile bioactive species [130] in the same way as that proposed by Margulies for cationic dye-montmorillonite compounds [131].

Novel organic-inorganic materials based on azobenzene-organoclay complexes reported by Fujita et al. [132] open the way to compounds showing an interesting photoresponse, changing in a reversible way their conformation (from *trans* to *cis*) by either UV irradiation or mild heat treatments (Fig. 2.12). Attractively, these changes are accompanied by variations of the basal spacing values of the layered solid, as revealed by XRD measurements, the photoresponse being repetitively maintained by the material over more than 15 cycles. This means that devices operating on mechanical behavior can be developed associated with photoactive organic-inorganic materials.

Multifunctional materials also based on intercalation compounds have been recently reported by Bénard et al. [133]. These authors have prepared transparent films consisting of spyropyran cations inserted into $MnPS_3$ by an ion-exchange process. Within the solid the spyropyran species are transformed in the stable merocyanin form exhibiting photochromic behavior, together with the magnetic properties inherent to the host lattice. Interestingly, the UV irradiation of these hybrid materials results in photomagnetic effects such as observed changes in the hysteresis loop, similarly to those reported for layered bimetallic oxalates [134]. In fact, these last systems are extremely interesting because of their possibility of exhibiting multivariate functional behavior coupling different transport phenomena. See, for instance, the simultaneous ferromagnetic and metal-conducting properties observed in Mn-Cr oxalates intercalated by TTF derivatives reported by Coronado et al. [135]. The preparation, by controlled modifications of inorganic substrates introducing organic groups (not only by intercalation but also following the grafting and sol-gel procedures above discussed), of designed organic-inorganic materials provided with fascinating multifunctional behavior appears of great interest for future new applications.

Another very promising class of organic-inorganic materials based on molecular dyes included into nanopores of microcrystals of inorganic solids such as AlPO-5 has been recently reported by Braun et al. [136]. This solid is optically transparent and presents low internal scattering losses, having a free pore diameter of 0.73 nm that allows the accommodation of linear cationic dyes such as 1-ethyl-4-(4-(*p*-dime-

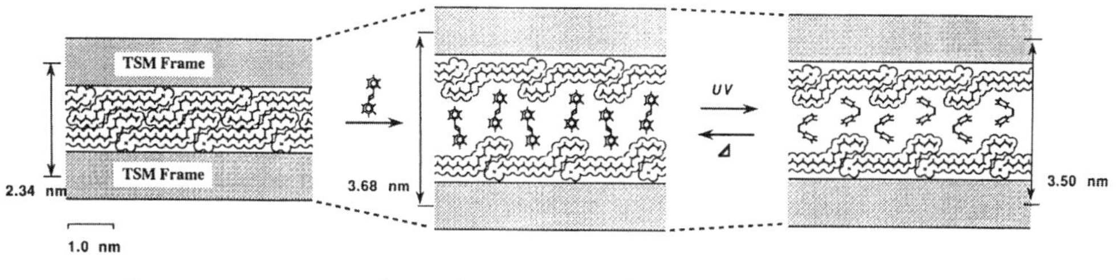

Fig. 2.12 Schematic representation of an azobenzene-organoclay material and its response to UV irradiation, after Fujita et al. [132]

thylaminophenyl)-1,3-butadienyl)-pyridinium. The resulting well-ordered materials contain the dye specie aligned along the c-axis of the AlPO monocrystals (Fig. 2.13) and exhibit interesting properties as microlasers. Sepiolite, having also channels of similar dimensions along the c-axis, shows the ability to accommodate cationic dyes such as methylene blue [137], and could be regarded as a host of potential interest for encapsulating suitable photoactive molecules for microlaser applications.

Polysesquiloxane bridges have also very interesting textural properties, derived from the presence of the bridging groups which could modulate the size of nanocavities. For instance, arylene-bridged polysilsesquioxane xerogels develop high BET specific surface areas (ca. 1000 m^2/g) with large pore volume (>0.5 cm^3/g) [85]. This behavior is not related only to the role of structural bridging groups as organic spacers, but also to the experimental conditions adopted during the synthesis and drying [74]. One of the most interesting features of these porous organic-inorganic solids is the possibility of using them as a confinement matrix for different compounds, such as nanosized CdS particles [138].

Grafting of –SH (*thiol or mercapto*) functions using appropriate organosilanes (*e.g.* 3-mercaptopropyltrimethoxysilane) on large porous solids such as MCM mesoporous silica, modifies the reactivity of their inorganic surfaces, allowing the uptake of metals such as Cd by thiol-cation interactions [139–140]. Further treatment of these complexes with H_2S gas or Na_2S solution gives nanoparticles of CdS, conferring peculiar properties on this organic-inorganic material, such as photocatalytic activity for H_2 generation from isopropanol [139]. Interestingly, the size of these nanoparticles can be controlled because the increase of the density of –SH groups decreases the CdS particle size as revealed by UV-Vis spectroscopy [139]. The controlled production of small semiconductor nanoparticles appears to be one of the most interesting issues for developing new nanoelectronic and optoelectronic devices. The reason arises from the very attractive quantum properties possessed by these kinds of nanostructures when they have been grown either in an insulating matrix or as a semi-insulating coating.

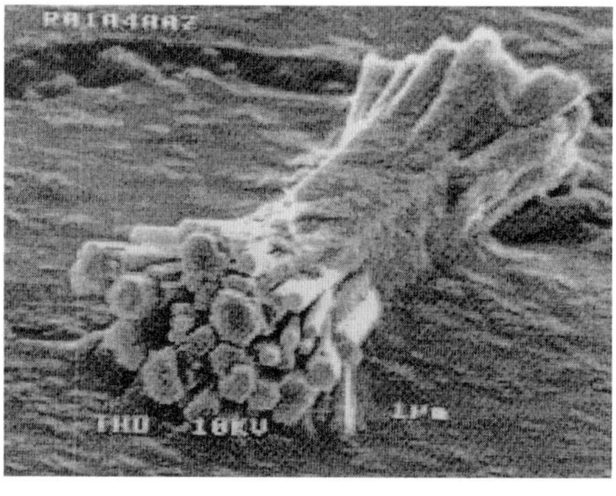

Fig. 2.13 SEM image showing the morphology of a pyridine-2 dye/AlPO-5 organic-inorganic material with lasing properties, after Braun et al. [136]

2.3.4
Electrical Behaviors: Ionic & Electronic Conductors

Intercalation of organic ions, molecules or polymers into layered solids can modify electrical behaviors inherent to the pristine host lattice. By contrast with phyllosilicates, whose layers could be considered as insulators, 2D solids containing transition metal ions, i.e. MS_2, M=Mo, Ti, Ta etc. dichalcogenides, exhibit electronic conductivity like semiconductors or metals. For instance, the *2H*-TaS_2 polytype where Ta is in prismatic tetrahedral coordination, has a typical metallic behavior. In addition, this solid, as well as other MS_2 compounds of Group V, shows superconductivity with a T_c value for *2H*-TaS_2 of 0.8 K. Intercalation of organic molecules in these solids modifies some of their electrical behavior measured over a wide temperature range. In this way, intercalations on 2H-TaS_2 give organic-inorganic superconducting materials in which the T_c value increases by a factor of about 5 with respect to the starting host [141–143]. Of course, these values are far away from the T_c corresponding to YBaCuOs and related superconducting mixed oxides, but their existence indicates significant changes in the intrinsic characteristics of 2D solids that could lead to understanding for instance host-guest charge transfer mechanisms.

PEO-clay nanocomposites based on homoionic montmorillonite and other smectites are anisotropic solid electrolyte materials presenting ionic conductivity values several orders of magnitude higher than that of the parent silicate. The thermal stability of this class of materials is strongly enhanced in comparison with conventional PEO-salt complexes. Moreover, a salient feature of clay nanocomposites compared to these last electrolytes, is that here the anion is the silicate layer, and therefore theoretically must exclusively involve a cationic transport mechanism. Recent experiments on solid-state polarization techniques using PEO/Na-montmorillonite confirm this hypothesis, affording transport number (t_+) values close to 0.99 (theoretical value of 1 for an ideally pure cationic solid electrolyte) [144]. Such a unique behavior is of importance with a view to potential applications of these nanocomposites. PEO intercalations in 2D host lattice solids showing electronically conducting properties, such as transition metal oxides (e.g. V_2O_5 xerogel: $V_2O_5 \cdot nH_2O$) [145–146], transition-metal dichalcogenides (MS_2, M=Mo, Ti, Ta) [47] [147–148] and phosphorus trichalcogenides (MPS_3, M=Cd, Mn) [149] offer the possibility of developing materials provided with both electronic and ionic conductivity, which is of potential interest for electrochemical devices, as discussed below (Sect. 2.3.5).

Sol-gel procedures involving TEOS, poly(ethylene glycol) and lithium salts give materials provided with ionic conductivity, measured by impedance spectroscopy at room temperature, in the 10^{-5} to 10^{-7} S/cm range [150]. Thus, these organic-inorganic materials exhibit higher conductivities than vitreous solids based on the same Li/silica composition and also than PEO-silicates based on intercalation methods. The use of poly(ethylene glycol) in place of PEO is due to solubility problems of this higher molecular weight polymer in the solvents used to make compatible the sol-gel starting reagents.

Grafting reactions of arylsilanes such as $Cl_2Si(CH_3)(CH_2)_2Ph$ on sepiolite or with polyorganosiloxane networks generated by sol-gel, give organic-inorganic materials that can be sulfonated producing arylsulfonic groups [65] [151]. As these groups contains H^+ species, this type of material could be regarded as proton conductors, and the corresponding cation-exchanged derivatives as ion-conductor hybrid solids. Sol-gel methods open the way to adequate coatings for electrodes and other electroactive elements.

2.3.5
Electroactivity & Electrochemical Devices

Although the intercalation of conducting polymers in 2D solids always gives less conductive nanocomposite materials compared to bulk polymers [29], the synergetic behavior between the guest polymer and the host solid is interesting for applications involving electrochemical devices. Promising uses of those nanocomposites, based on intercalations of electroactive polymers into transition metal oxides such as V_2O_5 xerogel, chalcogenides such as MoS_2, chlorides such as $RuCl_3$, or oxychlorides such as FeOCl, in modified positive electrodes for rechargeable lithium batteries, are actually in consideration. In fact, PEO can facilitate the mobility of Li^+ ions, whereas intercalation of electronically conducting polymers such as PPy and PANI assures a greater electronic conductivity to the organic-inorganic materials. Resulting electroactive nanocomposites appear as very attractive materials for designing efficient battery electrodes [152–159]. In this way, nanocomposites based on the intercalation of PANI and PPy conducting polymers into V_2O_5 xerogel show electrochemical Li insertion behaviors that are appropriate for their use as cathodes in rechargeable batteries [153] [157–159]. It should be remarked that these hybrids are efficient as electrodes for electrochemical devices only when the polymer is conditioned by a previous oxidation consisting in H_2O_2 or O_2 oxidative treatments [157]. The last procedure involves heating at 150 °C to assure water elimination from the electroactive materials, being usually assumed as a favorable procedure for improving the work of the cathodes, increasing their Li capacity and the cyclability of the devices (Fig. 2.14). The diffusion coefficient of Li^+ ions is increased by one order of magnitude in comparison with the pristine vanadium oxide. The increase of the electronic conductivity due to the molecular coupling between the conducting polymer and the electroactive inorganic host, as well as the enhancement of Li diffusion, are determining factors improving the overall behavior of the Li-cells, in particular at high current densities [158].

Hybrids based on clusters of transition metals (Mo, V, W etc.) and conducting polymers (PANI, PPy etc.) also appear of interest for applications in energy storage [3]. Other polyanions, such as the electroactive hexacyanoferrate $[Fe(CN)_6]^{3-/4-}$, can also form hybrid materials by their combination with PPy and PANI. These materials were tested as active cathode materials in reversible lithium batteries, showing good cyclability and relatively high specific charge (140 Ah/Kg) [160–162].

With the aim of developing ion-selective sensors for detection of ionic species in aqueous solutions, macrocyclic compounds such as crown-ethers, MAO and nonac-

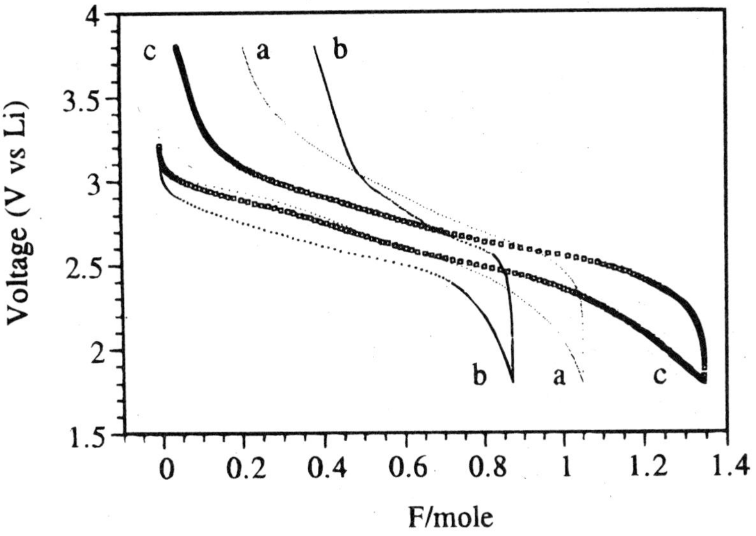

Fig. 2.14 Voltage *vs.* Li insertion degree in electrochemical Li-cells based on V_2O_5-PANI polymer nanocomposites, corresponding to starting V_2O_5 (a), PANI intercalated in V_2O_5 (b), and this last compound treated with O_2 (c), after Leroux et al. [157]

tine, acting as neutral carriers of cations, could be entrapped either in silica or in polysiloxane networks (Sect. 2.3.1). These hybrid materials were used to cover metal substrates that act as electronic collectors to construct modified electrodes. It has been clearly shown that the selectivity of the resulting devices is strongly related to the nature and concentration of both the macrocycle and the electrolyte (analyte). These electrodes are particularly sensitive to alkaline cations, lead, silver and ammonia ions, giving linear potentiometric responses with near-Nernstian slopes in wide cation concentration ranges [163].

Ion-sensing membranes also based in neutral carriers and prepared by sol-gel procedures have been reported by Kimura et al. [164–167]. These authors described the use of valinomycin and a bis-12C4-derivative for applications in ion-sensitive field-effect transistors (ISFETs), in which a voluminous anion such as tetraphenylborate was chemically bonded to the neutral carriers. On the other hand, Reinhoudt et al. [168] developed K^+-selective chemically-modified field-effect transistors (CHEMFETs) using functionalized polysiloxane membranes containing variable amounts of cyanopropylsiloxane groups to modulate the polarity of the hybrid system, incorporating either a free or a covalently attached ionophore such as valinomycin, also bonded to tetraphenylborate anions.

If the modification of electrodes by grafting reactions with organosilanes is interesting to modulate electrical responses towards ionic or molecular species, in some cases there is also of great interest to obtain biocompatible devices. This is for instance the case for iridium pacemaker electrodes which were functionalized by reaction with silanes, introducing epoxy groups able to immobilize PEO in a further reaction [169]. In this way, it significantly enhanced cell adhesion to microelectrodes applied to stimulation and/or recording, showing a new face of the utility of organic-inorganic materials.

2.4 Conclusions

Within the context of nanostructured solids, well-controlled design and preparation of organic-inorganic materials provided with suitable functions is now available [180]. Scientists and engineers have the possibility of modifying inorganic solids by introducing in various designed steps predetermined functions, following processes of grafting or intercalation using selected organic and organo-metallic compounds. Sol-gel procedures, conducted in *one-pot* processes, are also a realistic alternative for producing materials useful as components of devices based on a highly selective molecular adsorbent or catalyst, as well as on optical, opto-electronic and electrochemical properties. Complex hybrid systems exhibiting multifunctionality, i.e. the simultaneous ability to act physically or chemically in controlled ways, represent the opportunity to design new devices based on hybrid materials deriving from combinations at the molecular scale.

Acknowledgements

The author would like to express his sincere gratitude to those researchers and technicians belonging to the CSIC that directly or indirectly have made possible much of the progress reported in this chapter. Friendly encouragement from Prof. J. J. Fripiat and Prof. J. M. Serratosa to work on organic-inorganic functional materials is particularly acknowledged. I would like to thank Dr. P. Aranda and Dr. S. Letaïef for their kind assistance in the revision and management of the manuscript. Finally, financial support from CSIC, CYCIT, EU BRITE–EURAM Program, Comunidad Autónoma de Madrid, Ramon Areces Foundation and TOLSA S.A., is also acknowledged.

References

1 G. A. Ozin, *Adv. Mater.* **1992,** *4,* 612–649.
2 E. Ruiz-Hitzky, *Mol. Cryst. Liq. Cryst. Inc. Nonlin. Op.* **1988,** *161,* 433–452.
3 P. Gómez-Romero, *Adv. Mater.* **2001,** *13,* 163–174.
4 J. M. Thomas, W. J. Thomas: *Principles and Practice of Heterogeneous Catalysis,* VCH, Weinheim, 1997.
5 J. E. Gieseking, *Soil Sci.* **1939,** *47,* 1–13.
6 S. B. Hendricks, *J. Phys. Chem.* **1941,** *45,* 65–81.
7 D. M. C. MacEwan, *Nature,* **1944,** *154,* 577–578.
8 J. W. Jordan, *Mineralog. Mag.* **1949,** *28,* 598–605.
9 G. Lagaly, *Solid State Ionics* **1986,** *22,* 43–51.
10 J. M. Serratosa, *Clays Clay Miner.* **1968,** *16,* 93–97.
11 R. A. Schoonheydt: in *Advanced Techniques for Clay Mineral Analysis,* ed J. J. Fripiat, Elsevier, Lausanne, 1981, 163–189.

12 J. Sanz, J. M. Serratosa: in *Organo-Clay Complexes And Interactions,* eds S. Yariv and H. Cross, Marcel Dekker Inc., New York, 2002.

13 M. Ogawa, K. Kuroda, *Bull. Chem. Soc. Jpn.* **1997**, *70,* 2593–2618.

14 M. Martínez Lara, A. Jiménez López, L. Moreno Real, S. Bruque, B. Casal, E. Ruiz-Hitzky, *Mat. Res. Bull.* **198,** *5,* 549–555.

15 M. L. Rojas-Cervantes, B. Casal, P. Aranda, M. Savirón, J.C. Galván, E. Ruiz-Hitzky, *Coll. Polym. Sci.* **2001**, *279,* 990–1004.

16 W. M. R. Divigalpitiya, R. F. Frindt, S. R. Morrison, *Science* **1989**, *246,* 369–371.

17 D. O'Hare: in *Inorganic Materials,* 2nd edition, eds D.W. Bruce and D. O'Hare, John Wiley & Sons, Chichester, 1993.

18 S. Miyata, T. Kumora, *Chem. Lett.* **1973,** 843–848.

19 F. Leroux, J. P. Besse, *Chem. Mater.* **2001,** *13,* 3507–3515.

20 W.F. Bradley, *J. Am. Chem. Soc.* **1945,** *67,* 975–981.

21 B. K. G. Theng: *The Chemistry of Clay-Organic Reactions,* Adam Hilger Ltd., London, 1974.

22 A. Rausell-Colom, J. M. Serratosa: in *Chemistry of Clays and Clay Minerals,* ed A. C. D. Newman, The Mineralogical Society, London, 1987.

23 S. Yariv: in *Organo-clay Complexes and Interactions,* eds S. Yariv and H. Cross, Marcel Dekker Inc., New York, 2002.

24 M. S. Witthingham, A. J. Jacobson eds: *Intercalation Chemistry,* Academic Press, New York, 1982.

25a) E. Ruiz-Hitzky, B. Casal, *Nature* **1978,** *276,* 596–597.

25b) E. Ruiz-Hitzky, B. Casal, P. Aranda, J. C. Galván, *Rev. Inorg. Chem.* **2001,** *21,* 125–159.

26 J. J. Fripiat, J. Chaussidon, A. Jelly: *Chimie Physique des Phénomènes de Surfaces et Application aux Oxydes et aux Silicates,* Masson, Paris, 1970.

27 E. Gutiérrez, E. Ruiz-Hitzky, *Mol. Cryst. Liq. Cryst. Inc. Nonlin. Opt.* **1988,** *161,* 453–458.

28 H. E. Doner, M. M. Mortland, *Science* **1969,** *166,* 1406–1407.

29 E. Ruiz-Hitzky, *Adv. Mater.* **1993,** *5,* 334–340.

30 M. G. Kanatzidis, L. M. Tonge, T. J. Marks, H. O. Marcy, C. R. Kannewurf, *J. Amer. Chem. Soc.* **1986,** *109,* 3797–3799.

31 R. Schöllhorn: *Intercalation Chemistry,* eds M. S. Whittingham and A. J. Jacobson, Academic Press, New York, 1982.

32 A. Blumstein, *Bull. Soc. Chim. Fr.* **1961,** 899–905.

33 H. Z. Friedlander, *ACS Div. Polym. Chem. Reprints* **1963,** *4,* 300–306.

34 Y. Fukushima, A. Okada, M. Kawasumi, T. Kurauchi, O. Kamigaito, *Clay Miner.* **1988,** *23,* 27–34.

35 E. Ruiz-Hitzky, P. Aranda, *Adv. Mater.* **1990,** *2,* 545–547.

36 P. Aranda, E. Ruiz-Hitzky, *Chem. Mater.* **1992,** *4,* 1395–1403.

37 J. Wu, M. M. Lerner, *Chem. Mater.* **1993,** *5,* 835–838.

38 P. Aranda, E. Ruiz-Hitzky, *Acta Polymer.* **1994,** *45,* 59–67.

39 R. A. Vaia, H. Ishii, E. P. Giannelis, *Chem. Mater.* **1993,** *5,* 1694–1696.

40 R. A. Vaia, S. Vasudevan, W. Krawiec, L. G. Scanlon, E. P. Giannelis, *Adv. Mater.* **1995,** *7,* 154–156.

41 E. P. Giannelis, *Adv. Mater.* **1996,** *8,* 29–35.

42 P. Aranda, J. C. Galvan, E. Ruiz-Hitzky, in *Organic/Inorganic Hybrid Materials,* eds R. M. Laine, C. Sanchez, C. J. Brinker, E. Giannelis, MRS Symposium Proceedings, Warrendale, 1998, 375–380, Vol. 519.

43 V. Mehrota, E. P. Giannelis, *Solid State Ionics,* **1992,** *51,* 115–122.

44 E. Ruiz-Hitzky, P. Aranda: in *Polymer-Clay Nanocomposites,* eds T. J. Pinnavaia and G. W. Beall, John Wiley & Sons, Chichester, 2000.

45 T.-C. Chang, S.-Y. Ho, K.-J. Chao, *J. Chin. Chem. Soc.* **1992,** *39,* 209–212.

46 L. Wang, P. Brazis, M. Rocci, C. R. Kannewurf, M. G. Kanatzidis: in *Organic/ Inorganic Hybrid Materials,* eds R. M. Laine, C. Sanchez, C. J. Brinker, E. Giannelis, MRS Symposium Proceedings, Warrendale, 1998, 257–264, Vol. 519.

47 E. Ruiz-Hitzky, R. Jiménez, B. Casal, V. Manríquez, A. Santa Ana, G. González, *Adv. Mater.* **1993,** *5,* 738–741.

48 K. A. Carrado, L. Xu, *Chem. Mater.* **1998,** *10,* 1440–1445.
49 K. A. Carrado L. Xu, S. Seifert, R. Csencsits: in *Polymer-Clay Nanocomposites,* eds T. J. Pinnavaia and G. W. Beall, John Wiley & Sons, Chichester, 2000.
50 K. A. Carrado, *Appl. Clay Sci.* **2000,** *17,* 1–23.
51 F. Leroux, P. Aranda, J. P. Besse, E. Ruiz-Hitzky, *Eur. J. Inorg. Chem.* **2003,** 1242–1251.
52 R. K. Iler: *The Chemistry of Silica,* John Wiley & Sons, New York, 1979.
53 F. Aragón de la Cruz, F. Esteban, C. Vitón, in *Proc. Int. Clay Conf. Madrid,* ed J. M. Serratosa, División de Ciencias CSIC, Madrid, 1973.
54 a) H. P. Eugster, *Science* **1967,** *157,* 1177–1180; b) G. Lagaly, K. Beneke, A. Weiss, *Naturforsch.* **1973,** *28b,* 234–238.
55 J. M. Rojo, E. Ruiz-Hitzky, J. Sanz, J. M. Serratosa, *Rev. Chim. Miner.* **1983,** *20,* 807–816.
56 E. Ruiz-Hitzky, J. M. Rojo, *Nature* **1980,** *287,* 28–30.
57 E. Ruiz-Hitzky, J. M. Rojo, G. Lagaly, *Coll. Polym. Sci.* **1985,** *263,* 1025–1030.
58 M. Ogawa, S. Okutomo, K. Kuroda, *J. Am. Chem. Soc.* **1998,** *120,* 7361–7362.
59 M. Ogawa, M. Miyoshi, K. Kuroda, *Chem. Mater.* **1998,** *10,* 3787–3789.
60 C. W. Lentz, *Inorg. Chem.* **1964,** *3,* 574–579.
61 J. J. Fripiat, E. Mendelovici, *Bull. Soc. Chim. Fr.* **1968, 483–492.**
62 E. Ruiz-Hitzky, A. Van Meerbeek, *Coll. Polym. Sci.* **1978,** *256,* 135–139.
63 A. Van Meerbeek, E. Ruiz-Hitzky, *Coll. Polym. Sci.* **1979,** *257,* 178–181.
64 E. Ruiz-Hitzky, J. J. Fripiat, *Clays Clay Miner.* **1976,** *24,* 25–30.
65 A. J. Aznar, E. Ruiz-Hitzky, *Mol. Cryst. Liq. Cryst. Inc. Nonlin. Opt.* **1988,** *161,* 459–469.
66 B. Siffert, H. Biava, *Clays Clay Min.* **1976,** *24,* 303–311.
67 M. N. Fernández-Hernández, E. Ruiz-Hitzky, *Clay Miner.* **1979,** *14,* 295–305.
68 E. Ruiz-Hitzky, M. N. Fernández-Hernández, J. M. Serratosa, Spanish Patent 479518, 1979.
69 B. Casal, E. Ruiz-Hitzky, in *Proc. III European Clay Conference Oslo,* ed I. Th. Rosenqvist, Nordic Society for Clay Research, Oslo, 1977, Vol. 1.
70 B. Casal, E. Ruiz-Hitzky, *An. Quím.* **1984,** *80,* 315–320.
71 B. Casal, E. Ruiz-Hitzky, J. M. Serratosa, Spanish Patent 480550, 1980.
72 E. Ruiz-Hitzky, B. Casal, *J. Catal.* **1985,** *92,* 291–295.
73 S. Yamanaka, K. Yamanaka, M. Hattori, *J. Incl. Phenom.* **1984,** *2,* 297–304.
74 C. Sanchez, F. Ribot, *New J. Chem.* **1994,** *18,* 1007–1047.
75 P. Judeinstein, C. Sanchez, *J. Mater. Chem.* **1996,** *6,* 511–525; C. Sanchez, F. Ribot, B. Lebeau, *J. Mater. Chem.* **1999,** *9,* 35–44.
76 C. J. Brinker, G. W. Scherer: *Sol-Gel Science. The Physics and Chemistry of Sol-Gel Processing,* Academic Press, Boston, 1990.
77 J. S. Beck, U.S. Patent N° 5.057.57296, 1991.
78 J. S. Beck, J. C. Vartuli, W. J. Roth, M. E. Leonowicz, C. T. Kresge, K. D. Schmitt, C. T-W. Chu, D. H. Olson, E. W. Sheppard, S. B. Mc Cullen, J. B. Higgins, J. L. Schlenker, *J. Am. Chem. Soc.* **1992,** *114,* 10834–10843.
79 M. H. Lim, A. Stein, *Chem. Mater.* **1999,** *11,* 3285–3295.
80 F. De Juan, E. Ruiz-Hitzky, *Adv. Mater.* **2000,** *12,* 430–432.
81 E. Ruiz-Hitzky, S. Letaïef, V. Prévot, *Adv. Mater.* **2002,** *14,* 439–443.
82 K. J. Shea, D. A. Loy, O. W. Webster, *J. Am. Chem. Soc.* **1992,** *114,* 6700–6710.
83 R. J. P. Corriu, J. J. E. Moreau, P. Thepot, M. Wong Chi Man, *Chem. Mater.* **1992,** *4,* 1217–1224.
84 J. D. Lichtenhan, N. Q. Vu, J. A. Carter, J. W. Gilman, F. J. Feher, *Macromolecules* **1993,** *26,* 2141–2142.
85 D. A. Loy, K. J. Shea, *Chem. Rev.* **1995,** *95,* 1431–1442.
86 K. J. Shea, D. A. Loy, *Chem. Mater.* **2001,** *13,* 3306–3319.
87 R. M. Barrer, *Clays Clay Miner.* **1989,** *37,* 385–395.
88 T. González Carreño, J. A. Martin Rubi, *J. Chromatogr.* **1977,** *133,* 184–189.
89 A. Yamagishi: in *Proc. Int. Clay Conf. Denver,* eds L. G. Schulz, H. van Olphen, F. A. Mumpton, The Clay Mineral Society, Denver, 1985, 329–334.
90 W. F. Jaynes, S. A. Boyd, *Soil Sci. Soc. Am. J.* **1991,** *55,* 43–48.

91 S. Wu, G. Sheng, S. A. Boyd, *Adv. Agron.* **1997**, *59*, 25–62.

92 G. Sheng, X. Wang, S. Wu, S. A. Boyd, *J. Environ. Qual.* **1998**, *27*, 806–814.

93 Y. El Nahal, S. Nir, C. Serban, O. Rabinovitch, B. Rubin, *J. Agric. Food. Chem.* **2000**, *48*, 4791–4801.

94 Y. Yan, T. Bein, *Chem. Mater.* **1993**, *5*, 905–907.

95 J. H. Fendler: *Membrane Mimetic Chemistry*, John Willey & Sons, New York, 1982.

96 E. Ruiz-Hitzky, P. Aranda, B. Casal, J. C. Galván, *Adv. Mater.* **1995**, *7*, 180–184.

97 P. Aranda, J. C. Galván, B. Casal, E. Ruiz-Hitzky, *Coll. Polym. Sci.* **1994**, *272*, 712–720.

98 J. C. Galván, P. Aranda, J. M. Amarilla, B. Casal, E. Ruiz-Hitzky, *J. Mater. Chem.* **1993**, *3*, 687–688.

99 A. Jiménez-Morales, J. C. Galván, P. Aranda, E. Ruiz-Hitzky: in *Organic/Inorganic Hybrid Materials*, eds R. M Laine, C. Sánchez, C. J. Brinker, E. Giannelis, MRS Symp. Proc., Warrendale, 1998, 211–216, Vol. 519.

100 C. Guizard, P. Lacan, *New J. Chem.* **1994**, *18*, 1097–1107.

101 P. Lacan, C. Guizard, P. Le Gall, D. Wettling, L. Cot, *J. Membr. Sci.* **1995**, *100*, 99–109.

102 T. G. Waddell, D. E. Leyden, *J. Org. Chem.* **1981**, *46*, 2406–2407.

103 J. S. Bradshaw, R. L. Bruening, K. E. Krakowiak, B. J. Tarbet, M. L. Bruening, R. M. Izatt, J. J. Christensen, *J. Chem. Soc., Chem. Commun.* **1988**, 814–817.

104 M. Nakajima, K. Kimura, T. Shono, *Anal. Chem.* **1983**, *55*, 463–467.

105 T. Iwachido, H. Naito, F. Samukawa, K. Ishimaru, K. Tôe, *Bull. Chem. Soc. Jpn.* **1986**, *59*, 1475–1480.

106 E. Blasius, K. P. Janzen, W. Klein, H. Klotz, V. B. Nguyen; T. Nguyen-Tien, R. Pfeiffer, G. Scholten, H. Simon, H. Tockemer, A. Toussaint, *J. Chromatogr.* **1980**, *201*, 147–166.

107 E. Blasius, K. P. Janzen, *Israel J. Chem.* **1985**, *26*, 25–34.

108 K. Kimura, T. Shono, *J. Liq. Chromatogr.* **1982**, *5*, 223–255.

109 O. Lev, Z. Wu, S. Bharathi, V. Glezer, A. Modestov, J. Gun, L. Ravinovich, S. Sampath, *Chem. Mater.* **1997**, *9*, 2354–2375.

110 M. Barboiu, C. Guizard, N. Hovnanian, J. Palmeri, C. Reibel, L. Cot, C. Luca, *J. Membr. Sci.* **2000**, *172*, 91–103.

111 P. Aranda, A. Jiménez-Morales, J. C. Galván, B. Casal, E. Ruiz-Hitzky, *J. Mater. Chem.* **1995**, *5*, 817–825.

112 M. M. Mortland, V. Berkheiser, *Clays Clay Miner.* **1976**, *24*, 60–63.

113 T. J. Pinnavaia, *Science* **1983**, *220*, 365–371.

114 P. Hernan, C. Del Pino, E. Ruiz-Hitzky, *Chem. Mater.* **1992**, *4*, 49–55.

115 R. V. Parish, D. Habibi, V. Mohammadi, *J. Organomet. Chem.* **1989**, *369*, 17–28.

116 Ch. Merkckle, J. Blümel, *Chem. Mater.* **2001**, *13*, 3617–3628.

117 J. Barrios, G. Poncelet, J. J. Fripiat, *J. Catal.* **1981**, *63*, 362–370.

118 B. Casal, F. Bergaya, D. Challal, J. J. Fripiat, E. Ruiz-Hitzky, H. Van Damme, *J. Mol. Catal.* **1985**, *33*, 83–86.

119 J. Barrios-Neira, L. Rodrique, E. Ruiz-Hitzky, *J. Microsc. Spec. Electron.* **1974**, *20*, 295–298.

120 D. E. De Vos, I. F. J. Vankelecom, P. A. Jacobs: *Chiral Catalysts Immobilization and Recycling*, Wiley-VCH, Weinheim, 2000

121 A. Corma, M. Iglesias, C. del Pino, F. Sanchez, *J. Chem. Soc., Chem. Commun.* **1991**, 1253–1255.

122 A. Corma, M. Iglesias, C. del Pino, F. Sanchez, *J. Organomet. Chem.* **1992**, *431*, 233–246.

123 K. Bergman, C. T. O'Konski, *J. Phys. Chem.* **1963**, *67*, 2169–2177.

124 J. Cenens, R. Schoonheydt, *Clays Clay Miner.* **1988**, *36*, 214–224.

125 A. J. Aznar, B. Casal, E. Ruiz-Hitzky, I. Lopez-Arbeloa, F. Lopez-Arbeloa, J. Santaren, A. Alvarez, *Clay Miner.* **1992**, *27*, 101–108.

126 B. Ackermans, R. A. Schoonheydt, E. Ruiz-Hitzky, *Chem. Soc., Faraday Trans.* **1996**, *92*, 4479–4484.

127 T. Fujita, N. Iyi, T. Kosugi, A. Ando, T. Deguchi, T. Sota, *Clays Clay Miner.* **1997**, *45*, 77–84.

128 G. Rytwo, S. Nir, L. Margulies, B. Casal, J. Merino, E. Ruiz-Hitzky, J. M. Serratosa, *Clays Clay Miner,* **1998,** *46,* 340–348.
129 E. Ruiz-Hitzky, *J. Mater. Chem.* **2001,** *11,* 86–91.
130 B. Casal, J. Merino, J. M. Serratosa, E. Ruiz-Hitzky, *Appl. Clay Sci.* **2001,** *18,* 245–254.
131 L. Margulies, H. Rozen, E. Cohen, *Nature* **1985,** *315,* 658–659.
132 T. Fujita, N. Iyi, Z. Klapyta, *Mat. Res. Bull.* **1998,** *33,* 1693–1701.
133 S. Bénard, A. Léaustic, E. Rivière, P. Yu, R. Clément, *Chem. Mater.* **2001,** *13,* 3709–3716.
134 S. Bernard, P. Yu, E. Rivière, K. Nakatani, J. F. Delouis, *Chem. Mater.* **2001,** *13,* 159–162.
135 E. Coronado, J. R. Galanmascaros, C. J. Gomezgarcia, V. Laukhin, *Nature* **2000,** *408,* 447–449.
136 I. Braun, G. Ihlein, F. Laeri, J. U. Nöckel, G. Schulz-Ekloff, F. Schüth, U. Vietze, O. Weiss, D. Wöhrle, *Appl. Phys. B* **2000,** *70,* 335–344.
137 E. Ruiz-Hitzky, *J. Mater. Chem.* **2001,** *11,* 86–91.
138 K. M. Choi, K. J. Shea, *Chem. Mater.* **1993,** *5,* 1067–1069.
139 H. Wellmann, J. Rathousky, M. Wark, A. Zukal, G. Schulz-Ekloff, *Microporous & Mesoporous Mater.* **2001,** *44,* 419–425.
140 T. Hirai, H. Okubo, I. Komasawa, *J. Coll. Interf. Sci.* **2001,** *235,* 358–364.
141 F. R. Gamble, F. J. Disalvo, R. A. Klemm, T. H. Geballe, *Science* **1970,** *168,* 568–570.
142 F. R. Gamble, J. H. Osiecki, F. J. DiSalvo, *J. Chem. Phys.* **1971,** *55,* 3545–3530.
143 F. R. Gamble, A. H. Thompson, *Solid State Commun.* **1978,** *27,* 379–382.
144 P. Aranda, Y. Mosqueda, E. Pérez-Cappe, E. Ruiz-Hitzky, *J. Polym. Sci. B Polym. Phys.* (in press).
145 Y.-J. Liu, D. C. DeGroot, J. L. Schindler, C. R. Kannewurf, M. G. Kanatzidis, *Chem. Mater.* **1991,** *3,* 992–994.
146 E. Ruiz-Hitzky, P. Aranda, B. Casal, *J. Mater. Chem.* **1992,** *2,* 581–582.
147 R. Bissessur, M. G. Kanatzidis, J. L. Schindler, C. R. Kannewurf, *J. Chem. Soc., Chem. Commun.* **1993,** 1582–1585.
148 L. Wang, M. G. Kanatzidis, *Chem. Mater.* **2001,** *13,* 3717–3727.
149 I. Lagadic, A. Léaustic, R. Clément, *J. Chem. Soc., Chem. Commun.* **1992,** 1396–1397.
150 D. Ravaine, A. Seminel, Y. Charbouillot, M. Vincens, *J. Non-Cryst. Solids* **1986,** *82,* 210–219.
151 A. J. Aznar, J. Sanz, E. Ruiz-Hitzky, *Colloid & Polym. Sci.* **1992,** *270,* 165–176.
152 G. M. Kloster, J. A. Thomas, P. W. Brazis, C. R. Kannewurf, D. F. Shriver, *Chem. Mater.* **1996,** *8,* 2418–2420.
153 M. Lira-Cantú, P. Gomez-Romero, *J. Electrochem. Soc.* **1999,** *146,* 2029–2033.
154 M. Lira-Cantú, P. Gomez-Romero, *J. New Mater. Electrochem. Syst.* **1999,** *2,* 142–144.
155 M. Lira-Cantú, P. Gomez-Romero, *Int. J. Inorg. Mater.* **1999,** *1,* 111–116.
156 E. Shouji, D. A. Buttry, *Langmuir* **1999,** *15,* 669–673.
157 F. Leroux, B. E. Koene, L. F. Nazar, *J. Electrochem. Soc.* **1996,** *143,* L181–L183.
158 F. Leroux, G. Goward, W. P. Power, L. F. Nazar, *J. Electrochem. Soc.* **1997,** *144,* 3886–3895.
159 G. Goward, F. Leroux, W. P. Power, L. F. Nazar, *Electrochim. Acta* **1998,** *43,* 1307–1313.
160 G. Torres-Gomez, M. Lira-Cantu, P. Gomez-Romero, *J. New Mater. Electrochem. Syst.* **1999,** *2,* 145–150.
161 G. Torres-Gomez, K. West, S. Skaarup, P. Gomez-Romero, *J. Electrochem. Soc.* **2000,** *147,* 2513–2516.
162 G. Torres-Gomez, E. M. Tejada-Rosales, P. Gomez-Romero, *Chem. Mater.* **2001,** *13,* 3693–3697.
163 E. Ruiz-Hitzky, J. C. Galván, A. Jiménez-Morales, P. Aranda, Spanish Patent P. 9900956, 1999.
164 K. Kimura, T. Sunagawa, M. Yokoyama, *Chem. Commun.* **1996,** 745–746.
165 Y. Tsujimura, T. Sunagawa, A. Yokoyama, K. Kimura, *Analyst,* **1996,** *121,* 1705–1709.
166 K. Kimura, T. Sunagawa, M. Yokoyama, *Anal. Chem.* **1997,** *69,* 2379–2393.

167 K. Kimura, T. Sunagawa, S. Yajima, S. Miyake, M. Yokoyama, *Anal. Chem.* **1998**, *70*, 4309–4313.

168 D. N. Reinhoudt, J. F. J. Engbersen, Z. Brzózka, H. H. Van den Vlekkert, G. W. N. Honig, H. A. J. Holterman, U. H. Verkerk, *Anal. Chem.* **1994**, *66*, 3618–3623.

169 M. Stelzle, R. Wagner, W. Nisch, W. Jägermann, R. Fröhlich, M. Schaldach, *Biosensors & Bioelectron.* **1997**, *12*, 853–865.

170 R. K. Iller, US Patent 2,657.149, 1953.

171 F. Aragón, J. Cano Ruiz, D. M. C. MacEwan, *Nature* **1959**, *183*, 740–741.

172 E. Michel, A. Weiss, *Z. Naturforschg.* **1965**, *20b*, 1307–1308.

173 P. Hagenmuller, J. Portier, B. Barbe, P. Bouclier, *Z. Anorg. Allg. Chem.* **1967**, *355*, 209–218.

174 A. Weiss, R. Ruthard, *Z. Naturforschg.* **1969**, *24b*, 355.

175 S. Yamanaka, *Inorg Chem.* **1976**, *15*, 2811–2817.

176 G. Alberti, U. Costantino, S. Allulli, N. Tomassini, *J. Inorg Nucl. Chem.* **1978**, *40*, 1113–1117.

177 G. Lagaly, *Adv. Coll. Interf. Sci.* **1979**, *11*, 105–148.

178 H. Schmidt, G. Tünker, H. Scholze, German Patent 3,011.761, 1980.

179 P. Gómez-Romero, M. Lira-Cantú, *Adv. Mater.* **1997**, *9*, 144–147.

180 E. Ruiz-Hitzky, *Chem. Rec.* **2003**, *3*, 88–100.

3
Bridged Polysilsesquioxanes. Molecular-Engineering Nanostructured Hybrid Organic-Inorganic Materials

K. J. Shea, J. Moreau, D. A. Loy, R. J. P. Corriu, B. Boury

3.1
Introduction

Bridged polysilsesquioxanes are a family of hybrid organic-inorganic materials prepared by sol-gel processing of monomers (Fig. 3.1) that contain a variable organic bridging group and two or more trialkoxysilyl groups [1–3]. This arrangement offers exceptional opportunities to combine the important properties from both organic and inorganic realms and to create entirely new compositions with truly unique properties. This molecular versatility, coupled with the mild sol-gel conditions and the ability to prepare bulk, thin films and fibers, makes this group of chemically and thermally robust materials a key resource in advanced materials science and technology. This chapter reviews the syntheses of bridged monomers and their sol-gel polymerization to hyper-cross-linked networks, and how the polymerization conditions, methods for post-gelation processing and the nature of the bridging group affect the physical and chemical properties of the resulting materials.

The key to this versatility, and much of the unique properties of this class of materials, comes from the bridging organic group that is covalently attached to the polymerizable trialkoxysilyl groups through Si–C bonds. This organic group can be varied in length, rigidity, geometry of substitution, and functionality. Because the organic group remains an integral component of the material, this variability provides an opportunity to engineer bulk properties such as porosity, thermal stability, refractive index, optical clarity, chemical resistance, hydrophobicity, and dielectric constant without the threat of phase segregation at longer length scales. The fine degree of control over bulk chemical and physical properties has made these materials excellent candidates for applications ranging from optical device fabrication [4] to catalyst supports [5–7] and ceramics precursors [8].

A few representatives of bridged polysilsesquioxane monomers are shown in Fig. 3.2. The organic fragments in the building blocks range from rigid arylenic (Fig. 3.2, **1** and **2**) [9–13], acetylenic (Fig. 3.2, **3**, **4**)[12, 14–19], and olefinic (Fig. 3.2, **5–7**) [20–23] bridging groups to flexible alkylenes ranging from 1 to 14 methylene groups (Fig. 3.2, **8**, **9**) in length [11, 24–30]. They also include a variety of functio-

Gómez-Romero: Organic-Inorganic Materials. K. J. Shea

ISBN: 3-527-30484-3

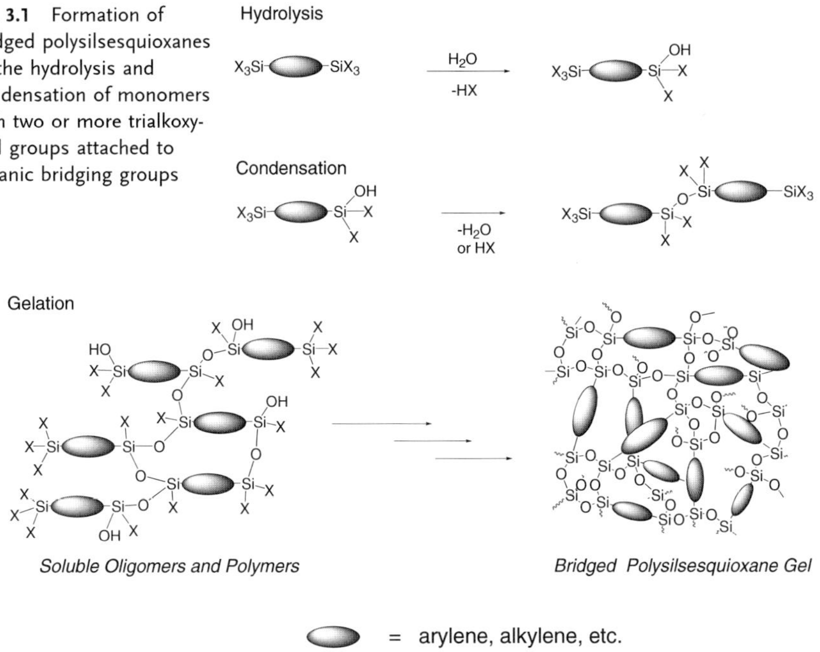

Fig. 3.1 Formation of bridged polysilsesquioxanes by the hydrolysis and condensation of monomers with two or more trialkoxysilyl groups attached to organic bridging groups

nalized groups such as amines (Fig. 3.2, **10**) [3, 31–36], ethers (Fig. 3.2, **11**) [34, 37], sulfides (Fig. 3.2, **12**) [33, 38, 39], phosphines (Fig. 3.2, **13**) [40–46], amides (Fig. 3.2, **14**) [47, 48], ureas (Fig. 3.2, **15**) [49–58], carbamates [31, 59–61], carbonates [62], viologens (Fig. 3.2, **16**) [63, 64] and azobenzenes (Fig. 3.2, **17**) [65]. In addition, bridging groups have included organometallics in which the metal atom is part of the bridge as in this ferrocenyl-bridged monomer (Fig. 3.2, **18**) [66–68] or pendant to the bridge as in this eta-6-arenechromiumtricarbonyl complex (Fig. 3.2, **19**) [69–71, 72, 73, 74]. At present there are a few commercially available monomers, mostly for surface modification or coating applications (Fig. 3.2, **3**, **5**, **8–10**, **15**) [75–87], but new bridged monomers are beginning to appear for opto-electronic applications [88].

Sol-gel polymerization of these molecular building blocks permits the rapid formation of bridged polysilsesquioxanes that irreversibly form gels. The organic group, which comprises approximately 40–60 wt percent of the material, is an integral part of the network architecture. Upon drying, these gels afford amorphous xerogels or aerogels whose surface areas can be tailored, through selection of the organic spacer, to be as high as 1800 m^2g^{-1} in supercritical processed aerogels [13, 26, 89, 90] or, at the other extreme, nonporous glassy xerogels [25, 30, 91, 92]. Bulk properties such as *pore size* may be controlled with a fidelity that is more reminiscent of surfactant-templated mesoporous molecular sieves. Optical properties can be manipulated by incorporating chromophores in the bridging organic component [93–95]. Most recently, functional bridged polysilsesquioxanes have been prepared for use as high-capacity adsorbents [58, 96–102].

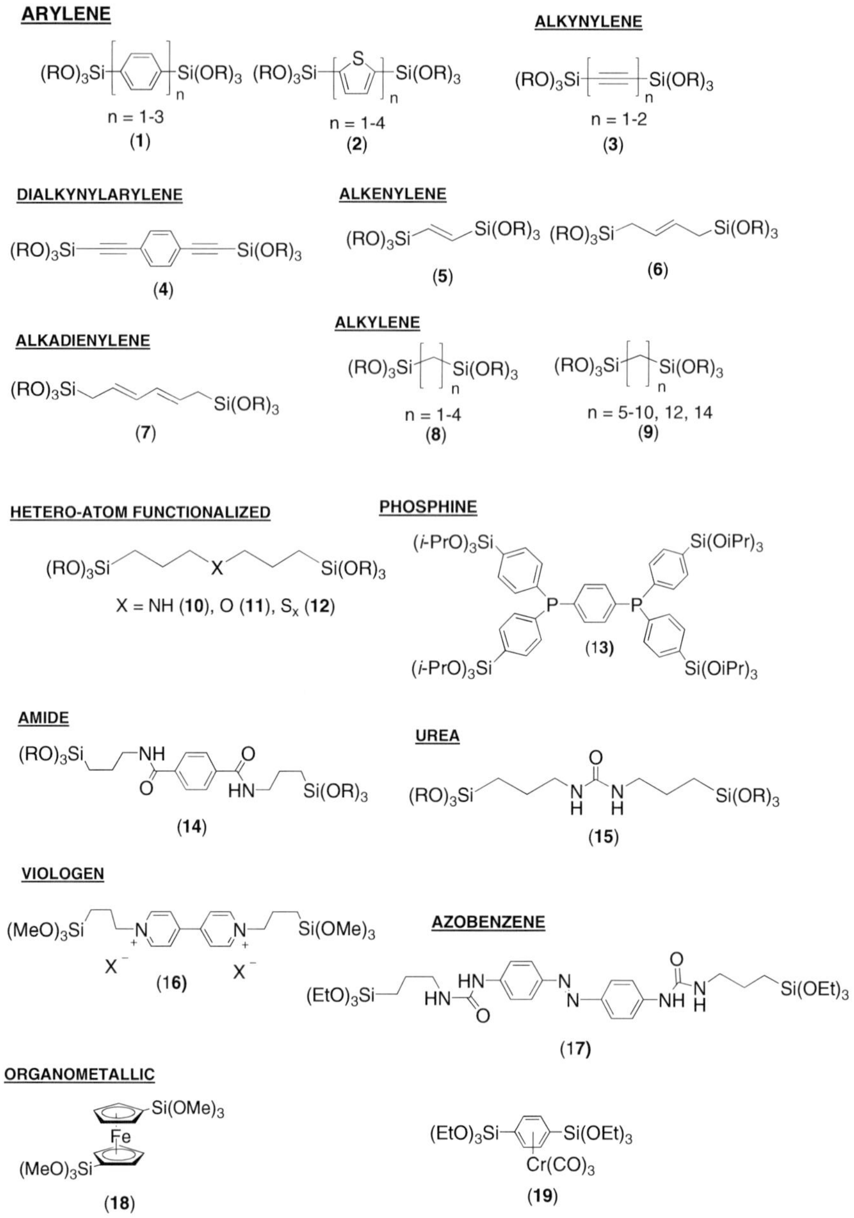

Fig. 3.2 Representative bridged polysilsesquioxane monomers

3.2 Historical Background

Sol-gel processable monomers containing two or more trichlorosilyl or trialkoxysilyl groups (Fig. 3.1) have been known for over 55 years [103]. Prior to the late 1980s, virtually all of these compounds were used as coupling agents, surface modifiers, or coatings, and as components of adhesive formulations. For example, the phenylene (Fig. 3.2, **1**) and acetylene (Fig. 3.2, **2**) [104] bridged monomers were first prepared in the 1950s for use as coatings on glass. The dipropyltetrasulfide-bridged monomer or "Si-69" (Fig. 3.2, **12**) was developed as a coupling agent for elastomers in the 1970s [33, 38, 39]. More than 30×10^6 kg of Si-69 is produced each year; much of this is used in silica-rubber composites.

Initial investigations of bridged polysilsesquioxanes with rigid arylene (Fig. 3.2, **1**) and acetylene (Fig. 3.2, **3**) bridging groups were undertaken to determine if the porosity of amorphous hybrid materials could be controlled at the molecular level [9–13]. Subsequently, bridged polysilsesquioxanes were prepared from monomers with trimethoxysilyl groups [105] and a rapidly growing number of new bridging groups [3, 106]. The primary focus was still control over porosity, based on the nature of the organic group or through its destruction as a template. Since then, the field has broadened in scope to include control of the optical [4], thermomechanical [34, 107], and chemical properties. As was mentioned earlier, a bridging organic group has also been reported as an expedient means for securely attaching organic functionalities, such as dyes or metals, to existing porous materials. The ease with which porous bridged polysilsesquioxane gels can be prepared also led to their application as encapsulants for biochemicals [108–111]. Recently, efforts to prepare functionalized materials have evolved into strategies for controlling the long-range order based on surfactant templating, hydrogen bonding, organometallic complexation, and mesogenic interactions. The rest of this chapter will focus first on how various monomers can be prepared, continue with a review of the basics for bridged polysilsesquioxane sol-gel polymerization and processing, and end with a survey of applications.

3.3 Monomer Synthesis

One of the attractive features of working with bridged polysilsesquioxanes is the relative ease with which the monomers can be prepared. This means that one can make monomers quickly and begin to study the construction of the hyper-cross-linked materials with little delay. There are a number of useful synthetic methods that permit bridged monomers to be prepared in one or two steps from readily available starting materials. The three most commonly used approaches are by (i) metallation of aryl, alkyl, and alkynyl precursors followed by reaction with a tetrafunctional silane, (ii) hydrosilylation of dienes (or polyenes) or, less commonly, diynes, and (iii) reaction of a bifunctional organic group with an organotrialkoxysilane bearing a reactive functional group.

3.3.1
Metallation

Metallation includes the reactions of arylene Grignards with tetralkoxysilanes (Fig. 3.3a) [12, 105] or from Grignard reagents bearing trialkoxysilyl groups (Fig. 3.3b) [112, 113], lithium-halogen exchange (Fig. 3.3c) [11, 12, 105], and deprotonation of acetylenes (Fig. 3.3d) [12, 16, 17, 104, 114], or treatment of bis(trialkoxysilyl)methane with a Grignard, organolithium, or metal hydride [115] (Fig. 3.3e). In each case, the resulting organometallic reagent is reacted with a tetraalkoxysilane or chlorotrialkoxysilane or other electrophiles to give the final products in moderate to good yields.

3.3.2
Hydrosilylation

Hydrosilylation (Fig. 3.4) is an efficient reaction for preparing bridged monomers in high yields from chemicals bearing two or more terminal olefins [25, 26, 116, 117]. The reaction has been used to create monomers with alkylene and hetero-

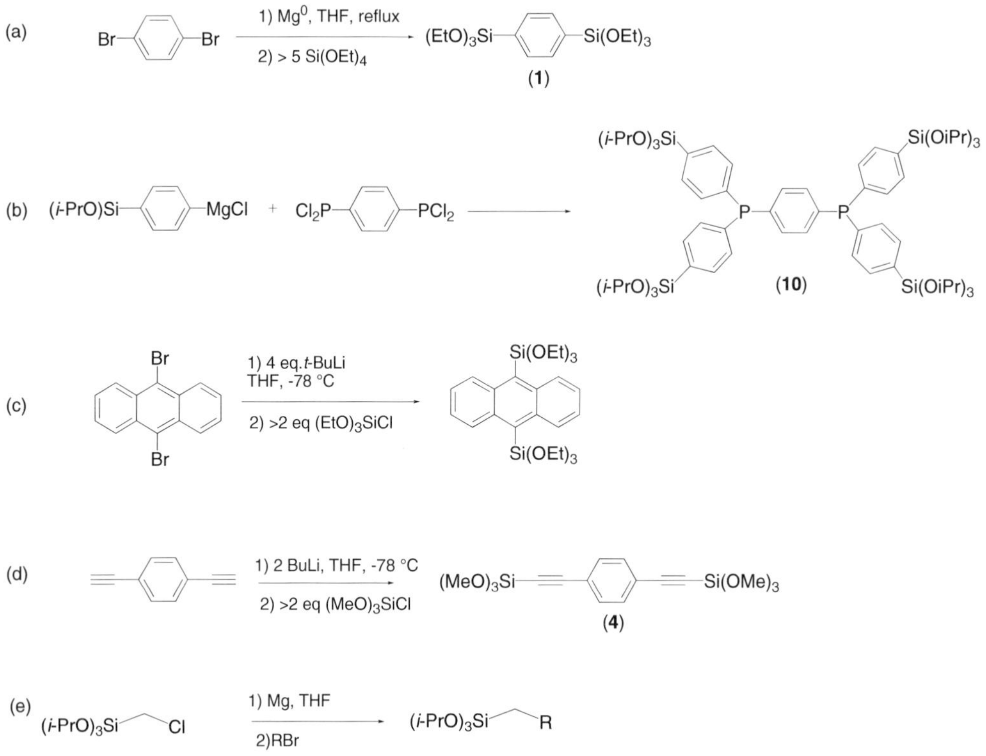

Fig. 3.3 Preparation of monomers by metal-halogen exchange or metallation chemistry

atom-functionalized bridging groups. Addition of the Si–H group in trichlorosilanes or trialkoxysilanes across carbon-carbon double bonds is most often catalyzed with a noble metal catalyst such as chloroplatinic acid or Karsted's or Spier's catalyst [118], usually placing the silicon at the terminal position of the double bond (Fig. 3.4a). Hydrosilylation of 1,5-hexadiene (Fig. 3.4b) affords the 1,6-bis(trimethoxysilyl)hexane (Fig. 3.4, **9**) as a clear colorless oil. The 1,4-butylene-bridged monomer (Fig. 3.4, **8**) can be prepared from butadiene by palladium-catalyzed hydrosilylation, *in situ* isomerization to afford the 3-butenyltrichlorosilane, followed by a second hydrosilylation (Fig. 3.4c) [25]. The resulting trichlorosilane is readily converted to trialkoxysilanes with trialkylorthoformates [25] or with alcohols and an amine [12].

3.3.3
Functionalization of an Organotrialkoxysilane

This synthetic route has become increasingly common because it permits a great number of bridging groups to be prepared from readily available starting materials (Fig. 3.5). For example, an electrophilic substituent on the organotrialkoxysilane can be reacted with any organic molecule with two or more nucleophilic groups to create a bridged monomer. Suitable electrophilic groups available in commercially available silane coupling agents include isocyanates, alkyl or benzyl halides, epoxides, and acrylates. Isocyanates, the most frequently used electrophile for preparing new bridged monomers, react readily with amines (Fig. 3.5a) to give urea linkages (Fig. 3.5, **15**) [50, 52–56, 119] with alcohols (Fig. 3.5b) in the presence of tin or acidic catalysts to give urethanes or carbamates (Fig. 3.5, **19**) [31, 61, 120–122], or with carboxylic acids to give, after decarboxylation, an amide linkage. Alkyl halide substituted organotrialkoxysilanes (Fig. 3.5c) have been used with diamines to give bridging groups with amino functionalities (Fig. 3.5, **20**) [7, 123–127]. Amines have proven to be one of the most useful starting materials for preparing bridged mono-

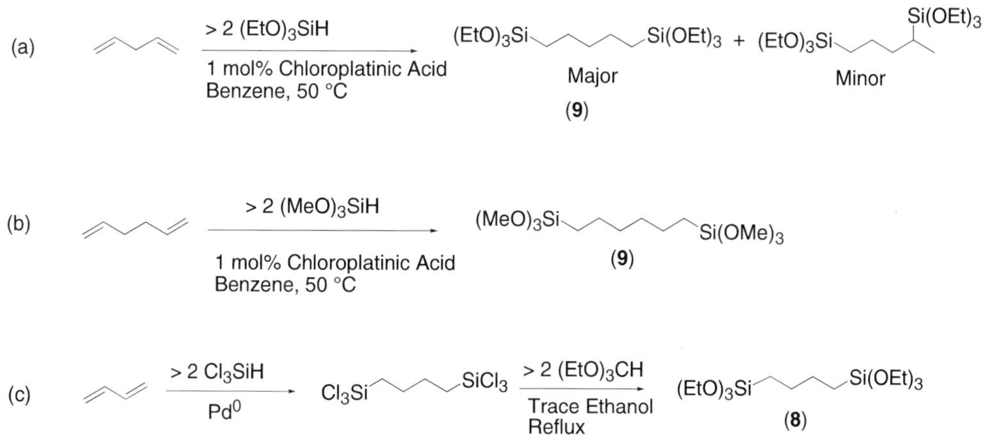

Fig. 3.4 Preparation of monomers by hydrosilylation reactions

mers. A number of amide-containing bridges (Fig. 3.5 d) have been prepared from precursors bearing two or more sulfonyl chlorides or acid chlorides (Fig. 3.5, **21**) [47]. Bridging groups based on Schiff bases (Fig. 3.5, **22**) have been prepared by reacting (aminopropyl)trialkoxysilanes with di- or trialdehydes (Fig. 3.5 e) [128–130].

3.3.4
Other Approaches

The reaction of the silyl anion of trichlorosilane with allyl or benzyl halides has been used to prepare 2,4-hexadienylene (**7**) [22], 2-butenylene (Fig. 3.6 a&b, **6**)[22, 23, 131], and xylylene (Fig. 3.6 c, **23**) [132–134] bridged monomers. Other approaches include ruthenium-catalyzed silylation/desilylation of vinyltriethoxysilane (Fig. 3.6 d) to afford a mixture of the *E*-isomer of the ethenylene-bridged monomer (Fig. 3.6, **5**) and the vinylidene isomer [20, 21, 135], photochemical isomerization of the *E*-isomer to the *Z* isomer of **5** (Fig. 3.6 e) [20, 21], Heck vinylation (Fig. 3.6 f) to afford crown- [136, 137] or oligoarylenevinylene-bridged monomers (Fig. 3.6, **24**) [18, 127, 138], and the Diels-Alder reaction (Fig. 3.6 g) of 1,2-bis(trichlorosilyl)ethene with cyclopentadiene (Fig. 3.6, **25**) [139]. A promising new approach for preparing arylene-bridged monomers is the Murai coupling reaction between vinyltrialkoxysilanes and carbonyl functionalized aryl compounds [140].

Another method for forming bridging groups is through the formation of a metal complex (Fig. 3.7) using Lewis base (electron donor) groups such as isonitriles (Fig. 3.7, **26**) [141, 142], phosphines (Fig. 3.7, **27**) [5, 40–42, 44–46, 57, 143–150], amines [6, 151], thiols [152], or diamines (Fig. 3.7, **28**)[126] as metal ligands in an

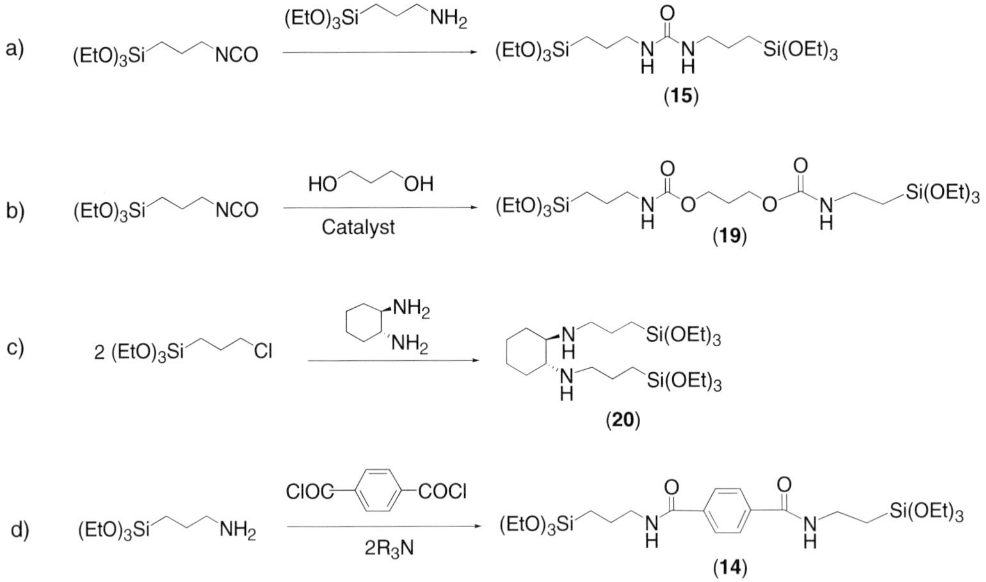

Fig. 3.5 Preparation of bridged monomers from organotrialkoxysilanes

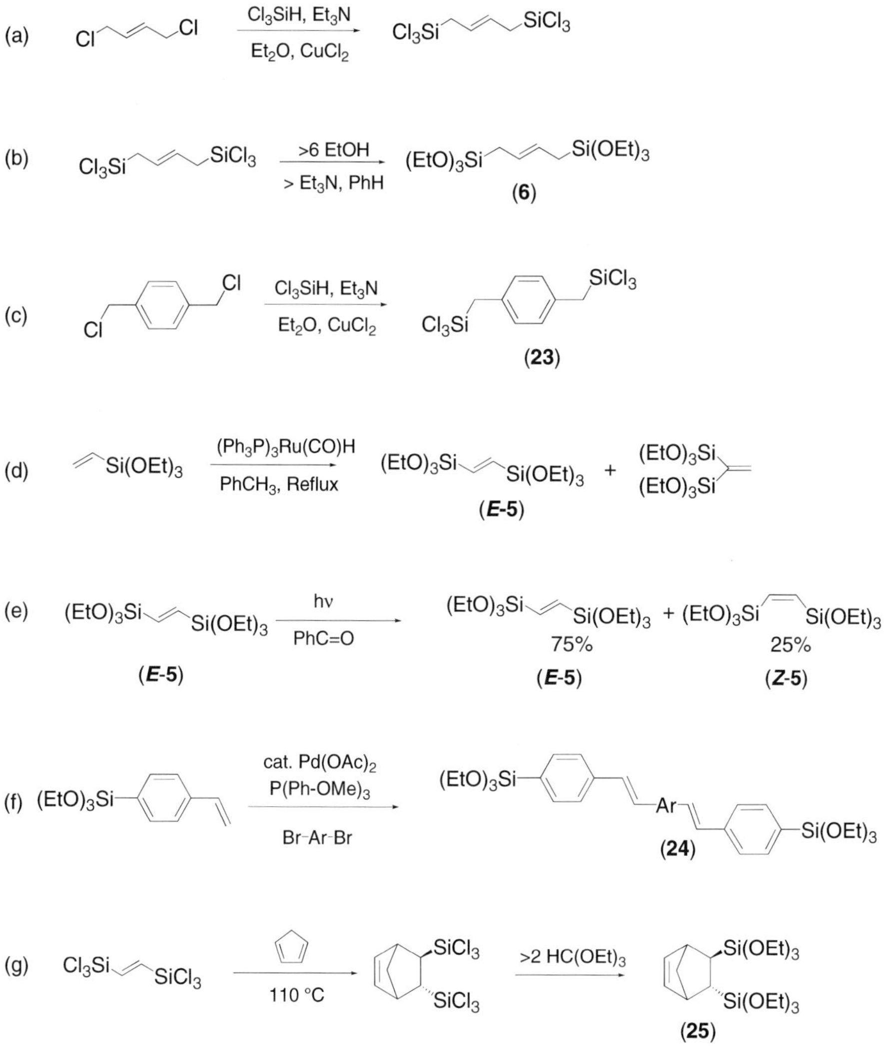

Fig. 3.6 Miscellaneous methods of synthesizing monomers

organometallic bridging group. Electrochemically active, ferrocenylene-bridged monomers are readily prepared by metallation and reaction with chlorotrialkoxysilane [66, 67]. Trialkoxysilyl-arene-chromiumtricarbonyl complexes (Fig. 3.5, **19**) readily form upon reaction of the corresponding trialkoxysilylaryl compound with chromium hexacarbonyl [69–71, 72, 73, 74].

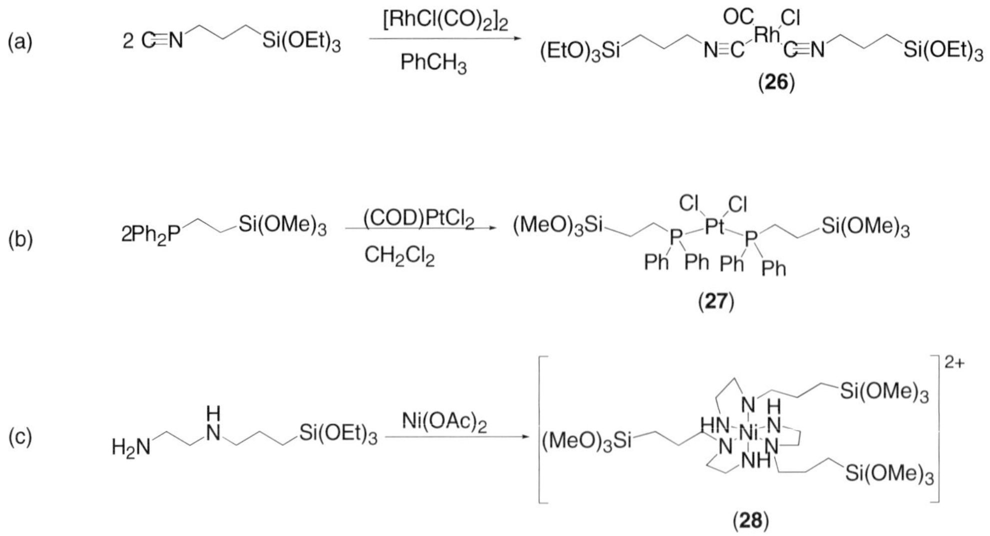

Fig. 3.7 Formation of organometallic bridged polysilsesquioxanes

3.4
Sol-Gel Processing of Bridged Polysilsesquioxanes

Sol-gel processing can be viewed as a series of stages (Fig. 3.8), hydrolysis and condensation chemistry, gelation, aging, and drying [153], through which an alkoxysilane monomer is converted into a hyper-cross-linked siloxane network.

3.4.1
Hydrolysis and Condensation

Sol-gel polymerization of bridged trialkoxysilanes proceeds by a series of hydrolysis and condensation reactions that afford up to three siloxane bonds to each silicon atom and produce three equivalents of alcohol [1, 154–156]. The dynamic result is a "sol" composed of the hydrolyzed and condensed species that will eventually grow into a percolating network or gel. The reactions are typically performed in the same alcohol generated by the monomer hydrolysis or in tetrahydrofuran. At least three equivalents of water as the co-reactant are added to the polymerization reaction. The sol-gel polymerization is generally acid or base catalyzed, although fluoride catalysts have also been used by several research groups [105, 157–160]. Hydrochloric acid is typically used as the acidic catalyst. Ammonium hydroxide, sodium hydroxide, and potassium hydroxide have been used as basic catalysts.

Hydrolysis and condensation rates of alkyl- and aryltrialkoxysilanes are significantly faster than tetraalkoxysilanes under acidic conditions and slower under basic conditions [154, 161]. For example, simple alkyltriethoxysilanes hydrolyze 6–10 times faster than tetraethoxysilane (TEOS) with acid catalysts. It has been shown for silica sol-gels derived from tetraethoxysilane that only silanol-silanol or water-producing condensation is observed, while tetramethoxysilane leads to both water and methanol producing condensation reactions. While no-one has verified this reactivity trend with silsesquioxanes, it is not unreasonable to suspect that a similar reactivity is observed. The resulting "sol" will contain, depending on the time, reaction conditions, catalyst and other experimental variables, monomeric species, cyclic and acyclic oligomers, polymers and colloids. As would be expected from the sol-gel chemistry of tetraalkoxysilanes, the condensation rates decrease with increasing size of the alkoxide substituent (MeO > EtO > *n*-PrO) [162].

3.4.2
Gelation

The next stage of the sol-gel process is gelation. Gelation is the manifestation of the percolation of colloidal polysilsesquioxanes throughout the liquid. The ease with which bridged polysilsesquioxanes form gels may be their single most distinguishing trait. The six reactive alkoxide groups result in rapid gelation times for both

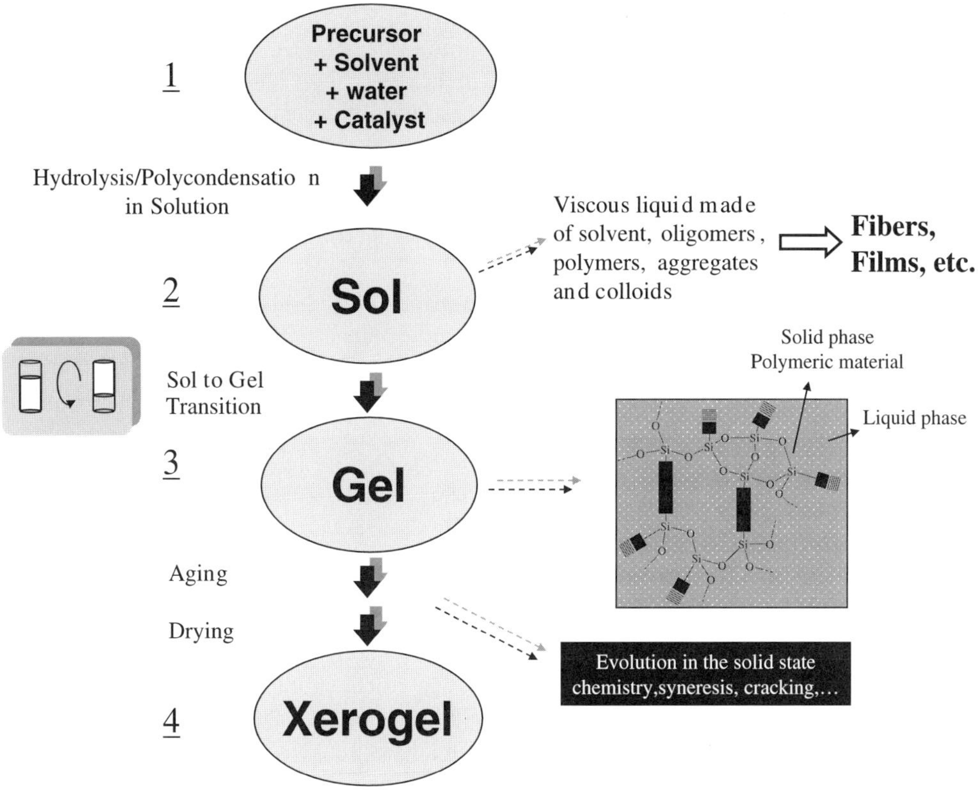

Fig. 3.8 The sol-gel process

traditional aqueous and anhydrous formic acid sol-gel procedures [1, 155, 156]. Gelation occurs in minutes to hours for many bridged polysilsesquioxanes at 0.4 M monomer concentration. This concentration is approximately one-fifth of the concentration of tetraethoxysilane used in typical sol-gel formulations that require days for gels to form. In contrast, the majority of organotrialkoxysilanes, $RSi(OR)_3$, will not form gels under any sol-gel conditions [156]. To date, only one type of bridged monomer, 5,6-bis(triethoxysilyl)norbornene (Fig. 3.6, **25**), has resisted forming gels under all sol-gel conditions tested [139]. Steric hindrance from the bridging norbornenylene group in (*E*)- and (*Z*)-5,6-bis(triethoxysilyl)norbornene monomers is apparently sufficient to impede condensation so that stable solutions of hydrolyzed monomers and oligomers are obtained even after standing for months. Gels have been prepared in nonpolar solvents such as toluene through a transesterification reaction between the triethoxysilyl groups and anhydrous formic acid [163]. The first reported sol-gel polymerization in supercritical carbon dioxide was used to prepare phenylene-bridged polysilsesquioxane aerogels in a single step from the anhydrous sol-gel polymerization of the phenylene monomer (Fig. 3.2, **1**) with 6 equivalents of formic acid [89].

Acid catalysis results in gels with less condensed networks (65–75%) and more residual alkoxide and silanol groups than those prepared under basic conditions (75–90%). The degree of condensation is directly related to the number of residual silanol and ethoxide groups at silicon. This, in turn, contributes to the overall polarity of the material and its surface properties. Less highly condensed materials, formed under acidic conditions, have a greater "presentation" of these polar groups. To attain similar hydrocarbon character in a silica gel can only be achieved through surface silylation of the accessible silanols, yielding a hydrophobic material. The "organophilicity" of bridged polysilsesquioxanes can be tuned through the selection of the appropriate organic bridging group while still maintaining a silanol-rich surface.

Substituent effects on the sol-gel formation of bridged polysilsesquioxanes include the steric sensitivity of hydrolysis and condensation of trialkoxysilyl groups to the size of the alkoxide or bridging group, electronic effects from the organic substituent, and the influence of flexible bridging groups on intramolecular condensations [155]. For example, the gelation rates for the sol-gel polymerization of hexylene-bridged monomers decrease as one goes from methoxide to ethoxide and to propoxide [117], just as the hydrolysis and condensation rates did. Gels obtained from monomers with the more reactive methoxide groups often display more mesopores (20 Å < pore diameter < 500 Å) and macropores (>500 Å) and fewer micropores when compared with those prepared from monomers bearing ethoxide groups.

It is important to note that cyclization reactions are pervasive in siloxane-based polymerization chemistry, and cyclization reactions during the polymerization of tetraalkoxysilane [164] and organotrialkoxysilanes can delay or even prevent the formation of gels [156, 165, 166]. What is interesting about bridged polysilsesquioxanes is that the formation of siloxane rings does not appear to be as important a factor in determining whether a gel will form as is the case with silica or other siloxanes. This is probably due to the greater functionality of the bridged monomers, hexafunctional as opposed to tetrafunctional (TEOS) and trifunctional (trialkoxysilanes), that allows for the ready formation of extensive networks. It is only when *intramolecular* condensations leading to carbosiloxane rings (Fig. 3.9) become significant contributors to the sol-gel chemistry of the bridged monomers that polymer growth dramatically slows. These intramolecular condensation reactions become important with monomers containing bridging groups between one and four methylene repeating units long (Fig. 3.9, **8**) [27] or with *cis* substitution geometries on a cycloalkane or a carbon-carbon double bond [21]. These cyclization reactions take place during the early stages of polymerization and alter the composition of the basic structural units that are subsequently incorporated into the silsesquioxane network.

The cyclic intermediates have been characterized by *in situ* ^{29}Si NMR, chemical ionization mass spectrometry, and actual isolation of the intermediates from the

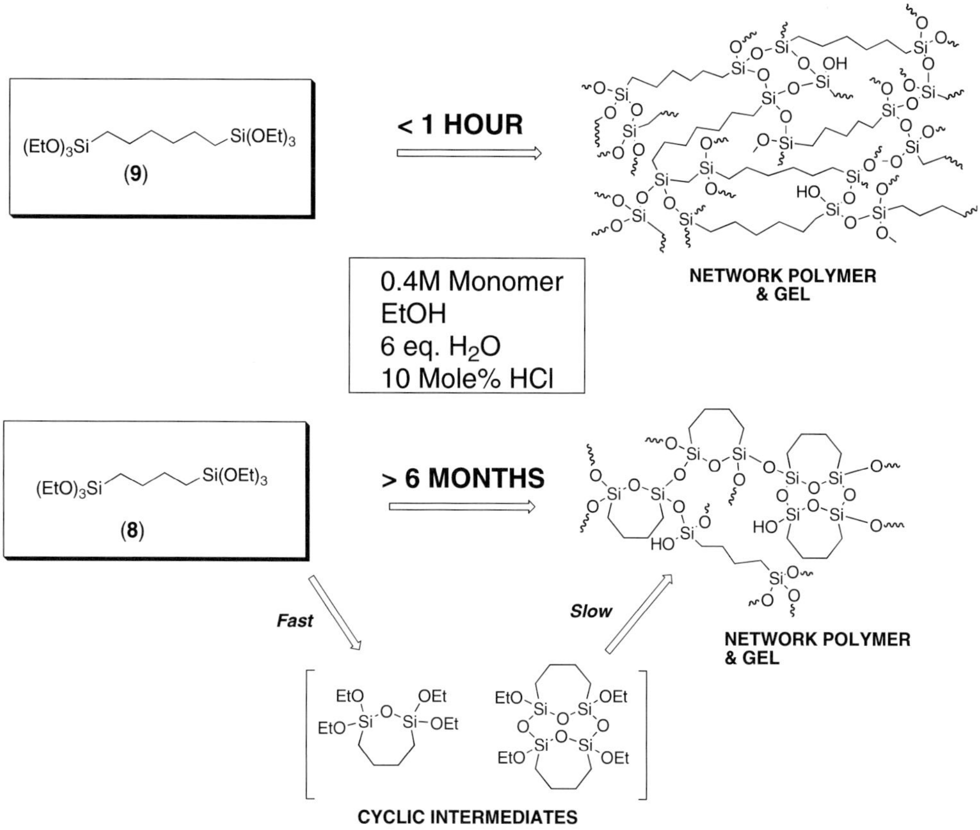

Fig. 3.9 Effect of monomer structure on gelation times: reducing the length of the alkylene-bridging group by *two methylenes* increases cyclization reactions to the extent that gelation is delayed by months

reaction solutions [21, 27, 131, 155]. One example is the sol-gel polymerization of 1,4-bis(triethoxysilyl)butane (Fig. 3.9, **8**). The gel time for this monomer is in excess of 6 months. Under the same "standard" conditions, the corresponding six-carbon, hexylene-bridged (Fig. 3.9, **9**) homologue gels in less than an hour.

These cyclic carbosiloxanes are local thermodynamic sinks that produce kinetic bottlenecks in the production of bridged polysilsesquioxanes with sufficiently high connectivity to become gels. The formation of cyclics results in retarding or, in some cases, preventing gelation. An additional finding is that the cyclic structures are incorporated intact into the final xerogel. Because cyclization alters the structure of the building block that eventually makes up the xerogel network, it is expected that this will contribute importantly to the bulk properties of the xerogel as well.

3.4.3 Aging and Drying

The next stages of sol-gel processing are aging and drying of the gels. While reaction rates and physical changes slow in the gel state, one must remember that the system is still dynamic. This can best be exemplified by the phenomenon known as "syneresis" which is often characterized by fairly abrupt shrinkage with expulsion of solvent from the gel. Thus, it is critical to control the aging time such that samples being compared have all proceeded to more or less the same extent through the aging process. With a single monomer and formulation, whether or not the samples are allowed to age through syneresis has a profound effect on the porosity and texture of the resulting xerogels [167].

Xerogels (Fig. 3.10, left) have been prepared by air-drying, washing with water and then drying [11], or exchanging the polymerization solvent with one with a lower dielectric than that from slow air-drying [12]. Because of the low concentration of monomers used in sol-gel processing, bridged polysilsesquioxane xerogels can lose as much as 80–95% of their volume upon air-drying. This shrinkage can result in the collapse of pores, resulting in nonporous materials [25, 28, 34, 62, 168, 169]. However, most bridged polysilsesquioxane xerogels remain porous with surface areas between 200 and 1200 m^2g^{-1}. Thin film coatings can be readily prepared by spin casting, dip coating or spraying. Gels can be processed as monolithic aerogels by replacing the original solvent with supercritical carbon dioxide extraction and then slowly venting (Fig. 3.10, right) [13, 26, 62, 168, 170, 171]. Alternatively, aerogels can be prepared by directly polymerizing the monomers with formic acid in supercritical carbon dioxide [13, 89].

3.5 Characterization of Bridged Polysilsesquioxanes

Characterization of hyper-cross-linked bridged polysilsesquioxanes is made difficult by the very physical properties of intractability and insolubility that make them attractive for many applications. X-ray powder diffraction [12, 172–175] as well as

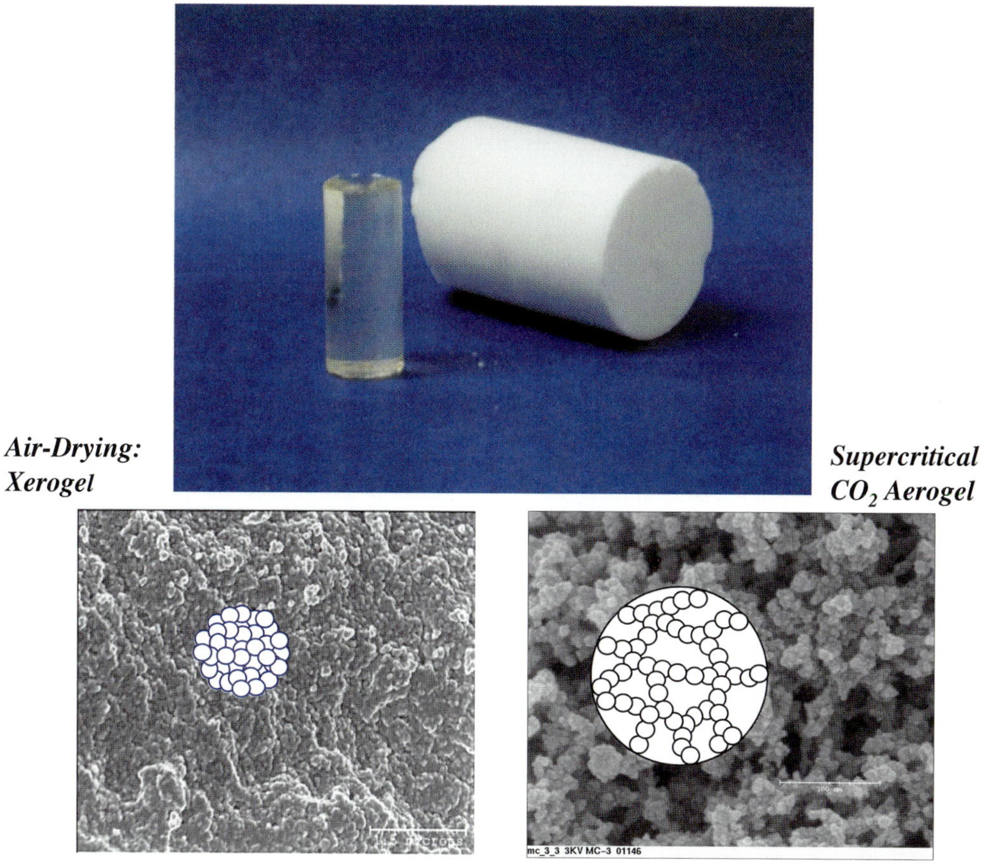

Fig. 3.10 Scanning electron micrographs of a hexylene-bridged polysilsesquioxane xerogel (left) and the analogous aerogel (right)

small-angle X-ray [17, 176] and neutron [177] scattering have been used to establish the composition of many bridged polysilsesquioxanes. Many of the materials are completely amorphous and exhibit fractal dimensions that are intermediate between a surface and mass material. The powder diffraction patterns from the amorphous bulk sol-gel materials reveal broad peaks consistent with an un-ordered arrangement of bridging groups and silicon centers. Bridging groups designed as self-assembling, and self-assembly additives, can be used to induce order in bridged polysilsesquioxanes under certain circumstances that will be discussed (*vide infra*).

Solid-state ^{13}C and ^{29}Si NMR spectroscopies are widely used to determine the molecular structure of the bridged polysilsesquioxanes [12, 21, 26, 156, 173, 178, 179]. In particular, the integrity of the bridging group and the extent of reaction is elucidated by deconvoluting the ^{29}Si NMR data (Fig. 3.11). The degree of condensation for the hexylene-bridged polysilsesquioxanes shows the general trend of more cross-linking in reactions prepared with base catalysts. It was shown that in most of the cases, cross-polarization magic angle spinning (CP MAS) sequences can be used more conveniently and at the same level of confidence as single pulse (HPDEC) experiments. The degree of condensation can have an important influence on how porous the final gels may be (*vide infra*) [179].

3.5.1
Porosity in Bridged Polysilsesquioxanes

Porosity is a key property of materials used for catalysts, chromatographic supports, membranes, and adsorbent materials [180]. High surface areas and control over the pore size are important goals for synthetic materials programs. The bridging organic group provides an opportunity to systematically vary the size, shape, geometry, and functionality of a molecular building block in order to probe how bulk structural properties such as porosity are affected. In amorphous bridged polysilsesquioxanes, the relationship between the bridging group and the porosity is subtle. Nevertheless, precise levels of control have been achieved. Important factors contributing to the porosity in these materials include the compliance of the network, which is a function of the degree of condensation at the silicon and the flexibility of the bridging group. For example, long flexible alkylene (Fig. 3.12, **9**) [25], fluoroalkylene (Fig. 3.12, **29**) [169], or even heteroatom-functionalized alkylene (Fig. 3.12, **10–12**) bridging groups [34, 168], particularly when polymerized under acidic con-

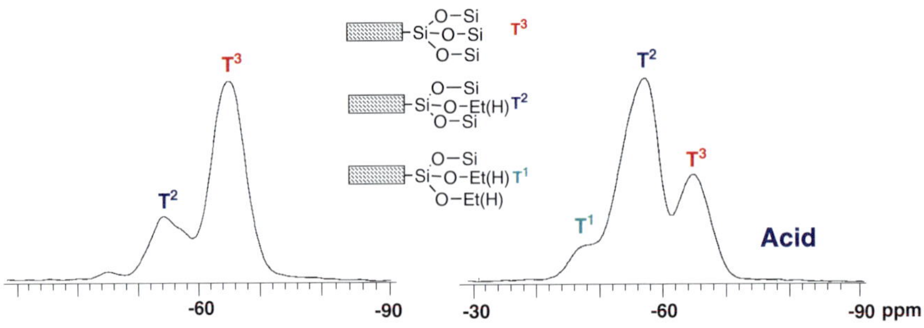

Degree of Condensation

Alkoxide Group	OH^-	H^+
Methoxy	72.7%	73.1%
Ethoxy	90.3%	75.2%
Propoxy	90.2%	76.0%

Fig. 3.11 ^{29}Si CP MAS NMR spectra of hexylene-bridged polysilsesquioxane aerogels prepared by base (left) and acid (right) catalyzed sol-gel polymerizations of 1,6-bis(triethoxysilyl)hexane (0.4 M solution in ethanol or tetrahydrofuran, 2.4 M H_2O, and 0.04 M catalyst) (*116*)

ditions, can lead to complete collapse of the porosity, resulting in nonporous xerogels or thin films. The ability to tailor nonporous, bridged polysilsesquioxanes may be useful for fabricating chemical barriers, dense membranes, or optical coatings. It is also important to note that, because of the sensitivity of the sol-gel process to factors such as pH, catalyst, temperature, solvent, and aging time, these factors must be carefully controlled to permit the structure-property effects of the bridging group to be determined reproducibly [15, 19, 30, 106, 116, 117, 157, 160, 167, 181–184].

3.5.2
Pore Size Control

Less compliant networks prepared with basic catalysts to give more condensed networks and/or more rigid bridging groups retain their porosity after drying. Arylene (Fig. 3.2, **1**, **2**) [9–13, 16, 66, 105, 115, 173, 185] and ethenylene (Fig. 3.2, **5**) [20, 21] bridged polysilsesquioxanes give rise to materials with surface areas as high as $1800\ m^2g^{-1}$. The high surface area contains large contributions from micropores with mean pore diameters < 20 Å. Alkylene-bridged polysilsesquioxanes prepared with base catalysts and bridging groups up to 10 carbons in length are in the form of mesoporous gels (20 Å < mean pore diameter < 500 Å) [26]. The mean pore diameter was shown to be roughly proportional to the length of the bridging groups. Introduction of unsaturated functionalities such as olefinic or aromatic groups into the organic bridges may decrease the flexibility and further prevent collapse of pores during drying. There is sufficient empirical data to predict with fair confidence if a given bridging group under a defined set of sol-gel polymerization conditions will be porous or not. This marks a fairly significant advance in our understanding of the molecular determinants of porosity.

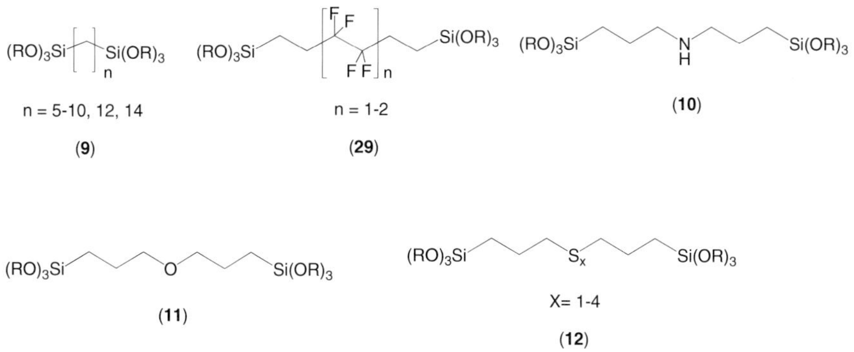

Fig. 3.12 Bridged monomers with flexible bridging groups capable of forming more compliant networks that can be engineered to afford nonporous thin films by reducing the degree of condensation at silicon

3.5.3 Pore Templating

Another strategy for creating porosity is to use the organic group as a template for porosity (Fig. 3.13). While a common strategy in the preparation of zeolites, this has only recently been applied in hybrid sol-gel materials. Templating relies on an organic group to occupy space until calcination, chemical oxidation, chemical rearrangements, or hydrolysis eliminates the template. This will leave a pore whose size and shape roughly corresponds to that of the organic molecule. Templating of porosity was first performed serendipitously with acetylene-bridged polysilsesquioxanes (Fig. 3.14, **3**) that lost acetylene during thermolysis (Fig. 3.14a) [12]. An alternative approach used a low-temperature, inductively coupled plasma to burn away organic bridging groups in arylene- (Fig. 3.14b) and alkylene- (Fig. 3.14c) bridged polysilsesquioxane xerogels [24, 186–188]. The resulting silica gels were porous. If the xerogels were porous before oxidation, the mean pore diameter shifted to larger sizes. When nonporous alkylene-bridged polysilsesquioxane xerogels were treated, mesoporous silica gels were obtained. The size of the pores increased as the length of the bridging group increased. Thermal oxidation was also used to convert nonporous, dendrimeric silsesquioxanes into porous silica [189]. Thermal oxidation of thin films of bridged polysilsesquioxanes was also used to template porosity in silica membranes [190]. A variety of chemical processes have also been explored for cleaving, chemically modifying, or removing portions of the organic bridging groups. These include the use of retro-Diels-Alder reactions (Fig. 3.14, **25**) to modify [139] or cleave the bridging group [191], decarboxylation of dialkyl carbonate bridging groups (Fig. 3.14, **30**) [168], and thermal decomposition of phenolic carbamate linkages in the bridging groups [191, 192]. A clever variant of carbamate templating was used to prepare imprinted gels with amine groups advantageously positioned about a pore for DDT detection [193].

Fluoride-catalyzed or thermally induced cleavage of Si–C bonds has been used in a number of acetylene- and alkynylarylene-bridged polysilsesquioxanes, such as Fig. 3.14, **3**, **4**, to template porosity in silica gels [17, 114, 158, 174, 176, 194–198]. This has been a particularly versatile template system that has been used to explo-

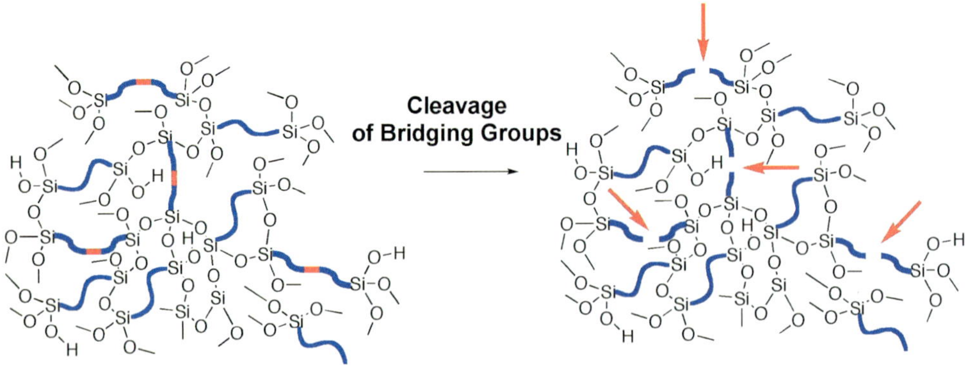

Fig. 3.13 Templating porosity in bridged polysilsesquioxanes

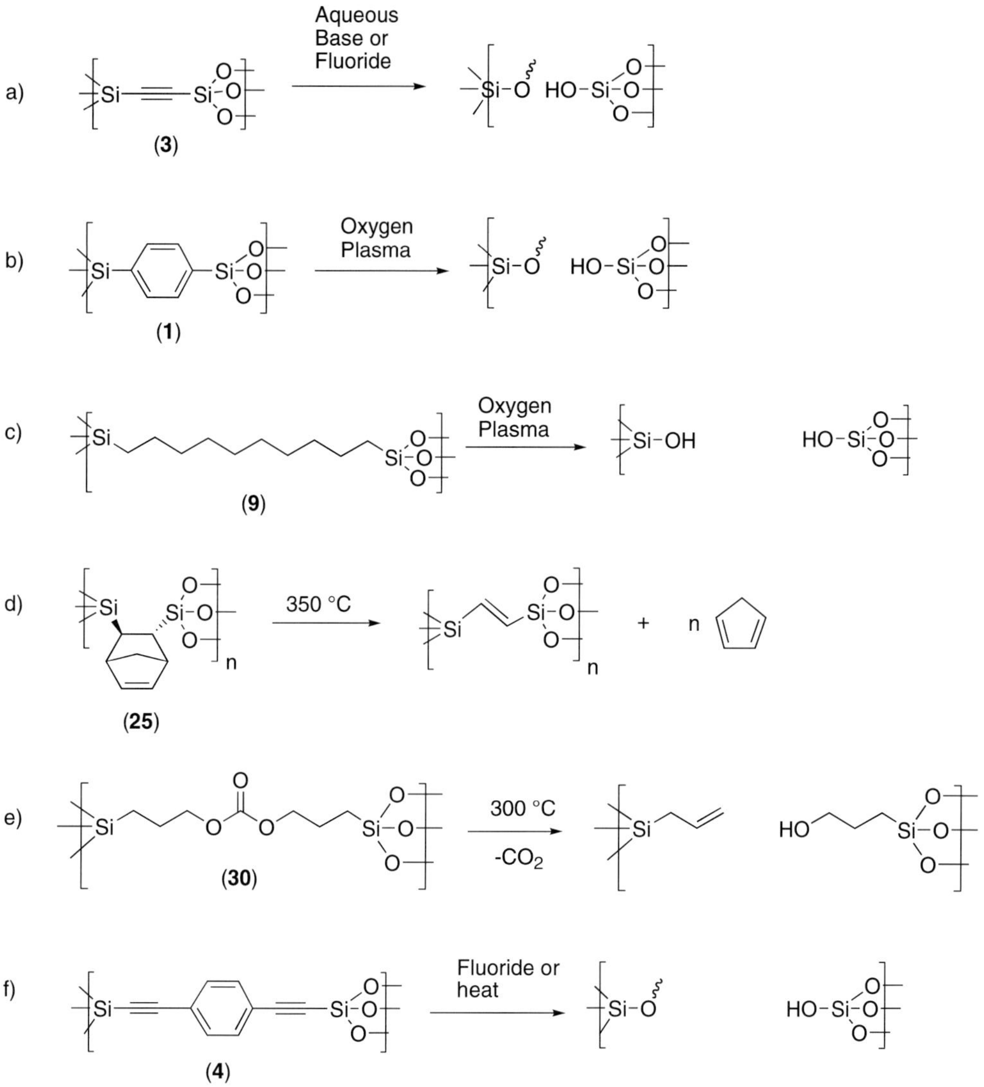

Fig. 3.14 Templating porosity in bridged polysilsesquioxanes with the organic bridging group

re how a number of readily accessible template geometries affect the porosity of the resulting silica gels. More recently, hydrolysis of carbamate linkages was used to prepare porous 3-aminopropyl-functionalized polysilsesquioxanes [192].

3.6
Influence of Bridging Group on Nanostructures

In bulk sol-gel polymerizations of bridged monomers, the influence of the organic bridging group on the long-range order is dwarfed by the dynamics of the sol-gel polymers that lead to fractal structures. However, under certain circumstances it is possible to observe what appear to be nanostructures whose construction appears to be directed by strong inter-bridging group interactions. In nearly all cases, creation of bridged polysilsesquioxane nanostructures requires some dimensional restriction to where the growing sol-gel polymer resides in an "interphase."

3.6.1
Surfactant Templated Mesoporous Materials

For example, bridged polysilsesquioxanes with long-range periodic structure have been prepared using surfactant templating techniques (Fig. 3.15) [199–223]. In these materials, the polysilsesquioxane makes up the skeleton of a geometric array of pores in hexagonal close-packed cylinders or tetrahedral networks. The size and geometry of the mesopores depend on the surfactant and processing conditions. The bridged polysilsesquioxane adds improved toughness and organophilicity that may make these materials valuable as adsorbents of organic chemicals. In the case of the surfactant templated, phenylene-bridged polysilsesquioxane, a well-organized material was obtained in which a periodical organization of the monomeric units of the framework of the wall was established [222, 224]. However, it is not clear that there are advantages to the more time-intensive surfactant templating, compared with the single-step sol-gel processing of bridged polysilsesquioxane xerogels with higher surface areas and a comparable degree of control over the pore size distributions.

3.6.2
Mesogenic Bridging Groups

Thin films of bridged polysilsesquioxanes with mesogenic bridging groups show significant birefringence (Fig. 3.16) that suggests that there is considerable ordering in the liquid thin films before gelation. In these materials, flat and rigid bridging groups such as bis-alkynylarylene [16, 19, 225–227] and arylene [227, 228] groups facilitate the formation of anisotropic structures. The birefringence, particularly noticeable around cracks that develop during the curing of the sol-gel films, is thought to arise from anisotropic structures whose formation is driven by the influence of the mesogenic bridging groups on cyclization and organization into "ordered colloids" (Fig. 3.17). Flexible alkylene groups do not exhibit birefringence when processed under identical conditions [227].

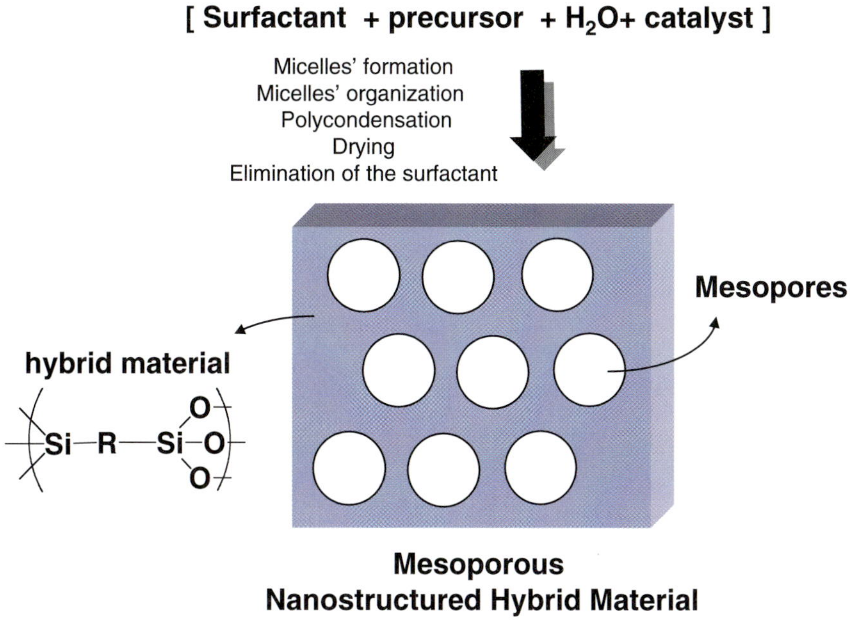

Fig. 3.15 Surfactant templating of mesoporous bridged polysilsesquioxanes

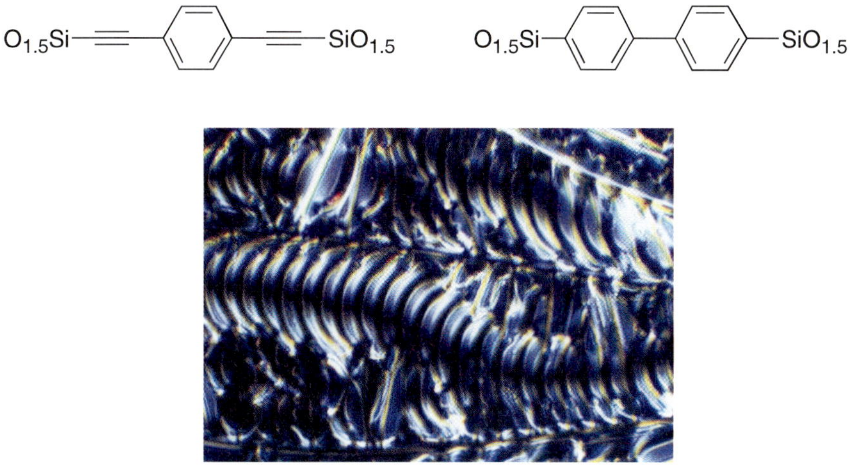

Fig. 3.16 Birefringent thin films of dialkynylarylene-bridged polysilsesquioxanes

In bulk samples, there is some chemical evidence, such as the *in situ* polymerization of thiophenylene (Fig. 3.2, **2**), diacetylene (Fig. 3.2, **3**), or diethynylbenzene (Fig. 3.2, **4**) bridging groups, as to the dispersion and long-range organization of the organic bridging groups in the hyper-cross-linked network [115, 195]. X-ray diffraction peaks show correlations with the length of the bridging groups [174]. These patterns rule out the long-range order observed in crystalline materials or even surfactant templated materials, but cannot exclude the possibility of shorter-range order in the form of anisotropic colloids. More studies that bridge between bonding information obtained from NMR and x-ray diffraction studies are necessary to resolve the structural details of these intractable materials.

3.6.3
Supramolecular Organization

Crystallization in polymers and self-assembly of amphiphiles based on polymethylene segments is an effective technique for introducing order into materials including polysilsesquioxanes with pendant alkyl chains. For such self-assembly, based on the organic bridging group, to occur in typical sol-gel solutions, the non-bonding interactions need to be relatively strong. To date, alkylene-bridging groups up to 14 methylenes in length have not demonstrated ordered structures based on crystallization of the alkylene chains [25, 26]. However, if hydrogen bonding, through bis-urea-alkylene groups or other organogelators [56, 229], is added to the bridging group, it is possible to apply supramolecular organization to introduce order into bridged polysilsesquioxanes. The introduction of bis-urea groups into the bridging group permits the formation of strong hydrogen bonds between the organic bridging groups that are relatively independent of the siloxane network. This appro-

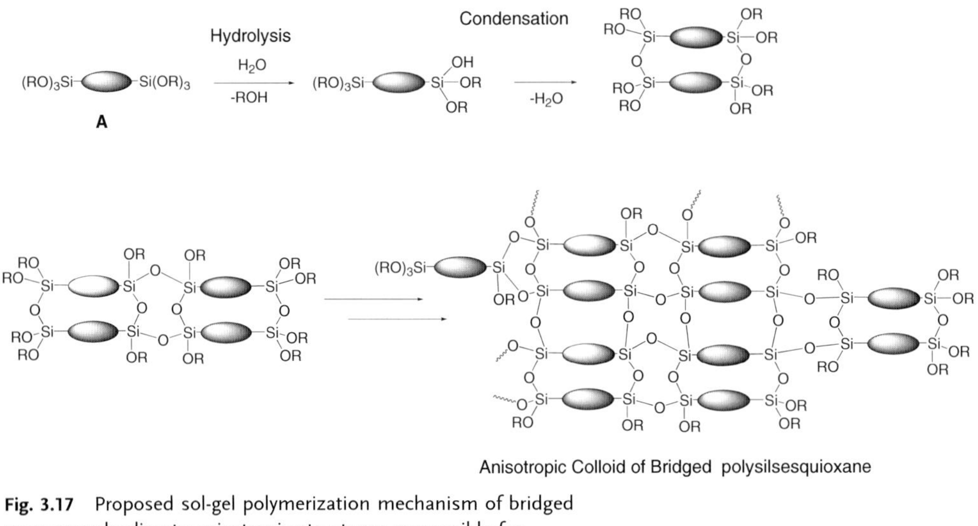

Fig. 3.17 Proposed sol-gel polymerization mechanism of bridged monomers leading to anisotropic structures responsible for birefringence of thin films

ach was first used successfully to prepare ordered materials by precipitation from the sol [229]. By removing solvent from the sol-gel chemistry and undertaking the solid-state polymerization of crystalline monomers, the non-bonding interactions and long-range structure can be preserved into the final material [230]. However, even in the solid-state polymerizations of bridged polysilsesquioxanes, anisotropic nanostructures are obtained only in the thin-film forms.

3.6.4
Metal Templating

It is possible to utilize metal ligand bonding to organize bridged polysilsesquioxanes into nanostructured materials with order mostly at relatively short length scales. This was first demonstrated with the *in situ* formation of palladium complexes during the sol-gel processing of 2-butenylene-bridged polysilsesquioxanes (Fig. 3.18) [22, 170]. In this case the metal is not required to form a gel, but the complex can conceivably increase the level of cross-linking. The coordination of the metal can affect the disposition of the organic bridging groups in space. Assembling the organometallic complex before sol-gel polymerization or *in situ* before gelation will result in ligand orientations based predominantly on coordination geometries that are distinctly different from those complexes that are formed after gelation [212, 231–233]. Other metal templating is performed by assembling an organometallic bridging group from a metal salt or compound and two or more organotrialkoxysilane bearing one or more Lewis basic functionalities such as a thiol [152], amine [6, 151], phosphine [5, 40–42, 44, 46, 57, 143–150], and isonitrile [141, 142]. In this form the metal is an integral part of the bridging group. The section on catalysts and catalyst supports will discuss the application of these organometallic systems to catalyzing chemical reactions.

3.7
Thermal Stability and Mechanical Properties

One of the most attractive features of bridged polysilsesquioxanes is their excellent thermal stability in inert atmospheres and in air. Arylene-bridged polysilsesquioxanes (Fig. 3.2, **1**) are stable to 500 °C [12]. Alkylene-bridged polysilsesquioxanes

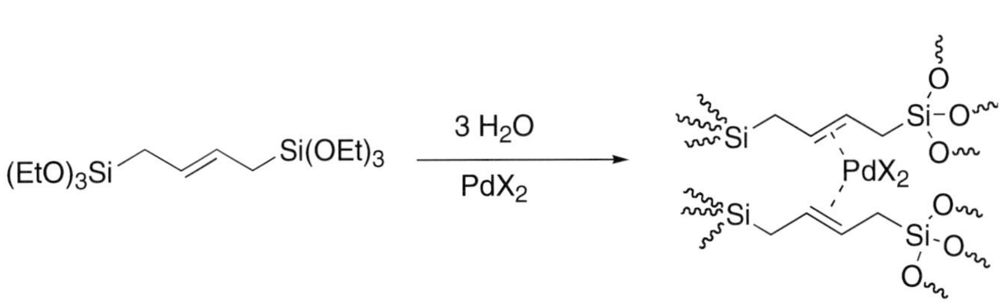

Fig. 3.18 In-situ metal templating of 2-butenylene-bridged polysilsesquioxanes

(Fig. 3.2, **8**, **9**) are stable to over 400 °C [11, 25, 30, 117]. The mechanical properties of bridged polysilsesquioxanes have been less well characterized than their thermal stability. Bis-imidoarylene-bridged polysilsesquioxanes (Fig. 3.19, **31**, **32**) can be prepared in the form of hard thin films that are thermally stable at 400 °C for over 40 hours (Fig. 3.19) [34]. The mechanical properties of some highly branched polysilsesquioxanes or "star-gels" (Fig. 3.19, **33**, **34**) have resulted in their description as compliant gels or glasses [234–236]. Compressive strain studies revealed that the star-gels are considerably tougher than epoxy resins or silica gel. More recently, mechanical analysis of thin films of surfactant-templated ethylene-bridged polysilsesquioxanes (Fig. 3.2, **8**) showed an increase in the modulus (4.3 GPa) and hardness (4.8 GPa) with increasing amounts of the ethylene bridging group [107].

3.8 Chemical Properties

The general physical and chemical stability of bridged polysilsesquioxanes relative to organic polymers has been touted as one of the advantages of these materials [1, 12]. However, the high level of functionality that can be incorporated without interference into the hyper-cross-linked network provides an enormous range of controlled chemical transformations that can be explored with a similarly expanded range of applications. The two distinctly different regions to consider when examining the chemical properties of bridged polysilsesquioxanes are the silicon centers and the organic bridging group.

The silicon atoms can potentially have three siloxane bonds to other silicons. However, they commonly possess residual alkoxide groups that can undergo fur-

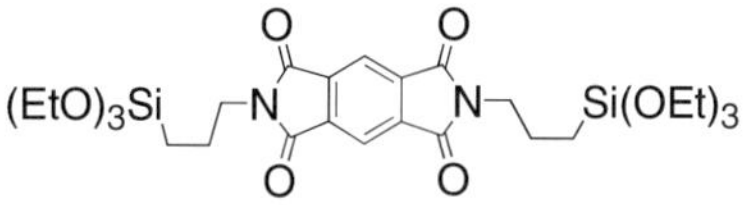

(31)

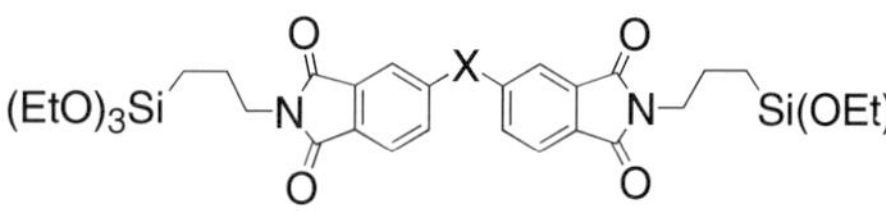

(32)

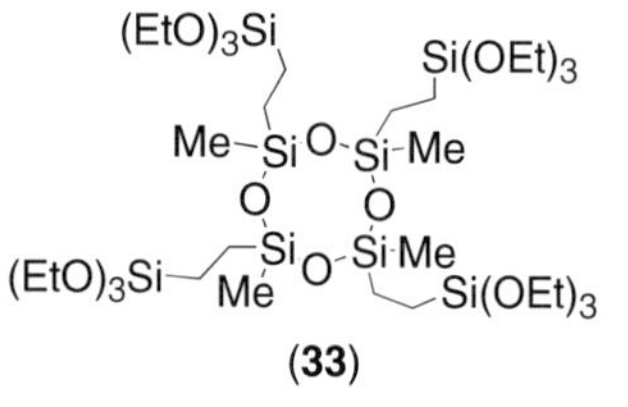

(33)

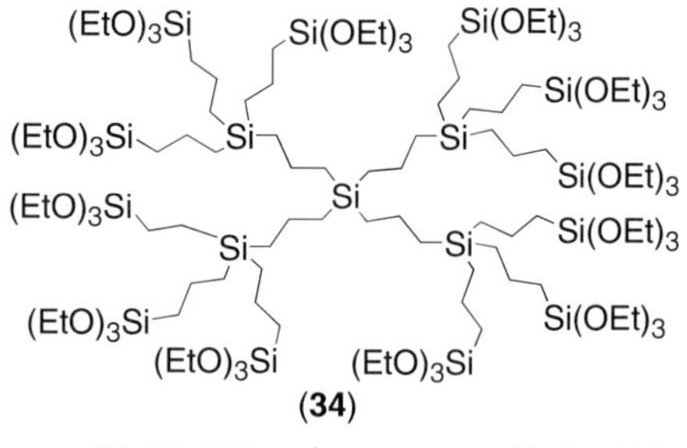

(34)

Fig. 3.19 Bis-imidoarylene bridged monomers (**31–32**) (*34*) and monomers with dendritic bridging groups (**33** and **34**) (*189, 234–240*) ther hydrolysis, silanols that can re-esterify, be deprotonated or even attack electrophilic species including other silicon centers. The equilibria for hydrolysis and condensation are known to strongly favor the hydrolyzate and condensate products, respectively. Furthermore, after gelation and especially after drying the reactivity of the bridged materials diminishes considerably. Drying can induce additional condensation reactions to afford a more hydrophobic material. But like silica gels, many of the siloxane bonds reopen upon exposure, suggesting that the siloxane bonds are strained. There have been some indications that fluoride and high pH can be used to catalyze the redistribution and reorganization of the siloxane network to a certain extent [241]. Fluoride, and to a lesser degree hydroxide, will also facilitate the cleavage of the silicon-carbon bond to the organic bridging group. This has been effectively applied to the intentional chemical removal of bridging groups with alkynyl C–Si bonds described in the section on templating. A few other Si–C bonds which need care with regard to Si–C bond cleavage include Si–Aryl linkages under acidic conditions and Si–benzyl or Si–allyl bonds under basic conditions. Some cleavage of Si–allyl bonds in bridged polysilsesquioxanes has been observed [23, 131, 198]. For the most part, however, the C–Si linkages seem to be relatively robust to aqueous acid or base. Perhaps the heterogeneous nature of the reactions between the intractable, insoluble network and the liquid reagents protects the network to a certain extent. One unusual reaction available to methylene-bridged polysilsesquioxanes is the deprotonation of the bridging methylene and the subsequent nucleophilic attack on another silicon center to afford a tris-silylated methine bridging group [242].

The second method for chemically modifying bridged polysilsesquioxanes is through direct chemical modification of the organic bridging group. One of the most widely used chemistries is the formation of metal complexes, a subject that will be treated in later sections. Other reactions include addition reactions with the bridging groups through bromination of unsaturated carbon-carbon double bonds [201], sulfonation of phenylene bridging groups [243], photochemical isomerizations [65, 244], binding of proteins [121], Diels-Alder chemistry [191, 245] and polymerization chemistry of diacetylene [195] and thiophene [115] bridging groups. Chemistry directed at the organic bridging group also includes cleavage through retro Diels Alder reactions [191], decarboxylation [62, 168], thermal degradation of urethanes [191, 192], and complete oxidation of the bridging group [24, 187, 188]. Thermal or photochemical decomposition of organometallic bridging groups has proven to be an effective method for preparing highly dispersed metallic or semiconducting nanoclusters [4].

3.9 Applications

The primary application of bridged polysilsesquioxanes has been for surface modification as coupling agents such as those with sulfides in the bridging group (Fig. 3.2, **12**) that are used with silica-filled rubber [38]. In addition, bridged polysilsesquioxanes have been widely applied in coating formulations. Materials with alkylene (Fig. 3.2, **8**, **9**) [107, 246], ether (Fig. 3.2, **11**) [247], and urea (Fig. 3.2, **15**) [248] functionalized bridging groups can result in the formation of tough and relatively hard films to protect easily scratched surfaces. Bridged polysilsesquioxane coatings can also be used as protective layers for metals [249] or for microelectronic applications such as low *k* dielectrics [246] and photoresists [250]. A more recent application took advantage of the more hydrophobic, yet readily gelled, ethylene and hexylene-bridged polysilsesquioxanes as hydrophobic, encapsulating sol-gels for enzymes [109–111].

3.9.1 Optics and Electronics

3.9.1.1 Dyes

Incorporation of dyes into the sol-gel matrix is a useful strategy for preparing waveguides, lasers, sensors, light-emitting diodes, and nonlinear optical (NLO) materials [4]. Using a dye molecule as a bridging group provides an exceptionally high loading of chromophores and ensures against leaching and/or phase separation of the dye molecule (Fig. 3.20).

Dyes that have been incorporated into bridging groups include terphenylene (Fig. 3.2, **1**) [12], coumarins (Fig. 3.20, **35**) [93, 251], fullerenes [252, 253], oligothiophenes (Fig. 3.2, **2**) [11, 115, 254, 255], viologens (Fig. 3.2, **16**) [63, 64], lanthanide complexes [232, 256–258], triarylamines (Fig. 3.20, **36**) [88], nitroaromatics

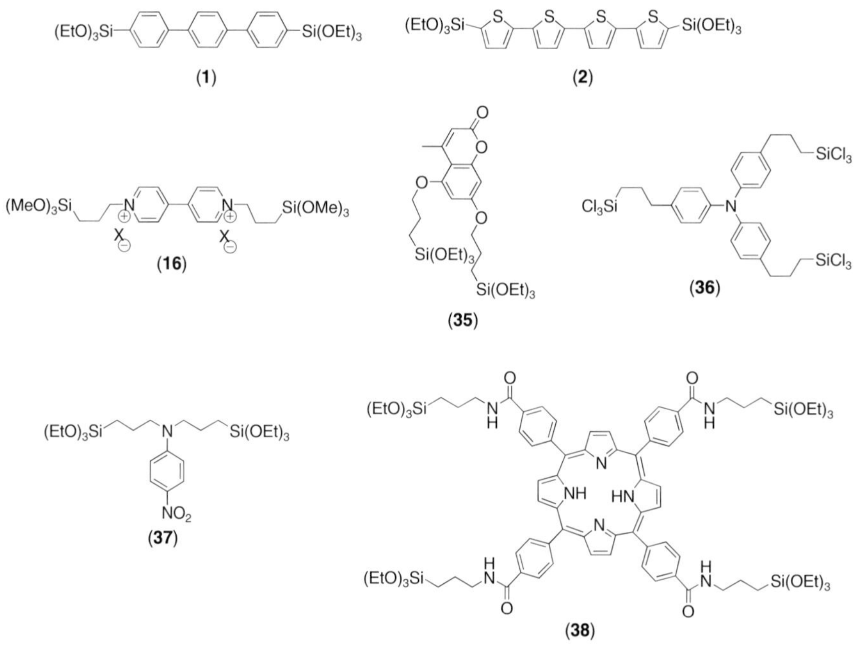

Fig. 3.20 Bridged polysilsesquioxane monomers with chromophore bridging groups

(Fig. 3.20, **37**) [259–261], phthalocyanines [262, 263], porphyrins (Fig. 3.20, **38**) [53, 264–266], and phenylenevinylene [138, 267–270] functionalities. Dithienylethene- [271] and azobenzene- (Fig. 3.2, **17**) [65, 244] bridged polysilsesquioxanes are photochromic materials whose refractive index can be reversibly photoswitched. These materials have significant potential for optical components for wave guiding. This approach has also been used for the preparation of second-order NLO bridged polysilsesquioxanes containing 4-nitro-*N*,*N*-bis[(3-triethoxysilyl)propyl]aniline (Fig. 3.20, **37**) in each monomer unit [259].

3.9.1.2 **Nano- and Quantum Dots in Bridged Polysilsesquioxanes**

It is desirable to prepare nano- and quantum dots in high modulus matrices to prevent aggregation. To be useful the matrices should be transparent for optical applications, possessing a high surface area for catalysis, and inert to the intended chemical processes. High-surface-area bridged polysilsesquioxanes were used to encapsulate quantum dots of gold that were prepared through inverse micelle techniques [272]. Similarly, nanosized CdS particles have been deposited by *external* doping, treating porous bridged polysilsesquioxanes with successive washes of $CdNO_3$ and Na_2S [69, 71, 72, 273]. The CdS particles are distributed uniformly throughout the material. Although not perfectly monodisperse, the average CdS particle size correlates with the average pore size of the xerogel. However, it is the ability to use the functionality of the bridged polysilsesquioxanes to facilitate and control the formation of these small structures that is unique. Gold nanoparticles have been prepared through site-selective reduction of auric chloride in thiol-functionalized, surfactant-templated, bridged polysilsesquioxanes [274, 275].

It is also possible to use the bridged polysilsesquioxanes as molecular, organometallic precursors that can be photochemically or thermally converted into the nanoclusters *in situ*. Nanosized irregularly shaped Cr, Fe, Co, and Pt particles (10–90 nm) have also been prepared in xerogels by an *internal* doping method. Zerovalent transition-metal complexes [69–72, 276] were incorporated in bridged polysilsesquioxanes by copolymerization of the organometallic functionalized monomers, such as eta-6-arenechromium carbonyl monomer (Fig. 3.2, **18**) with a dilutant monomer, 1,4-bis(triethoxysilyl)benzene (Fig. 3.2, **1**). In many cases, discrete metal particles can be formed either by heating or by irradiating with light the dried xerogel under vacuum. In both internal and external methods, the deposition of nanoparticles slightly reduced the surface area but the overall porosity remained intact. Both methods are employed to fabricate intimate nanosized composites of Cr^0/CdS.

3.9.2
Separations Media

The combination of high surface area with chemical functionality makes bridged polysilsesquioxanes ideal chromatographic supports for HPLC. The suitability of phenylene (Fig. 3.2, **1**) bridged polysilsesquioxanes as packing materials for HPLC columns to separate aromatic compounds compared favorably with similar silica-gel-filled columns [1]. Bridged polysilsesquioxanes have also been used as pore templates in inorganic molecular sieving membranes [190] and as a structural material [107]. Nonporous gas separation membranes based on bis-amido-arylene bridged polysilsesquioxanes (Fig. 3.2, **14**) demonstrated hydrogen:methane selectivities as high as 120:1 as flow rates similar to organic polymer based membranes [47]. Separations are due to differences in gas solubility and diffusion parameters rather than molecular sieving. A number of groups are investigating selective transport across bridged polysilsesquioxane membranes by taking advantage of the ability to introduce very high levels of functionality in a robust thin film. For example, silver or other chalcophilic metal selective membranes have been prepared with sulfide groups in the organic bridging groups [32], and membranes based on polysilsesquioxanes with copper complexed with amine functionalized bridges have been recently applied to gas separations [277]. More recently, proton conducting membranes for fuel cells based on clusters of phosphotungstenic acid dispersed in various nonporous bridged polysilsesquioxanes have been characterized to have suitable ion conductivities (0.03 S cm^{-1}) at 160 °C [278–280].

3.9.3 Catalyst Supports and Catalysts

The high surface area and chemical functionality make bridged polysilsesquioxanes excellent candidates for catalyst supports. The basic concept is either to incorporate Lewis basic groups that can coordinate to metals. The metal can be pendent (Fig. 3.2, **19**) to the bridging group or it can be part of the bridging group (Fig. 3.2, **18**). The high surface area and high relative loadings of metals in organometallic-bridged polysilsesquioxanes make them attractive for supported catalysts with tailored reactivities and selectivities. A series of bridged polysilsesquioxanes with ruthenium (Fig. 3.21, **39**) [281], iridium [282], and rhodium (Fig. 3.21, **40**) [141, 142, 283] complexes as part of the bridging group have been prepared and used for hydrogenation of aldehydes, olefins, or arenes with increased reactivity, presumably because of prevention of less reactive dimers seen with the homogeneous catalysts. Bridged polysilsesquioxanes with rhodium [49, 57, 284] have been used for hydroformylation of olefins with enhanced selectivities for the terminal carbonyl. The degree of condensation and flexibility of the network were used to engineer the materials for their applications. Chiral rhodium catalysts based on an asymmetric diaminocyclohexyl-functionalized bridge (Fig. 3.21, **41**) as the metal ligand have been prepared and used in asymmetric reductions [123, 124, 126]. A bridged polysilsesquioxane with a chiral binaphthyl bridging group coordinating rhodium was shown to provide moderate enantiomeric excesses in the asymmetric reduction of prochiral ketones [125]. More often, the bridged polysilsesquioxane is used as an agent for attaching a highly dispersed catalyst to a solid support of another material. For example, a chromium(III)-bridged polysilsesquioxane, coordinated by two hydroxyaryl imines tethered to triethoxysilyl groups, was used to prepare a heterogeneous catalyst by surface silylation of silica gels [285]. The resulting heterogeneous catalyst was effective at gas-phase oxidation of alkylaromatics to carboxylic acid functionalized aromatics.

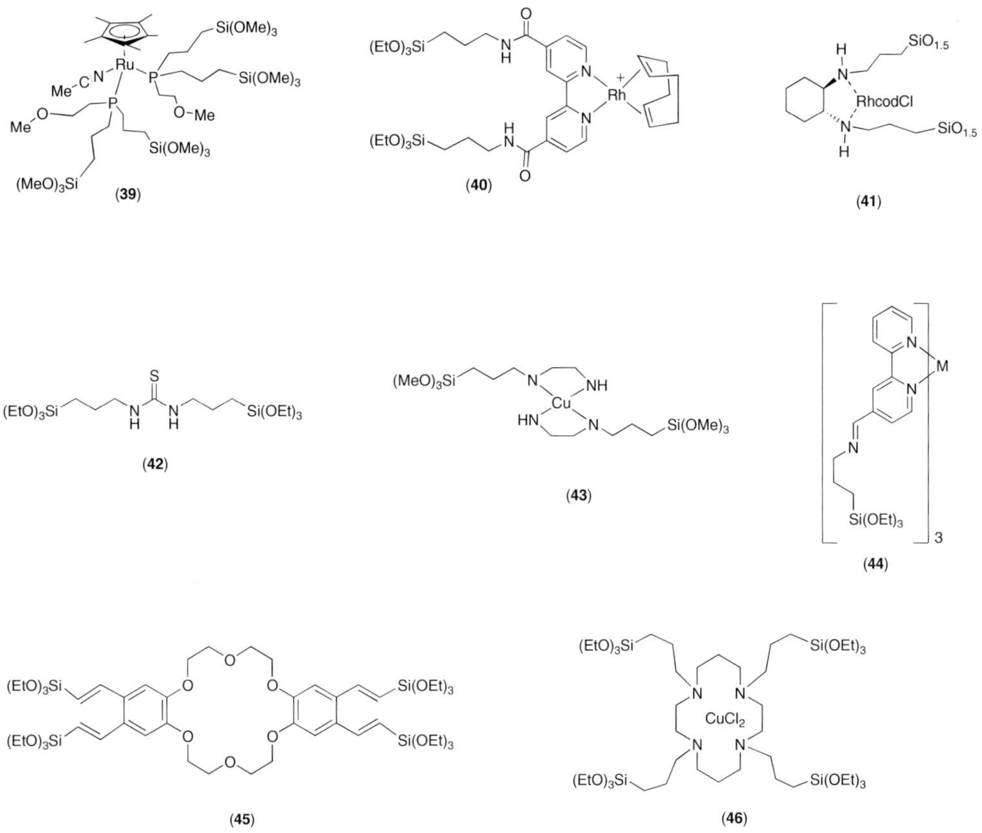

Fig. 3.21 Organometallic bridged polysilsesquioxanes for nanocluster synthesis, catalysts, and metal adsorption

3.9.4 Metal and Organic Adsorbents

The same functionalities that make good ligands for catalysts can also be used to bind metals in adsorbent materials. Bridged polysilsesquioxanes with thiocarbamate groups (Fig. 3.21, **42**) have been used extensively as porous, high surface area adsorbents for metals from solution [58, 97–102, 286, 287]. Dipropylamine bridged polysilsesquioxanes were used to adsorb metals from solvents and water [288–290]. The copper complex with two equivalents of ethylenebis(aminopropyl)triethoxysilane (Fig. 3.21, **43**) has been used to functionalize the interior of pores in a surfactant-templated “MCM-41” silica [211]. Copper was washed out of the bridging group, leaving the ethylenediamine groups positioned to preferentially sequester copper(II) from aqueous solutions. Assembly of bridged silsesquioxanes around a metal, such as that with the bipyridyl-based system (Fig. 3.21, **44**), can be used to imprint surface metal selective coordination sites [130]. Crown ether bridged polysilsesquioxanes (Fig. 3.21, **45**) have been prepared that can bind alkali-metal cations

[136, 137]. Copper can also be incorporated into tetraazamacrocycle-bridged polysilsesquioxanes (Fig. 3.21, **46**) through both external and internal doping methods [231]. Internal doping occurs without a loss of copper because of the strong binding from the chelating ligand. Furthermore, the presence of copper in the macrocycle during the sol-gel polymerization appears to rigidify the bridging group, providing a new tool for modifying the structural features of these hybrids.

Because of the homogeneous distribution of organics throughout high surface area materials, bridged polysilsesquioxanes have great potential as adsorbents for volatile organic contaminants including phenols [291]. Templating based on cleavable bridging groups that generate a recognition site and requisite chemical functionalities has also been developed for DDT adsorption and sensing [193].

3.10
Summary

In a relatively short period of time, bridged polysilsesquioxanes have emerged as a versatile class of materials. These hybrid organic-inorganic materials are built from precursors that integrate organic and inorganic groups at molecular length scales. A considerable body of experimental data has established that this configuration permits engineering of the physical properties of the resulting xerogels and aerogels. The materials can be fabricated into a variety of forms that range from nonporous films to high surface area microporous monolithic structures. More importantly, this process can be achieved in a single step. These features, coupled with the ability to prepare materials with organic functionality at every repeat unit without perturbing the formation of the gels or with adverse effects to the porosity, make these materials attractive for separations, catalysis, optics, sensors, and dielectric coatings.

References

1 D. A. Loy, K. J. Shea, *Chem. Rev.* **95**, 1431–1442 (1995).

2 R. J. P. Corriu, D. Leclercq, *Angew. Chem., Int. Ed. Engl.* **35**, 1421–1436 (1996).

3 K. J. Shea, D. A. Loy, *Chemistry of Materials* **13**, 3306–3319 (2001).

4 M. Choi, K. J. Shea, *Plast. Eng. (N. Y.)* **49**, 437–480 (1998).

5 E. Lindner, T. Schneller, F. Auer, H. A. Mayer, *Angew. Chem., Int. Ed.* **38**, 2155–2174 (1999).

6 U. Schubert, *New J. Chem.* **18**, 1049–1058 (1994).

7 J. J. E. Moreau, M. Wong Chi Man, *Coordination Chemistry Reviews* **178–180**, 1073–1084 (1998).

8 R. J. P. Corriu, *Angew. Chem., Int. Ed.* **39**, 1376–1398 (2000).

9 K. J. Shea, D. A. Loy, O. W. Webster, *Chem. Mater.* **1**, 572–574 (1989).

10 K. J. Shea, O. Webster, D. A. Loy, *Mater. Res. Soc. Symp. Proc.* **180**, 975–980 (1990).

11 J. H. Small, K. J. Shea, D. A. Loy, *J. Non-Cryst. Solids* **160**, 234–246 (1993).

12 K. J. Shea, D. A. Loy, O. Webster, *J. Am. Chem. Soc.* **114**, 6700–6710 (1992).

13 D. A. Loy, K. J. Shea, E. M. Russick, *Mater. Res. Soc. Symp. Proc.* **271**, 699–704 (1992).

14 G. Cerveau *et al.*, *Tailor-Made Silicon-Oxygen Compd., [Lect. Workshop]*, 273–293 (1996).

15 R. Corriu, *Polyhedron* **17**, 925–934 (1998).
16 B. Boury, R. J. P. Corriu, V. Le Strat, P. Delord, M. Nobili, *Angew. Chem., Int. Ed.* **38**, 3172–3175 (1999).
17 B. Boury, R. J. P. Corriu, V. Le Strat, P. Delord, *New J. Chem.* **23**, 531–538 (1999).
18 G. Cerveau, S. Chappellet, R. J. P. Corriu, B. Dabiens, *Journal of Organometallic Chemistry* **626**, 92–99 (2001).
19 B. Boury, R. J. P. Corriu, H. Muramatsu, *New Journal of Chemistry* **26**, 981–988 (2002).
20 J. P. Carpenter *et al., Polym. Prepr. (Am. Chem. Soc., Div. Polym. Chem.)* **39**, 589–590 (1998).
21 D. A. Loy *et al., Chem. Mater.* **10**, 4129–4140 (1998).
22 R. J. P. Corriu, J. J. E. Moreau, P. Thepot, M. W. C. Man, *J. Mater. Chem.* **4**, 987–989 (1994).
23 R. M. Shaltout, D. A. Loy, J. P. Carpenter, K. Dorhour, K. J. Shea, *Polym. Prepr. (Am. Chem. Soc., Div. Polym. Chem.)* **40**, 906–907 (1999).
24 D. A. Loy, R. J. Buss, R. A. Assink, K. J. Shea, H. Oviatt, *Polym. Prepr. (Am. Chem. Soc., Div. Polym. Chem.)* **34**, 244–245 (1993).
25 H. W. Oviatt, Jr., K. J. Shea, J. H. Small, *Chem. Mater.* **5**, 943–950 (1993).
26 D. A. Loy *et al., J. Non-Cryst. Solids* **186**, 44–53 (1995).
27 D. A. Loy *et al., J. Am. Chem. Soc.* **118**, 8501–8502 (1996).
28 S. A. Myers, R. A. Assink, D. A. Loy, K. J. Shea, *Perkin* 2, 545–549 (2000).
29 K. J. Shea, D. A. Loy, *Polym. Prepr. (Am. Chem. Soc., Div. Polym. Chem.)* **41**, 505 (2000).
30 M. R. Minke, K. J. Shea, D. A. Loy, *Polym. Prepr. (Am. Chem. Soc., Div. Polym. Chem.)* **42**, 887–888 (2001).
31 C. Li, G. L. Wilkes, *Journal of Inorganic and Organometallic Polymers* **8**, 33–45 (1998).
32 O. Villamo, C. Barboiu, M. Barboiu, W. Yau-Chun-Wan, N. Hovnanian, *Journal of Membrane Science* **204**, 97–110 (2002).
33 D. Zhu, W. J. Van Ooij, *Journal of Adhesion Science and Technology* **16**, 1235–1260 (2002).
34 S. T. Hobson, K. J. Shea, *Chem. Mater.* **9**, 616–623 (1997).
35 L. R. Dalton *et al., Chem. Mater.* **7**, 1060–1081 (1995).
36 H. W. Oviatt, Jr. *et al., Chem. Mater.* **7**, 493–498 (1995).
37 B. Fell, B. Meyer, *Chem.-Ztg.* **115**, 39–43 (1991).
38 F. Thurn, S. Wolff, *Kautsch. Gummi, Kunstst.* **28**, 733–739 (1975).
39 S. Kohjiya, Y. Ikeda, *Rubber Chemistry and Technology* **73**, 534–550 (2000).
40 J.-P. Bezombes, C. Chuit, R. J. P. Corriu, C. Reye, *J. Mater. Chem.* **8**, 1749–1759 (1998).
41 J.-P. Bezombes, C. Chuit, R. J. P. Corriu, C. Reye, *J. Mater. Chem.* **9**, 1727–1734 (1999).
42 J.-P. Bezombes, C. Chuit, R. J. P. Corriu, C. Reye, *Can. J. Chem.* **78**, 1519–1525 (2000).
43 R. J. P. Corriu, F. Embert, Y. Guari, A. Mehdi, C. Reye, *Chemical Communications* 1116–1117 (2001).
44 F. Embert, A. Mehdi, C. Reye, R. J. P. Corriu, *Chemistry of Materials* **13**, 4542–4549 (2001).
45 E. Lindner *et al., Chem. Mater.* **9**, 1524–1537 (1997).
46 E. Lindner *et al., Chemistry of Materials* **11**, 1833–1845 (1999).
47 C. Guizard, P. Lacan, *New J. Chem.* **18**, 1097–1107 (1994).
48 G. C. Gemeinhardt, S. K. Young, K. A. Mauritz, *Polymer Preprints (American Chemical Society, Division of Polymer Chemistry)* **39**, 379–380 (1998).
49 D. Cauzzi *et al., J. Organomet. Chem.* **541**, 377–389 (1997).
50 L. D. Carlos, V. de Bermudez, R. A. Sa Ferreira, L. Marques, M. Assuncao, *Chemistry of Materials* **11**, 581–588 (1999).
51 V. de Bermudez, L. D. Carlos, L. Alcacer, *Chemistry of Materials* **11**, 569–580 (1999).
52 C. Li, T. Glass, G. L. Wilkes, *Journal of Inorganic and Organometallic Polymers* **9**, 79–106 (1999).
53 J. C. Biazzotto *et al., Journal of Non-Crystalline Solids* **273**, 186–192 (2000).
54 D. A. Loy, C. R. Baugher, D. A. Schneider, A. Sanchez, F. Gonzalez, *Polym. Prepr. (Am. Chem. Soc., Div. Polym. Chem.)* **42**, 180–181 (2001).

55 C. Li, G. L. Wilkes, *Chemistry of Materials* **13**, 3663–3668 (2001).
56 J. J. E. Moreau *et al.*, *Journal of the American Chemical Society* **123**, 7957–7958 (2001).
57 E. Lindner *et al.*, *Journal of Organometallic Chemistry* **641**, 165–172 (2002).
58 N. N. Vlasova, O. Y. Raspopina, Y. N. Pozhidaev, M. G. Voronkov, *Russian Journal of General Chemistry* (translation of *Zhurnal Obshchei Khimii*) **72**, 55–57 (2002).
59 S. Cuney *et al.*, *Journal of Applied Polymer Science* **65**, 2373–2386 (1997).
60 C. Li, G. L. Wilkes, *J. Inorg. Organomet. Polym.* **7**, 203–216 (1997).
61 V. de Zea Bermudez, M. C. Goncalves, L. D. Carlos, *Ionics* **5**, 251–260 (1999).
62 D. A. Loy *et al.*, *Mater. Res. Soc. Symp. Proc.* **576**, 99–104 (1999).
63 D. C. Bookbinder, M. S. Wrighton, *J. Am. Chem. Soc.* **102**, 5123–5125 (1980).
64 J. G. Gaudiello, P. K. Ghosh, A. J. Bard, *J. Am. Chem. Soc.* **107**, 3027–3032 (1985).
65 N. Liu *et al.*, *Journal of the American Chemical Society* **124**, 14540–14541 (2002).
66 G. Cerveau, R. J. P. Corriu, N. Costa, *J. Non-Cryst. Solids* **163**, 226–235 (1993).
67 P. Audebert, P. Calas, G. Cerveau, R. J. P. Corriu, N. Costa, *J. Electroanal. Chem.* **372**, 275–277 (1994).
68 P. Audebert, G. Cerveau, R. J. P. Corriu, N. Costa, *J. Electroanal. Chem.* **413**, 89–96 (1996).
69 K. M. Choi, K. J. Shea, *Chem. Mater.* **5**, 1067–1069 (1993).
70 K. M. Choi, K. J. Shea, *J. Am. Chem. Soc.* **116**, 9052–9060 (1994).
71 K. M. Choi, K. J. Shea, *Mater. Res. Soc. Symp. Proc.* **346**, 763–771 (1994).
72 K. M. Choi, J. C. Hemminger, K. J. Shea, *J. Phys. Chem.* **99**, 4720–4732 (1995).
73 G. Cerveau, R. J. P. Corriu, C. Lepeytre, *J. Mater. Chem.* **5**, 793–795 (1995).
74 G. Cerveau, R. J. P. Corriu, C. Lepeytre, *Chem. Mater.* **9**, 2561–2566 (1997).
75 T. P. Chou *et al.*, *Journal of Non-Crystalline Solids* **290**, 153–162 (2001).
76 J.-I. Chen, R. Chareonsak, V. Puengpipat, S. Marturunkakul, *Journal of Applied Polymer Science* **74**, 1341–1346 (1999).
77 W. T. Ferrar *et al.*, *Polym. Prepr. (Am. Chem. Soc., Div. Polym. Chem.)* **41**, 503–504 (2000).
78 T. E. Gentle, R. H. Baney, *Mater. Res. Soc. Symp. Proc.* **274**, 115–119 (1992).
79 S. B. Hamilton, Jr., *Adhes. Age* **14**, 23–27 (1971).
80 K. Jordens, G. Wilkes, *Journal of Macromolecular Science, Pure and Applied Chemistry* **A38**, 185–207 (2001).
81 K. Jordens, G. Wilkes, *Polym. Mater. Sci. Eng.* **73**, 290–291 (1995).
82 J. M. Klosowski, G. A. L. Gant, *ACS Symp. Ser.* **113**, 113–127 (1979).
83 S. J. Landon, N. B. Dawkins, B. A. Waldman, *Journal of the Adhesive and Sealant Council* **1996**, 21–36 (1996).
84 K. L. Marshall *et al.*, *J. Appl. Phys.* **64**, 2279–2285 (1988).
85 W. J. van Ooij, T. Child, *Chemtech* **28**, 26–35 (1998).
86 J. Wen, K. Jordens, G. L. Wilkes, *Mater. Res. Soc. Symp. Proc.* **435**, 207–213 (1996).
87 G. L. Witucki, *J. Coat. Technol.* **65**, 57–60 (1993).
88 W. Li *et al.*, *Advanced Materials* **11**, 730–734 (1999).
89 D. A. Loy, E. M. Russick, S. A. Yamanaka, B. M. Baugher, K. J. Shea, *Chem. Mater.* **9**, 2264–2268 (1997).
90 G. W. Scherert, *Advances in Colloid and Interface Science* **76–77**, 321–339 (1998).
91 P. J. Barrie, S. W. Carr, D. L. Ou, A. C. Sullivan, *Chem. Mater.* **7**, 265–270 (1995).
92 D. A. Loy *et al.*, *Chem. Mater.* **8**, 656–663 (1996).
93 T. Suratwala *et al.*, *Chemistry of Materials* **10**, 199–209 (1998).
94 J. E. Mark, *Heterog. Chem. Rev.* **3**, 307–326 (1996).
95 S. Hofacker, G. Schottner, *Journal of Sol-Gel Science and Technology* **13**, 479–484 (1998).
96 N. N. Vlasova, G. Y. Zhila, M. G. Voronkov, *Zh. Obshch. Khim.* **66**, 1952–1954 (1996).
97 N. N. Vlasova *et al.*, in *U.S.S.R.* ((Irkutskij Institut Organicheskij Khimii So An Sssr, USSR). Su, 1996).
98 N. N. Vlasova, G. Y. Zhila, A. I. Kirillov, E. P. Khalikova, M. G. Voronkov, *Dokl. Akad. Nauk* **348**, 777–779 (1996).
99 M. G. Voronkov, N. N. Vlasova, A. E. Pestunovich, *Russian Journal of General Chemistry* (translation of *Zhurnal Obshchei Khimii*) **68**, 770–774 (1998).

100 N. N. Vlasova, Y. N. Pozhidaev, O. Y. Raspopina, L. I. Belousova, M. G. Voronkov, *Russian Journal of General Chemistry* (translation of *Zhurnal Obshchei Khimii*) **69**, 1391–1394 (1999).

101 N. N. Vlasova *et al.*, *Doklady Akademii Nauk* **364**, 492–494 (1999).

102 A. I. Kirillov *et al.*, *Zhurnal Prikladnoi Khimii (Sankt-Peterburg)* **73**, 520–521 (2000).

103 A. J. D. Barry, D. E. Hook, British Patent 635,645 (1950).

104 E. M. Boldebuck. U.S. Patent 2,551,924 (1951).

105 R. J. P. Corriu, J. J. E. Moreau, P. Thepot, M. W. C. Man, *Chem. Mater.* **4**, 1217–1224 (1992).

106 G. Cerveau, R. J. P. Corriu, E. Framery, *Chemistry of Materials* **13**, 3373–3388 (2001).

107 Y. Lu *et al.*, *J. Am. Chem. Soc.* **122**, 5258–5261 (2000).

108 M. T. Reetz, K.-E. Jaeger, *Chemistry and Physics of Lipids* **93**, 3–14 (1998).

109 M. T. Reetz, *Adv. Mater.* **9**, 943–954 (1997).

110 M. T. Reetz, A. Zonta, J. Simpelkamp, A. Rufinska, B. Tesche, *J. Sol-Gel Sci. Technol.* **7**, 35–43 (1996).

111 M. Reetz, A. Zonta, J. Simpelkamp, *Angew. Chem., Int. Ed. Engl.* **34**, 301–303 (1995).

112 D. J. Brondani, R. J. P. Corriu, S. El Ayoubi, J. J. E. Moreau, M. Wong Chi Man, *J. Organomet. Chem.* **451**, C1–C3 (1993).

113 R. J. P. Corriu, M. Granier, G. F. Lanneau, *J. Organomet. Chem.* **562**, 79–88 (1998).

114 B. Boury, R. J. P. Corriu, V. L. Strat, *Chem. Mater.* **11**, 2796–2803 (1999).

115 R. J. P. Corriu *et al.*, *Chem. Mater.* **6**, 640–649 (1994).

116 B. M. Baugher *et al.*, *Mater. Res. Soc. Symp. Proc.* **371**, 253–259 (1995).

117 D. A. Loy, B. M. Baugher, S. Prabakar, R. A. Assink, K. J. Shea, *Mater. Res. Soc. Symp. Proc.* **371**, 229–234 (1995).

118 I. Ojima, *Chem. Org. Silicon Compd.* **2**, 1479–1526 (1989).

119 D. M. Gara, D. A. Loy, *Polymer Preprints (American Chemical Society, Division of Polymer Chemistry)* **43**, 676–677 (2002).

120 H. Wolter, W. Storch, H. Ott, *Mater. Res. Soc. Symp. Proc.* **346**, 143–149 (1994).

121 C. Kim, E. K. Kim, I.-J. Chin, K. D. Park, Y. H. Kim, *Pollimo* **19**, 240–246 (1995).

122 F. Schapman, J. P. Couvercelle, C. Bunel, *Polymer* **39**, 965–971 (1998).

123 A. Adima, J. J. E. Moreau, M. W. C. Man, *J. Mater. Chem.* **7**, 2331–2333 (1997).

124 A. Adima, J. J. E. Moreau, M. W. Chi Man, *Chirality* **12**, 411–420 (2000).

125 P. Hesemann, J. J. E. Moreau, *Tetrahedron: Asymmetry* **11**, 2183–2194 (2000).

126 C. Bied, D. Gauthier, J. J. E. Moreau, M. W. C. Man, *J. Sol-Gel Sci. Technol.* **20**, 313–320 (2001).

127 A. Brethon, P. Hesemann, L. Rejaud, J. J. E. Moreau, M. Wong Chi Man, *J. Organomet. Chem.* **627**, 239–248 (2001).

128 D. C. Tahmassebi, T. Sasaki, *Journal of Organic Chemistry* **59**, 679–681 (1994).

129 K.-o. Hwang, T. Sasaki, *Journal of Materials Chemistry* **8**, 2153–2156 (1998).

130 K.-O. Hwang, Y. Yakura, F. S. Ohuchi, T. Sasaki, *Mater. Sci. Eng., C* **C3**, 137–141 (1995).

131 D. A. Loy *et al.*, *J. Am. Chem. Soc.* **121**, 5413–5425 (1999).

132 S. W. Carr, M. Motevalli, D. L. Ou, A. C. Sullivan, *Journal of Materials Chemistry* **7**, 865–872 (1997).

133 S. W. Carr, L. Courtney, A. C. Sullivan, *Chemistry of Materials* **9**, 1751–1756 (1997).

134 S. W. Carr, D. Li Ou, A. C. Sullivan, *Journal of Sol-Gel Science and Technology* **13**, 31–36 (1998).

135 B. Marciniec, *Silicon for the Chemical Industry I (Conf. Proc.)*, 119–130 (1992).

136 C. Chuit, R. J. P. Corriu, G. Dubois, C. Reye, *Chem. Commun.* 723–724 (1999).

137 G. Dubois, C. Reye, R. J. P. Corriu, C. Chuit, *J. Mater. Chem.* **10**, 1091–1098 (2000).

138 C. Carbonneau, R. Frantz, J.-O. Durand, G. F. Lanneau, R. J. P. Corriu, *Tetrahedron Lett.* **40**, 5855–5858 (1999).

139 M. D. McClain, D. A. Loy, S. Prabakar, *Mater. Res. Soc. Symp. Proc.* **435**, 277–282 (1996).

140 F. Kakiuchi, S. Murai, *Accounts of Chemical Research* **35**, 826–834 (2002).

141 H. Gao, R. J. Angelici, *Journal of the American Chemical Society* **119**, 6937–6938 (1997).
142 H. Gao, R. J. Angelici, *Organometallics* **18**, 989–995 (1999).
143 J.-P. Bezombes *et al.*, *J. Organomet. Chem.* **535**, 81–90 (1997).
144 E. Lindner, A. Jaeger, T. Schneller, H. A. Mayer, *Chemistry of Materials* **9**, 81–90 (1997).
145 E. Lindner, T. Salesch, F. Hoehn, A. Mayer, *Z. Anorg. Allg. Chem.* **625**, 2133–2143 (1999).
146 E. Lindner, A. Enderle, A. Baumann, *Journal of Organometallic Chemistry* **558**, 235–237 (1998).
147 E. Lindner, S. Brugger, S. Steinbrecher, E. Plies, H. A. Mayer, *Z. Anorg. Allg. Chem.* **627**, 1731–1740 (2001).
148 T. Salesch *et al.*, *Advanced Functional Materials* **12**, 134–142 (2002).
149 F. H. Carre *et al.*, *Eur. J. Inorg. Chem.*, *4*, 647–653 (2000).
150 C. Chuit *et al.*, *J. Organomet. Chem.* **511**, 171–175 (1996).
151 U. Schubert, N. Huesing, A. Lorenz, *Chem. Mater.* **7**, 2010–2027 (1995).
152 M. G. Voronkov *et al.*, *Zhurnal Obshchei Khimii* **52**, 2751–2754 (1982).
153 C. Brinker, G. Scherer, *Sol-Gel Science: The Physics and Chemistry of Sol-Gel Processing* (1990). Academic Press
154 F. D. Osterholtz, E. R. Pohl, *J. Adhes. Sci. Technol.* **6**, 127–149 (1992).
155 K. J. Shea, D. A. Loy, *Acc. Chem. Res.* **34**, 707–716 (2001).
156 D. A. Loy, B. M. Baugher, C. R. Baugher, D. A. Schneider, K. Rahimian, *Chem. Mater.* **12**, 3624–3632 (2000).
157 G. Cerveau, R. J. P. Corriu, C. Lepeytre, P. H. Mutin, *J. Mater. Chem.* **8**, 2707–2713 (1998).
158 P. Chevalier, R. J. P. Corriu, P. Delord, J. J. E. Moreau, M. Wong Chi Man, *New J. Chem.* **22**, 423–433 (1998).
159 G. Cerveau, R. J. P. Corriu, E. Framery, *Journal of Materials Chemistry* **11**, 713–717 (2001).
160 G. Cerveau, R. J. P. Corriu, C. Fischmeister-Lepeytre, *J. Mater. Chem.* **9**, 1149–1154 (1999).
161 R. J. Hook, *J. Non-Cryst. Solids* **195**, 1–15 (1996).
162 M. G. Voronkov, V. I. Lavrent'yev, *Top. Curr. Chem.* **102**, 199–236 (1982).
163 K. G. Sharp, G. W. Scherer, *J. Sol-Gel Sci. Technol.* **8**, 165–171 (1997).
164 L. W. Kelts, N. J. Armstrong, *J. Mater. Res.* **4**, 423–433 (1989).
165 D. A. Loy, B. M. Saugher, D. A. Schneider, *Polym. Prepr. (Am. Chem. Soc., Div. Polym. Chem.)* **39**, 418–419 (1998).
166 D. A. Schneider, B. M. Baugher, D. A. Loy, K. Rahimian, T. Alam, *Mater. Res. Soc. Symp. Proc.* **628**, CC6.35.1–CC6.35.6 (2001).
167 G. Cerveau, R. J. P. Corriu, E. Framery, S. Ghosh, H. P. Mutin, *Journal of Materials Chemistry* **12**, 3021–3026 (2002).
168 D. A. Loy *et al.*, *Chemistry of Materials* **11**, 3333–3341 (1999).
169 B. Ameduri, B. Boutevin, J. J. E. Moreau, H. Moutaabbid, M. W. Chi Man, *J. Fluorine Chem.* **104**, 185–194 (2000).
170 S. A. Yamanaka, J. P. Carpenter, M. D. McClain, D. A. Loy, *Int. SAMPE Tech. Conf.* **27**, 568–577 (1995).
171 N. Huesing, U. Schubert, K. Misof, P. Fratzl, *Chemistry of Materials* **10**, 3024–3032 (1998).
172 R. Corriu, *C. R. Acad. Sci., Ser. IIc: Chim.* **1**, 83–89 (1998).
173 F. Ben, B. Boury, R. J. P. Corriu, V. Le Strat, *Chem. Mater.* **12**, 3249–3252 (2000).
174 B. Boury, R. J. P. Corriu, P. Delord, V. Le Strat, *J. Non-Cryst. Solids* **265**, 41–50 (2000).
175 B. Boury, F. Ben, R. J. P. Corriu, *Angewandte Chemie, International Edition* **40**, 2853–2856 (2001).
176 B. Boury *et al.*, *Chem. Mater.* **11**, 281–291 (1999).
177 D. W. Schaefer *et al.*, *Mater. Res. Soc. Symp. Proc.* **435**, 301–306 (1996).
178 C. Bonhomme *et al.*, *Mater. Res. Soc. Symp. Proc.* **435**, 437–442 (1996).
179 F. Babonneau, J. Maquet, *Polyhedron* **19**, 315–322 (2000).
180 T. J. Barton *et al.*, *Chem. Mater.* **11**, 2633–2656 (1999).
181 B. M. Baugher, D. A. Schneider, D. A. Loy, K. Rahimian, *Mater. Res. Soc. Symp. Proc.* **576**, 105–110 (1999).
182 G. Cerveau, R. J. P. Corriu, C. Lepeytre, *J. Organomet. Chem.* **548**, 99–103 (1997).

183 G. Cerveau, R. J. P. Corriu, *Coord. Chem. Rev.* **178–180**, 1051–1071 (1998).

184 G. Cerveau, R. J. P. Corriu, E. Framery, *Chem. Commun.* 2081–2082 (1999).

185 K. J. Shea, D. A. Loy, O. W. Webster, *Polym. Mater. Sci. Eng.* **63**, 281–285 (1990).

186 D. A. Loy, K. J. Shea. U.S. Patent 5,321,102 (1994).

187 D. A. Loy, K. J. Shea, R. J. Buss, R. A. Assink, *ACS Symp. Ser.* **572**, 122–133 (1994).

188 D. A. Loy, R. J. Buss, R. A. Assink, K. J. Shea, H. Oviatt, *Mater. Res. Soc. Symp. Proc.* **346**, 825–829 (1994).

189 B. Boury, R. J. P. Corriu, R. Nunez, *Chem. Mater.* **10**, 1795–1804 (1998).

190 C. J. Brinker *et al.*, *J. Membr. Sci.* **94**, 85–102 (1994).

191 R. M. Shaltout *et al.*, *Polym. Prepr. (Am. Chem. Soc., Div. Polym. Chem.)* **41**, 508–509 (2000).

192 A. Katz, M. E. Davis, *Nature (London)* **403**, 286–289 (2000).

193 A. L. Graham, C. A. Carlson, P. L. Edmiston, *Analytical Chemistry* **74**, 458–467 (2002).

194 P. Chevallier, R. Corriu, J. Moreau, C. M. M. Wong, (Rhone Poulenc Chimie, Fr.). Fr 2,728,572 (1996).

195 R. J. P. Corriu, J. J. E. Moreau, P. Thepot, M. W. C. Man, *Chem. Mater.* **8**, 100–6 (1996).

196 P. M. Chevalier, R. J. P. Corriu, J. J. E. Moreau, M. W. C. Man, *J. Sol-Gel Sci. Technol.* **8**, 603–607 (1997).

197 B. Boury, R. J. P. Corriu, *Adv. Mater. (Weinheim, Ger.)* **12**, 989–992 (2000).

198 G. Cerveau, R. J. P. Corriu, B. Dabiens, *J. Mater. Chem.* **10**, 1113–1120 (2000).

199 S. Inagaki, S. Guan, Y. Fukushima, T. Ohsuna, O. Terasaki, *Journal of the American Chemical Society* **121**, 9611–9614 (1999).

200 N. Igarashi, Y. Tanaka, S.-I. Nakata, T. Tatsumi, *Chemistry Letters*, 1–2 (1999).

201 M. H. Lim, C. F. Blanford, A. Stein, *J. Am. Chem. Soc.* **119**, 4090–4091 (1997).

202 M. Kruk, T. Asefa, M. Jaroniec, G. A. Ozin, *Journal of the American Chemical Society* **124**, 6383–6392 (2002).

203 M. Kruk, M. Jaroniec, S. Guan, S. Inagaki, *Journal of Physical Chemistry B* **105**, 681–689 (2001).

204 M. Koya, H. Nakajima, *Studies in Surface Science and Catalysis* **117**, 243–248 (1998).

205 M. J. MacLachlan, T. Asefa, G. A. Ozin, *Chemistry–A European Journal* **6**, 2507–2511 (2000).

206 J. R. Matos *et al.*, *Chemistry of Materials* **14**, 1903–1905 (2002).

207 B. J. Melde, B. T. Holland, C. F. Blanford, A. Stein, *Chemistry of Materials* **11**, 3302–3308 (1999).

208 A. Sayari, S. Hamoudi, Y. Yang, I. L. Moudrakovski, J. R. Ripmeester, *Chem. Mater.* **12**, 3857–3863 (2000).

209 K. Yu, B. Smarsly, C. J. Brinker, *Advanced Functional Materials* **13**, 47–52 (2003).

210 C. Yoshina-Ishii, T. Asefa, N. Coombs, M. J. MacLachlan, G. A. Ozin, *Chem. Commun. (Cambridge)*, 2539–2540 (1999).

211 S. Dai *et al.*, *Journal of the American Chemical Society* **122**, 992–993 (2000).

212 R. J. P. Corriu *et al.*, *Journal of Materials Chemistry* **12**, 1355–1362 (2002).

213 R. J. P. Corriu, A. Mehdi, C. Reye, C. Thieuleux, *Chemical Communications (Cambridge, United Kingdom)*, 1382–1383 (2002).

214 R. J. P. Corriu *et al.*, *Chemical Communications (Cambridge, United Kingdom)*, 763–764 (2001).

215 H. Dong, J. Xu, K.-Y. Qiu, S. A. Jansen, Y. Wei, *Polymer Preprints (American Chemical Society, Division of Polymer Chemistry)* **41**, 194–195 (2000).

216 H. Fan, Y. Lu, R. A. Assink, G. P. Lopez, C. J. Brinker, *Materials Research Society Symposium Proceedings* **628**, CC6.41.1–CC6.41.7 (2001).

217 S. Guan, S. Inagaki, T. Ohsuna, O. Terasaki, *Microporous and Mesoporous Materials* **44–45**, 165–172 (2001).

218 S. Guan, S. Inagaki, T. Ohsuna, O. Terasaki, *Journal of the American Chemical Society* **122**, 5660–5661 (2000).

219 A. Fukuoka *et al.*, *Journal of the American Chemical Society* **123**, 3373–3374 (2001).

220 T. Asefa, M. J. MacLachlan, H. Grondey, N. Coombs, G. A. Ozin, *Angewandte Chemie, International Edition* **39**, 1808–1811 (2000).

221 T. Asefa, C. Yoshina-Ishii, M. J. MacLachlan, G. A. Ozin, *Journal of Materials Chemistry* **10**, 1751–1755 (2000).

222 T. Asefa, M. J. MacLachlan, N. Coombs, G. A. Ozin, *Nature (London)* **402**, 867–871 (1999).
223 R. J. P. Corriu, A. Mehdi, C. Reye, *C. R. Acad. Sci., Ser. IIc: Chim.* **2**, 35–39 (1999).
224 S. Inagaki, S. Guan, T. Ohsuna, O. Terasaki, *Nature (London, United Kingdom)* **416**, 304–307 (2002).
225 R. Corriu *et al.*, *Polym. Prepr. (Am. Chem. Soc., Div. Polym. Chem.)* **41**, 510–511 (2000).
226 B. Boury, F. Ben, R. J. P. Corriu, P. Delord, M. Nobili, *Chemistry of Materials* **14**, 730–738 (2002).
227 A. Vergnes *et al.*, *Journal of Sol-Gel Science and Technology* **26**, 621–624 (2003).
228 F. Ben, B. Boury, R. J. P. Corriu, *Advanced Materials (Weinheim, Germany)* **14**, 1081–1084 (2002).
229 J. J. E. Moreau, L. Vellutini, M. W. C. Man, C. Bied, *J. Am. Chem. Soc.* **123**, 1509–1510 (2001).
230 H. Muramatsu, R. Corriu, B. Boury, *Journal of the American Chemical Society* **125**, 854–855 (2003).
231 G. Dubois *et al.*, *Angewandte Chemie, International Edition* **40**, 1087–1090 (2001).
232 R. J. P. Corriu, F. Embert, Y. Guari, C. Reye, R. Guilard, *Chemistry–A European Journal* **8**, 5732–5741 (2002).
233 R. J. P. Corriu, C. Hoarau, A. Mehdi, C. Reye, *Chem. Commun. (Cambridge)*, 71–72 (2000).
234 K. G. Sharp, M. J. Michalczyk, *J. Sol-Gel Sci. Technol.* **8**, 541–546 (1997).
235 K. G. Sharp, M. J. Michalczyk, *Mater. Res. Soc. Symp. Proc.* **435**, 105–112 (1996).
236 M. J. Michalczyk, K. G. Sharp, *Tailor-Made Silicon-Oxygen Compd., [Lect. Workshop]*, 295–303 (1996).
237 J. W. Kriesel, T. D. Tilley, *Chemistry of Materials* **11**, 1190–1193 (1999).
238 J. W. Kriesel, T. D. Tilley, *Polymer Preprints (American Chemical Society, Division of Polymer Chemistry)* **41**, 566–567 (2000).
239 J. W. Kriesel, T. D. Tilley, *Chemistry of Materials* **12**, 1171–1179 (2000).
240 T. D. Tilley, *Journal of Molecular Catalysis A: Chemical* **182–183**, 17–24 (2002).
241 G. Cerveau, R. J. P. Corriu, E. Framery, *Comptes Rendus de l'Academie des Sciences, Serie IIc: Chimie* **4**, 79–83 (2001).
242 K. Yamamoto, Y. Sakata, Y. Nohara, Y. Takahashi, T. Tatsumi, *Science (Washington, DC, United States)* **300**, 470–472 (2003).
243 T. A. Brandvold, J. S. Holmgren, T. P. Malloy, (UOP Inc.) Cont.-in-part of U.S. Patent 5,371,154 (1995).
244 N. Liu *et al.*, *Angewandte Chemie, International Edition* **42**, 1731–1734 (2003).
245 R. M. Shaltout, D. A. Loy, *Polym. Mater. Sci. Eng.* **80**, 195–196 (1999).
246 C. Jin *et al.*, *Materials Research Society Symposium Proceedings* **443**, 99–104 (1997).
247 T. Kobayakawa, S. Tanaka, T. Imura, (Tokuyama Soda Kk, Japan). Jpn. Kokai Tokkyo Koho. 05,140,549 (1993).
248 C. Li, G. L. Wilkes, *J. Macromol. Sci., Pure Appl. Chem.* **A37**, 549–571 (2000).
249 B. Arkles, *Chemtech* **29**, 7–14 (1999).
250 N. Koike, H. Tsukagoshi, (Toshiba Corp., Japan). Jpn. Kokai Tokkyo Koho 60, 169, 847 (1985).
251 T. Suratwala *et al.*, *Chemistry of Materials* **10**, 190–198 (1998).
252 A. Kraus, M. Schneider, A. Gugel, K. Mullen, *Journal of Materials Chemistry* **7**, 763–765 (1997).
253 H. Xia *et al.*, *Fullerene Science and Technology* **5**, 1621–1626 (1997).
254 H. Nakashima, M. Irie, *Macromol. Chem. Phys.* **200**, 683–692 (1999).
255 C. D. Ki, J. K. Kim, S. S. Hwang, S. I. Hong, *Polymer Preprints (American Chemical Society, Division of Polymer Chemistry)* **41**, 600–601 (2000).
256 R. E. Taylor-Smith, K. M. Choi, *Mater. Res. Soc. Symp. Proc.* **576**, 433–438 (1999).
257 Y. H. Li *et al.*, *Thin Solid Films* **385**, 205–208 (2001).
258 V. de Bermudez *et al.*, *Journal of Physical Chemistry B* **105**, 3378–3386 (2001).
259 S. T. Hobson, J. Zieba, P. N. Prasad, K. J. Shea, *Mater. Res. Soc. Symp. Proc.* **561**, 21–28 (1999).
260 Z. Yang *et al.*, *Chem. Mater.* **6**, 1899–901 (1994).
261 C. Sanchez, F. Ribot, B. Lebeau, *Journal of Materials Chemistry* **9**, 35–44 (1999).

262 A. O. Ribeiro, J. C. Biazzotto, O. A. Serra, *Journal of Non-Crystalline Solids* **273**, 198–202 (2000).

263 S. Mangematin, A. B. Sorokin, *Journal of Porphyrins and Phthalocyanines* **5**, 674–680 (2001).

264 J. C. Biazzotto *et al.*, *Journal of Non-Crystalline Solids* **247**, 134–140 (1999).

265 K. J. Ciuffi *et al.*, *Journal of Non-Crystalline Solids* **273**, 100–108 (2000).

266 H. C. Sacco *et al.*, *Journal of Non-Crystalline Solids* **273**, 150–158 (2000).

267 E. Sugiono, T. Metzroth, H. Detert, *Advanced Synthesis & Catalysis* **343**, 351–359 (2001).

268 R. J. P. Corriu, P. Hesemann, G. F. Lanneau, *Chem. Commun. (Cambridge)*, 1845–1846 (1996).

269 C. Carbonneau *et al.*, *New Journal of Chemistry* **25**, 1398–1402 (2001).

270 R. Frantz *et al.*, *Tetrahedron Letters* **43**, 6569–6572 (2002).

271 J. Biteau *et al.*, *Chemistry of Materials* **10**, 1945–1950 (1998).

272 A. Martino, S. A. Yamanaka, J. S. Kawola, D. A. Loy, *Chem. Mater.* **9**, 423–429 (1997).

273 K. M. Choi, K. J. Shea, *J. Phys. Chem.* **98**, 3207–14 (1994).

274 Y. Guari *et al.*, *Chemical Communications (Cambridge, United Kingdom)*, 1374–1375 (2001).

275 Y. Guari *et al.*, *Chemistry of Materials* **15**, 2017–2024 (2003).

276 K. M. Choi, K. J. Shea, *J. Sol-Gel Sci. Technol.* **5**, 143–57 (1995).

277 R. Corriu, A. Mehdi, C. Reye, H. Ledon, (Centre National de la Recherche Scientifique, Fr.) WO 0,276,991 (2002).

278 I. Honma, H. Nakajima, O. Nishikawa, T. Sugimoto, S. Nomura, *Journal of the Electrochemical Society* **150**, A616–A619 (2003).

279 I. Honma, H. Nakajima, O. Nishikawa, T. Sugimoto, S. Nomura, *Journal of the Electrochemical Society* **149**, A1389–A1392 (2002).

280 I. Honma, H. Nakajima, O. Nishikawa, T. Sugimoto, S. Nomura, *Electrochemistry (Tokyo, Japan)* **70**, 920–923 (2002).

281 E. Lindner, A. Jager, F. Auer, W. Wielandt, P. Wegner, *Journal of Molecular Catalysis A: Chemical* **129**, 91–95 (1998).

282 E. Fache, C. Mercier, N. Pagnier, B. Despeyroux, P. Panster, *J. Mol. Catal.* **79**, 117–31 (1993).

283 M. Berry, R. K. Champaneria, J. A. S. Howell, *Journal of Molecular Catalysis* **37**, 243–52 (1986).

284 E. Lindner *et al.*, *Journal of Molecular Catalysis A: Chemical* **157**, 97–109 (2000).

285 I. C. Chisem *et al.*, *Chemical Communications (Cambridge)*, 1949–1950 (1998).

286 M. G. Voronkov, N. N. Vlasova, Y. N. Pozhidaev, *Zh. Prikl. Khim. (S.-Peterburg)* **69**, 705–718 (1996).

287 Y. N. Pozhidaev, S. A. Bol'shakova, A. E. Pestunovich, N. N. Vlasova, M. G. Voronkov, *Dokl. Akad. Nauk* **355**, 653–655 (1997).

288 P. Panster, K. H. Koenig, E. Schopenhauer-Gehrmann, P. Kleinschmit, (Degussa A.-G., Fed. Rep. Ger.) DE 3,706,523 (1988).

289 P. Panster, R. Gradl, P. Kleinschmit, (Degussa A.-G., Germany). DE 3,837,418 (1990).

290 P. Panster, S. Wieland,. (Degussa A.-G., Germany). DE 3,925,359 (1991).

291 M. C. Burleigh, M. A. Markowitz, M. S. Spector, B. P. Gaber, *Environmental Science and Technology* **36**, 2515–2518 (2002).

4
Porous Inorganic-Organic Hybrid Materials

Nicola Hüsing and Ulrich Schubert

4.1
Introduction

In recent years the emphasis of material design has dramatically changed towards a general understanding and control of the fundamental connections between the chemistry on a molecular level and the materials properties on the macroscopic scale. Looking towards the 21st century, advances in information processing, communication (microelectronics), medicine etc. require miniaturized materials with superior properties and performance. "Nanotechnology", which is essentially material science on a nanometer scale, is a subject that combines the efforts of scientists in interdisciplinary research (physicists, chemists and material scientists). Developing and designing materials from a molecular/atomic level requires the ability to control deliberately the positioning of the molecular building blocks within a material. This is also an essential requirement in the synthesis of porous materials in which the arrangement of different building blocks forming the solid framework determines the pores' size, shape and arrangement.

A wide variety of porous inorganic frameworks is known. Following the classification by IUPAC, they can be grouped by the size of their pores: microporous solids with pore diameters up to 2 nm with zeolites as the most prominent example, mesoporous solids with pore sizes between 2 and 50 nm, for example aerogels, pillared clays, M41S materials, and macroporous solids such as glasses and foams. In addition, these materials can be distinguished by the arrangement of the pores – periodic or random – and the pore radii distribution, which can be either narrow or quite broad.

For most of these porous materials, modifications of the inorganic backbone are required to provide a certain specific surface chemistry or active sites on the inner pore surface. This makes the materials more viable for applications in catalysis, sensing or separation technologies, for example. In principle, the materials can be functionalized by simple inclusion of active species in the confined pore spaces, but in many cases covalent anchoring of functional moieties (mostly organic or organometallic) is required to avoid leaching etc.

Gómez-Romero: Organic-Inorganic Materials. Nicola Hüsing

ISBN: 3-527-30484-3

Materials composed of both inorganic and organic entities are called hybrid materials. They combine, to some extent, the properties of both constituents in *one* material. The general idea behind these materials is similar to that of composites, where two or more materials are combined that differ in form and (mostly) chemical composition. While macroscopic constituents with defined phase boundaries are used for composite materials, molecular building blocks of different composition (inorganic and organic/organometallic/biological) are instead combined in hybrid materials. The combination of different building blocks allows the generation of materials with new properties or combinations of properties not accessible otherwise. In the case of porous materials, the goal is to achieve a synergy between the properties originating from the porous inorganic framework and the properties of the organic entities.

In the following, we will focus the discussion on the chemical functionalization of two different types of hybrid mesoporous materials, that is aerogels and M41S materials. Both classes of materials exhibit high surface areas and porosities, aerogels with a random orientation of the pores and a broad pore size distribution, and M41S materials with a narrow pore size distribution and a periodic and highly regular arrangement of the pores. Each of these materials has a distinct way of preparation to produce the pore structure. In this article, we show the possibilities but also the problems associated with a modification of the materials by organic groups.

Following a short section on the formation of oxidic inorganic networks in general (sol-gel chemistry), we will discuss the synthesis and properties of aerogels as well as M41S materials in detail. In the second part, different alternatives for the functionalization and modification of both types of materials are presented, starting with some general considerations on how the functional organic moieties can be introduced, followed by a discussion on different strategies such as grafting of organic groups, co-condensation reactions, incorporation of metal complexes, biological entities etc. It will be shown that, in principle, the general strategies for the modification of both types of porous materials are the same, but that different problems are encountered, which means that the applicability of the various methods is very different.

4.2 Inorganic Network Formation

The sol-gel process is particularly well suited to create inorganic-organic hybrid materials, because the solids are formed at low temperatures, at which organic groups are not degraded. It is a method for converting soluble molecular or oligomeric compounds into extended inorganic (oxidic) networks, and amorphous or microcrystalline solids are obtained. In the first stage of the process, colloidal particles – a sol – are formed which then aggregate to an extended gel network. There are two possibilities for the formation of gels: Solid gels may be dissolved (peptized) and the obtained sols re-aggregated under different conditions (usually at a different pH), or sol particles may be formed by chemical reactions from molecular precursors, such as metal or semi-metal alkoxides, or hydrolyzable metal salts. The most important precursors for porous oxide materials, especially when organically modified materials are prepared, are alkoxides. The chemical reactions during sol-gel processing of metal or semi-metal alkoxides can be formally described by three equations (given for an alkoxysilane in Eq. 1).

$$\text{Si–OR} + H_2O \rightarrow \text{Si–OH} + \text{ROH} \qquad \text{Hydrolysis}$$

$$\left.\begin{array}{l}\text{Si–OH} + \text{Si–OR} \rightarrow \text{Si–O–Si} + \text{ROH}\\ \text{Si–OH} + \text{Si–OH} \rightarrow \text{Si–O–Si} + H_2O\end{array}\right\} \text{Condensation} \qquad (1)$$

In the first part of the overall reaction, Si–OR groups are hydrolyzed, and reactive Si–OH groups are created. The Si–OH groups then undergo condensation reactions (by elimination of water or alcohol), which result in a stepwise build-up of the oxide network.

Formation of sol-gel materials is kinetically controlled, and therefore the reaction conditions (solvent, pH, additives, concentration, temperature, etc.) influence the network structure to a very high degree. One of the most important parameters is the pH, and different types of silica networks are obtained when $Si(OR)_4$ is processed under acidic or basic conditions, respectively (see below).

Metal alkoxides are much more reactive towards water than alkoxysilanes. The reactivity of many metal alkoxides towards water is so high that precipitates are spontaneously formed. While the reactivity of alkoxysilanes has to be promoted by acid or base catalysts, the reaction rates of metal alkoxides must be moderated to get gels instead of precipitates. The most common method is to substitute part of the alkoxy groups by chelating or bridging ligands. The new molecular compound $M(OR)_{n-x}L_x$ (L = bidentate anionic ligand) shows a different reactivity, structure and functionality compared to the unsubstituted alkoxide $M(OR)_n$. During hydrolytic polycondensation of such precursors, the groups L are largely retained. The most common bidentate ligands are acetate or β-diketonates.

A *gel* consists of a sponge-like, three-dimensional solid network, the pores being filled with another substance, usually a liquid. When gels are prepared by hydrolysis and condensation reactions of hydrolyzable metal or semi-metal compounds (via the sol stage), the pore liquid mainly consists of water and/or alcohol. The resulting "wet" gels are therefore called *aquagels, hydrogels,* or *alcogels.* When the pore liquid is replaced by air without destroying the gel structure and without major shrinkage of the gel body, *aerogels* are obtained. A *xerogel* is formed instead, when the wet gels are conventionally dried, that is by temperature increase or pressure decrease, with concomitant large shrinkage (and mostly destruction) of the initially uniform gel body. The reason for the large shrinkage of a gel body upon evaporation of the pore liquid is the action of capillary forces on the pore walls as the liquid retreats into the gel body. This eventually results in the collapse of the filigrane, highly porous inorganic networks of the aquagels or alcogels. Therefore, special drying methods had to be developed to prepare aerogels, such as the supercritical drying technique.

Two different approaches can be used for the incorporation of organic groups into an inorganic network, namely embedding of organic molecules into gels without chemical bonding, and incorporation of organic molecules via covalent bonding.

Embedding of organic molecules or polymers into the inorganic host is mostly achieved by dissolving them in the precursor solution. The gel matrix is formed around them and traps them. A variety of organic or organometallic molecules can be employed, such as dyes, catalytically active metal complexes, sensor compounds, biomolecules or polymers (see below).

Very important modifications of sol-gel materials are based on a covalent linkage of the organic groups. In silicate systems, it is possible to use organotrialkoxysilanes, $R'Si(OR)_3$ or $(RO)_3Si–R'–Si(OR)_3$, as precursors for sol-gel processing in the same way as tetraalkoxysilanes. In the organically substituted derivatives, the groups R′ are bonded through Si–C links to the network-forming inorganic part of the molecule. Since Si–C bonds are hydrolytically stable, the organic groups are retained in the final material. The choice of the organic group R′ is nearly unlimited, and many functional organic groups can be used. Because of the difficulties in obtaining three-dimensional gel networks from organically substituted trialkoxysilanes, they are typically used as mixtures with the corresponding tetraalkoxysilane.

Unlike Si–C bonds, metal-carbon bonds are usually cleaved by water. However, the organic groups can be introduced via the bidentate ligands mentioned above. Usually, these complexing ligands are used to moderate the reactivity of the alkoxide precursors (see above), but they can also be used for an organic functionalization.

4.3 Preparation and Properties

4.3.1 Aerogels

Aerogels are characterized by an extremely high porosity with well accessible, cylindrical, branched mesopores. The resulting bulk densities are typically in the range 0.003–0.500 g cm^{-3}. Aerogels can be prepared as monoliths, granulates or powders. There are some review articles on various aspects of aerogels [1]. The focus of this article is on a sub-class of aerogels, which are composed of both inorganic and organic entities, that is inorganic-organic hybrid aerogels. The organic groups may either be part of the gel network or, more often, modify the surface of an otherwise purely inorganic network.

The filigrane solid network and, as a result, the pore structure of aerogels is formed by condensation of primary particles which have diameters in the lower nanometer range. Fig. 4.1 shows the structure of a SiO_2 aerogel schematically. Generation and aggregation of the particles is controlled by chemical processes, mostly the sol-gel process. The primary particles can be chain-like, cyclic or three-dimensional primary oligomeric species formed by polycondensation reactions. Condensation reactions also lead to the formation of sol particles from the primary particles, and the aggregation of the sol particles to form the oxidic gel network.

The kind of inorganic network formed by the condensation reactions not only depends on the absolute rates of the individual reactions but also on the relative rate of the hydrolysis and condensation reactions [2]. Colloidal and polymeric gels constitute the two extremes with regard to both the microstructure and the resulting properties. In colloidal gels, formed under basic conditions, dense colloidal particles are interconnected like a string of pearls. The resulting network has a particulate character with large particles and large pores. Polymeric silica gels are formed under acidic conditions (pH 2–5). Their primary particles consist of linear or branched polymer chains. Preparation of aerogels from polymeric gels is more difficult, because diffusion processes are strongly inhibited by the smaller pores. Complete removal of the pore liquid is therefore more difficult and results in a larger shrinkage during drying [3]. For this reason, silica aerogels are usually prepared by base-catalyzed reaction of $Si(OMe)_4$ (TMOS) or $Si(OEt)_4$ (TEOS), mostly with ammonia as the catalyst.

The basic chemical processes are similar for metal alkoxides, although their reactivity is much higher. Therefore, modified precursors are often employed, as mentioned above. It should be pointed out that aerogels prepared from these modified precursors are initially inorganic-organic hybrid aerogels. However, the properties originating from the presence of the organic groups in the as-synthesized aerogels, used to control the reactivity of the precursors, were rarely noticed. Because the goal is usually to prepare metal-oxide aerogels, the organic groups are removed by cal-

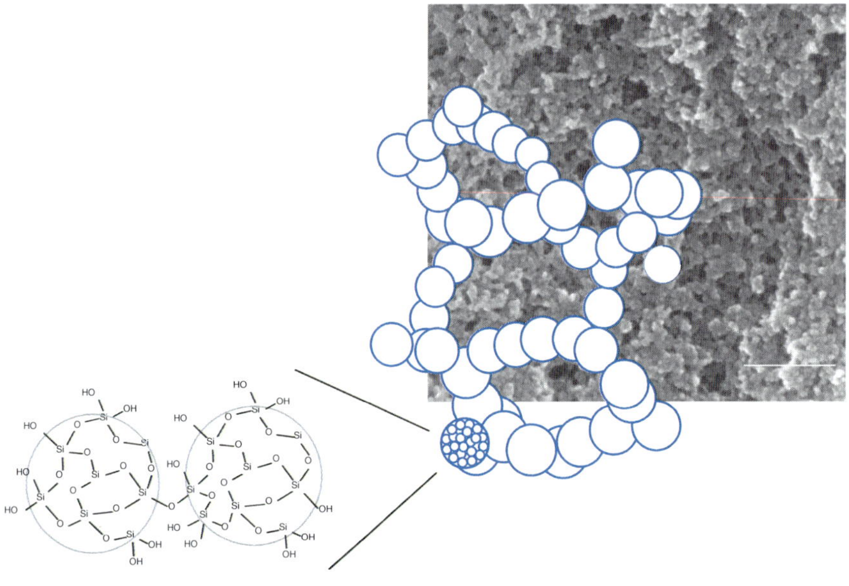

Fig. 4.1 Schematic structure and SEM image of a silica aerogel [Scalebar = 200 nm]

cination. For example, an alumina aerogel was prepared from ethylacetylacetonate-modified $Al(O^sBu)_3$ [4], a titania/silica aerogel from $Si(OMe)_4$ and acac-modified $Ti(O^iPr)_4$ (acac = acetylacetonate) [5] or acetate-modified $Ti(OBu)_4$,[6] a vanadia-silica aerogel from $Si(OMe)_4$ and $VO(acac)_2$ [7] and a mullite aerogel from TEOS and β-diketonate-modified $Al(O^sBu)_3$. [8]

The solvent plays an important role in sol-gel processes. It not only serves to homogenize the precursors in the initial stage, but also influences the particle and network forming reactions to a very high degree by its polarity and viscosity. In the preparation of aerogels it has an additional function. Since gelation results in only a marginal volume change, and the drying process during the aerogel production is performed, by definition, in such a way that shrinkage is minimal, the volume of the aerogel body and thus its density is determined by the volume of the reaction solution. Therefore, the density of aerogels is simply modified by variation of the solvent volume, an alcohol in most cases.

For technical applications in which large amounts of aerogels are needed, alkoxysilanes are too expensive. Therefore, *water glass*, an aqueous sodium silicate solution, is used as the silica source. The sodium silicate solution is ion-exchanged and the resulting silicic acid solution gelled by changing the pH to acidic conditions [9]. Ion exchange can also be performed after gelation [10].

The sol-gel transition (gel point) is reached when a continuous network is formed. An extremely important aspect of the preparation of inorganic gels is that the chemical reactions are not completed with the formation of the solid gel network. The pore liquid still contains condensable particles, and structural rearrangements of the gel network take place (aging, Ostwald ripening). Therefore, the gels need to be aged for a certain period of time before they can be dried.

When the liquid is evaporated from a wet gel, the gel initially shrinks by the volume that was previously occupied by the liquid. If the network is compliant, as it is in alkoxide-derived gels, the gel deforms. Upon shrinkage, OH groups at the inner surface approach each other and can react to form new M–O–M bridges. As drying proceeds, the network becomes increasingly stiffer and the surface tension in the liquid rises correspondingly because the pore radii become smaller. When the surface tension is no longer capable of deforming the network, the gel body becomes too stiff for further shrinkage. In this stage the tension in the gel is so large that the probability of cracking is highest.

Two processes are important for the collapse of the network upon drying: First, the slower shrinkage of the network in the interior of the gel body results in a pressure gradient which causes cracks. Second, larger pores will empty faster than smaller during drying, that is if pores with different radii are present, the meniscus of the liquid drops faster in larger pores (Fig. 4.2). The walls between pores of different size are therefore subjected to uneven stress, and crack. Strategies were developed to avoid cracking upon shrinkage. However, this is not sufficient for the production of aerogels. Methods must be applied which conserve the pore structure of the wet gels, that is which also avoid shrinkage.

The most common method is still supercritical drying. In this procedure, the pore liquid is put into the supercritical state. Above the critical point, liquid/gas

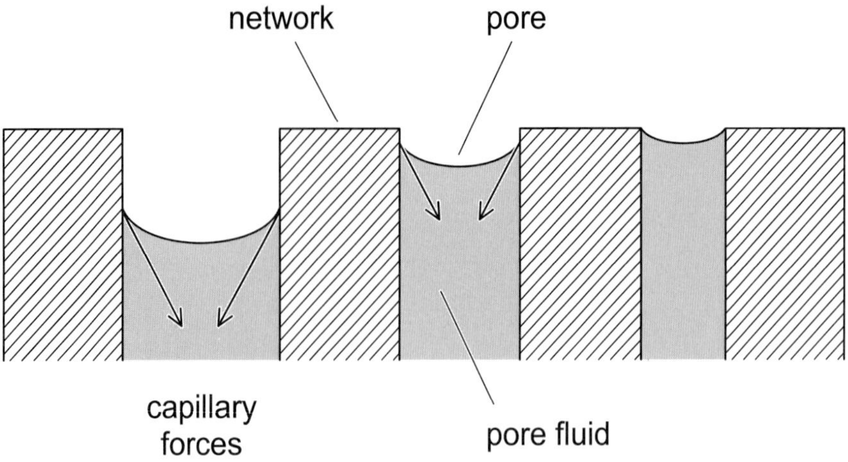

Fig. 4.2 Representation of the contracting surface forces in pores of different size during drying. At the same pressure, the curvature of all menisci in the pores is the same. For this reason the larger pores empty first (after Ref. [2])

interfaces no longer exist, and capillary forces are thus avoided. The supercritical fluid is then slowly vented without crossing the phase boundary between the liquid and the gas. One possibility is to put the original pore liquid, which is usually an alcohol, in the supercritical state. This method is also called the high-temperature method, because about 250 °C (and 5–8 MPa) are necessary to do so. Problems arise from the combination of high temperatures and high pressure, and the easy flammability of the organic solvents. In addition, severe structural and chemical changes occur ("accelerated aging") that level off the structural and chemical characteristics of the wet gels and thus favor the formation of "thermodynamically stabilized" aerogels. Organically modified aerogels are particularly sensitive to the high temperatures, and functional organic groups may be destroyed during drying [11]. An alternative is the use of liquid carbon dioxide. This has the advantage of a very low critical temperature at a moderate critical pressure. However, a time-consuming exchange of the original pore liquid for liquid CO_2 is necessary. Another requirement is the miscibility of the pore liquid with carbon dioxide. For example, water and CO_2 are immiscible; therefore, an intermediate solvent exchange (water against acetone, for example) is necessary [12]. Structural changes of the network take place to a much smaller extent than in the high-temperature method.

Ambient-pressure methods for preparing aerogels are just emerging. They include both surface modification and network strengthening. Additionally or alternatively, the contact angle between the pore liquid and the pore walls has to be influenced by deliberate modification of the inner surface and variation of the solvent to minimize the capillary forces.

Smith et al. published the first synthesis of ambient-pressure-dried SiO_2 aerogels in 1992 [13]. The procedure involves, in principle, the passivation of the inner surface by silylation of the Si–OH groups (that is conversion in Si–OSiR_3 groups). The gel shrinks strongly during evaporation of the solvent from the pores at ambient pressure, as expected. However, no irreversible narrowing of the pores by formation of Si–O–Si bonds is possible because of the silylation. Therefore, the gel expands to nearly its original size after the pores have been emptied. This is called the "spring-back" effect [14].

Another strategy is to increase the strength and stiffness of the network by aging the wet gels in solutions of tetraalkoxysilanes in aqueous alcohols [15]. Shrinkage during drying is thus completely avoided. The monomers added during aging mainly condense in the smallest pores and at the particle necks. The microporosity of the gels is thus lost. The gels can also be dried at ambient pressure at temperatures between 20–180 °C because of the stiffer network.

4.3.2
M41S materials

M41S materials are mesoporous solids with a periodic and regular arrangement of pores, well-defined and tunable pore sizes between 2 and 50 nm, and (normally) amorphous inorganic framework structures. They share characteristics of both gels and zeolites, and are typically characterized by a high specific surface area.

In the early 1990s, researchers discovered that in addition to single molecules such as tetramethylammonium bromide used for the preparation of zeolites, molecular assemblies, as found in liquid crystals, can also be used for templating inorganic matrices [16–17]. With this discovery, research in the field of templating and patterning inorganic structures to get perfectly periodic, regularly sized and shaped channels, layers and cavities has expanded dramatically. This supramolecular templating relies on the ability of amphiphilic molecules such as surfactants to self-assemble into micellar structures that, when concentrated in aqueous solutions, undergo a second stage of self-organization resulting in lyotropic liquid crystalline mesophases. Molecular inorganic species can cooperatively co-assemble with these structure-directing agents (templates) eventually to condense and form the mesoscopically ordered inorganic backbone of the final material (Fig. 4.3 [17]). The mesostructured nanocomposite is typically either calcined, ozonolyzed or solvent-extracted to obtain a porous inorganic material. The pore dimension in the porous material relates to the chain length of the hydrophobic tail of the template molecule.

The initial focus was mainly on the preparation of more stable materials, expansion of the pore sizes and structures, variable framework compositions, the use of new template molecules from ionic to non-ionic surfactants and block copolymers, and the processing of the materials as powders, fibers [18], thin films [19], monoliths [20] and other interesting shapes [21, 22]. As a result of this research effort it is currently possible to produce mesoporous materials with pore sizes from 2–50 nm [23, 24], and to synthesize mesoporous transition metal oxides [25, 26], metal sulfides [27, 28] and hybrid silica/organic frameworks in addition to purely siliceous or aluminosilicate materials [29, 30, 31]. Furthermore, the materials can

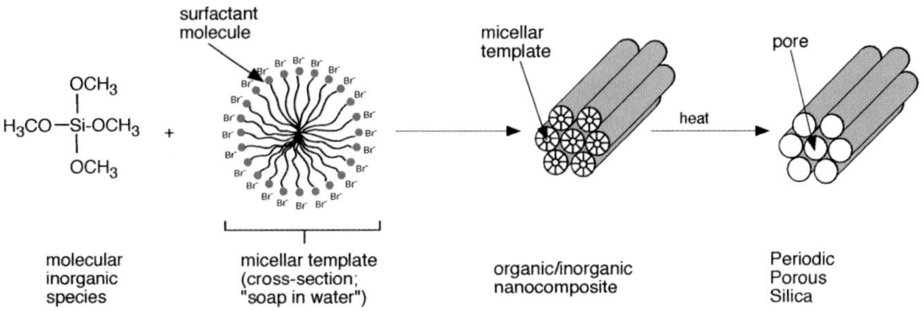

Fig. 4.3 Scheme of silica surfactant self-assembly and formation of mesoporous silica (reprinted with friendly permission of Nature Publishing Group from Ref. [17])

be patterned by using lithography in combination with photoacids [32] and soft lithography techniques to produce hierarchically patterned materials [33]. There are some excellent review articles on different aspects of mesostructured materials [34, 35, 36, 37].

Different preparative pathways were chosen for the synthesis of mesoporous silica. The original synthesis protocol involved the use of a strong base with cationic surfactant molecules (CTAB) resulting in powders referred to as MCM (Mobil composition of matter) materials. Two mechanistic possibilities were proposed: The first assumed preformed liquid crystals from the template molecule in which the inorganic species condense in the hydrophilic domains of the lyotropic liquid crystal phases to form the inorganic framework. The second is based on a cooperative mechanism in which the inorganic species induce the formation of the liquid crystal-like phases. In both cases strong interactions between the inorganic species, which are typically negatively charged at high pH, and the cationic surfactant molecules can be assumed. Pathway one was considered less likely because the synthesis was performed with surfactant concentrations far below the critical micelle concentration. Therefore, the cooperative mechanism was favored. However, it was shown that under certain synthesis conditions pathway one is also feasible.

A different approach was taken by a Japanese group which transformed a layered hydrated silicate "Kanemite" into a hexagonal material (referred to as FSM = folded sheet mechanism materials) by intercalation of ammonium surfactants into the layers, and folding and condensation of the silicate through an interlayer cross-linking process (Fig. 4.4) [38].

Following the original synthesis, a variety of protocols for mesostructured materials was developed. The most prominent examples for materials prepared under acidic conditions are the so-called SBA (Santa Barbara Amorphous) materials, which were synthesized with a broad variety of pore sizes and structures [39].

Other modifications of the synthesis relate to the choice of the template molecule which can be anionic (for example, sodium dodecylsulfate [40, 41]), cationic (for example, alkyltrimethylammonium halides), or even neutral (for example, amines [42], poly(ethylene oxides) (EOx) [43], alkyl poly(ethylene oxides) [44], water-soluble block copolymers [23]), resulting in different interactions between the inorga-

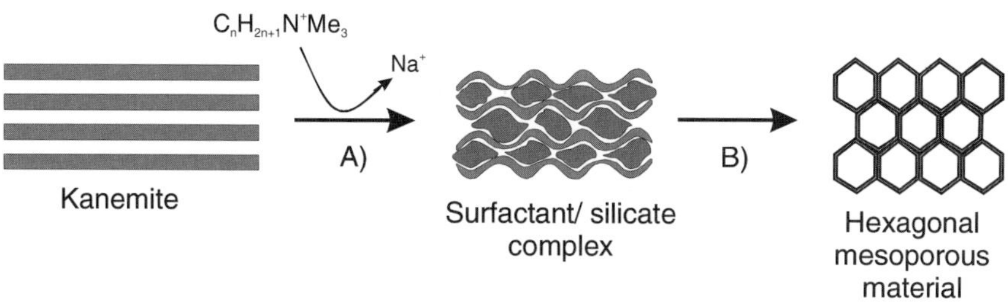

Fig. 4.4 Folding of a layered silicate around intercalated surfactant molecules. (A) Ion-exchange; (B) calcination

nic network-former and the template molecule as shown in Fig. 4.5. Covalent interactions between the template molecule and the inorganic network former are mainly used in the synthesis of mesostructured transition metal oxides, for example in niobia or tantala materials [26]. The reaction mechanism depends strongly on the choice of surfactant and counterions, and the type of interaction between the template and the inorganic source, the pH, and the concentration of the precursors and molar compositions. However, it is clear that the inorganic source can have a distinct influence on the lytropic liquid crystal phase behavior of the surfactant template.

Many different liquid crystal mesophases were used for the templating of periodic porous materials. The most common is the hexagonal phase with *p6mm* symmetry, for example MCM-41 or SBA-3, in which closed-packed cylinders of surfactants form the mesophase. Lamellar phases can be found in MCM-50, for example. However, lamellar phases are not stable upon template removal (template removal results in collapse of the layers and dense silica is formed). For many applications,

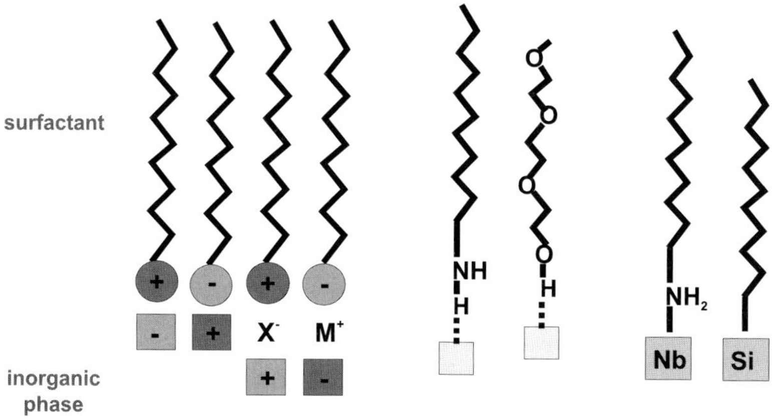

Fig. 4.5 Scheme of the different interaction possibilities between inorganic components and surfactant head group: electrostatic: a) S^+I^-, b) S^-I^+, c) $S^+X^-I^+$, d) $S^-M^+I^-$; hydrogen bonding interactions: e) S^0I^0 and f) N^0I^0; covalent bonding g) S–I

for example in catalysis or for sensors, a three-dimensional mesostructure is favorable, which can be found, for example, in the bicontinuous gyroid cubic phase of MCM-48 with a *Ia*3*d* symmetry or in the cubic close-packed arrangement of SBA-1 with a *Pm*3*n* symmetry. In these phases, all pores are interconnected and also connect to the surface, which allows a rapid transport of molecular compounds within the inorganic structure.

The focus of this article will be on the modification of the mesostructured material by covalently linked functional moieties under retention of the highly periodic pore and framework structure. Most of the work is focused on the modification of silica-based M41S materials.

4.4 Methods for Introducing Organic Groups into Inorganic Materials

The favorable physical and derived materials properties of porous materials, for example the extremely low thermal conductivity or unusual sound-propagation characteristics of aerogels, are a consequence of their highly porous structure. Therefore, any chemical modification of the materials must retain this structure. In the case of mesostructured porous materials, the periodicity of the structure must also be retained. The purpose of modifying porous materials by organic groups is to extend the spectrum of properties without adversely affecting the existing positive properties. For example, unmodified silica aerogels or M41S materials are rather hydrophilic. These properties can be improved by organic groups. The incorporation of functional organic groups allows the implementation of new properties and makes the materials viable for a broader spectrum of applications, for example in catalysis, for sensors or even sieving and separations.

Chemical modification of porous materials in general, and covalent modification by organic entities in particular, can be achieved at various stages of the preparation process (Fig. 4.6):

1. post-synthesis modification of the final porous product by gaseous, liquid or dissolved organic or organometallic species by making use of the reactivity of surface OH groups;
2. liquid-phase modification in the wet gel stage or – for mesostructured materials – prior to removal of the template;
3. addition of molecular, but non-reactive, compounds to the precursor solution;
4. use of organically substituted co-precursors.

Examples, and the relative importance of these methods, will be discussed in the following sections.

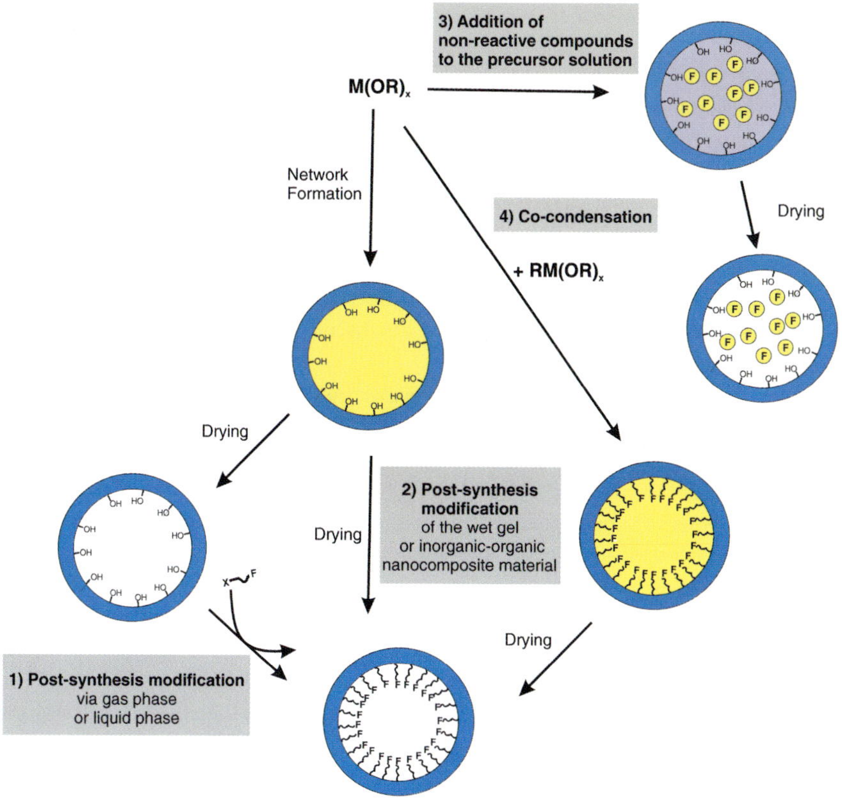

Fig. 4.6 Different options to modify a porous material (aerogel or M41S)

4.5 Porous Inorganic-Organic Hybrid Materials

In this section, some groups of porous inorganic-organic hybrid materials will be discussed. The classification is roughly based on the nature of the incorporated organic or organometallic entities, but is somewhat arbitrary. We will always start with the discussion of aerogels, followed by M41S materials.

4.5.1 Functionalization of Porous Inorganic Materials by Organic Groups

4.5.1.1 Post-synthesis Modification

One of the particular problems for the technical use of unmodified SiO_2 aerogels is their long-term stability in a humid atmosphere. The large number of silanol groups on the inner surface results in adsorption and capillary condensation of water, and eventually in the cracking of the gel body by the resulting capillary forces. Approaches to hydrophobize aerogels permanently were therefore investigated quite early. The possibility of reacting an aerogel post-synthesis with a hydrophobizing reagent, such as trimethylchlorosilane, appears attractive due to the high porosity of aerogels. However, post-treatment with a liquid phase is difficult because wet gels are formed again, and the laborious drying process has to be repeated. Another more general possibility for the subsequent modification of aerogels is to use their reaction with gaseous compounds. For example, gaseous dimethyldichlorosilane or hexamethyldisilazane was used to remove the reactive silanol groups and thus to hydrophobize the aerogel permanently [45]. In another method, the aerogels were treated in methanol vapor at 240 °C to convert the Si–OH groups in Si–OMe groups [46]. However, this method does not result in a permanent hydrophobization, because the Si–OMe groups can be re-converted to Si–OH groups in a humid atmosphere.

While most of the above-mentioned experiments were aimed at converting Si–OH groups into Si–OSiR_3 groups, modification of the porous materials can, in principle, also be achieved by adsorbing organic molecules post-synthesis on the inner surface of the porous material. For the reasons mentioned above, this way of modifying aerogels is only practical via the gas phase. One of the rare examples is a porphyrine derivative sublimed into a silica aerogel [47].

Grafting procedures are more common in the modification of periodic mesoporous materials, since this can be done via the gas phase but also easily via the liquid phase (unlike aerogels, the problem of destroying the porous structure of the material does not exist for M41S materials). It was shown that surface modifications done by the liquid phase are advantageous, since the number of surface silanol groups can be increased by a pretreatment with water. This eventually results in a higher concentration of grafted molecules on the pore walls [48]. The post-synthesis grafting process is limited by the fact that the functional groups may concentrate near the opening of the channels and by the limited diffusion into the pores. Furthermore, it is strongly influenced by the number of accessible and reactive surface silanol groups. Post-synthetic grafting of molecules onto the pore walls leads to a significant reduction in pore size due to the formed monolayer of condensed molecules. Therefore, it is advisable to use a material with larger pores (more than 3 nm), in which the smaller curvature of the pore also results in less steric hindrance.

Research on post-synthesis grafting is mostly focused on MCM-41, with its simple hexagonal arrangement of parallel, one-dimensional channels. But FSM-type materials, SBA-15, or MCM-48 with cubic structures, were also used. The grafting process itself is not limited to simple organic groups. Functional entities can also be incorporated and in further synthesis steps can even be transformed to create new functionalities. A huge amount of examples can be found in the literature, but only a selection can be given within the scope of this article.

As for aerogels, hydrophobization of the inner surface is important for later applications. An enhanced surface hydrophobicity in MCM materials was achieved by silylation of the pore walls with an excess of trimethylchlorosilane and hexamethyldisiloxane under reflux [49, 50]. A similar procedure has been published for other chlorosilanes such as triethyl-, tripropyl-, ethyldimethyl-, octyldimethyl- or cyclohexyldimethylchlorosilane [51].

The use of disilazanes ($HN(SiR_3)_2$) allows the modification of the surface under mild conditions [52]. Methyl groups can also be introduced by esterification reactions of surface silanol groups with butanol to form Si–OBu, followed by reaction with the Grignard reagent CH_3MgI, leading to $Si–CH_3$ groups [53].

An example of a mesoporous material modified with functional groups is a thiol-functionalized MCM-41 which shows a high adsorption efficiency for heavy metal ions from water [54, 55]. Another example is the modification with amino groups, for example by reaction with alkoxysilylpropyl(ethylenediamine), -(diethylenetriamine) or -(ethylenediaminetriacetic) acetate which allows the complexation of metal ions [56].

Anchoring of functional groups by post-synthesis grafting and their subsequent transformation into new functionalities was shown for chloropropyl groups, for example, by reaction with piperidine to yield grafted piperidinopropyl groups [57] or glycidyl functionalities which were converted to triazacyclononanes [58]. Stable heterogeneous catalysts with adjustable base strength were synthesized by functionalization of an MCM-41 material with glycidyl moieties which were reacted in a second step with different amines such as piperidine, pyrrolidine, or 1,5,7-triazabicyclo[4,4,0]dec-5-ene [59]. The walls of a cubic mesoporous silicate MCM-48 were grafted with ferrocenyl end groups by a ring-opening reaction of the ansa-ferrocene $[Fe((\eta\text{-}C_5H_4)_2SiMe_2)]$ [60].

In order to generate acidic sites within the pore structure of mesostructured materials, phenyltrimethoxysilane was grafted onto the surface and subsequently sulfonated with chlorosulfonic acid [61].

A novel twist was introduced by Liu and coworkers who used the grafting method to synthesize a hierarchical porous material containing a rigid mesoporous oxide frame coated by a soft, "microporous" organic monolayer [62]. Fig. 4.7 shows schematically the different synthesis steps. First, template molecules of a certain shape (dipod and tripod) were synthesized and modified with aminopropyltrimethoxysilane (APS) allowing for covalent binding to the calcined mesostructured silica. In a second step, octadecyltrimethoxysilane (OTS) was added to the solution to form a monolayer, similar to self-assembled monolayers, on uncoated areas of the meso-

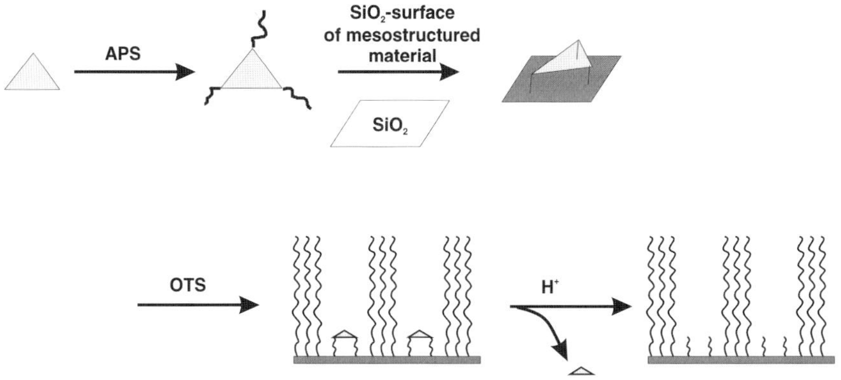

Fig. 4.7 Creation of microcavities by a molecular imprinting approach (after Ref. [62])

structured material. The organic template molecules (dipod, tripod) were then removed under acidic conditions resulting in a material in which empty spaces mimic the shape of tri- or dipod template molecules.

4.5.1.2 Liquid-Phase Modification in the Wet Gel Stage or Prior to Surfactant Removal

Modification of a gel in the alcogel or hydrogel stage is time-consuming, because a limiting factor is the diffusion rate of the modifying reagent into the pores. This process is nevertheless a key step in the ambient pressure drying of aerogels (see above), where the wet gels are treated with a silylating reagent, such as Me_3SiCl. It should be noted that aerogels prepared by this method are inherently hydrophobic. The original method involved at least one solvent exchange. A cost- and time-efficient variation of the process was developed in which the silylation is carried out directly in the water phase of hydrogels obtained from waterglass solutions. When Me_3SiCl is added to the hydrogel, the surface Si–OH groups are silylated and, at the same time, hydrophobic hexamethyldisiloxane, $Me_3Si–O–SiMe_3$, is formed which expels the aqueous phase from the pores [63].

More recently, silica aerogels were similarly functionalized. When the silica hydrogel obtained from waterglass solutions was treated with a mixture of HCl, $Me_3Si–O–SiMe_3$ and $(CH_2{=}CH)Me_2Si–O–SiMe_2(CH{=}CH_2)$, Me_3SiCl and $(CH_2{=}CH)Me_2SiCl$ are formed *in situ* by cleavage of the Si–O–Si bonds of the disiloxanes. The chlorosilanes then react with the surface Si–OH groups and thus silylate and functionalize the surface at the same time [64]. Use of the vinyl functionalized aerogels as reactive fillers for silicone rubbers resulted in a shift to a higher modulus of elasticity when the vinyl content was increased. The vinyl content also impacted the stiffness and hardness of the cured rubbers.

As-synthesized, non-dried mesostructured materials can be treated similarly to wet alcogels. Grafting of functional moieties after removal of the surfactant in the calcined and porous product is time-consuming and often results in low loadings of the functional silanes. An efficient approach is based on a simple ion exchange reaction. When the mesoporous silica is synthesized by the use of cationic surfactants under alkaline conditions, anionic SiO^- groups are located on the silica surface, balanced by positively charged surfactants in the channels. It was shown that it is possible to attach surface ligands such as aminopropyl-, octyl- or mercaptopropyl groups by a reaction of the surfactant-silica composite with the corresponding chlorosilane [65]. The driving force for this reaction is the formation of covalent Si–O–Si bonds, and the surfactant is efficiently removed during this reaction (Fig. 4.8). A similar approach was described by using cationic metal complexes such as copper(II) 3-(2-aminoethylamino)propyltrimethoxysilane complexes, which also led to substitution and repulsion of the surfactant from the mesostructured material [66]. This process was successfully extended to the incorporation of metal porphyrin complexes into MCM-41 [67].

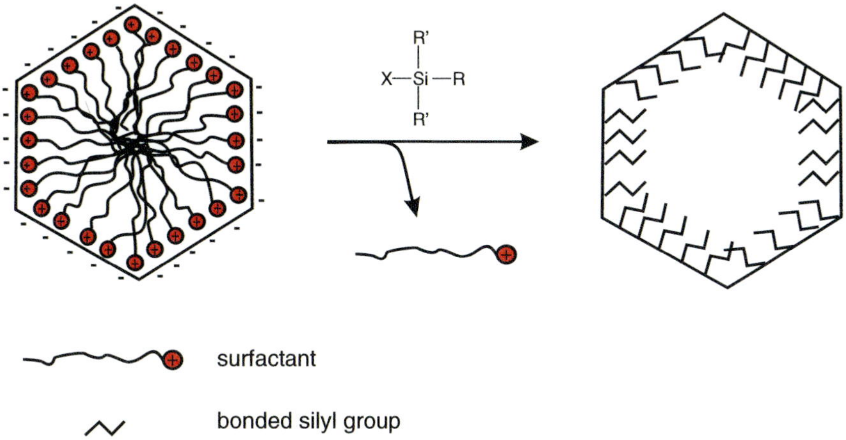

Fig. 4.8 Template displacement by chloroorganosilanes (after Ref. [65])

4.5.1.3 **Addition of Non-Reactive Compounds to the Precursor Solution**

In this approach, organic molecules are physically entrapped in gel networks without chemical bonding. This is achieved by dissolving the guest compound in the precursor solution; the host network is then formed around the guest during sol-gel processing. This approach is routinely used for inorganic-organic hybrid xerogels. The doped wet gels can, in principle, also be converted to aerogels. However, the probability is very high that the guest compound is leached during supercritical drying and the concomitant rinsing processes. Thus when a fluorescent dye (POPOV) was trapped in the gel network, the portion leached during aging and drying was relatively high [68]. C_{60}, or C_{60}/C_{70} mixtures, were similarly incorporated into SiO_2 aerogels [69]. This procedure mainly produced fullerene clusters that gave rise to quantum confinement effects. Fullerene-host matrix interactions seem to play a very important role.

For mesostructured materials entrapment of guest species was demonstrated in a huge number of examples. By contrast with aerogels, leaching is not a dominant problem due to the different synthesis procedure. This type of inclusion chemistry into M41S materials is reviewed elsewhere in detail [36].

A different approach for the modification of mesostructured materials is to use the template molecule as the modifying agent. This was demonstrated by the application of ferrocenyl surfactants (for example 11-ferrocenylundecylammonium bromide) in the synthesis of mesostructured thin films. These surfactants allow for the synthesis of functional mesostructured films with redox-active compounds incorporated in a one-step procedure (Fig. 4.9) [70].

Polymerizable surfactants (for example with diacetylene moieties) fulfil two different functions – as structure-directing agents and as monomers for polymerization reactions – in the formation of robust and functional mesostructured nanocomposites [71].

The surfactant can also be used as modifying agent for covalently bonded functions. This was shown for amphiphilic block copolymers which consist of a hydro-

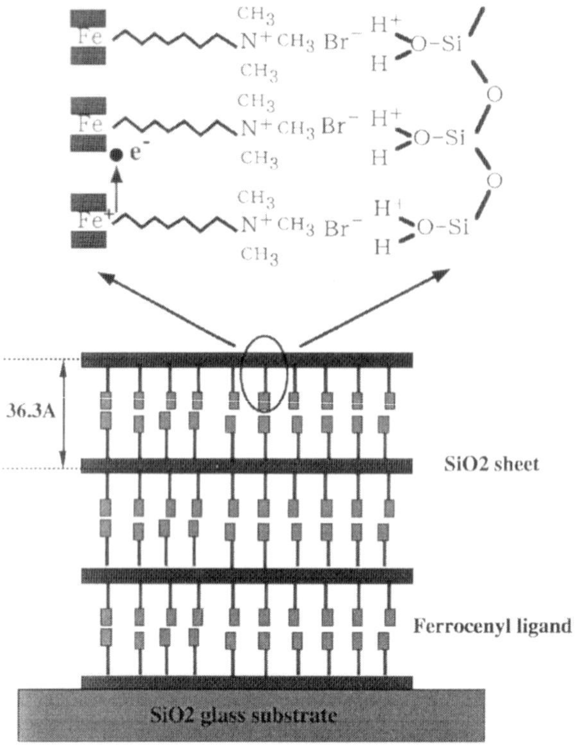

Fig. 4.9 Schematic diagram of a mesostructured ferrocenyl-surfactant/silica material (reprinted with friendly permission of Wiley-VCH from Ref. [70])

phobic silicone part and a hydrophilic poly(ethylene oxide) part. A lamellar silica composite material was produced with the aid of this template. Lamellar materials normally collapse upon template removal. However, in the case of the inorganic-organic template, the silicone part acts as buffer in between the layers even after calcination up to 450 °C because the silicone is partly oxidatively degraded, forming a silsesquioxane which covers the silica layers with a hydrophobic Si–CH_3 coating [72].

4.5.1.4 **Use of Organically Substituted Co-precursors**

The most general approach is to use organically substituted precursors, such as organically substituted trialkoxysilanes $R'Si(OR)_3$, where R′ can be a non-functional or functional organic group, or metal alkoxides substituted by non-functional or functional bidentate ligands. The use of organically substituted precursors results in the covalent bonding of the organic groups to the oxide network.

Because the connectivity of the building blocks is reduced by the organic substituents, such precursors are normally processed together with unsubstituted alkoxides. Sol-gel processing of mixtures of $Si(OR)_4$ and $R'Si(OR)_3$ (R′ = alkyl, aryl or $(CH_2)_nA$ ($n = 2, 3$; A = SH, OCH_2–epoxy, OC(O)C(Me)=CH_2, NCO, Cl, NHC(O)OMe PPh_2)) under alkaline conditions proceeds in two steps [73, 74] because the hydrolysis and condensation rate of $R'Si(OR)_3$ or $(RO)_3Si$–X–A is slower than that of the corresponding tetraalkoxysilane $Si(OR)_4$. In the first stage of the reaction, the gel network is formed only by hydrolysis and condensation of $Si(OR)_4$, while $R'Si(OR)_3$ is a kind of co-solvent. Hydrolysis and condensation of $R'Si(OR)_3$ is strongly delayed under these conditions. Once $R'Si(OR)_{3-n}(OH)_n$ is formed, the organically substituted units condense to the already-existing silica network formed from the tetraalkoxysilane in the first stage. The R'Si≡ groups modify the inner surface while the typical aerogel network and thus the typical aerogel properties are retained.

Surprisingly, the structure of aerogels (porosity, specific inner surface) is much more influenced by the $Si(OR)_4/R'Si(OR)_3$ ratio than by the properties of the group R'. Because the trialkoxysilane basically acts as a co-solvent in the first stage of the reaction, the tetraalkoxysilane is more diluted in this stage (resulting in a decrease of the network density), and a larger relative amount of water and catalyst is available for its reaction (resulting in larger particles).

Permanently hydrophobic aerogels (Fig. 4.10) were prepared in a one-step process using a mixture of $Si(OR)_4$ and $R'Si(OR)_3$ (R' = alkyl or aryl group) under alkaline reaction conditions [11, 75, 76]. The elastic properties of SiO_2 aerogels were also improved by chemical modifications of the precursors. The reason for the improved elastic properties of hybrid aerogels is probably that the presence of alkyl or aryl groups at the surface of the secondary particles impairs stiffening of the particle necks by Ostwald ripening [75].

Another hydrophobic aerogel was prepared from $TMOS/CF_3CH_2CH_2Si(OMe)_3$ mixtures. The CF_3-substituted aerogel completely adsorbs oil from oil-water mixtu-

Fig. 4.10 After hydrophobization, silica aerogels even float on water for long periods of time (reprinted with friendly permission of Wiley-VCH from Ref. [1 f])

res at oil/aerogels ratios up to 3.5, producing a dry solid when separated from the water [77].

Sol-gel processing of TEOS/$(MeO)_3Si(CH_2)_3NR'_2$ mixtures (NR'_2=NH_2 or $NHCH_2$-CH_2NH_2), followed by drying of the wet gels with supercritical CO_2, also resulted in organically functionalized aerogels. However, due to the basic amino groups (which can undergo hydrogen bonding), both precursor silanes appear to be involved in the build-up of the aerogel network, by contrast with alkoxysilanes $R'Si(OR)_3$ without basic properties [78]. Aerogel-like materials ("low-density xerogels") were obtained from TEOS/$(MeO)_3Si(CH_2)_3NR'_2$ (NR'_2=NH_2 or $NHCH_2CH_2NH_2$) mixtures even without supercritical drying. It was shown that the aminopropylsilane reacts first, and TEOS reacts with the thus-formed nucleus. This nucleation mechanism occurs only when a large reactivity difference between the aminopropylsilane and $Si(OR)_4$ exists. Due to the nucleation mechanism, the particles and aggregates increase in size as the TEOS/$(MeO)_3Si(CH_2)_3NR'_2$ ratio increases, and therefore the pores between the particles and aggregates also increase [79]. Aerogels were also prepared from TEOS/$(MeO)_3Si(CH_2)_3NR'_2$ mixtures under acidic conditions, but the network structure was not investigated in this case [80].

The co-condensation approach can also be used in the synthesis of organically modified M41S materials. Mixtures of tetraalkoxysilanes and organotrialkoxysilanes in the presence of surfactant template molecules were employed in either hydrothermal synthesis of mesostructured hybrid powders or in sol-gel based approaches towards mesoporous thin films and bulk materials. Systematic studies of the hydrolyis and condensation kinetics of the different precursors were not performed, but the incorporation of the organic groups was in most cases confirmed by ^{29}Si MAS-NMR spectra. This approach is similar to the one described in the aerogel section; the tetraethoxysilane acts as primary silica source to form the network.

Major issues are whether the periodic pore structure can be retained even if part of the network-forming tetraalkoxysilane is substituted by a trialkoxysilane, and whether the surfactant template can be removed without destroying the organic functionality.

This approach is viable for the incorporation of phenyl [81], n-octyl [81], thiol [81, 82, 83], amino [83, 84], epoxy [81], imidazole [81], allyl [81], cyano [84], vinyl groups [85] or even dyes [83, 86, 87] such as [3-(2,4-dinitrophenylamino)propyl], 4-[(4-dimethylaminophenyl)azo] benzoic acid and sulfo-rhodamine B functionalities. The materials are typically prepared at room temperature from alkaline mixtures containing varying molar ratios of TEOS and the organofunctional trialkoxysilane, but neutral, acidic, or hydrothermal conditions can also be used resulting in materials with a different degree of ordering of the mesostructure.

Since the typically applied calcination process to remove the template would result in degradation of the organic groups, an acidic solvent extraction technique is used to remove the surfactant but to retain the organic moieties covalently attached to the pore wall. This extraction procedure can be done for almost all functionalized materials; only the imidazole, epoxy and allyl modified materials show a significant loss of order in the mesostructure upon acidic extraction of the surfactant. The thermally very stable Si–phenyl bond (up to 450 °C) is the only one to allow for a thermal degradation of the surfactant and to retain the phenyl moieties in the final material [88].

The accessibility and reactivity of the functional groups in the mesostructured materials was tested in several experiments, for example by bromination of vinyl groups [85], uptake of mercury by thiol-modified MCM, or oxidation of the thiol to sulfonic acid groups [82].

The modification of mesostructured materials by organic groups via this co-condensation approach is not limited to a single functional group, but also binary mixtures were applied [89].

Another approach based on co-condensation of various alkoxysilanes is the use of alkoxysilane-modified surfactants, where the trialkoxysilyl-modified molecule acts both as structure-directing agent and as covalent tether of organic groups to the inorganic framework (Fig. 4.11) [90, 91].

4.5.2
Bridged Silsequioxanes

A different network structure is obtained when the organic groups link two or more $Si(OR)_3$ groups, that is when silanes of the type $(RO)_3Si–X–Si(OR)_3$ are employed as precursors. The use of such precursors permits the organic groups to become an integral part of the network rather than just a pendant functionality. For example, aerogels with $X = (CH_2)_n$ [92] or $CH_2–CHR–CH_2–OC(O)O–CH_2–CHR–CH_2$ (R = H, Me) [93] were prepared.

When precursors of the type $(RO)_3Si(CH_2)_nSi(OR)_3$ with $n = 2, 6, 8, 10, 14$ were processed under either basic or acidic conditions, the length of the spacer (n) had a decisive influence on the later aerogel structure. Most of the alkylene-bridged aerogels were mesoporous, high-surface area materials. However, the specific surface area decreased with increasing n; the base-catalyzed gel with $n = 14$ was nonporous.

A bridged mixed-metal precursor was obtained by reaction of 3-(propyltrimethoxysilyl)acetylacetone with $Ti(O^iPr)_4$. Silica/titania aerogels were prepared by sol-gel processing of the resulting single source precursor followed by supercritical drying [94].

A special type of precursor substituted by multiple $Si(OR)_3$ groups is the dendrimer, an example of which is shown in Fig. 4.12. Carbosilane dendrimer precursors with 36, 108 or 216 terminal Si–OEt groups resulted in high-surface area aerogels after supercritical drying [95].

A functional aerogel was obtained from $[(EtO)_3Si(CH_2)_3]_2S_4$ and sodium dodecyl sulfate as a surfactant at ambient drying conditions. A metastable lamellar surfactant-silica composite was formed first instead of an intermediate alcogel. The sur-

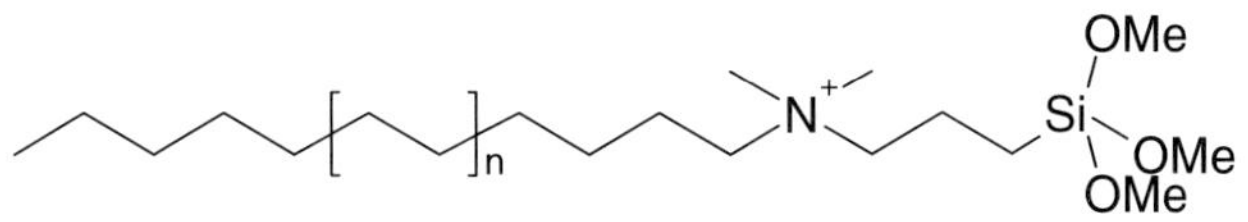

Fig. 4.11 Alkoxysilyl-modified surfactants

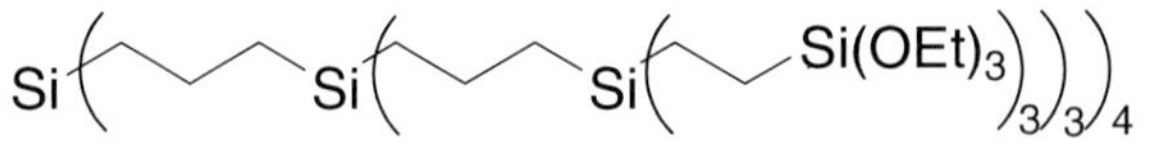

Fig. 4.12 A third-generation carbosilane dendrimer with 108 Si–OEt groups

factant mesostructure facilitates the formation of monodispersed silica nanoparticles that aggregate into a highly porous solid. The porous structure was maintained after removal of the surfactant [96].

The synthesis of materials in which bridging organic groups are integrated within the framework of mesostructured materials (designated as HMM [30, 97], UOFMN [29], PMO [31, 98], or BSQM [99]) was also reported. A variety of spacer groups was incorporated such as C_2H_4, C_6H_4 or even functional groups such as –CH=CH– [31] or cyclam [100] moieties. Furthermore, different periodic pore geometries (cubic, hexagonal) are accessible [101].

The largest pores (cagelike pores of 10 nm) in a well-ordered mesostructured silica material with integrated organic groups were reported by using a large block copolymer as surfactant [102] and bis(triethoxysilyl)ethane as the framework-forming component. In contrast, super-microporous organic-integrated silica with periodic and uniform pore sizes of 1–2 nm were prepared from alkylamine surfactants and bis(triethoxysilyl)ethane [103].

Typically, the framework of the MCM-type mesostructured materials is amorphous. However, an interesting feature was discovered when 1,2-bis(triethoxysilyl)benzene was used as network former. Reaction of a mixture of octadecyltrimethylammonium salt, sodium hydroxide and water resulted in a material possessing a crystal-like pore-wall structure [104]. This additional periodicity was attributed to a regular arrangement of $O_{1.5}Si–C_6H_4–SiO_{1.5}$ units in the pore walls due to an interaction between the phenylene groups (Fig. 4.13).

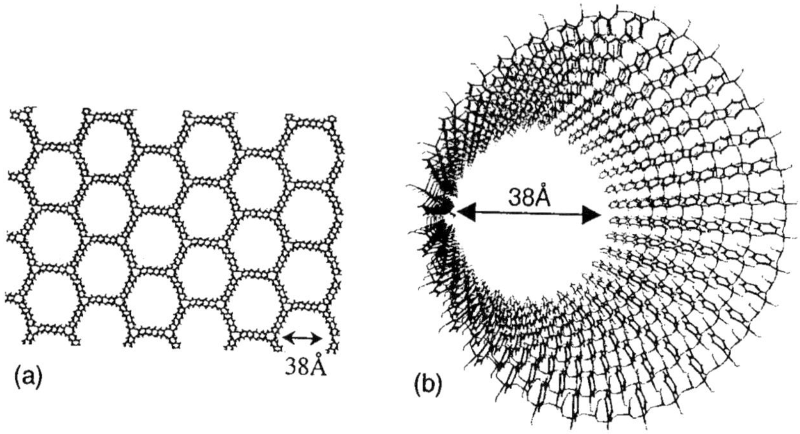

Fig. 4.13 Structural model of a mesostructured hybrid material containing phenylene groups. (a) 2D-hexagonal arrangement of the pores; (b) structure of the pore walls with a periodic stacking of the phenylene groups (reprinted with friendly permission of the MRS society from Ref. [104])

Additional organic functionalization of mesostructured thin films was achieved by co-condensation of bis(triethoxysilyl)ethane with a series of alkoxysilanes with quaternary ammonium groups [105].

4.5.3
Incorporation of Metal Complexes for Catalysis

Organically substituted alkoxysilanes with Lewis-basic groups A can be used to tether metal complexes to the aerogel network (Fig. 4.14). Such materials are interesting for the extraction of metal ions, for getting colored aerogels or, most important, to obtain catalytic materials when the tethered metal complex has catalytic properties.

Ni^{2+} and Cu^{2+} ions were incorporated into silica aerogels by two methods [80]. In the first one, the wet gel obtained by hydrolytic polycondensation of TEOS/$(MeO)_3Si(CH_2)_3NHCH_2CH_2NH_2$ mixtures was impregnated with alcoholic solutions of metal salts. In the second method, the amino-substituted alkoxysilane was pre-reacted with the nickel or copper salts, and the obtained $[(MeO)_3Si(CH_2)_3NHCH_2CH_2NH_2]_nM^{2+}$ complex (Fig. 4.14) used as a co-precursor for sol-gel processing with TEOS. No substantial leaching of the metal was observed during supercritically drying with CO_2. The second approach was also used to bind Pd^{2+} to silica aerogels. Aerogels doped by monodispersed 2.2 nm Pd crystals were obtained after supercritical drying and calcination at 400 °C [106]. The Pd crystallites were completely accessible inside the silica particles of the aerogel structure and were stable against sintering during supercritical drying and calcination, unlike samples made by Pd^{2+} impregnation of pre-formed silica.

In theory, heterogenization of homogeneous catalysts promises a combination of the engineering advantages of heterogeneous catalysts and the advantageous features of homogeneous catalysts, such as well-defined structures that can be correlated with reactivity. Sol-gel processing is a particularly useful method for heterogenizing catalytic metal complexes [107]. The general approach is to derivatize the catalytic metal complexes with pendant $(CH_2)_nSi(OR)_3$ groups, and to use the

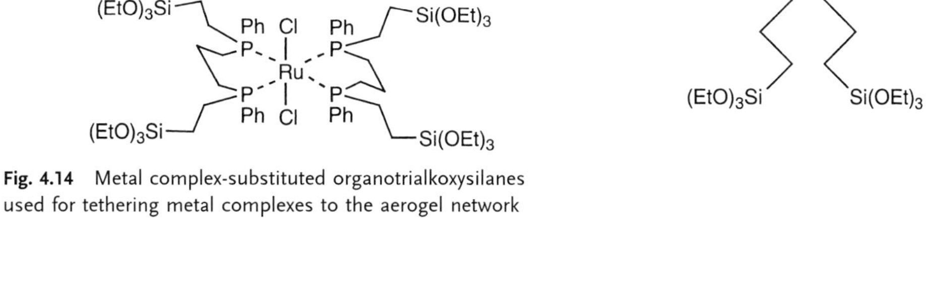

Fig. 4.14 Metal complex-substituted organotrialkoxysilanes used for tethering metal complexes to the aerogel network

$Si(OR)_3$-substituted metal complexes thus obtained as (co-)precursors for sol-gel processing. While silica xerogels containing covalently bonded catalytic metal complexes have already been investigated for some time, this approach was only recently extended to silica aerogels. The first examples are a ruthenium phosphine catalyst [108] and salicylideneamidato complexes of Co^{2+} and Cu^{2+} [109] (Fig. 4.14). The Ru complex-containing aerogel had a much higher catalytic activity than the corresponding xerogel, which was attributed to the favorable textural properties of the aerogel. A Mn(III) porphyrine-doped silica gel, as a catalyst for the epoxidation of olefins, was obtained by co-hydrolysis and condensation of $Si(OEt)_4$ and a $(CH_2)_3Si(OEt)_3$-substituted porphyrine derivative, followed by supercritical drying. No leaching of the porphyrine was observed. In contrast, when an aerogel was prepared under the same conditions from the porphyrine lacking the $(CH_2)_3Si(OEt)_3$ group, that is when the porphyrine was just entrapped in the aerogel matrix without covalent bonding, the porphyrine was completely leached [110].

Silica-titania mixed oxide aerogels are highly active catalysts for the epoxidation of olefins by peroxides, because of the pronounced mesoporosity and the high dispersion of titania in the silica matrix. However, the catalyst is deactivated by aqueous hydrogen peroxide due to leaching of the active titanium species. The leaching was explained by the hydrophilicity of unmodified silica-titania mixed oxides. Hydrophobization of the aerogel surface was achieved by partial replacement of the silica precursor TMOS by $MeSi(OMe)_3$ [111]. Although leaching of titanium was thus prevented, selectivities and activities were not satisfying. Acidity of the mixed oxides is crucial for activating the peroxide, but it can also result in acid-catalyzed side reactions. Tuning the acidity with weakly basic additives, especially amines, was shown to improve considerably the performance of mesoporous silica-titania catalysts. The immobilization of amines by employing $TMOS/(MeO)_3Si(CH_2)_3NR_2$ mixtures (NR_2=NH_2, NHMe, NHPh, NEt_2, NMe_2) led to recyclable, selective and active catalysts for the epoxidation of olefins and allylic alcohols [112]. The significant selectivity enhancement is probably due to neutralization of surface Si–OH groups via hydrogen bond interactions, and thus the suppression of acid-catalyzed side reactions. Surface modification by covalently bonded functional organic groups has a much broader application range than tuning the polarity alone by methyl or phenyl groups. For this reason, the studies were also extended to $(MeO)_3Si(CH_2)_3Cl$ and various acetoxyalkyl(trialkoxy)silanes as modifiers [113].

In comparison to zeolites, the MCM materials have larger pores, and therefore offer the option of incorporating even larger organometallic and transition metal compounds for catalytic applications. The enormous surface area of these materials makes them very interesting for solid catalyst supports.

As for aerogels, the covalent attachment of the organometallic or coordination compounds to the pore walls of mesoporous silica can be effected either by direct grafting or via a spacer ligand (tethering). The attachment of guests to the pore walls by direct grafting, which is typically achieved by a direct reaction of surface hydroxyl groups with reactive molecules, was demonstrated, for example, using titanocene complexes [114], aluminum chloride [115], tetrabutyl tin [116], $VO(O^iPr)_3$ [117] or the molybdenum complex $[Cp(CO)_3MoSnMe_3]$ [118]. An impressive example for direct grafting processes is the grafting of rare-earth silylamide complexes such as $M[N(SiHMe_2)_2]_3$ (M = Ln, Y, La, Er, Lu, Nd) via surface organometallic chemistry on mesoporous aluminosilicates, followed by Brønsted acid/base reactions with different ligands XY resulting in materials for asymmetric catalysis (Fig. 4.15) [119, 120]. The acidic surface silanol groups that are not coordinated to a metal center are silylated by the corresponding silane, providing the material with hydrophobicity.

Besides the direct grafting of metal centers onto the surface of MCM materials, the metal centers can be reacted with a pre-functionalized surface. This pre-functionalization can be performed as described above either by post-synthesis modification or by co-condensation of different alkoxysilanes. The surface thus modified can be reacted further to yield the desired metal-doped material. For example, 3-chloropropylsilane-modified MCM was used to anchor a pentadentate ligand 3-[N,N′-bis-3-(3,5-di-*tert*-butylsalicylidenaminopropyl]amine (salpr), which was reacted with manganese acetylacetonate to give the corresponding Schiff base complex (Fig. 4.16) [121]. Several other examples prove the viability of this procedure, for example the incorporation of chiral catalysts for the enantioselective addition of diethylzinc to benzaldehyde [122, 123], or the reaction of cobalt(II) salts [56] or [*meso*-tetrakis(4-chlorophenyl)porphyrine Ru(II)] [124] with different amine-modified surfaces. Several other examples have already been mentioned in the grafting section of this article. Metal nanostructures were prepared in MCM-41 and MCM-48 by functionalization of the intrachannel surface of the silica host with positive charges for the accommodation of highly concentrated and negatively charged metal complexes, which formed Au and Pt nanowire bundles after reduction [125].

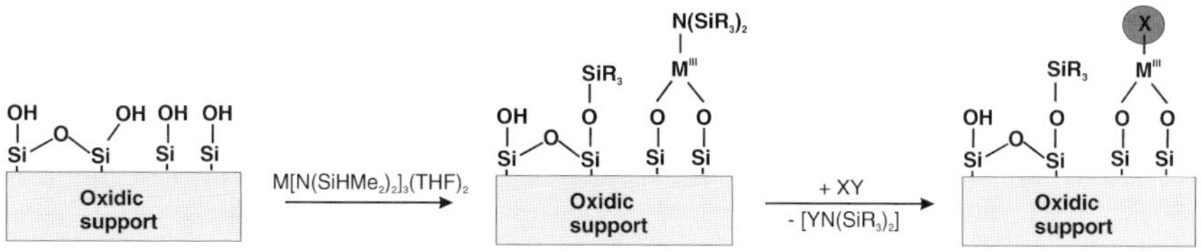

Fig. 4.15 Immobilization of metal silylamides onto mesoporous silica with simultaneous surface silylation and subsequent ligand exchange

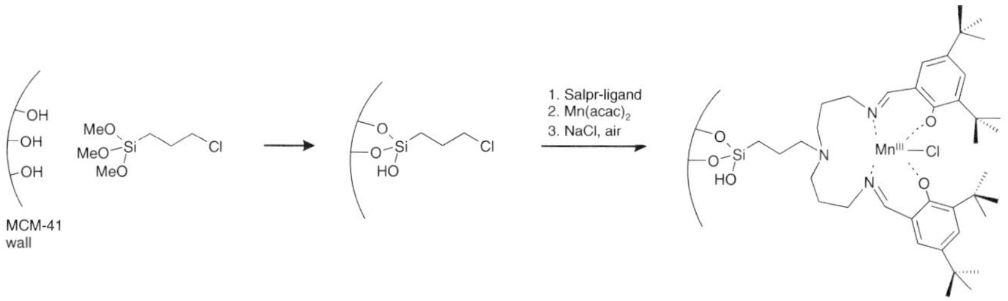

Fig. 4.16 Multistep synthesis for the internal attachment of a manganese (III) Schiff-base complex

4.5.4
Incorporation of Biomolecules

Sol-gel derived silica-based materials hold promise as biocompatible scaffolds for the immobilization of enzymes, proteins, cells etc. due to their biocompatibility, porosity, chemical inertness and mechanical stability [126]. In this short section some examples of encapsulated biomolecules are presented. The immobilization technique should allow the biomolecule to maintain its catalytic activity while diminishing other processes that are detrimental to the enzyme such as autolysis. However, one has to keep in mind that the interactions of these molecules with the silica matrix are typically not by covalent bonding but by hydrogen bonding or electrostatic interactions.

Aerogels are very suitable for the immobilization of large entities such as cells due to their characteristic high surface area in combination with ultrahigh porosity and large pores. This makes them viable as biosensors for the detection of chemicals and organisms within the environment, or even as biocatalysts, for which it is often advantageous to use immobilized enzymes because it is easier to separate them from the reaction products.

Different approaches were taken for the immobilization of bioactive species within the silica aerogel network. The biomolecules can either be incorporated by a post-synthesis treatment or *in situ* during sol-gel processing. The sol-gel processing parameters can be adjusted in such a way that enzymes and other bioactive species tolerate the gel synthesis conditions. In the preparation of aerogels another critical step – drying in supercritical solvents – has to be considered as harmful for biomolecules.

The lipase *Pseudomonas cepacia* tolerates the conditions of supercritical drying with carbon dioxide and can thus be incorporated into the aerogel *in situ* during sol-gel processing followed by supercritical drying. Its catalytic activity is even higher than in the corresponding xerogels [127]. Hybridization of the inorganic network with an organic additive such as poly(vinyl alcohol) (PVA) improves the catalytic activity of the enzyme [128].

A comparison of the *in situ* approach for the incorporation of bioactive species with the post-synthesis treatment was performed for penicillin G amidase, thermolysin and α-chymotrypsin. It was found that the activity of penicillin G amidase is much higher after loading an existing aerogel with the enzyme (post-synthesis treatment) compared to incorporating it already in the sol-gel process, probably due to hindered diffusion of substrates to the enzyme [129].

Macroporous, supercritically dried and re-wetted gels were reported as a colonization scaffold for bacteria for an aerosol collection device and a biosensor system [130]. Storing the re-wetted aerogel discs in a bacterial growth medium allowed the colonization of *Escherischia coli*. The immobilized bacterial cells elicit a measurable response as a result of an environmental trigger. This proof-of-principle study showed that organisms can survive in a wetted macroporous silica gel.

Incorporation of enzymes in periodic mesoporous silica materials has also been reported. The uniform pore diameters of 2–50 nm approximate the molecular diameter of enzymes and offer therefore a good scaffold for biomolecules. However, the same questions apply here, whether the enzyme can tolerate the synthesis conditions such as high temperatures etc.

Different enzymes were immobilized in MCM-41 by a post-synthesis treatment to enhance their stability while exploiting the catalytic activity of these hybrid materials [131]. Enzymes ranging from cytochrome C to papain and trypsin were adsorbed into porous siliceous MCM-41. The enzymes still showed catalytic activity. Leaching of trypsin from the host can be prevented by silylation with different organoalkoxysilanes, thus narrowing the pore openings. Cytochrome C was also successfully immobilized in MCM-48 and SBA-15 and the layered transition metal oxide Nb-TMS4 by impregnation of the porous material with a solution of the enzyme [132]. Horseradish peroxidase is another example of an enzyme adsorbed in different mesostructured materials with various pore sizes [133].

Biomolecules were not only incorporated into mesostructured materials by simple impregnation techniques, but they were also used as templates in the synthesis of periodic mesoporous silica (DAM-1, Dallas amorphous material-1). For example, vitamin E-TPGS (α-tocopheryl polyethylene glycol 1000 succinate) was found to be an efficient template for the synthesis of hexagonal mesostructured silica [134].

4.5.5 Incorporation of Polymers

The combination of porous oxide materials as hosts and organic polymers as guests offers interesting possibilities for new types of hybrid materials. One obvious application is the improvement of the mechanical stability of porous materials; another, which is especially important for periodic mesoporous materials, is the option of forming aligned polymers which can be conducting, or have interesting optical properties due to the space confinement.

The diffusion of polymers into preformed highly porous systems is very difficult, probably because of the associated loss of entropy of the polymer chain. However, there are several other options for how polymers can be incorporated into the porous host.

1. The porous network can be formed around preformed polymers, that is, the sol-gel process can be performed in solutions of organic polymers.
2. Organic monomers can be added to the precursor mixture for the preparation of the inorganic host. Subsequent polymerization can either be performed simultaneously with the formation of the inorganic framework or after the inorganic network has been formed. A variation of the latter approach is the use of precursors with polymerizable organic groups such as 3-(methacryloxypropyl)trimethoxysilane. The silane part of these bifunctional molecules becomes part of the network structure during the hydrolytic polycondensation reaction and the unsaturated groups can be polymerized with the organic co-monomers. This results in a covalent tethering of the polymer to the inorganic framework.

3. The preformed porous material can be doped with organic monomers, which can then be polymerized inside the pore systems. This option is only used for M41S-type materials.
4. For mesostructured materials there is an additional option: Because of the templating mechanism in the synthesis of the mesostructured materials, compartments of different polarity are formed. The hydrophobic core made by the template molecule can be used as monomer and can be polymerized after formation of the mesostructure.

Experiments to prepare aerogel/polymer composites by performing the sol-gel process in solutions of organic polymers mostly failed, because the polymer was washed out during subsequent supercritical drying. This approach is only successful when the organic polymers are capable of forming hydrogen bonds to the surface silanol groups of the inorganic matrix. Recent examples are chitosan-silica hybrid aerogels [135]. The gels were formed by adding TEOS to an aqueous solution of chitosan, a polysaccharide obtained by alkaline deacetylation of chitin. Because of the hydrogen bonds between the silica network and chitosan, shrinkage of the gels during supercritical drying was particularly low.

A polymer-doped aerogel with interpenetrating, but not connected, networks was obtained when the organic polymer was generated *in situ* during the sol-gel reaction by radical polymerization of *N,N´*-dimethylacrylamide (in the presence of *N,N´*-methylenebisacrylamide) followed by supercritical drying with CO_2 [136].

The covalent bonding of an organic polymer to the aerogel structure was achieved as follows. Poly(benzobisthiazole), a temperature-stable polymer, was sulfonated to obtain a coupling site with the silica structures. The obtained sulfobiphenylpolybenzobisthiazole (SBPPBT, Fig. 4.17), was first reacted with $(MeO)_3Si(CH_2)_3$-NEt_2 as the coupling agent, followed by sol-gel processing with $Si(OMe)_4$ as the silica source and supercritical drying with CO_2. The obtained aerogels had low densities and low values of the longitudinal sound velocity, but were substantially less brittle than unmodified silica aerogels [137].

Aerogels with very good elastic properties were prepared by co-condensation of flexible polydimethylsiloxanes (PDMS) with TEOS. An elastic compression up to 30% was observed for aerogels in which 20 wt% PDMS was incorporated [138]. Another inorganic-organic hybrid polymer was prepared by crosslinking the alumina network with hexamethylenediisocyanate. This hybrid material had a lower density and a broader pore radii distribution than an unmodified alumina aerogel [139].

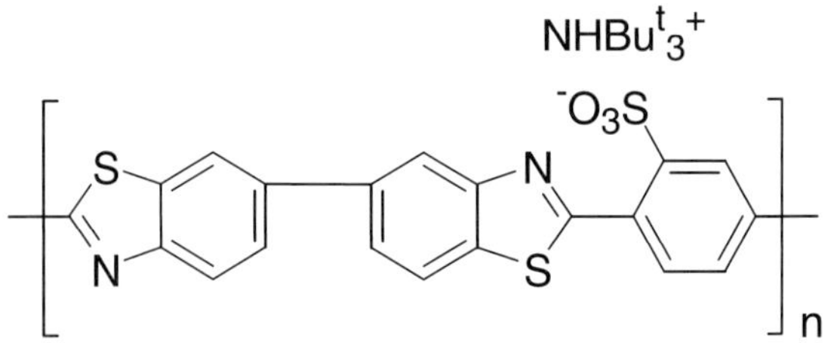

Fig. 4.17 Structure of the SBPPBT polymer

The main interest in inorganic-organic hybrid materials based on transition-metal aerogels is for electrically conducting materials, especially the polypyrrole (PPy)-vanadia combination for electrodes in lithium batteries [1g]. The objective is to attain two interpenetrating networks between the metal oxide (to achieve lithium intercalation) and a conducting polymer (to provide an electrically conducting network). Nanocomposite PPy/V_2O_5 aerogels were prepared by three approaches [140]. One approach was a co-synthesis route where the organic and inorganic networks were prepared simultaneously from a common solvent. The second approach involved impregnation of a vanadia gel with the pyrrole monomer, which then polymerized as initiated by V(V) centers. The aerogels were then formed by either supercritical or ambient-pressure drying. In the third approach a "microcomposite" was prepared by forming the gel in the presence of preformed PPy particles. The electrochemical properties of the hybrid aerogels critically depend on controlling the microstructure of the hybrid material. The second (post-gelation) synthesis method led to relatively poor electrochemical properties because the inhomogeneous distribution of PPy limits access to the V_2O_5 phase. The organic particles in the "microcomposite" were too isolated to form a conductive pathway. In comparison, co-synthesis offers excellent microstructural control, and a homogeneous distribution of the PPy phase was achieved. Reversible lithium capacity of 3Li/V_2O_5 was obtained, although the hybrid material did not exhibit improved macroscopic conductivity.

Recent results with another organic/inorganic aerogel system showed that the conductivity can indeed be enhanced. PPy/MoO_3 hybrid aerogels were synthesized by a co-solvent method; their electrical conductivity was approximately one hundred times higher than that of pristine MoO_3 gels, and ten times higher than that of a material in which the conducting polymer (polyaniline) was intercalated into the MoO_3 structure [141].

The preparation of a polymer-silica mesoporous nanocomposite was reported by using acid-catalyzed sol-gel reactions of TEOS with poly(styrene-*co*-styrylethyltrimethoxysilane) in the presence of dibenzoyltartaric acid (DBTA) as a non-surfactant pore-forming agent. After removal of DBTA by extraction, a mesoporous material with a polymer covalently bonded to the silica walls was obtained [142].

Inorganic-polymer hybrid mesostructured materials with a polymer covalently attached to the inorganic framework were prepared by co-condensation of TMOS with 3-methacryloxypropyltrimethoxysilane in the presence of a template. The template molecule was extracted in a second step, resulting in an ordered MCM hybrid material with covalently bonded methacrylate units. In a further step, MMA was adsorbed into the material and polymerized resulting in a covalently coupled inorganic-organic composite material [143]. A similar material was prepared in a one-pot synthesis using a quite complex mixture of a hydrolyzable and condensable inorganic precursor, water, hydrochloric acid, ethanol, surfactant, 3-methacryloxypropyltrimethoxysilane as coupling agent, dodecylmethacrylate as monomer and hexanedioldimethacrylate as crosslinker. A layered polymer-silica nanocomposite, again with covalently bonded methacrylate units, was formed by a simple dip coating procedure relying on self-assembly driven by solvent evaporation (Fig. 4.18). The organic polymer was formed by UV curing in a second step [144]. This example clearly shows that organic molecules can be placed in a very controlled fashion in mesostructured inorganic materials [145].

The feasibility of forming polymer networks within the pore channels of preformed MCM type materials by polymerizing adsorbed monomers was demonstrated by Llewellyn et al. [146]. They adsorbed styrene, vinyl acetate, and methyl methacrylate monomers into dehydrated and evacuated MCM hosts and studied the effect of confinement on polymer growth. The most prominent example is the polymerization of methyl methacrylate within the preformed mesoporous hosts MCM-41 and MCM-48, which was studied in detail. Polymerization proceeded via radical initiation with high yield and resulted in nanocomposites with high polymer content [147, 148]. Poly(vinyl ether)s and poly(vinylcarbazole)s were also incorporated, and detailed investigations of the polymerizations under confinement in the restricted geometry were performed [149].

An interesting new concept was presented by using vapor phase impregnation techniques. The well-defined channels of mesoporous silica were used as a mold for the synthesis of poly(phenolformaldehyde) fibers with a high aspect ratio ($>10^4$) and controlled diameter. These fibrous copies of the mesochannels were extracted by treatment with HF to remove the silica, and exact replicas of the diameter and length of the mould pores were obtained [150]. An extension of this work was demonstrated by the vapor phase impregnation of metal-containing monomers such as [1]-silaferrocenophanes into the channels of MCM-41 and subsequent ring-opening polymerization of these precursors. The resulting poly(ferrocenylsilane) was converted by pyrolysis into magnetic iron nanoparticles, which are confined to the channels of the silica-based material [151].

Having a well-defined channel system, especially in MCM-41 with its straight hexagonal channels, the incorporation of electrically conducting polymers is of special interest with regard to designing nanometer-scaled electronic devices. The conducting polymer polyaniline was successfully stabilized as filaments in the channels of MCM-41 by adsorption of aniline vapor into the dehydrated host, followed by reaction with peroxydisulfate [152, 153]. In addition to polyaniline, other conducting species such as graphitic carbon wires generated from organic polymers were incorporated in MCM type materials. In this approach, the monomer acrylo-

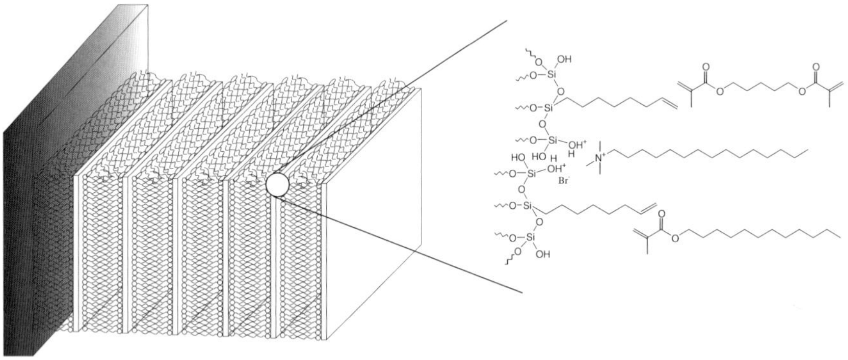

Fig. 4.18 Assembly mixture for a poly(dodecylmethacrylate)/silica nanocomposite

nitrile was introduced into the host through vapor or solution transfer, and polymerized in the channels. Pyrolysis of the long chains of polyacrylonitrile thus formed led to the formation of carbonized material in the channels of the host [154].

In another example, crystalline nanofibres of linear polyethylene with an ultra-high molecular weight and a defined diameter of 30–50 nm were grown catalytically inside the channels of an MCM material with active transition metal complexes grafted to the silica walls [155]. This concept was extended to a variety of metal complexes [156, 157, 158].

An example in which the structure-directing agent forms the polymeric material inside the mesoporous channel is the use of an oligoethylene glycol functionalized diacetylenic surfactant as structure directing agent. It also acts as monomeric precursor for the conjugated polymer polydiacetylene because of its polymerizable groups [71].

4.5.6 Creation of Carbon Structures

Silica aerogels are very interesting thermal insulation materials. Due to their low density and small pore radii, the heat transport via the solid aerogel skeleton and the gas phase is low. The radiative transport below 20 °C is also low, because silica aerogels absorb heat sufficiently. However, use of silica aerogel for heat insulation at medium temperatures (50–500 °C) requires reduction of the radiative heat transport. The reason for this is that the radiation maximum at these temperatures is in a range where silica has a low specific extinction. Carbon black is very well suited for infrared opacification due to its broad absorption band in the relevant range.

A better possibility than mixing aerogels with soot is to generate carbonaceous structures in silica aerogels, either by pyrolysis of organically modified aerogels or by pyrolysis of gaseous organic compounds. In the latter method, silica aerogels were reacted with acetylene, methane, propane, xylol or furfuryl alcohol in a stream of an inert gas at high temperatures. The temperatures necessary for pyrolysis depend on the gas and vary between 350 and 850 °C. The aerogels strongly favor the decomposition of the organic compounds. Carbon is deposited as graphite on the inner surface of the aerogels. Even nanometer-sized carbon fibers, rings and tubes were found [159].

Pyrolysis of organic groups of inorganic-organic hybrid aerogels is another possibility to subsequently modify aerogels by carbon structures. Pyrolysis of organically substituted aerogels obtained from $R'Si(OR)_3/Si(OR)_4$ mixtures resulted in the coating of the aerogel network with carbon nanostructures [160]. This was particularly efficient for $R' =$ phenyl. The chitosan-silica hybrid aerogels mentioned above were also used to prepare carbon-silica composite aerogels [135]. C_{60}/C_{70} was incorporated in silica aerogels by adding the fullerenes to the sol-gel precursor mixture [69].

MCM-type materials can also be used as host for carbonaceous structures. One example was already mentioned in the case of poly(acrylonitrile), which was converted to carbon nanostructures by pyrolysis [154]. In addition, to aid the generation of carbon nanostructures within the silicate matrix, fullerenes such as C_{60} or C_{70} were incorporated within the mesoporous host by post-synthetic washing with toluene/C_{60} or C_{70} solutions [161, 162].

The MCM material was also used as template for the preparation of mesoporous carbon. The first synthesis of such mesoscopically ordered carbon materials was reported by Ryoo et al. by infiltration of the inorganic template with sucrose and carbonizing the organic sugar at 900 °C inside the pores of MCM-48 mesoporous silica [163]. The carbon (CMK-1) obtained after subsequent removal of the silica template with HF or KOH solution showed a highly ordered pore arrangement. Several CMKs with different pore sizes and different structural characteristics were synthesized [164, 165, 166, 167].

Carbon species can also be infiltrated into mesoporous silica by chemical vapor infiltration using propylene as carbon precursor. Carbothermal reduction of the as-prepared SiO_2/C material at 1250–1450 °C in an inert atmosphere led to an almost complete conversion into high surface area SiC material [168].

References

1 a) R. J. Ayen, P. A. Iacobucci, *Rev. Chem. Eng.* **1988**, *5*, 157–198; b) H. D. Gesser, P. C. Goswami, *Chem. Rev.* **1989**, *89*, 765–788; c) G. M. Pajonk, *Appl. Catal.* **1991**, *72*, 217–266; d) J. Fricke, A. Emmerling, *Struct. Bonding*, **1992**, *77*, 37–87; e) M. Schneider, A. Baiker, *Catal. Rev. – Sci. Eng.* **1995**, *37*, 515–556; f) N. Hüsing, U. Schubert, *Angew. Chem.* **1998**, *110*, 22–47; *Angew. Chem. Int. Ed. Engl.* **1998**, *37*, 22–45; g) D. R. Rolison, B. Dunn, *J. Mater. Chem.* **2001**, *11*, 963–980.

2 D.W. Schaefer, K. D. Keefer, *Mat. Res. Soc. Symp. Proc.* **1984**, *32*, 1–14; K. D. Schaefer, *Mat. Res. Soc. Symp. Proc.* **1984**, *32*, 15–24; J. Livage, M. Henry, C. Sanchez, *Prog. Solid State Chem.* **1988**, *18*, 259–341; C. J. Brinker, G. W. Scherer, *Sol-Gel Science*, Academic Press, New York, 1990.

3 T. Heinrich, U. Klett, J. Fricke, *J. Porous Mat.* **1995**, *1*, 7–17.

4 Y. Mizushima, M. Hori, *J. Non-Cryst. Solids* **1994**, *167*, 1–8.

5 D. C. Dutoit, M. Schneider, A. Baiker, *J. Catal.* **1995**, *153*, 165–176; R. Hutter, T. Mallat, A. Baiker, *J. Catal.* **1995**, *153*, 177–189.

6 J. J. Calvino, M. A. Cauqui, G. Cifredo, J. A. Perez, J. M. Rodriguez-Izquiredo, *Mat. Res. Soc. Symp. Proc.* **1994**, *346*, 685–690; S. Bernal, J. J. Calvino, M. A. Cauqui, J. M. Rodriguez-Izquiredo, H. Vidal, *Stud. Surf. Sci. Catal.* **1995**, *91*, 461–470.

7 U. Klett, J. Fricke, *J. Non-Cryst. Solids* **1998**, *225*, 188–191.

8 T. Heinrich, F. Raether, H. Marsmann, *J. Non-Cryst. Solids* **1994**, *168*, 14–22.

9 R.-M. Jansen, B. Kessler, J. Wonner, A. Zimmermann, DE 4316540 A1, **1993** [*Chem. Abstr.* **1995**, *122*, 65331r]; R.-M. Jansen, A. Zimmermann, EP 690023 A2, **1993** [*Chem. Abstr.* **1996**, *124*, 150025e].

10 G. Herrmann, R. Iden, M. Mielke, F. Teich, B. Ziegler, *J. Non-Cryst. Solids*, **1995**, *186*, 380–387.

11 N. Hüsing, F. Schwertfeger, W. Tappert, U. Schubert, *J. Non-Cryst. Solids* **1995**, *186*, 37–43.

12 C. J. Brinker, K. J. Ward, K. D. Keefer, E. Holupka, P. J. Bray, R. K. Pearson in *Aerogels*, ed J. Fricke, *Springer Proc. Phys.*, Springer, Berlin, 1986, 57–67, Vol. 6.

13 P. J. Davis, C. J. Brinker, D. M. Smith, *J. Non-Cryst. Solids* **1992**, *142*, 189–196; P. J. Davis, C. J. Brinker, D. M. Smith, R. A. Assink, *J. Non-Cryst. Solids* **1992**, *142*, 197–207; R. Deshpande, D. W. Hua, D. M. Smith, C. J. Brinker, *J. Non-Cryst. Solids* **1992**, *144*, 32–44; D. M. Smith,

R. Deshpande, C. J. Brinker, *Mat. Res. Soc. Symp. Proc.* **1992**, *271*, 567–572.
14 D. M. Smith, D. Stein, J. M. Anderson, W. Ackerman, *J. Non-Cryst. Solids* **1995**, *186*, 104–112.
15 S. Hæreid, M. Dahle, S. Lima, M. A. Einarsrud, *J. Non-Cryst. Solids* **1995**, *186*, 96–103.
16 T. Yanagisawa, T. Shimizu, K. Kuroda, C. Kato, *Bull. Chem. Soc. Jpn.* **1990**, *63*, 988–992.
17 C. T. Kresge, M. E. Leonowicz, W. J. Roth, J. C. Vartuli, J. S. Beck, *Nature* **1992**, *359*, 710–712.
18 Q. S. Huo, D. Y. Zhao, J. L. Feng, K. Weston, S. K. Buratto, G. D. Stucky, S. Schacht, F. Schuth, *Adv. Mater.* **1997**, *9*, 974–978.
19 Y. Lu, R. Ganguli, C. A. Drewien, M. T. Anderson, C. J. Brinker, W. Gong, Y. Guo, H. Soyez, B. Dunn, M. H. Huang, J. I. Zink, *Nature* **1997**, *389*, 364–368; M. Ogawa, T. Kikuchi, *Adv. Mater.* **1998**, *10*, 1077–1080.
20 G. S. Attard, J. C. Glyde, C. G. Goltner *Nature* **1995**, *378*, 366–368; C. G. Göltner, S. Henke, M. C. Weisenberger, M. Antonietti, *Angew. Chem.* **1998**, *110*, 633–636, *Angew. Chem. Int. Ed. Eng.* **1998**, *37*, 613–616.
21 I. Sokolov, H. Yang, G. A. Ozin, C. T. Kresge, *Adv. Mater.* **1999**, *11*, 636–642.
22 S. M. Yang, I. Sokolov, N. Coombs, C. T. Kresge, G. A. Ozin, *Adv. Mater.* **1999**, *11*, 1427–1431.
23 D. Y. Zhao, J. L. Feng, Q. S. Huo, N. Melosh, G. H. Frederickson, B. F. Chmelka, G. D. Stucky, *Science* **1998**, *279*, 548–552.
24 P. Schmidt-Winkel, W. W. Lukens, D. Y. Zhao, P. D. Yang, B. F. Chmelka, G. D. Stucky, *J. Am. Chem. Soc.* **1999**, *121*, 254–255.
25 U. Ciesla, D. Demuth, R. Leon, P. Petroff, G. Stucky, K. Unger, F. Schuth, *J. Chem. Soc. Chem. Commun.* **1994**, 1387–1388; P. D. Yang, D. Y. Zhao, D. I. Margolese, B. F. Chmelka, G. D. Stucky, *Nature* **1998**, *396*, 152–155.
26 D. M. Antonelli, J. Y. Ying, *Angew. Chem. Int. Ed. Engl.* **1996**, *35*, 426–430; D. M. Antonelli, A. Nakahira, J. Y. Ying, *Inorg. Chem.* **1996**, *35*, 3126–3136.
27 P. V. Braun, P. Osenar, S. I. Stupp, *Nature* **1996**, *380*, 325–328.
28 M. J. MacLachlan, N. Coombs, G. A. Ozin, *Nature* **1999**, *397*, 681–684.
29 B. J. Melde, B. T. Holland, C. F. Blanford, A. Stein, *Chem. Mater.* **1999**, *11*, 3302–3308.
30 S. Inagaki, S. Guan, Y. Fukushima, T. Ohsuna, O. Terasaki, *J. Am. Chem. Soc.* **1999**, *121*, 9611–9614.
31 T. Asefa, M. J. MacLachlan, N. Coombs, G. A. Ozin, *Nature* **1999**, *402*, 867–871.
32 D. A. Doshi, N. K. Hüsing, M. C. Lu, H. Y. Fan, Y. F. Lu, K. Simmons-Potter, B. G. Potter, A. J. Hurd, C. J. Brinker, *Science* **2000**, *290*, 107–111.
33 M. Trau, N. Yao, E. Kim, Y. Xia, G. M. Whitesides, I. A. Aksay, *Nature* **1997**, *390*, 674–676.
34 J. Y. Ying, C. P. Mehnert, M. S. Wong, *Angew. Chem.* **1999**, *38*, 58–82; *Angew. Chem. Int. Ed. Engl.* **1999**, *38*, 56–77.
35 A. Corma, *Chem. Rev.* **1997**, *97*, 2373–2419.
36 K. Moller, T. Bein, *Chem. Mater.* **1998**, *10*, 2950–2963.
37 K. J. Edler, S. J. Roser, *Int. Rev. Phys. Chem.* **2001**, *20*, 387–466.
38 S. Inagaki, Y. Fukushima, K. Kuroda, *J. Chem. Soc. Chem. Commun.* **1994**, 680–681.
39 Q. Huo, D. I. Margolese, G. D. Stucky, *Chem. Mater.* **1996**, *8*, 1147–1160.
40 M. H. Huang, B. S. Dunn, J. I. Zink, *J. Am. Chem. Soc.* **2000**, *122*, 3739–3745.
41 Q. Huo, D. I. Margolese, U. Ciesla, D. G. Demuth, P. Feng, T. E. Gier, P. Sieger, A. Firouzi, B. F. Chmelka, F. Schüth, G. D. Stucky, *Chem. Mater.* **1994**, *6*, 1176–1191.
42 P. T. Tanev, T. J. Pinnavaia, *Science* **1995**, *267*, 865–867; P. T. Tanev, T. J. Pinnavaia, *Science* **1996**, *271*, 1267–1269.
43 S. A. Bagshaw, E. Prouzet, T. J. Pinnavaia, *Science* **1995**, *269*, 1242–1244.
44 D. Zhao, Q. Huo, J. Feng, B. F. Chmelka, G. D. Stucky, *J. Am. Chem. Soc.* **1998**, *120*, 6024–6036.
45 M. Mielke, private communication, quoted in [1d].
46 K. H. Lee, S. Y. Kim, K. P. Yoo, *J. Non-Cryst. Solids* **1995**, *186*, 18–22.

47 S. Reul, W. Richter, D. Haarer, *J. Non-Cryst. Solids* **1992**, *145*, 149–153.
48 B. Lindlar, M. Lüchinger, A. Röthlisberger, M. Haouas, G. Pirngruber, A. Kogelbauer, R. Prins, *J. Mater. Chem.* **2002**, *12*, 528–533.
49 K. A. Koyano, T. Tatsumi, Y. Tanaka, S. Nakata, *J. Phys. Chem. B* **1997**, *101*, 9436–9440.
50 X. S. Zhao, G.Q. Lu, *J. Phys. Chem. B* **1998**, *102*, 1556–1561.
51 V. Antochshuk, M. Jaroniec, *J. Phys. Chem. B* **1999**, *103*, 6252–6261.
52 R. Anwander, I. Nagl, M. Widenmeyer, G. Engelhardt, O. Groeger, C. Palm, T. Röser, *J. Phys. Chem. B* **2000**, *104*, 3532–3544.
53 K. Yamamoto, T. Tatsumi, *Chem. Lett.* **2000**, *6*, 624–625.
54 X. Fen, G. E. Fryxell, L.-Q. Wang, A. Y. Kim, J. Liu, K. M. Kemner, *Science* **1997**, *276*, 923–926.
55 J. Liu, X. Feng, G. E. Fryxell, L.-Q. Wang, A. Y. Kim, M. Gong, *Adv. Mater.* **1998**, *10*, 161–165.
56 J. F. Diaz, K. J. Balkus Jr., F. Bedioui, V. Kurshev, L. Kevan, *Chem. Mater.* **1997**, *9*, 61–67.
57 A. Cauvel, G. Renard, D. Brunel, *J. Org. Chem.* **1997**, *62*, 749–751.
58 Y. V. Subba Rao, D. E. De Vos, T. Bein, P. A. Jacobs, *J. Chem. Soc. Chem. Commun.* **1997**, 355–356.
59 S. Jaenicke, G. K. Chuah, X. H. Lin, X. C. Hu, *Microporous Mesoporous Mater.* **2000**, *35–36*, 143–153.
60 P. Ferreira, I. S. Goncalves, F. Mosselmans, M. Pillinger, J. Rocha, A. Thursfield, *Eur. J. Inorg. Chem.* **2000**, 97–102.
61 B. Lindlar, M. Lüchinger, M. Haouas, A. Kogelbauer, R. Prins, *Stud. Surf. Sci. Catal.* **2001**, *135*, 4828–4834.
62 Y. Shin, J. Liu, L.-Q. Wang, Z. Nie, W. D. Samuels, G. E. Fryxell, G. J. Exarhos, *Angew. Chem.* **2000**, *112*, 2814–2819; *Angew. Chem. Int. Ed. Engl.* **2000**, *39*, 2702–2707.
63 F. Schwertfeger, D. Frank, M. Schmidt, *J. Non-Cryst. Solids* **1998**, *225*, 24–29.
64 J. K. Floess, R. Field, S. Rouanet, *J. Non-Cryst. Solids* **2001**, *285*, 101–108.
65 V. Antochshuk, M. Jaroniec, *Chem. Mater.* **2000**, *12*, 2496–2501; V. Antochshuk, M. Jaroniec, *J. Chem. Soc. Chem. Commun.* **1999**, 2373–2374.
66 S. Dai, Y. Shin, Y. Ju, M. C. Burleigh, J.-S. Lin, C. E. Barnes, Z. Xue, *Adv. Mater.* **1999**, *11*, 1226–1230.
67 B. T. Holland, C. Walkup, A. Stein, *J. Phys. Chem. B* **1998**, *102*, 4301–4309.
68 M. Bockhorst, K. Heinloth, G. M. Pajonk, R. Begag, E. Elaloui, *J. Non-Cryst. Solids* **1995**, *186*, 388–394.
69 L. Zhu, Y. Li, J. Wang, J. Shen, *Chem. Phys. Lett.* **1995**, *239*, 393–398; L. Zhu, Y. Li, J. Wang, J. Shen, *J. Appl. Phys.* **1995**, *77*, 2801–2803; J. Shen, J. Wang, B. Zhou, Z. Deng, Z. Weng, L. Zhu, L. Zhao, Y. Li, *J. Non-Cryst. Solids* **1998**, *225*, 315–318; L. Zhu, P. P. Ong, J. Shen, J. Wang, *J. Phys. Chem. Solids* **1998**, *59*, 819–824.
70 H. Zhou, I. Honma, *Jap. J. Appl. Phys.* **1999**, *38*, L958-L960; H. S. Zhou, H. Sasabe, I. Honma, *J. Mater. Chem.* **1998**, *8*, 515–516; I. Honma, H. S. Zhou, *Adv. Mater.* **1998**, *10*, 1532–1536.
71 Y. Lu, Y. Yang, A. Sellinger, M. Lu, J. Huang, H. Fan, R. Haddad, G. Lopez, A. R. Burns, D. Y. Sasaki, J. Shelnutt, C. J. Brinker, *Nature* **2001**, *410*, 913–917.
72 N. Hüsing, B. Launay, J. Bauer, G. Kickelbick, D. Doshi, *J. Sol-Gel Sci. Technol.* **2003**, *26*, 609–613.
73 B. Riegel, S. Plittersdorf, W. Kiefer, N. Hüsing, U. Schubert, *J. Mol. Struct.* **1997**, *410–411*, 157–160; N. Hüsing, U. Schubert, B. Riegel, W. Kiefer, *Mat. Res. Soc. Symp. Proc.* **1996**, *435*, 339–344.
74 N. Hüsing, U. Schubert, K. Misof, P. Fratzl, *Chem. Mater.* **1998**, *10*, 3024–3032.
75 F. Schwertfeger, W. Glaubitt, U. Schubert, *J. Non-Cryst. Solids* **1992**, *145*, 85–89; F. Schwertfeger, N. Hüsing, U. Schubert, *J. Sol-Gel Sci. Technol.* **1994**, *2*, 103–108; U. Schubert, F. Schwertfeger, N. Hüsing, E. Seyfried, *Mat. Res. Soc. Symp. Proc.* **1994**, *346*, 151–162.
76 A. Venkateswara Rao, D. Haranath, *Microporous Mesoporous Mat.* **1999**, *30*, 267–273; A. Venkateswara Rao, G. M. Pajonk, D. Haranath, *Mater. Sci. Technol.* **2001**, *17*, 343–348; A. Venkateswara Rao, G. M. Pajonk, *J. Non-Cryst. Solids* **2001**, *285*, 202–209.

77 J. G. Reynolds, P. R. Coronado, L. W. Hrubesh, *Energy Sources* **2001**, *23*, 831–843.
78 N. Hüsing, U. Schubert, R. Mezei, P. Fratzl, B. Riegel, W. Kiefer, D. Kohler, W. Mader, *Chem. Mater.* **1999**, *11*, 451–457.
79 C. Alié, R. Pirard, A. J. Lecloux, J. P. Pirard, *J. Non-Cryst. Solids* **1999**, *246*, 216–228; C. Alié, A. Benhaddou, R. Pirard, A. J. Lecloux, J. P. Pirard, *J. Non-Cryst. Solids* **2000**, *270*, 77–90; C. Alié, R. Pirard, A. J. Lecloux, J. P. Pirard, *J. Non-Cryst. Solids* **2001**, *285*, 135–141.
80 W. Cao, A. J. Hunt, *Mat. Res. Soc. Symp. Proc.* **1994**, *346*, 631–636.
81 S. Burkett, S. D. Sims, S. Mann, *J. Chem. Soc. Chem. Commun.* **1996**, 1367–1368; C. E. Fowler, S. L. Burkett, S. Mann, *J. Chem. Soc. Chem. Commun.* **1997**, 1769–1770.
82 M. H. Lim, C. F. Blanford, A. Stein, *Chem. Mater.* **1998**, *10*, 467–470.
83 H. Fan, Y. Lu, A. Stump, S. T. Reed, T. Baer, R. Schunk, V. Perez-Luna, G. P. Lopez, C. J. Brinker, *Nature* **2000**, *405*, 56–60.
84 D. J. Macquarrie, *J. Chem. Soc. Chem. Commun.* **1996**, 1961–1962.
85 M. H. Lim, C. F. Blanford, A. Stein, *J. Am. Chem. Soc.* **1997**, *119*, 4090–4091.
86 B. Lebeau, C. E. Fowler, S. R. Hall, S. Mann, *J. Mater. Chem.* **1999**, *9*, 2279–2281.
87 M. Ganschow, M. Wark, D. Wöhrle, G. Schulz-Ekloff, *Angew. Chem.* **2000**, *112*, 167–170; *Angew. Chem. Int. Ed. Engl.* **2000**, *39*, 160–163.
88 F. Babonneau, L. Leite, S. Fontluptx, *J. Mater. Chem.* **1999**, *9*, 175–178.
89 S. R. Hall, C. E. Fowler, B. Lebeau, S. Mann, *J. Chem. Soc. Chem. Commun.* **1999**, 201–202.
90 Q. Huo, D. I. Margolese, G. D. Stucky, *Chem. Mater.* **1996**, *8*, 1147–1160.
91 E. Ruiz-Hitzky, S. Letaief, V. Prévot, *Adv. Mater.* **2002**, *14*, 439–443.
92 D. A. Loy, G. M. Jamison, B. M. Baugher, E. M. Russick, R. A. Assink, S. Prabakar, K. J. Shea, *J. Non-Cryst. Solids* **1995**, *186*, 44–53.
93 D. A. Loy, J. V. Beach, B. M. Baugher, R. A. Assink, K. J. Shea, J. Tran, J. H. Small, *Chem. Mater.* **1999**, *11*, 3333–3341.
94 W. Rupp, N. Hüsing, U.Schubert, *J. Mater. Chem.*, **2002**, *12*, 2594–2596.
95 J. W. Kriesel, T. D. Tilley, *Adv. Mater.* **2001**, *13*, 1645–1648.
96 Y. Guo, A. R. Guadalupe, *J. Chem. Soc. Chem. Commun.* **1999**, 315–316.
97 S. Guan, S. Inagaki, T. Ohsuna, O. Terasaki *J. Am. Chem. Soc.* **2000**, *122*, 5660–5661.
98 C. Yoshina-Ishii, T. Asefa, M. J. Maclachlan, N. Coombs, G. A. Ozin, *J. Chem. Soc. Chem. Commun.* **2000**, 2539–2540.
99 Y. Lu, H. Fan, N. Doke, D. A. Loy, R. A. Assink, D. A. LaVan, C. J. Brinker, *J. Am. Chem. Soc.* **2000**, *122*, 5258–5261.
100 R. J. P. Corriu, A. Mehdi, C. Reyé, C. Thieuleux, *J. Chem. Soc. Chem. Commun.* **2002**, 1382–1383.
101 S. Guan, S. Inagaki, T. Ohsuna, O. Terasaki, *Microporous Mesoporous Mater.* **2001**, *44–45*, 165–172.
102 J. R. Matos, M. Kruk, L. P. Mercuri, M. Jaroniec, T. Asefa, N. Coombs, G. A. Ozin, T. Kamiyama, O. Terasaki, *Chem. Mater.* **2002**, *14*, 1903–1905.
103 M. D. McInall, J. Scott, L. Mercier, P. J. Kooyman, *J. Chem. Soc. Chem. Commun.* **2001**, 2282–2283.
104 S. Inagaki, S. Guan, T. Ohsuna, O. Terasaki, *Nature* **2002**, *416*, 304–307; S. Inagaki, S. Guan, *Mat. Res. Soc. Symp. Proc.* **2002**, *726*, 225–233.
105 E. M. Wong, M. A. Markowitz, S. B. Qadri, S. Golledge, D. G. Castner, B. P. Gaber, *J. Phys. Chem. B* **2002**, *106*, 6652–6658.
106 B. Heinrichs, F. Noville, J. P. Pirard, *J. Catal.* **1997**, *170*, 366–376.
107 U. Schubert, *New J.Chem.* **1994**, *18*, 1049–1058.
108 L. Schmid, M. Rohr, A. Baiker, *J. Chem. Soc. Chem. Commun.* **1999**, 2303–2304.
109 E. F. Murphy, L. Schmid, T. Bürgi, M. Maciejewski, A. Baiker, D. Günther, M. Schneider, *Chem. Mater.* **2001**, *13*, 1296–1304.
110 M. Bonnet, L. Schmid, A. Baiker, F. Diederich, *Adv. Funct. Mater.* **2002**, *12*, 39–42.
111 S. Klein, W. F. Maier, *Angew. Chem.* **1996**, *108*, 2376–2379; H. Kochkar, F. Figueras, *J. Catal.* **1997**, *171*, 420–430; C. A. Müller, M. Maciejewski, T. Mallat, A. Baiker, *J. Catal.* **1999**, *184*, 280–293.

112 C. A. Müller, M. Schneider, T. Mallat, A. Baiker, *Appl. Catal. A* **2000**, *201*, 253–261.
113 C. A. Müller, M. Schneider, T. Mallat, A. Baiker, *J. Catal.* **2000**, *189*, 221–232; A. Gisler, C. A. Müller, M. Schneider, T. Mallat, A. Baiker, *Topics Catal.* **2001**, *15*, 247–255.
114 T. Maschmeyer, F. Rey, G. Sankar, J. M. Thomas, *Nature* **1995**, *378*, 159–162.
115 R. Ryoo, S. Jun, J. M. Kim, M. J. Kim, *J. Chem. Soc. Chem. Commun.* **1997**, 2225–2226.
116 R. D. Oldroyd, G. Sankar, J. M. Thomas, D. Özkaya, *J. Phys. Chem. B* **1998**, *102*, 1849–1855.
117 M. Morey, A. Davidson, H. Eckert, G. Stucky, *Chem. Mater.* **1996**, *8*, 486–492.
118 C. Huber, K. Moller, T. Bein, *J. Chem. Soc. Chem. Commun.* **1994**, 2619–2620.
119 R. Anwander, O. Runte, J. Eppinger, G. Gerstberger, E. Herdtweck, M. Spiegler, *J. Chem. Soc. Dalton Trans.* **1998**, 847–858.
120 R. Anwander, R. Roesky, *J. Chem. Soc. Dalton Trans.* **1997**, 137–138.
121 P. Sutra, D. Brunel, *Chem. Commun.* **1996**, 2485–2486.
122 S. Xiang, Y. Zhang, Q. Xin, C. Li, *Angew. Chem.* **2002**, *114*, 849–852; *Angew. Chem. Int. Ed. Engl.* **2002**, *41*, 821–824.
123 M. Laspéras, N. Bellocq, D. Brunel, P. Moreau, *Tetrahedron Asymmetry* **1998**, *9*, 3053–3064.
124 C. J. Liu, S. G. Li, W. Q. Pang, C. M. Che, *J. Chem. Soc. Chem. Commun.* **1997**, 65–66.
125 C.-M. Yang, H.-S. Sheu, K.-J. Chao, *Adv. Funct. Mater.* **2002**, *12*, 143–148.
126 G. Carturan, R. Campostrini, S. Dire, V. Scardi, E. de Alteris, *J. Mol. Catal.* **1989**, *57*, L13-L16; S. Braun, S. Rappoport, R. Zusman, D. Avnir, M. Ottolenghi, *Mater. Lett.* **1990**, *10*, 1–5; L. M. Ellerby, C. R. Nishida, F. Nishida, S. Yamanaka, B. Dunn, J. S. Valentine, J. I. Zink, *Science* **1992**, *255*, 1113–1115.
127 P. Buisson, C. Hernandez, M. Pierre, A. C. Pierre, *J. Non-Cryst. Solids* **2001**, *285*, 295–302.
128 S. Maury, P. Buisson, A. C. Pierre, *Langmuir* **2001**, *17*, 6443–6446.
129 A. Basso, L. Dc Martin, C. Ebert, L. Gardossi, A. Tomat, M. Casarci, O. Li Rosi, *Tetrahedron Letters* **2000**, *41*, 8627–8630.
130 M. Power, B. Hosticka, E. Black, C. Daitch, P. Norris, *J. Non-Cryst. Solids* **2001**, *285*, 303–308.
131 J. F. Diaz, K. J. Balkus Jr., *J. Mol. Catal. B: Enzymatic* **1996**, *2*, 115–126.
132 L. Washmon-Kriel, V. L. Jiminez, K. J. Balkus Jr., *J. Mol. Catal. B: Enzymatic* **2000**, *10*, 453–469.
133 H. Takahashi, B. L. Toshiya Sasaki, C. Miyazaki, T. Kajino, S. Inagaki, *Chem. Mater.* **2000**, *12*, 3301–3305.
134 K. J. Balkus Jr., D. Coutinho, J. Lucas, L. Washmon-Kriel, *Mat. Res. Soc. Symp. Proc.* **2000**, *628*, CC10.7.1-CC10.7.6.
135 M. R. Ayers, A. J. Hunt, *J. Non-Cryst. Solids* **2001**, *285*, 123–127.
136 B. M. Novak, D. Auerbach, C. Verrier, *Chem. Mater.* **1994**, *6*, 282–286.
137 J. K. Premachandra, C. Kumudinie, J. E. Mark, T. D. Dang, F. E. Arnold, *J. Macromol. Sci.- Pure Appl. Chem.* **1999**, *A36*, 73–83.
138 S. J. Kramer, F. Rubio-Alonso, J. D. Mackenzie, *Mat. Res. Soc. Symp. Proc.* **1996**, 435, 295–300.
139 Y. Mizushima, M. Hori, *J. Non-Cryst. Solids* **1994**, *170*, 215–222.
140 B. C. Dave, B. Dunn, F. Leroux, L. F. Nazar, H. P. Wong, *Mat. Res. Soc. Symp. Proc.* **1996**, *435*, 611–616; H. P. Wong , B. C. Dave, F. Leroux, J. H. Harreld, B. Dunn, L. F. Nazar, *J. Mat. Chem.* **1998**, *8*, 1019–1027; J. H. Harreld, H. P. Wong, B. C. Dave, B. Dunn, L. F. Nazar, *J. Non-Cryst. Solids* **1998**, *225*, 319–324; J. H. Harreld, B. Dunn. L. F. Nazar, *Int. J. Inorg. Mater.* **1999**, *1*, 135–146.
141 W. Dong, B. Dunn, *Mat. Res. Soc. Symp. Proc.* **1999**, *576*, 269–274.
142 Q. Feng, J. Xu, H. Dong, S. Li, Y. Wei, *J. Mater. Chem.* **2000**, *10*, 2490–2494.
143 K. Moller, T. Bein, R. X. Fischer, *Chem. Mater.* **1999**, *11*, 665–673.
144 A. Sellinger, P.M. Weiss, N. Anh, Y. Lu, R.A. Assink, W. Gong, C.J. Brinker, *Nature* **1998**, *394*, 256–260.
145 R. Hernandez, A.-C. Franville, P. Minoofar, B. Dunn, J. I. Zink, *J. Am. Chem. Soc.* **2001**, *123*, 1248–1249.

146 P. L. Llewellyn, U. Ciesla, H. Decher, R. Stadler, F. Schüth, K. K. Unger, *Stud. Surf. Sci. Catal.* **1994**, *84*, 2013–2020.

147 K. Moller, T. Bein, R. X. Fischer, *Chem. Mater.* **1998**, *10*, 1841–1852.

148 S. M. Ng, S. Ogino, T. Aida, K. A. Koyano, T. Tatsumi, *Macromol. Rapid Commun.* **1997**, *18*, 991–996.

149 S. Spange, A. Gräser, A. Huwe, F. Kremer, C. Tintemann, P. Behrens, *Chem. Eur. J.* **2001**, *7*, 3722–3728.

150 S. A. Johnson, D. Khushalani, N. Coombs, I. Manners, G. A. Ozin, *J. Mater. Chem.* **1998**, *8*, 13–14.

151 M. J. MacLachlan, P. Aroca, N. Coombs, I. Manners, G. A. Ozin, *Adv. Mater.* **1998**, *10*, 145–149.

152 C. G. Wu, T. Bein, *Science* **1994**, *264*, 1757–1759.

153 C. G. Wu, T. Bein, *Chem. Mater.* **1994**, *6*, 1109–1112.

154 C. G. Wu, T. Bein, *Science* **1994**, *266*, 1013–1015.

155 K. Kageyama, J. I. Tamazawa, T. Aida, *Science* **1999**, *285*, 2113–2115.

156 R. Ramachandra Rao, B. M. Weckhysen, R. A. Schoonheydt, *J. Chem. Soc. Chem. Commun.* **1999**, 445–446.

157 B. M. Weckhysen, R. Ramachandra Rao, J. Pelgrims, R. A. Schoonheydt, P. Bodard, G. Debras, O. Collart, P. Van Der Voort, E. F. Vansant, *Chem. Eur. J.* **2000**, 6, 2960–2970.

158 H. Balcar, J. Sedlacek, J. Cejka, J. Vohlidal, *Macromol. Rapid Commun.* **2002**, *23*, 32–37.

159 A. J. Hunt, W. Cao, *Mat. Res. Soc. Symp. Proc.* **1994**, *346*, 451–455. W. Cao, X.-Y. Song, A. J. Hunt, *Mat. Res. Soc. Symp. Proc.* **1994**, *349*, 87–92; X.-Y. Song, W. Cao, A. J. Hunt, *Mat. Res. Soc. Symp. Proc.* **1994**, *349*, 269–274; X.-Y. Song, W. Cao, M. R. Ayers, A. J. Hunt, *J. Mater. Res.* **1995**, *10*, 251–254; A. J. Hunt, M. R. Ayers, W. Cao, *J. Non-Cryst. Solids* **1995**, *185*, 227–232; D. Lee, P. C. Stevens, S. Q. Zeng, A. J. Hunt, *J. Non-Cryst. Solids* **1995**, *186*, 285–290.

160 F. Schwertfeger, U. Schubert, *Chem. Mater.* **1995**, *7*, 1909–1914; J. Kuhn, V. Bock, M. C. Arduini-Schuster, E. Seyfried, U. Schubert, J. Fricke, Thermal Conductivity **1994**, *22*, 589–598.

161 G. Gu, W. Ding, Y. Du, H. Huang, S. Yang, *Appl. Phys. Lett.* **1997**, *70*, 2619–2621.

162 A. Govindaraj, M. Nath, M. Eswaramoorthy, *Chem. Phys. Lett.* **2000**, *317*, 35–39.

163 R. Ryoo, S.H. Joo, S. Jun, *J. Phys. Chem. B* **1999**, *103*, 7743–7746.

164 M. Kruk, M. Jaroniec, R. Ryoo, S. H. Joo, *J. Phys. Chem. B*, **2000**, *104*, 7960–7968.

165 S. Jun, S. H. Joo, R. Ryoo, M. Kruk, M. Jaroniec, Z. Liu, T. Ohsuna, O. Terasaki, *J. Am. Chem. Soc.* **2000**, *122*, 10712–10713.

166 J. Lee, S. Yoon, S. M. Oh, C.-H. Shin, T. Hyeon, *Adv. Mater.* **2000**, *12*, 359–362.

167 J. S. Lee, S. H. Joo, R. Ryoo, *J. Am. Chem. Soc.* **2002**, *124*, 1156–1157.

168 J. Parmentier, J. Patarin, J. Dentzer, C. Vix-Guterl, *Ceram. Int.* **2002**, *28*, 1–7.

5
Optical Properties of Functional Hybrid Organic-Inorganic Nanocomposites

Clément Sanchez, Bénédicte Lebeau, Frédéric Chaput and Jean-Pierre Boilot

5.1
Introduction

The concept of "hybrid organic-inorganic" nanocomposites exploded in the eighties with the expansion of soft inorganic chemistry processes [1–4]. Indeed the mild synthetic conditions offered by the sol-gel process (metallo-organic precursors, organic solvents, low processing temperatures, processing versatility of the colloidal state) allow the mixing of inorganic and organic components at the nanometric scale [5–8]. The so-called functional hybrids are thus nanocomposite materials that lie at the interface of the organic and inorganic realms, whose high versatility offers a wide range of possibilities for preparing tailor-made materials in terms of chemical and physical properties [5–8]. Because they present several advantages for designing materials for optical applications (versatile and relatively facile chemistry, easy shaping and patterning, good mechanical integrity and excellent optical quality), numerous silica- and/or siloxane-based hybrid organic-inorganic materials have been developed in the past few years [9–13]. Fig. 5.1 illustrates the diversity of siloxane-based monoliths that can be obtained by embedding or by grafting organic chromophores in sol-gel derived hybrid matrices.

During the past twenty years, optical studies performed on organic/inorganic nanocomposites have evolved towards different objectives: investigation of the fundamental spectroscopy of the dye molecule isolated in the sol-gel environment, study of dye energy transfer in solid matrices, use of luminescent molecules as probes of the sol-gel processing and finally development of materials with specific optical properties based on the properties of organic or inorganic chromophores [3, 8–14].

Some hybrid products have already entered the applied field. Examples include the one million TV sets sold annually by Toshiba, the screens of which are coated with hybrids made of indigo dyes embedded in a silica/zirconia matrix [15], organically doped sol-gel glassware sold by Spiegelau (Fig. 5.2) [16, 17], sol-gel entrapped enzymes sold by Fluka [18].

As illustrated in the three following examples [19–21], many applications of hybrid materials in the field of optics exploit combinations of properties such as

Gómez-Romero: Organic-Inorganic Materials. Clément Sanchez

ISBN: 3-527-30484-3

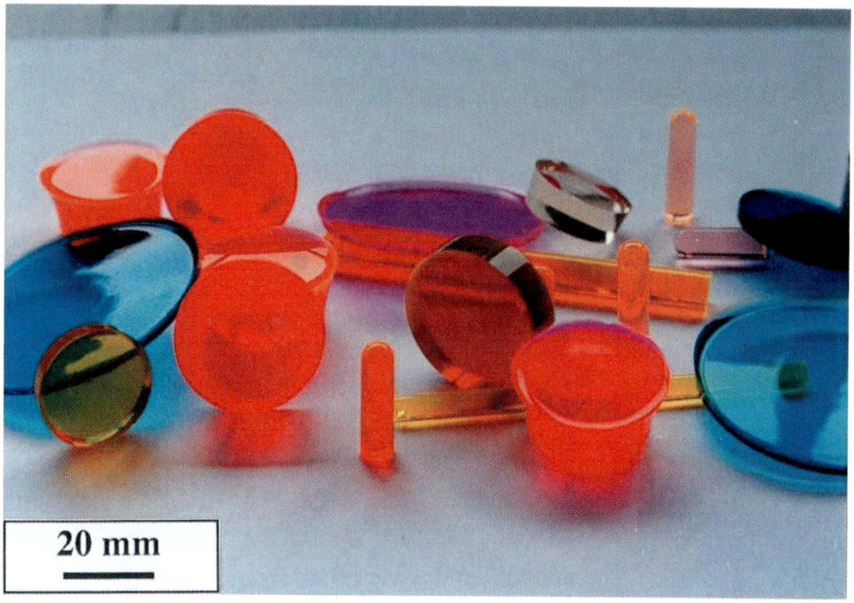

Fig. 5.1 Hybrid organic-inorganic materials containing organic chromophores prepared at the Laboratoire de Physique de la Matière Condensée (Ecole Polytechnique) [10]

transparency, good adhesion, barrier effect and corrosion protection, easy tuning of the refractive index, adjustable mechanical properties, and decorative properties.

- The megajoule class pulsed Nd: glass laser devoted to inertial confinement fusion research designed by the CEA/DAM requires 240 cavity end mirrors [19]. The approved laser design needs highly reflective (HR) substrates obtained by interference quarterwave stacks. Such HR properties can be obtained by coating large-area dielectric mirrors (about 50 cm in diameter, see Fig. 5.3) with silica and hybrid films made of ZrO_2-polyvinylpyrolidone (PVP) nanocomposites [19]. The low-refractive-index material is based on nanosized silica, while the high-refractive-index film is made from a hybrid system in which the nano-zirconia/PVP ratio has been optimised with respect to refractive index and laser damage threshold.
- Sol-gel derived hybrid coatings are also used in art conservation. One of the most striking examples concerns the long-term protection of the 14th century "Last Judgment Mosaic" situated above the gates of the St. Vitus cathedral in Prague castle [20]. The selected coating for treatment of the entire 13 m $\times$ 10 m mosaic is a multilayer system composed of a hybrid organic-inorganic functional layer made from organo alkoxysilanes and oxide nanoparticles, placed between the multicolored glass substrate and a fluoropolymer coating (see Fig. 5.4). This coating combination is a transparent efficient anti-corrosion barrier, much better than all the tested organic polymers [20].

Fig. 5.2 Glassware demonstrating the high quality optical appearance of crystal glass and other types of glasses coated with dye colored hybrid coatings (Courtesy of G. Schottner) [16, 17]

- Another kind of application which demonstrates simply the interest of using multifunctional hybrids concerns sol-gel derived hybrid organic-inorganic coatings developed by several glass packaging industries in Japan [21] and in Europe. These hybrid coatings (illustration in Fig. 5.5) [21] can not only give access to a wide variety of colors, thus enhancing the consumer appeal, but also allow an improvement of the mechanical properties of glass bottles. Moreover, these dye-colored bottles are easy to recycle as uncolored glass (these materials do not need color classification recycling), because, unlike conventional glass bottles, their coloration does not arise from transition metals that are very difficult to remove on re-melting.

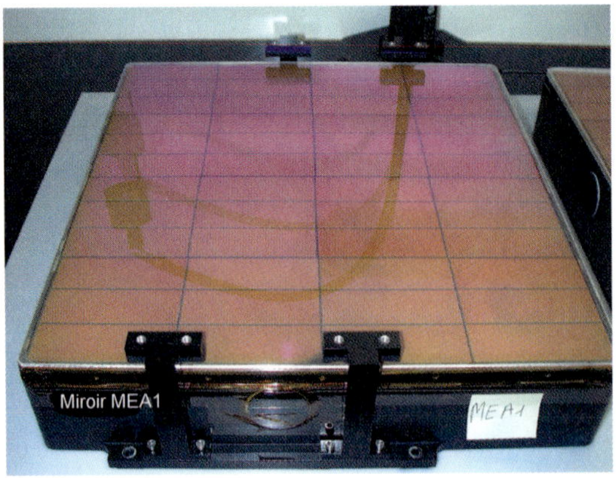

Fig. 5.3 Dielectric mirror coated with a stack of anti-reflective films. The combination between nano-silica and hybrid nanocomposite layers allows matching of the convenient reflection coefficient with a minimum of layers, a better pass-band and improved mechanical properties (courtesy of P. Belleville [19])

(a)

(b)

Fig. 5.4 The Last Judgment Mosaic in Prague is a large outdoor panel made from 1 million multicolored glass, or tesserae, embedded in mortar. This high potassium glass is chemically unstable and since the glass is exposed to harsh weather conditions (high SO_2 levels, rain, temperatures varying between −30 °C and +65 °C), the alkali reacts with the atmosphere and water to form salt deposits that blurs and corrodes the masterpiece. All organic polymers used in the previous protection attempts have failed to stop corrosion, because of their poor durability, poor adhesion to the glass, and large diffusion coefficient for SO_2 and water. After reviewing many possible choices the most efficient coating was made from a combination of a hybrid nanocomposite and a fuoropolymer. (a) Before, (b) after hybrid coating. © Getty Conservation Institute, photograph Jaroslav Zastoupil (courtesy of Eric Bescher [20])

Fig. 5.5 Easy to recycle colored glass bottles coated by hybrid organic-inorganic materials [21]

Hybrid nanocomposites present a paramount advantage for facilitating integration, miniaturisation and multifunctionalisation of devices, opening a land of opportunities for applications in the field known as "nanophotonics". As a consequence, there is widespread agreement in the scientific community that active optical applications of hybrid nanocomposites might present the most attractive field to realise applications for the 21st century. Indeed, the exploitation of active optical properties of photoactive coatings and systems is strongly emerging. For example, hybrid materials having an excellent laser efficiency and good photostability [22], very fast photochromic response [23], very high and stable second-order non-linear optical response [24], or acting as original pH sensors [25], electroluminescent diodes [26] or hybrid liquid crystals [27] have been reported in the past five years.

In this chapter, the most striking examples of functional hybrids exhibiting emission properties (solid-state dye lasers, rare-earth doped hybrids, electroluminescent devices), absorption properties (photochromic), nonlinear optical (NLO) properties (2nd order NLO, photochemical hole burning (PHB), photorefractivity), sensing properties, as well some examples of hybrid-based integrated optical devices are presented.

5.2
Hybrids with Emission Properties

5.2.1
Solid-State Dye-Laser Hybrid Materials

Laser dyes have been used for many years as a versatile source of tunable coherent optical radiation and are commonly used both in lasers and in optically pumped amplifiers. But most of the dye lasers are solvent based and require pumping of the dye solution through a resonator to slow down photodecomposition. Therefore, there has long been an interest in fabricating solid-state gain media containing organic laser dyes because it provides ease of use and replacement. Moreover, the encapsulation of the dye in a solid matrix is favorable for decreasing health and environmental hazards.

Avnir and co-workers [3, 28] first synthesized luminescent dye-doped sol-gel matrices and demonstrated the possibility of using them as solid-state dye lasers. However, it was only a few years later that the first laser emissions were reported [29–32] in sol-gel materials. The main reasons for this delay were the several drastic conditions required to prepare efficient solid-state organic dye lasers. The laser material must exhibit an excellent transparency without scattering and a well-polished surface (surface roughness < few nanometers). Other important parameters related to the laser emission are slope efficiency, output-energy pumping, lifetime and stability to aging. As a consequence of these several exigencies most of the studies in this area have been focused on the tailoring of the dye/matrix system and the demonstration of laser emission. Class II hybrid materials were quickly found to be well appropriate for constructing such solid-state dye lasers [8i, 33]. Indeed, the organic component could improve matrix characteristics such as hydrophobicity (to repel any residual polar solvent and thus protect the organic dye from chemical degradation), density (to decrease optical inhomogeneities) and mechanical resistance (necessary to stand the surface polishing). It was observed that hybrid laser materials exhibit substantially longer lifetimes and operate at much higher repetition rates than either inorganic xerogel systems or organic polymer hosts. These hybrid materials are able to provide a local environment in order to retain the strong fluorescent quantum yield of the dye and improve its photostability. Moreover, a high dye loading can be achieved with minimized dye aggregation [3, 34]. The development of a controlled hydrolysis process of hybrid precursors has allowed the synthesis of very dense and chemically stable matrices. The resulting materials exhibit a total absence of scattering losses all through the visible spectrum that clearly indicates the dramatic diminution of large pores. This not only affects the optical transparency of the laser material but also impacts its thermal conductivity and therefore the thermal problems that arise when dissipating heat internally through optical excitation.

Performance of solid-state dye laser materials is not only based on the properties of the host matrix but also on the choice of the organic dye. The dye must exhibit a good laser gain and be chemically, photochemically and thermally stable under the conditions of use of the laser.

Different xanthene, coumarin and perylene dyes have lased in a variety of matrices including pure inorganic, sol-gel composites and class II [6c] hybrid matrices [8i, 33]. The first significant laser characteristics were reported for rhodamine derivative-doped hybrid materials. However, the photochemical stability of this dye is far from being sufficient for use in solid-state dye lasers. For this purpose more stable and efficient dyes such as perylenes and pyrromethenes have been identified [22, 35–38].

The huge optimization potential of hybrid solid-state dye-laser materials is illustrated by the strong improvement of their performance over the past few years. Slope efficiency increased from a few percent to more than 80%. The maximum output energies increased simultaneously from a few μJ to a few mJ. At the same time, lifetimes progressed from about 10^4 pulses in the μJ range to more than 10^6 pulses in the mJ range. The development of a stable pyrromethene derivative-doped hybrid material was recently reported [22] with a lifetime of 5×10^5 pulses at 1 mJ with a repetition rate of 20 Hz. Moreover, it was observed that this lifetime could be increased to about 2×10^6 pulses by removing oxygen from the same sample prior to using it (see Fig. 5.6). The combination of a stable and efficient laser dye such as a pyrromethene derivative and the improved optical and thermal matrix properties allows the achievement of good laser emission in the 550–650 nm range.

It is now clear that the efficiency (with a conversion slope up to 86% and a threshold of 100 μJ), the tunability range (up to 60 nm) and the spatial beam quality are then similar to those obtained in equivalent liquid solutions. Experiments were also performed in a configuration for which the sample was continuously moved. These indicate that the laser system could be operated at about 1 mJ output energy at 20 Hz with a single sample for several months. These results are very promising and it may be forecast that effective tunable solid-state dye lasers will replace liquid dye laser systems in certain applications.

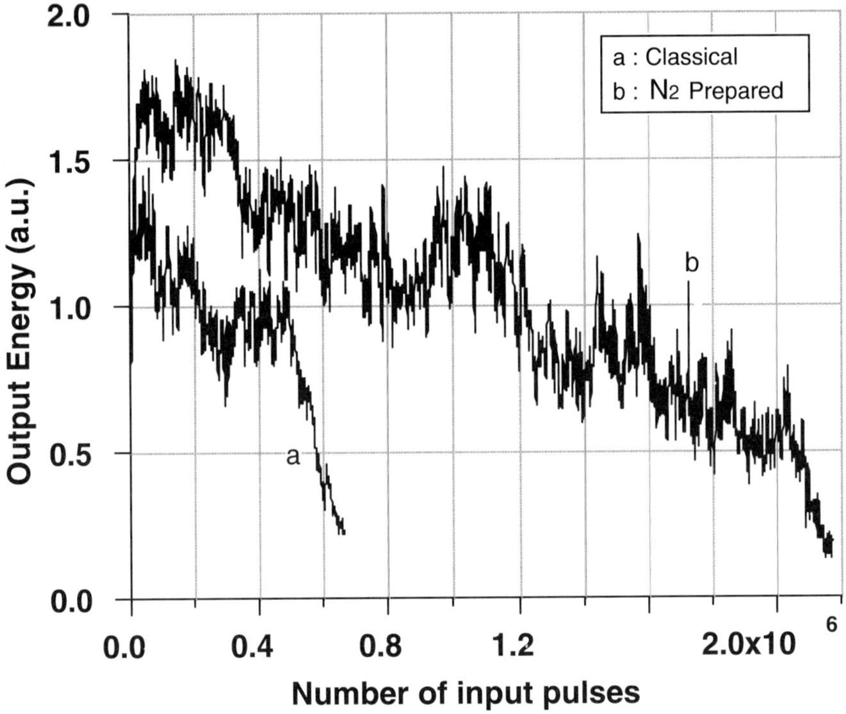

Fig. 5.6 Output energy evolution of the laser when using a pyrromethene 597 doped sample prepared (a) in "classical" conditions and (b) in oxygen-free conditions. The pumping conditions are 1 mJ/pulse at 20 Hz (from [22], reproduced with permission of the Optical Society of America)

Some studies [39–42] in the dye-doped sol-gel laser field were also devoted to the development of infrared (IR) dye-laser materials. The incorporation of infrared emitting dyes in sol-gel matrices not only extends the wavelength covered by sol-gel lasers but also relates to the possibility of developing a compact laser system that uses a diode laser as a pump. Most infrared emitting dyes belong to the cyanine family. The synthesis of IR dye-doped sol-gel materials has received relatively little attention because of the low stability of cyanine dyes. It was first shown that photostability of the IR dye, 1,1′,3,3,3′,3′-hexamethylindotricarbocyanine (HITC), was highly improved by its incorporation in a hydrophobic environment provided by an acid-free synthesized hybrid matrix. Dunn and co-workers [40] reported laser action of HITC and other IR dyes embedded in hybrid matrices and thus provided evidence that sol-gel lasers can be fabricated throughout the visible spectrum and into the near infrared. The future prospects for these materials are extremely promising, as hybrid solid-state dye lasers exhibit lower threshold powers for laser action (a few mJ), laser slope efficiency close to theoretical values, long laser lifetimes (10^6 pulses) and operation at high repetition rates (10 Hz).

Recent studies concern amplified spontaneous emission (ASE), a type of mirrorless lasing in which spontaneously emitted photons are amplified by stimulated emission as they travel through a gain medium [43]. These studies constitute a step towards device realization for integrated optical circuits. Yang and co-workers [44] showed the fabrication of mesostructured waveguides doped with the laser dye rhodamine 6G on low-refractive index mesoporous silica claddings. The waveguides exhibited ASE with a low pumping threshold of 10 kilowatts per square centimeter (instead of 200 kW cm^{-2} in silica sol-gel glasses for the same dye concentration) attributed to the mesostructure's ability to prevent aggregation of the dye molecules. ASE was also observed with a waveguide configuration based on a conjugated polymer blended with a zirconium – organosilicon glassy matrix [45].

5.2.2 Electroluminescent Hybrid Materials

Since the report of bright emission from thin layer molecular dye films, there is a growing interest in using organic electroluminescent devices in display applications because of their low operating voltage, large viewing angle and high efficiency and brightness [46]. A variety of materials have been proved to be useful for light emitting diodes (LEDs): low molecular weight molecules, conjugated polymers, main chain or side chain functional polymers and molecules dispersed in polymeric matrices. However, even if promising these materials present some drawbacks. The main difference between small molecules and polymers is in their manufacturing technology; the former are evaporated under vacuum whereas the latter are spin coated. Because of the low glass transition temperature (*Tg*), the small molecules tend to crystallize during the LED working (joule heating), while polymers, despite a good and low-cost processability, are in some cases difficult to synthesize. Furthermore, the processing of multilayer devices needed for a well-balanced double injection of the charge carriers using different polymers remains tricky because of dissolution. Sol-gel processes provide the opportunity to design new hybrid LEDs based on functionalized alkoxysilanes that are able to transport charges and emit light. Such grafted active units present some advantages over organic polymers and molecules. They allow:

- versatile material design and processing through spin coating techniques;
- efficient cross-linking through condensation reactions providing chemical stability and mechanical strength;
- the formation of insoluble films that favor the deposition of multilayer devices without swelling and leaching.

Hybrid organic-inorganic light emitting diodes (HLED) were developed for the first time by using new silylated precursors with hole transporting units prepared by modification of different active molecules (carbazole, oxadiazole, and tetraphenylphenylenediamine derivatives) (Fig. 5.7) [26, 47, 48]. Absorption and photoluminescence data show that the electronic structures of the chromophores are not significantly modified by the functionalization. The hole transporting hybrid layers generally exhibit about the same charge mobility as organic polymers having equivalent active units. The electric field dependence of mobility was described from the Poole-Frenkel formalism and the highest hole mobility was observed for the layer having tetraphenylphenylenediamine (Si-TPD in Fig. 5.7): 5.7×10^{-5} $cm^2\ V^{-1}\ s^{-1}$ at a field strength of $E = +5 \times 10^5$ V cm^{-1}. For the best carbazole compound (Si–KH in Fig. 5.7), the mobility is found to be about twentyfold lower at the same field [48].

In a preliminary experiment, the electroluminescent properties of sol-gel materials were demonstrated by manufacturing hybrid diodes emitting in the orange [26]. The emissive layer was prepared using a silylated precursor (SiDCM, Fig. 5.8) with a fluorescent molecule belonging to the DCM family: 4-dicyanomethylene-2-methyl-6-[p-(dimethyl-amino)styryl]-4H-pyran [49]. An oxadiazole derivative (PBD) was incorporated in a sol-gel layer as electron-transport guest molecule. Si-DCM and PBD molecules were incorporated or copolymerized with methyltriethoxysilane (MTEOS) as a crosslinking agent. The single-layer structure consisting of an emitting layer (Si-DCM doped with 50% PBD) shows light emission at about 15 V and a maximum luminance of 440 cd/cm^{-2} is obtained at 24 V.

The devices emitting in the green are formed from two hybrid thin layers (35–75 nm) prepared from silane precursors modified with hole transporting units and light emitting naphthalimide (NABUP, Fig. 5.9) based components, sandwiched between ITO and metallic electrodes. The maximum external quantum efficiency of the best diode using a LiF:Al cathode is about 1% at 100 cd m^{-2} and the luminance reaches 4000 cd m^{-2} at 27 V (Fig. 5.10) [47].

Electroluminescent silylated coumarine molecules with emission wavelengths that can be tailored by changing the substituent pattern have also been tested in HLED devices built with commercial hole and electron transport materials [50]. However, the efficiency of these HLEDs remained low.

All these first results on HLED are promising [26, 47, 48, 50], and further experiments are required to test these new materials as hole transporting layers in photorefractive and electroluminescent devices. It is important to point out that the final

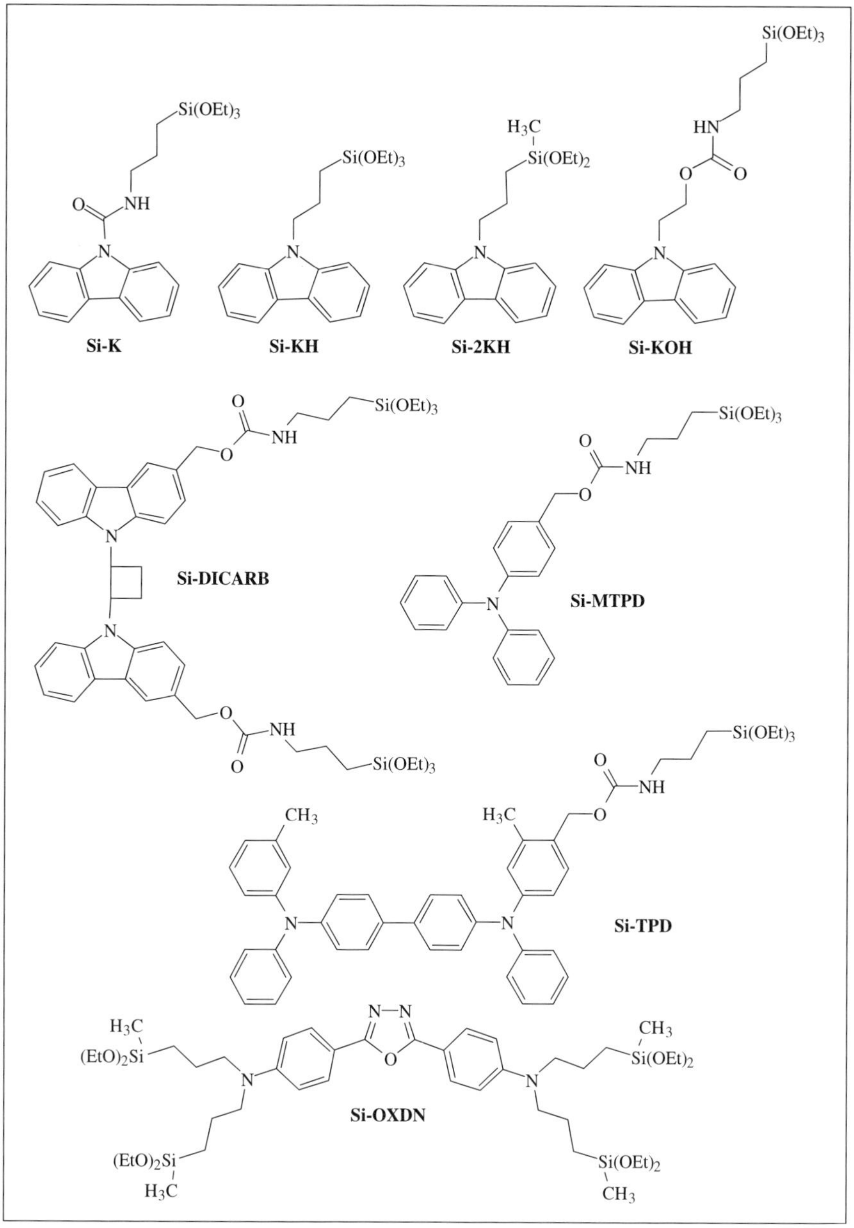

Fig. 5.7 Molecular structures of the silylated precursors with hole transporting units [48]

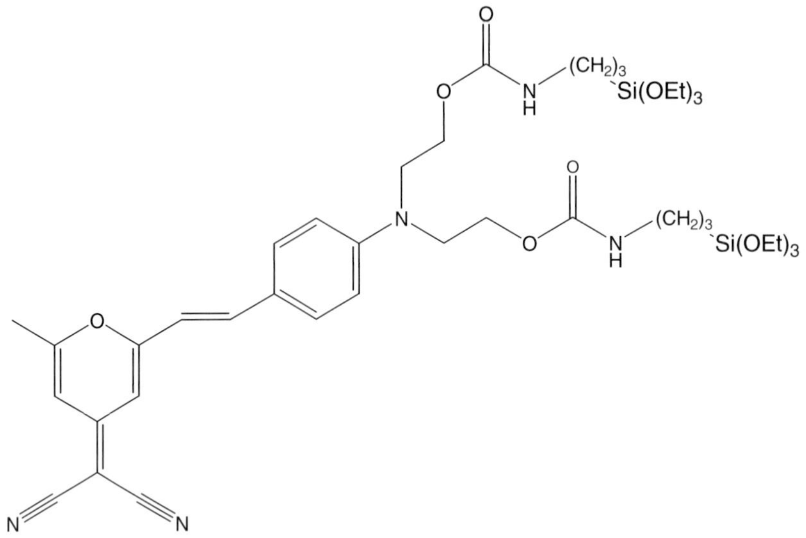

Fig. 5.8 Silylated precursor with light-emitting DCM (Si-DCM) used to prepare electroluminescent devices emitting in the orange [26]

efficiency of LEDs depends not only on the intrinsic physico-chemical quality and response of the different transporting carriers and light emitting layers, but also on a carefully controlled deposition process. This latter is, in many cases, of paramount importance for optimizing LED performance.

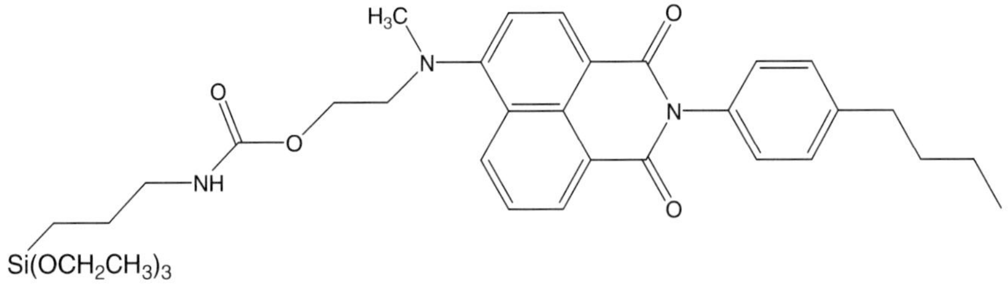

Fig. 5.9 Silylated precursor with light-emitting naphthalimide (NABUP) used to prepare electroluminescent devices emitting in the green [47]

5.2.3
Optical Properties of Lanthanide Doped Hybrid Materials

Due to their advantages, such as low temperature processing and shaping, higher sample homogeneity and purity, and the opportunity to prepare non-crystalline solids, sol-gel matrices have been investigated for some years as potential matrices for lanthanide luminescence [13]. The great interest of rare-earth ion (e.g. Nd^{3+},

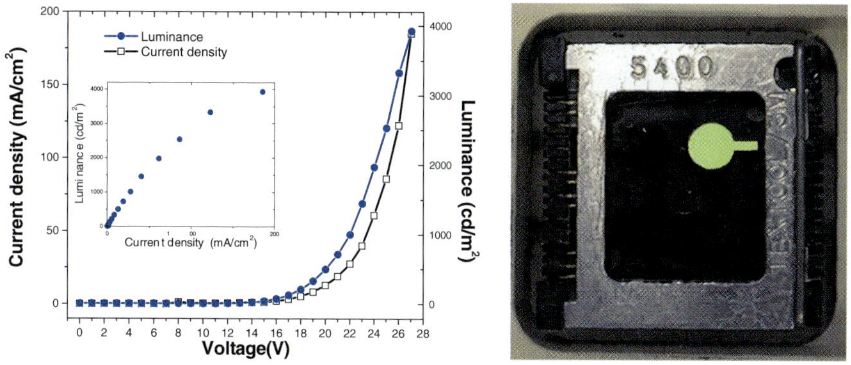

Fig. 5.10 Current density and luminance versus voltage for a bilayer device (PVK//Si-NABUP) emitting in the green with LiF/Al as a cathode material [47]

Eu^{3+}, Er^{3+}, Tr^{3+}, Ce^{3+}, Eu^{2+}) doped materials and devices in optics has been well known for a long time. Indeed, these materials offer a wide range of optical applications such as tunable lasers, luminescent displays and amplifiers for optical communication [51]. Lanthanide-doped hybrids are also promising materials for optical storage based photochemical hole-burning processes [51]. Moreover, Ce^{3+} doped glassy hosts present a lot of interest for applications in the scintillator field and in the search for a tunable laser in the UV and visible range [52].

The conventional processing of sol-gel derived matrices synthesized at room temperature, through the hydrolysis of tetravalent metal alkoxides $M(OR)_4$ (M = Si, Ti, Zr etc.) and inside which the lanthanide doping is simply performed by dissolving a rare-earth salt in the initial sol, usually yields matrices containing a large amount of hydroxyl groups and does not allow minimization of the segregation of the rare-earth ions. The presence of hydroxyl groups and the formation of lanthanide aggregates are both responsible for the decrease or even the quenching of the rare-earth luminescence, these effects being particularly drastic for Nd^{3+} ions. Therefore, the design of new sol-gel derived hybrid matrices with improved luminescent properties remains an important challenge.

In order to be able to achieve these objectives, main strategies consist in

1. The encapsulation of lanthanide ions within nanoparticles before embedding the doped nanoluminophore inside a hybrid matrix [53–56].
2. One-pot sol-gel synthesis performed through a careful tuning of sol-gel processing conditions (reactive hydrophobic precursors, hydrolysis ratio, catalysts, curing at moderate temperatures etc.) [57–69] to allow the minimization of the hydroxyl content. This one-pot strategy can be combined with the protection of the rare-earth ions via complexation or cryptation [70–87]. The luminescent ions can be sequestered through the simple embedding of rare-earth complexes within sol-gel hosts [70–81], or by using complexing ligands grafted to the sol-gel host through the condensation of trialkoxysilyl units [82–87].

3. Finally, the use of non-hydrolytic sol-gel processes permits the generation of hybrid matrices free from OH groups, but their main drawback is related to the difficulty of obtaining transparent monolithic pieces or films (i.e. powders are usually obtained) from such processes [88].

5.2.3.1 Encapsulation of Nano-Phosphors inside Hybrid Matrices

Er(III) oxalate nanoparticles (10–40 nm) dispersed in titania/(3-glycidoxypropyl)trimethoxysilane (GPTMS) hybrid materials were prepared by combining a microemulsion technique with the sol-gel process. A relatively strong room temperature green upconversion at 543 nm has been measured in the resulting hybrid nanocomposite films [53, 54]. Following a similar synthetic route, titania-hybrid composites containing neodymium oxide nanoparticles were prepared and exhibited a strong room temperature yellow to violet upconversion [55]. In addition to this violet emission, UV emission (372 nm) and blue emission (468 nm) were also observed in these hybrid matrices.

A luminescence quantum yield of 15% has been obtained from Eu or Nd doped YVO_4 nanoparticles. This quantum yield, lower than in bulk compounds, arises from residual crystalline defects and non-radiative relaxations from residual hydroxy surface groups. It can be increased up to 38% after annealing and deuteriation treatments [56]. After dispersion of these rare-earth doped YVO_4 nanoparticles in transparent hybrid matrices made from hydrolysis and condensation of vinyltriethoxysilane (VTES) and 3-(methacryloxypropyl)trimethoxysilane (MPTMS), efficient luminescent properties are observed [56]. These strategies open a promising area which has not been so far thoroughly explored. Such a kind of two-step process may allow synthesizing of well-defined and efficient nanophosphors, whose properties can be tuned in a second step during their incorporation inside transparent and easy-to-shape functional hybrid hosts.

5.2.3.2 One-pot Synthesis of Rare-Earth Doped Hybrid Matrices

Kozlova and co-workers [57] demonstrated fluorescence emission for Nd^{3+}, Sm^{3+}, Dy^{3+}, Er^{3+} and Tm^{3+} ions doped inside hybrid siloxane-oxide matrices. Re^{3+}-doped hybrid coatings were obtained via hydrolysis and co-condensation of diethoxymethylsilane (DEMS = $SiH(OEt)_2(CH_3)$) and a mixed metallic alkoxide synthesized by reacting zirconium tetrapropoxide ($Zr(OPr^n)_4$) and the rare-earth alkoxide. The use of a reactive hydridosilane precursor and the presence of methyl hydrophobic components on the siloxane backbone allowed elimination of residual solvent and reduction of the amount of hydroxyl groups in the final material. The zirconium alkoxide precursor was used to generate crosslinking zirconium oxo-polymers which are not photoreducible and can accommodate the high coordination demand of the rare earth. The rare-earth emission lifetime can be optimized by lowering the lanthanide concentration, and by tuning the sol-gel processing conditions and the thermal curing to minimize the amount of residual hydroxyl groups. As a result, broad emissions with relatively high quantum efficiencies (lifetimes between 5 μs and 160 μs for Nd^{3+}) were observed in these rare-earth ion-doped hybrid matrices [58, 59]. Optimization of such hybrid materials needs a better basic knowledge of the rare-earth localisation and environment. Eu^{3+} can be used as a probe because its emission properties are very sensitive to the rare-earth surroundings [60–62]. In particular the shift of the $^5D_0 \rightarrow ^7F_0$ transition and the ratio between the $^5D_0 \rightarrow ^7F_2$/$^5D_0 \rightarrow ^7F_1$ transitions are hypersensitive to the covalency, coordinence and symmetry of the Eu(III) first coordination shell. Such effects are particularly striking in Eu(III) doped siloxane-oxide matrices ($(CH_3)_2$–$(SiO)_n$–MOx), M = Ti, Zr, Al, Ta, Nb, Y, Sn, Ge etc.), the study of which should shed more light on the structural and chemical parameters allowing higher luminescent responses in rare-earth doped hybrid materials [61].

Optical amplification is a very important field of application. However, the existence of optical amplification at 1.55 μm wavelength in channel waveguides using erbium doping and a 980 nm laser pump involves matrices inside which the OH-group quantity should remain extremely low. Moreover, a large difference in refractive index between the guide and the cladding is necessary to induce a maximum confinement and thus increase the doping efficiency. This field was until recently limited to pure inorganic materials or semiconductors. Hybrid materials because of their versatility have recently allowed both requirements to be satisfied. For the first time, active erbium-doped organic-inorganic waveguides based on hydrophobic sol-gel precursors to reduce the OH content, and on a diacrylate monomer whose reactivity allows a high refractive index change under UV light exposure, have been made [63]. These results are quite encouraging and open many new possibilities in the field of amplifiers.

Transparent hybrid inorganic-organic monoliths doped with Ce (III) were prepared from co-condensation between $Si(OMe)_4$, $CH_3Si(OMe)_3$ and various functional trialcoxysilane, APTMS, GPTMS, TFTMS, CPTMS carrying aminopropyl, glycidyloxypropyl, trifluoropropyl, chloropropyl groups, respectively [64]. Different affinities of the silane functional group for Ce^{3+} ions were observed. Optical properties such as absorption spectra, emission spectra and fluorescence quantum yield were strongly affected by the Ce^{3+} environment. The emission intensities of the hybrid materials prepared from TFTM and CPTM were approximately 1000 times as much as those of materials prepared from APTMS and GPTMS. The fluorescence quantum yield was highest (11%) for the hybrid material prepared from the TFTMS precursor [64]. Such inorganic hydrogeneous materials can be used as neutron detectors. Of course due to their low density these materials could not be used as high-energy neutron detectors. In comparison to organic scintillators, they present higher hydrogen density and very high scintillation efficiency.

The versatility of sol-gel chemistry and the large choice of precursors allow the *in situ* generation of lower valence states such as Eu (II) or Ce (III) from Eu(III) and Ce(IV) salts respectively, within hybrid hosts [65–67]. Both undoped and Eu(II) or Ce(III) doped hybrid organic-inorganic matrices were prepared at room temperature from hydrolysis and co-condensation of organohydridosilanes $SiH(CH_3)(OEt)_2$, $SiH(OEt)_3$, alkylalkoxysilanes $Si(CH_3)(OEt)_3$ and zirconium propoxide in the presence of $EuCl_3$ or $Ce(NO_3)_4$. The cleavage of the Si–H bonds, catalyzed by transition metal alkoxide, is used to reduce *in situ* at room temperature the rare-earth cations. Depending on the chemical strategy, the resulting hybrid materials have been processed as transparent bulks or coatings, which exhibit a good transparency in the UV-visible domain (cut-off at 250 nm). Both the undoped and the Eu(II) or Ce(III) doped matrices exhibit a strong blue emission. The natures of the different emitting species have been clearly assigned by their very different kinetics of fluorescence ($\tau = 50$ ns for Ce^{3+}, $\tau < 1 \mu s$ for Eu^{2+} and $\tau = 4\ \mu s$ for photoinduced defects on the undoped matrices) and through excitation experiments [67]. As previously mentioned, defects in the silica component hybrid matrices can also be a source of luminescence which can be tuned by the presence of inorganic chromophores [68, 69]. Indeed, silica-polyethylene oxide or silica-polypropylene oxide hybrids exhibit an intense room temperature luminescence arising from electron-hole recombination on delocalized states provided by the silica nanophase. These hybrids, doped with divalent cations or with rare-earth ions, have enhanced emission properties because the cations solubilising close to the silica clusters decrease surface defects and the ensuing quenching mechanism [68, 69].

Direct inclusion of chelated or crypted lanthanides or lanthanide porphyrins inside sol-gel derived matrices, or *in situ* formation of the rare-earth complexes, leads to the formation of hybrid materials with generally improved luminescent properties [70–81]. For example, $[Eu(phen)_2]Cl_3$ and $[Tb(bpy)_2]Cl_3$ doped SiO_2–ZrO_2 or SiO_2–Ta_2O_5 sol-gel derived matrices (9Si/1Zr or Ta) with lanthanide concentration up to 20 mol% have been achieved. The resulting transparent hybrid bulks yield maximum emission intensities close (75%) to those reported for $Y(P,V)O_4$:Eu and $LaPO_4$:Ce phosphors [76].

In situ synthesis techniques have also been used for the first time to synthesize coordination compounds of europium beta-diketonates (2-thenoylfluoroacetonate, benzoyltrifluoroacetonate) in transparent organically modified silicates (made through acid co-hydrolysis of GPTMS and TEOS). These hybrid materials exhibit good monochromaticity and enhanced red-light emission under ultraviolet excitation, and the chelates reveal better photostability [77].

Polyhydroxyethylmethacylate-silica hybrid nanocomposites doped with Eu(trifluoroethoxy-acetone phenathrolin)$_3$ or Tb(salicylate)$_3$ complexes also exhibit high fluorescence intensities. Quantum efficiencies as high as 20% have been measured for Tb^{3+} emission under 244 nm excitation in hybrid materials made from Tb-benzoate complexes embedded in a silica gel [80].

Another interesting possibility consists in the complexation of the lanthanides by ligands that are covalently anchored to the hybrid networks [82–87]. Complexation of rare-earth ions (Eu^{3+}, Gd^{3+}, Tb^{3+} etc.) can be performed by using tri-alkoxysilyl functionalised organic groups carrying phosphine oxide [86] or bipyridine ligands [87] or carrying organic chromophores derived from dipicolinic acid or melinic acid (see Fig. 5.11) [82–84]. Condensation of these tri-alkoxysilyl functionalized chromophores in the presence of lanthanide salts leads to the formation of nanostructured hybrid organic-inorganic materials consisting of efficiently sequestered rare-earth species linked to pendant organic chromophores grafted to the inorganic host. The resulting hybrid can be shaped as monoliths or as transparent crack-free films giving rise to strong red or green emission, even at low lanthanide ion concentration [82–84]. Variations of the ligand structure lead to different coordinating properties and to variable absorption edges. As a result, the absorption efficiency or the ability of the chelate to transfer the absorbed energy to the lanthanide ion, and thus the emission quantum yield, can be modulated.

5.2.3.3 Rare-earth Doped Hybrids made via Non-hydrolytic Processes [88, 89]

Yuh and co-workers [88] used a non-hydrolytic process to prepare erbium doped organically modified materials with materials of extremely low hydroxyl content. The presence of methyl groups in erbium-doped methylsilsesquioxane was found to be beneficial to rare-earth emission because it gave a hydrophobic character to the matrix. Unfortunately the presence of the C–H groups was detrimental because it quenched some transitions of Er^{3+}. The use of organofluorosiloxane could be a way to synthesize more efficient luminescent hybrid materials free from both O–H and C–H groups.

5.2.3.4 Energy Transfer Processes between Lanthanides and Organic Dyes

The sol-gel process provides also the unique opportunity to introduce in the same solid network both organic and inorganic chromophores [90–92]. Moreover, the location of the organic dyes can *a priori* be tuned through dye-matrix interactions (covalent bonds, hydrogen bonds, ionic and van der Waals bonds). Only a few studies report the simultaneous incorporation of rare-earth ion and organic dyes in sol-gel derived hybrid hosts [58, 90–92]. Energy transfer between Nd^{3+} and rhodamine 6G dyes incorporated inside hybrid siloxane oxide coatings was demonstrated, opening a land of possibilities for making new optical devices.

Up to now, only a little work has been carried out in the area of hybrid matrices doped or co-doped with rare-earth cations. However, these first results show the high potential of room-temperature-processed hybrid materials. Efficient emission from lanthanides in the trivalent or divalent state, photochemical hole burning, efficient energy transfer between organic dyes and rare earth are among the properties that will open routes for the development of new optical materials in the future.

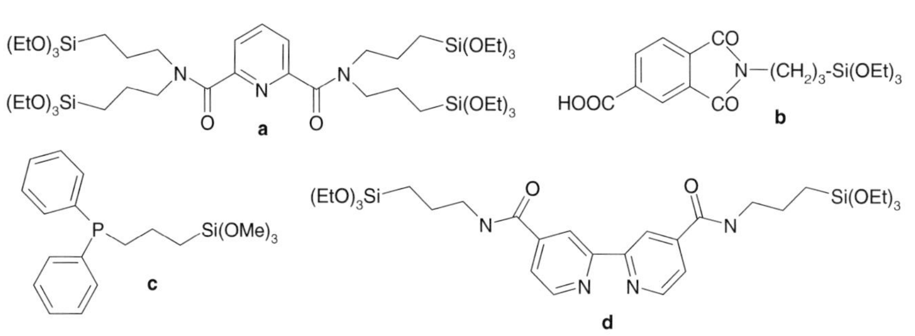

Fig. 5.11 Molecular structures of tri-alkoxysilyl functionalised organic groups able to complex rare-earth ions: (a) from [84], (b) from [85], (c) from [86] and (d) from [87]

5.3
Hybrid with Absorption Properties : Photochromic Hybrid Materials

Photochromic materials can be used for the design of optical switches, optical data storage devices, energy-conserving coatings, eye-protection glasses and privacy shields [93–95]. Photochromic materials and systems have several important uses depending on the rates of the optical transformations [93–95]. For example, very slow transformations are useful for optical data storage media, while fast transformations are required for optical switches. Nowadays, two main families of organic dyes have been coupled with solid host matrices to produce photochromic hybrid materials, namely (diarylethene or dithienylethene) and (spirooxazines or spiropyrans). Much less work has been devoted to the study of photochromic hybrids based on diazo dyes. The different kinds of photochromic hybrid materials will be successively presented.

5.3.1
Photochromic Hybrids for Optical Data Storage

Photochromic compounds such as diarylethene (DE) (see Fig. 5.12a), dithienylethene (DT) (see Fig. 5.12c) and furylfulgide (FF) (see Fig. 5.12b) derivatives [96–99] exhibit an improvement in photofatigue resistance and a good color stability. Photochromism of DE, DT and FF derivatives is based on a reversible isomerization, which does not exhibit thermal fading at room temperature. Therefore, a photochemically reversible change is observed upon alternate UV/visible irradiation. As these photochromic compounds are thermally stable and resistant to photochemical side reactions, they can be mainly used in optical information storage. For the first time, photochromic quantum yields of DE and FF embedded in hybrid matrices were reported [97]. Unlike DE doped silica gels, when DE and FF are embedded in hybrid matrices they both exhibit good quantum yields close to those measured in solution.

The grafting of DT dyes to the matrix allowed a strong improvement of the chromophore loading. Moreover, Boilot and co-workers processed from the trialcoxy-funtionalized DT precursors (see Fig. 5.12c) thin films a few μm thick which exhibit very interesting optical properties [96]. These class II [6c] photochromic hybrid materials exhibit large refractive index changes under UV-radiation ($\Delta n \sim 4 \times 10^{-2}$ at 785 nm, $\Delta n = 0.12$ at 1300 nm) [98]. Such photochromic properties have been used for writing channel waveguides (Fig. 5.13), gratings and others patterns utilized in integrated optics [99].

In the past years photochromic materials were proposed for application to optical data storage [100, 101]. But these materials suffer from a major deficiency of destructive readout, since illumination in the absorption band stimulates a photochemical reaction and erases the recorded information. As previously mentioned for DT hybrid gels, the photoinduced electrocyclization of the molecules produces both an important modification of the absorption spectrum and a concomitant change of the refractive index in the transparency region of the photochromic mate-

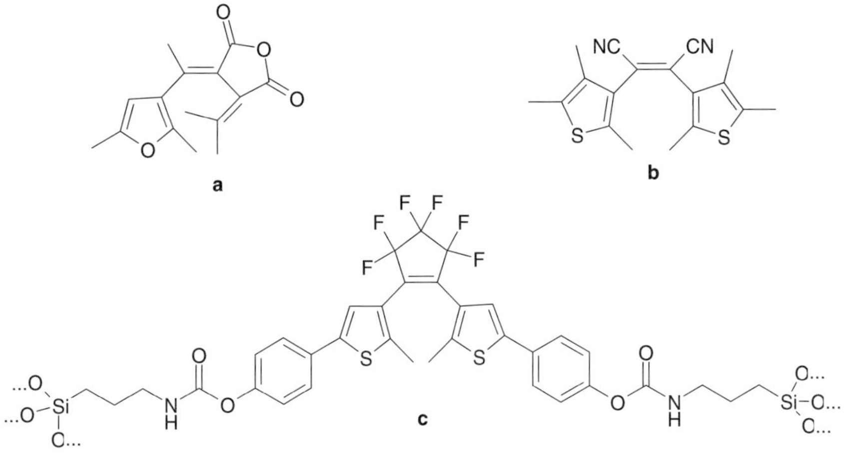

Fig. 5.12 Molecular structures of photochromic DE (a), FF (b) and DT functionalized alkoxysilane (c) dyes

rials. This allows the use of a wavelength out of the absorption bands for the non destructive readout. Moreover, DT in hybrid gels exhibits additional characteristics including thermally irreversible photochemical transition, high optical anisotropy and fixed orientation for the individual DT molecules due to their covalent grafting of molecules to the gel matrix. These properties ensure the stability (or remanence) of the material's optical properties in the absence of illumination in the absorption bands. Therefore, the photoinduced discoloration of DT sol-gel films (previously colored by UV light) with linearly polarized visible light induces a macroscopic optical anisotropy which results in a linear dichroism in the visible absorption band and in an important birefringence in the near-infrared transparency region which remains constant when the discoloring light is turned off.

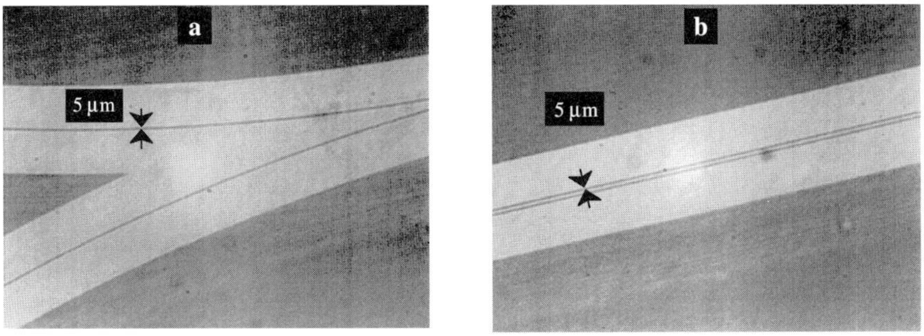

Fig. 5.13 Details of a directional coupler inscribed on DT gels after UV irradiation through a photomask (from [99], reproduced with permission of the American Chemical Society)

In these conditions, the following original writing-reading process was proposed in DT grafted hybrid gels [102]:

1. initialization by UV photocoloration;
2. writing the information by selective discoloration in the absorption band using polarized visible light;
3. reading the information in the infrared transparency range by measurement of the polarization change resulting from the photoinduced birefringence;
4. erasing the information by UV photocoloration.

For the reading process, the birefringence value was compatible with the reading speed of commercial systems like magneto-optical devices. Nevertheless, the writing energy was still too high; a speed of 1 MHz requires an incident power density of 12.5 mW μm^{-2}. This energy could be significantly reduced in the future by using photochromic molecules having larger photochemical quantum yields compared with that of the DT molecule used (0.4%).

Because during the course of isomerisation diazo dyes (for example Disperse Red 1 derivatives) undergo a large structural change which affects the medium around the molecules, new applications based on the photochromic behavior of azo dyes have been proposed in the form of laser-induced surface gratings [103]. Hybrids made from silica-based materials containing grafted azo-chromophores allowed the efficient inscribing of stable surface relief gratings having a high modulation depth on the hybrid films. High diffraction efficiencies of the gratings (30%) were observed when the hybrid films were exposed to *p*-polarized writing beams [98].

The possibility of patterning thin azobenzene photochromic hybrid films at nanometric resolution has recently been demonstrated by Landraud and co-workers using aperture near-field optical techniques [105]. The sample containing functionalized azobenzene species was locally illuminated through the aperture of a metallized tapered optical fiber. The surface topography imaged by *in situ* shear-force microscopy revealed the formation of nanometric surface relief whose lateral size was defined by the tip aperture diameter and whose protrusion height was direct-

Fig. 5.14 SEM photograph of the surface profile of sol-gel films after grating inscription [104]

ly proportional to the irradiation dose. This is due to the matter migration induced by repeated photoisomerization cycles of the azobenzene derivatives grafted to the gel matrix. As the illumination is strictly confined under the tip, compact arrays of bumps with nanometric lateral size can be inscribed without neighbor erasing effects (Fig. 5.15). This provides potential applications to optical nanolithography and high density optical data storage with a capacity of at least 1 Gbyte cm^{-2}.

5.3.2 Photochromic Hybrids for Fast Optical Switches

Photochromic molecules such as spirooxazines (SO) and spiropyrans (SP) (see Fig. 5.16), which have been embedded into sol-gel glasses, undergo large structural changes after absorption of a photon. The rates both of light-driven structural changes and of the back reactions are sensitive to the environment in the matrix. Pioneering work in this field by Levy and co-workers [12] showed the important role played by the dye-matrix interactions on the photochromic response of spiropyrans and the advantages of class I hybrids [6c] for photochromic devices.

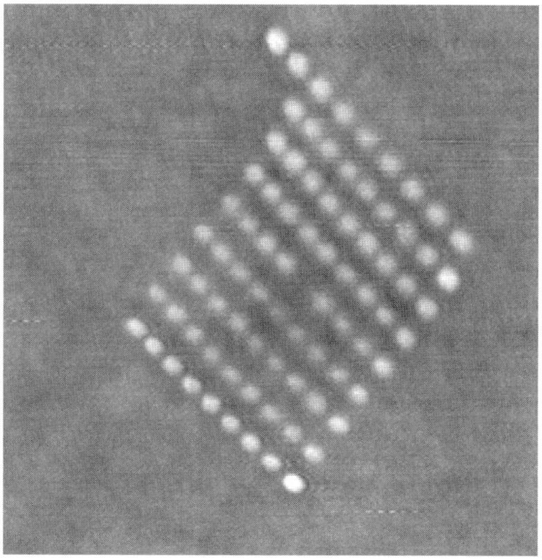

Fig. 5.15 Array of bumps inscribed in near-field on 20 nm-thick azo-hybrid film. The lateral size of each bump is 55 nm. The mean height of the bump is 5 nm (adapted from [105])

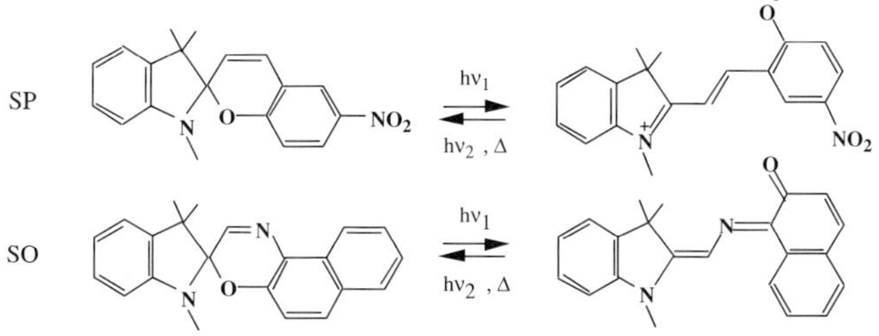

Fig. 5.16 Molecular structures of spiropyran (SP) and spirooxazine (SO)

SP and SO dyes retained their photochromic activity even at the dry stage when incorporated in a hybrid matrix. The advantages of hybrid systems have mainly been related to a high photochemical stability, the flexibility of the host matrix that allows molecular rearrangements of dyes, and the control of the polarity [12b, 106]. The main results concerning the photochromic properties of SP or SO doped sol-gel processed matrices are reported in Table 5.1.

For hybrid organic-inorganic matrices containing different chemical environments a competition between direct and reverse photochromisms can be observed [12, 23, 107]. Several authors introduced the hydrophobic/hydrophilic balance (HHB) of the hybrid matrix as an important factor which controls the competition between direct and reverse photochromism [12, 23, 107]. Usually spiropyran- or spirooxazine-doped sol-gel matrices or even spirooxazine-doped polymeric matrices exhibit slow thermal fading (see Table 5.1) because of steric constraints and/or the stabilization of the polar open form of the molecules through dye/matrix interactions. Hybrids synthesized from poly(N,N-dimethylacrylamide) having spiropyran on the side chain and cross-linked by silica produced through the hydrolysis of TMOS exhibit direct photochromism upon irradiation [118]. Their photoisomerisation behavior indicates a preferred solvation of the spiropyran moieties by the organic polymer host rather than by the silica component [118].

As far as photochromic devices are concerned, tuning between strong and fast photochromic coloration (high variation of the optical density) and a very fast thermal fading is needed. This can be performed through a careful adjustment of the interface experienced by the dye. The work of Schaudel and co-workers [23] evidenced that a very fast thermal bleaching (<0.5 s) can be observed in soft and strongly hydrophobic matrices, fully free from OH groups, synthesized from co-condensation between $MeHSi(OR)_2$ and $HSi(OR)_3$ precursors (see Fig. 5.17). The SO dye concentration can be increased by two orders of magnitude without affecting the kinetics of the photochromic effect by using precursors carrying aromatic groups such as $(phenyl)HSi(OR)_2$ [116]. Mesostructured hybrid materials made from a silica source and amphiphilic block copolymer templates (Pluronic®) are also excellent hosts for photochromic dyes. Inside these hybrid materials, SP and SO exhibit direct photochromism and quite fast kinetics [114]. Recently, photochromic nanocomposite coatings based on an epoxysilane (GPTS) network former, and different bisepoxides (cyclohexanedimethanoldiglycidyl (CHMG-nonpolar) or poly(phenylglycidylether)co-formaldehyde (PCF-polar)) as spacers, an organic amine (isophoronediamine) as thermal cross-linker and spirooxazines as photochromic dyes, have been developed [115]. In such composites the dye-spacer interactions have a noticeable influence on the switching kinetic behavior. Indeed, for these hybrid materials, the half-darkening and half-fading times of the SO dyes depend on the nature of the spacers. The nonpolar one (CHMG) did not influence the switching kinetics of the SO dyes significantly and fast switching times of 2–4 s were obtained, while the polar spacer (PCF) showed interactions with the SO dye which led to an increase of the switching time up to 25 s [115]. This effect, which increases with increasing spacer content and with increasing the polar character of the dye molecules, may be used to tune the photochromic behavior of the spyrooxazines for different kind of applications [115].

Tab. 5.1 Photochromism (D = direct, R = reverse) and bleaching constants k of spiropyrans and spirooxazines in selected matrices and solvents

Chromophores	Matrix	Effect	Characteristics k (s^{-1})	$t_{0.5}$ (s)	References
Spiropyran	SiO_2	D → R	D: 6.7×10^{-5} R: 1.7×10^{-5}		12a
	$EtSiO_{1.5}$	D	1.7×10^{-4}		12b
	SiO_2/Me_2SiO	D → R	D: 6.7×10^{-5} R: 5×10^{-5}		
	Me_2SiO/ZrO_2	D → R			23
	$MeHSiO/HsiO_{1.5}$	D	5×10^{-3}		23
	PMMA	D	$k_1 = 7 \times 10^{-4}$ $k_2 = 10^{-4}$		108
	Ethanol	D	3.7×10^{-4}		109
	SiO_2	D/R		2.3×10^5	110
	$MeSiO_{1.5}$	D			
	SiO_2 Cationic SP	D/R	D: 10_{+2} R: $4 \times 10^{+5}$		
Spirooxazine	$MeSiO_{1.5}$	D	$k_1 = 1.15 \times 10^{-2}$ $k_2 = 1.4 \times 10^{-3}$	2	107 112
	$RSiO_{1.5}/EtSiO_{1.5}/Et\ SiO_{1.5}$				
	$MeSiO_{1.5}/RsiO_{1.5}/R'SiO_{1.5}/$ SiO_2/Me_2SiO	D		2	113
	Me_2SiO/ZrO_2	D → R	$k_1 = 3.1 \times 10^{-2}$ $k_2 = 2 \times 10^{-3}$		23
	$MeHSiO/HsiO_{1.5}$	D	0.2		23
	PMMA	D	$k_1 = 4 \times 10^{-2}$ $k_2 = 4 \times 10^{-3}$		108
	Ethanol	D	0.2		95
	SiO_2	D/R	1.6×10^{-3}		110
	$MeSiO_{1.5}$	D	1.2×10^{-2}		
	Meso SiO_2/ Pluronic	D	0.15		114
	GPTS/PCT	D		3	115
	GPTS/CHMG	D		23	115
	$PhenylHSiO/HSiO_{1.5}$	D		1	116
	$[(BuSn)_{12}O_{14}(HO)_6]$ –OOC–PEG–COO–	D		1	117

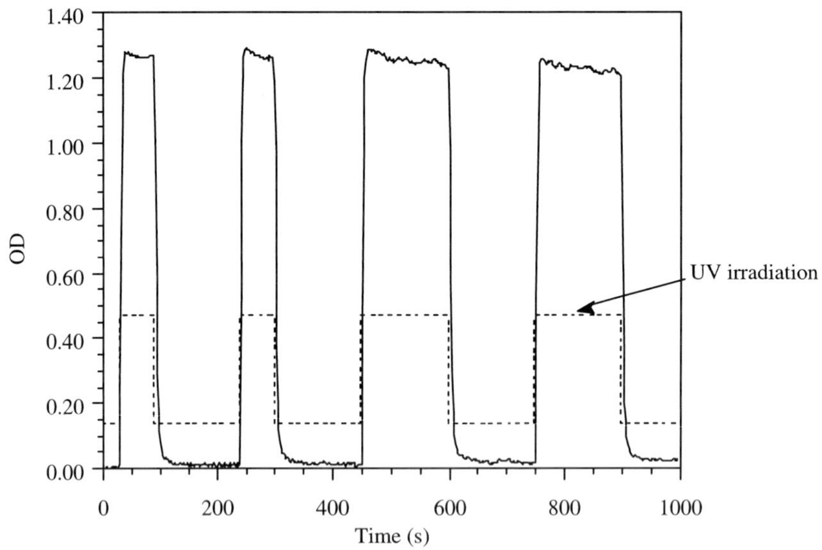

Fig. 5.17 Optical density versus time of SO embedded in strongly hydrophobic matrices, made from $MeHSi(OR)_2$ and $HSi(OR)_3$ [23])

In order to increase the dye content without aggregation, the possibility of grafting the photochromic dyes on the silica or siloxane based backbone has also been investigated [119, 120]. Strong interactions between the dye and the host matrix reduce the dye mobility and thus the thermal decoloration rate. In such systems, the spacer length between the dye and the matrix network [121] could optimize the rate of photochromic reactions.

One of the most important properties of photochromic compounds in long-term technological applications is their stability in the final product [122]. Schmidt and co-workers [123] have first examined the stability of spiropyrans, spirooxazine dyes embedded in hybrid matrices, with respect to their photochromic properties and matrix effects. They showed that the combination of two organic additives, methylimidazole and perfluoroalkylsiloxane, improved the photofatigue resistance.

5.3.3
Non-Siloxane-Based Hosts for the Design of New Photochromic Hybrid Materials

New photochromic hybrid matrices made with spyrooxazine or cationic spyropyran dyes embedded in non-siloxane based hosts have been reported very recently [117, 124].

Taking into account that the two main factors affecting the kinetics of thermal fading of SO are steric constraints and the stabilization of the polar open form of SO, an original non-silica based host has recently been tailor-made [117]. The hybrid matrix host was made of telechelic $[\alpha\text{-}\omega(COO)PEG]^{2-}$ chains cross-linked or connected by $[(BuSn)_{12}O_{14}(HO)_6]^{2+}$ tin-oxo clusters (see Fig. 5.18). SO doped

($[(BuSn)_{12}O_{14}(HO)_6]$(OOC-PEG-COO)) hybrid films exhibit a direct and intense photochromism with a fast thermal reponse, close to those reported for the best photochromic hybrid matrices [23, 114, 120]. Two combined effects can account for the observed properties. The soft PEG component allows enough flexibility to avoid mechanical and steric constraints of the dye and also permits a high dye solubility. Moreover the blurring of the polar poles of the cluster (carrying the Sn-OH bonds) by the carboxylate anions of the PEG units prevents accessibility of the polar open form of the SO dyes, thus avoiding its stabilisation [11].

Interplay between magnetic properties and photochromic behavior has also recently been demonstrated in new multifunctional hybrids [124]. N-methylated pyridospiropyran SP-R+ has been inserted by using exchange reactions inside lamellar $MnPS_3$ to yield $Mn_{1-x}PS_3(SP\text{-}R)_{2x}$ hybrid intercalates. The thin transparent films processed from these hybrid materials exhibit photochromism and acquire a spontaneous magnetization below 40 K. Irradiation of the hybrid film does not modify the critical temperature but affects considerably the hysteresis loop, increasing remanence and coercivity [124].

Hybridization of clay minerals with dicationic diarylethene molecules has recently been reported to produce highly functional photochromic materials. Transparent thin films were prepared by dispersing fine clay particles intercalated by organic compounds within a gelatin matrix [125].

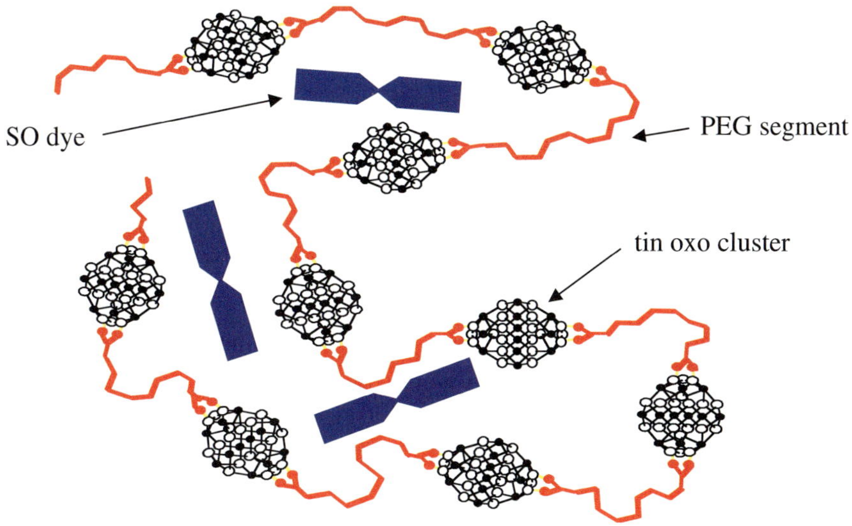

Fig. 5.18 Cartoon of the hybrid matrix host made of telechelic $[\alpha\text{-}\omega\ (COO)PEG]^{2-}$ chains cross-linked or connected by $[(BuSn)_{12}O_{14}(HO)_6]^{2+}$ tin-oxo clusters and doped with SO dye (adapted from [117])

5.4
Nonlinear Optics

5.4.1
Second-Order Nonlinear Optics in Hybrid Materials

Quadratic nonlinear optics (NLO) is a domain of "sol-gel optics" where the optical response and its temporal stability are strongly dependent on the nature of the bonding between the organic dye and the matrix. The random orientation of organic dyes in sol-gel matrices rules out second harmonic generation (SHG). Second-order nonlinearities can only be observed after poling in order to orient the organic dyes within the sol-gel matrix. Sol-gel processed hybrid materials being usually amorphous, they do not fulfill easily the phase-matching conditions required for frequency doubling. As a consequence, most targeted applications are related to the electro-optic effect (electro-optic modulator, photorefractive materials etc.).

The first SHG studies concerning hybrid materials prepared through sol-gel processing have started by using systems with embedded NLO dyes, because such systems can be easily synthesized [126]. However, guest/host (G/H) hybrid systems with second-order NLO properties have rapidly shown the weakness observed for G/H polymeric systems [127–129]: fast chromophore reorientation leading to the relaxation of poling-induced order, and low chromophore concentration in the composite due to poor solubility of the chromophore in the host matrix.

The achievement of these materials is related to the quadratic susceptibility of the dye and the thermal stability of the polymeric hybrid matrix. Therefore, hybrid materials built from many matrix-NLO chromophore couples started to be optimized in terms of intrinsic NLO response, matrix rigidity and dye-matrix interactions. Jeng and co-workers [128] have shown that some of the processing problems associated with *in situ* sol-gel approaches can be improved by using partially polymerized, soluble pre-polymer precursors known as spin-on glasses. The vitrification of the host matrix during the poling process allowed optimizing the orientation of dyes and the stability of the poling-induced anisotropy. However, the problem of low chromophore concentration was not yet resolved. Hybrid organic-inorganic materials inside which the NLO dyes are grafted to the inorganic network allow an easier optimization of the dye concentration and a better control of chemical and processing parameters. As a consequence, in the past years most of the sol-gel work performed on hybrid materials with quadratic NLO properties was devoted to dye-grafted hybrids. The grafting of the dyes to sol-gel matrices has been performed by using trialcoxy terminated organic spacers connected to the NLO chromophore.

The first hybrid side-chain systems (S–C) [130–132] have shown that covalent coupling of the NLO chromophore to the lattice could improve chromophore loading without any crystallization effects. The first hybrid materials which were siloxane-silica nanocomposites with nitroaniline type chromophores showed better NLO efficiency, but still too weak ($d_{33} = 11$ pm V^{-1}) for electro-optics applications. Several strategies have been used to improve the NLO response of the dye-grafted hybrid coatings: (i) the intrinsic NLO response of the dye can be increased by using chromophores such as N-(4-nitrophenyl)-L-prolinol (NPP) (see Fig. 5.19a) or disperse red one (DR1) (see Fig. 5.19b) derivatives which exhibit higher optical nonlinearities than nitroaniline ones; (ii) increasing the matrix rigidity can control the chromophore relaxation. The modification of the binary composition (siloxane-crosslinker), the nature of the $M(OR)_4$ crosslinking alkoxide ($SiR'_x(OR)_{4-x}/M(OR)_4$: R′ = any NLO chromophore; M = Zr, Si, Ti etc.), and the processing of these hybrid materials in the presence of polymers with well-known mechanical properties such as polymethylmethacrylates or polyimides, were the most commonly used strategies to minimize dye relaxation.

Tripathy and co-workers [133] have investigated such kinds of composites by curing polyamic acid precursors in the presence of functionalized trimethoxysilanes. A d_{33} value of 13.7 pm V^{-1} was measured for the cured film. Only 27% reduction in optical nonlinearity was observed after 168 h at 120 °C. No signal loss was observed at ambient temperatures over the same period of time. Using these hybrid side-chain systems, an interesting result has been obtained by D. Rhiel and co-workers [134] (d_{33} = 55 pm V^{-1} at λ = 1.064 μm) for an alkoxysilane-silica system functionalized with DR1. No NLO signal loss was observed at ambient temperature one week after poling. The good temporal NLO response stability of this system seems to be due to the strategy of synthesis. After the hydrolysis of precursors with excess of water in an acidic medium, a base was added to favor condensation reactions, thus improving crosslinking of the polymeric inorganic network.

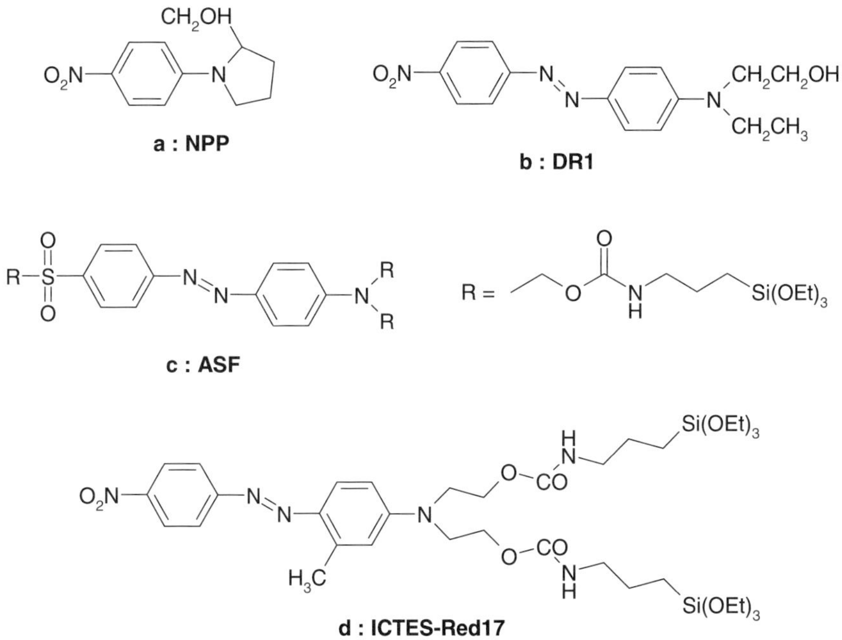

Fig. 5.19 Molecular structures of NLO dyes functionalized with trialkoxysilane groups

Second-order NLO responses were improved by increasing the rigidity of the amorphous sol-gel matrix (*i.e.* the glass transition temperature T_g) and the number of covalent connections between each dye and the metal oxo network. It has been shown that the covalent incorporation of a trialkoxysilane functionalized diazobenzene derivative, ASF (see Fig. 5.19c), as a main-chain (M-C) constituent of a sol-gel matrix leads to a large second-harmonic coefficient (27 pm V^{-1} at 1.32 μm) [135, 136], and a high thermal stability at 100 °C, close to that of polyimide/sol-gel interpenetrating networks [137].

The grafting of the NLO chromophores as side-chain or main-chain constituents has been performed by numerous research groups [126a, 129–132, 134–138]. The resulting high second-harmonic coefficients from 1.6 to 55 pm V^{-1} demonstrated the potential of this approach. Because siloxane based hybrids behave as thermosets, the processing conditions are very important. Care must be taken so that the hybrid matrix does not become too hard before a reasonable degree of chromophore anisotropy is obtained by electric field poling. Systematic investigations [24, 139, 140] on the incorporation of chromophores with high quadratic hyperpolarizability (μβ) values into inorganic-organic hybrid matrices obtained from hydrolysis and condensation of ICTES-Red 17 (see Fig. 5.19d) and $Si(OMe)_4$ precursors led to very fruitful results. Indeed, the coupling of structural investigations and *in situ* temperature-dependent NLO measurements allowed tuning of large nonlinearities (150 pm V^{-1} at 1.34 μm) and stability at room temperature as well as at higher temperatures (80 °C) through the control of chemical composition, aging of the sols, thermal curing and poling procedures [24]. These high nonlinearities [24, 140] are now quite competitive with those reported for inorganic solids and organic polymeric materials. Since these very promising results in terms of NLO efficiency, recent studies in this area have been mainly focused on the dye relaxation. The most recent researches have reported hybrid sol-gel materials with higher induced order stability but weaker NLO efficiency [141–145].

First results on new (non-azo) chromophores are also very promising. By molecular engineering towards large electro-optic molecular figures of merit (using a simple two-state two-level model) and suitable thermal stability, and by controlling the intermolecular dipole-dipole interactions by tuning push-pull chromophore concentration, M. Blanchard-Desce and co-workers [146] have recently shown the ability to prepare hybrid thin-films with very large electro-optic coefficients. For instance, using push-pull phenylpolyene chromophores grafted to the silica matrix with a μβ eight times larger than that of DR1 and carbazole units to avoid dipolar chromophore interactions, this strategy has led to a very high electro-optic coefficient (r_{33}) of 48 pm V^{-1} at 831 nm [147] which is suitable for EO based applications.

In summary, better results are generally obtained when NLO dyes are grafted to the oxide matrix via several covalent bonds (hybrids of class II). Moreover, class II hybrid organic-inorganic materials allow a better control of the film thickness, the film hydrophobicity, the matrix rigidity, densification and NLO chromophore concentrations. It was also observed that dye-grafting leads to higher surface quality films and corona poling can be optimized to preserve this quality [148].

However, as far as NLO devices are concerned measurements of electro-optical efficiency, wave-guiding properties and evaluation of optical losses must complete these promising results.

5.4.2
Hybrid Photorefractive Materials

The large potential of these NLO hybrid materials in the field of photorefractive materials has been demonstrated [149–151]. These materials exhibit spatial modulation of the refractive index due to redistribution of photogenerated holes in an optically nonlinear medium. As such, they have the ability to manipulate light and are potentially important for optical applications, including image processing, optical storage, programmable optical interconnects and simulation of neural networks. Photorefractive effects have been reported in hybrid sol-gel materials containing a second-order nonlinear chromophore (Disperse Red 1 (DR1) or 4-[N,N-bis (β-hydroxyethyl)amino]4′ nitrostilben (DHS) derivatives) and a charge-transporting molecule (carbazole unit: CB) which was used to form a charge transfer complex with the photosensitizer (Trinitrofluorenone : TNF) to facilitate photocarrier generation at visible wavelengths. In both cases, the carbazole derivatives and the NLO chromophore were grafted to the inorganic backbone. Electro-optical effects and photoconductivity have been observed. Holographic four-wave mixing and two-beam coupling experiments proved the photorefractive nature of the hybrid made with DHS-CB-TNF. These hybrids which exhibit stable photorefractive memory effects with a net internal gain of about 200 cm^{-1} open a promising approach towards the preparation of new photorefractive materials.

5.4.3
Photochemical Hole Burning in Hybrid Materials

Because of inhomogeneous broadening due to the disorder, the absorption spectrum of chromophores diluted in amorphous networks is broad even at low temperature (4 K). However, the absorption patterns due to isolated dye molecules are very sharp (homogeneous broadening). Some dye molecules can be photochemically bleached under selective irradiation and at sufficiently low temperatures the back reaction can be slow enough to maintain a persistent photochemical narrow hole. Therefore, this molecular absorption can be selectively revealed at low temperatures by using a monochromatic laser, opening the possibility of designing frequency-selective optical memories.

Photochemical Hole Burning (PHB) has been attracting attention in recent years because of its scientific interest and the possibility of its application to high-density optical storage. Furthermore, PHB is a useful method for estimating the local structure around the organic dyes. Among several requirements for practical application, high hole efficiencies and high temperature ($T > 77$ K) persistent spectral hole burning with long term stability are the most important. PHB was first achieved in various organically modified silica matrices doped with different porphyrins or quinizarin derivatives [152, 153]. A comparison with pure inorganic systems showed that low PHB efficiency of hybrid systems was mainly related to the mobility of the guest molecule in matrix pores. However, the presence of organic groups covering the inner surface of the pores allowed reduction of guest-host interaction and of electron-phonon coupling. A more complete study [154, 155] of these hybrid systems showed that the inorganic carpet provided by an organic group attached to the inorganic network has a screening effect on the acidic properties of the hydroxyl groups and helps the preservation of the chemical identity of the dopant. Moreover, the bigger the organic groups, the weaker is the linear electron-phonon interaction between the guest and the silicate network. Increasing the size of the organic side group could also reduce the contribution of librational movements of the guest molecule to optical dephasing and thus improve hole-burning efficiency. Looking for persistence of spectral hole burning at high temperatures implies a good choice for the host and a good control of the coupling between the host and the guest molecule.

Tetrakis(4-carboxyphenyl)porphine (TCPP) was cross-linked to an amorphous silica gel through the aminopropyl group of aminopropyltriethoxysilane (APTES) [156]. In these TCPP cross-linked silica gels, properties of photochemical holes, such as hole width, quantum efficiency and irreversible broadening under cyclic annealing experiments, were improved. The cross-linking process allows reduction of the guest mobility, inhibiting dimerization or aggregation and suppressing structural change.

Efficient hole burning was observed in hybrid matrices in which protoporphyrin IX was grafted by two chemical bonds to the inorganic network [157]. As shown in Fig. 5.20, several holes were burnt at 5 K using different burning times. Previously burnt holes were not erased when a new hole was burnt on their "blue" or "red" side. Also, no phonon side bands in the hole-burnt spectra could be recorded, which means that the electron-phonon coupling was very low in this system. The hole burning quantum efficiency was high ($\Phi = 1.2 \times 10^{-2}$). Holes are easily burnt (in 1 s with a power of 0.2 mW cm^{-2}) and they remain quite narrow ($\Gamma w < 1$ cm^{-1}) even at temperatures as high as 20 K. For the first time persistent spectral hole burning was also observed in these hybrid materials at 120 K where detectable holes were burnt with moderate fluences (2–5 mW cm^{-2} during 5 min). However, at this temperature, the hole was quite broad ($\Gamma w = 120$ cm^{-1}). The photochemical event responsible for PSHB is the rotation of the inner protons of the porphyrin cycle. As expected, holes disappeared at 150 K because the inner protons begin to rotate freely. This limitation should easily be overcome by using other photochemical mechanisms like, for instance, a photoinduced electron transfer reaction between a donor molecule and an electron trap, which should be deep enough to prevent detrapping at room temperature. These hybrids of class II look promising for high temperature persistent hole-burning materials.

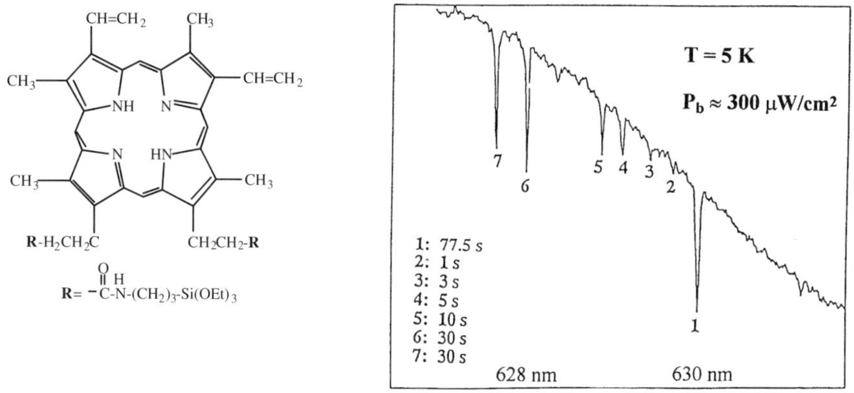

Fig. 5.20 Holes burnt at 5 K in hybrid matrices in which protoporphyrin IX was grafted by two chemical bonds to the inorganic network. Burning times are shown in the insert. P_b is the burning power. Hole widths are laser limited (adapted from [157])

5.4.4 Optical Limiters

The interest of hybrid matrices for preserving optical properties of a molecule was also shown with dyes known to exhibit strong reverse saturable absorption (RSA). The excited state absorption of such dye molecules is stronger in some spectral ranges than the ground state absorption. As a consequence, such chromophores become less transparent as the incident intensity increases. Such a property can be used to realize optical limiters, transmitting light at low powers and becoming opaque for increasing light inputs. Efficient solid-state reverse saturable absorbers were made by entrapment of phthalocyanines and fullerene C_{60} in hybrid matrices [158–161]. First, pristine fullerene C_{60} molecules were embedded within SiO_2 sol-gel matrices, but RSA was limited by the low chromophore concentration [162]. Because of the low solubility of C_{60} in water, increase in the amount of chromophore led to C_{60} aggregation that caused decrease of RSA. A better molecular dispersion was obtained after inclusion of water-soluble fullerenes in silica matrices [163, 164]. Optical limiting properties were also observed in C_{60}-SiO_2 and C_{60}-SiO_2-TiO_2 nanocomposites with a large laser damage threshold [165]. However, the final amount of fullerene molecules in the matrix is limited by the still low solubility of the complex. Improved OL performances were obtained by the use of grafting of C_{60} to the matrix backbone. Fullerene C_{60} derivative has been functionalized with Si tri-alkoxyde groups (see Fig. 5.21) to increase its concentration in the final matrix without any clustering effects [160, 161, 166–168]. The resulting material, processed as thick film, was used to fabricate efficient multilayered optical limiter devices in the red to near infrared range. Fullerene nanocomposites with optical limiting properties are now very close to commercial application (see Fig. 5.22).

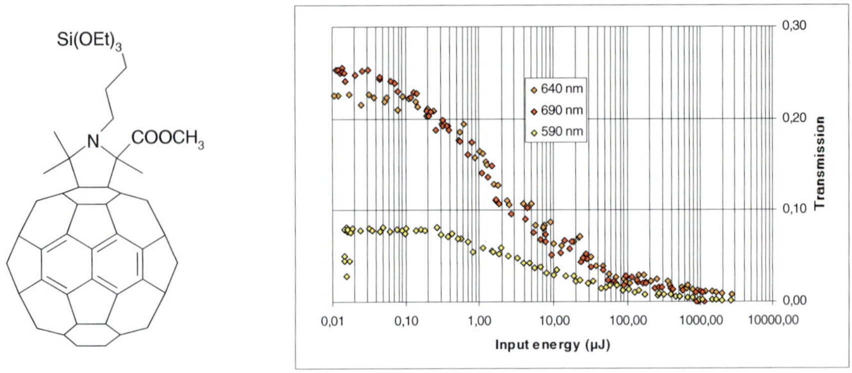

Fig. 5.21 Molecular structure of the Fullerene C_{60} functionalized with a propyltrialkoxysilane group and transmission spectrum of hybrid C_{60}-silica based matrix [166–168]

Fig. 5.22 A nonlinear hybrid C_{60}-silica based material replaces the reticule in one part of the binocular, right photograph [166–168] (courtesy Plinio Innocenzi)

5.5 Hybrid Optical Sensors

Due to their attractive optical properties, particularly intensive activity has recently characterized the area of reactive organically doped sol-gel materials for sensor applications [169]. Their ease of processing, along with several other inherent advantages of sol-gel materials such as tunability of physical properties, high photochemical and thermal stability, chemical inertness and negligible swelling both in aqueous and organic solvents, has resulted in an abundance of sol-gel optical sensors. Sol-gel entrapment of molecules capable of responding to chemical changes in the environment has evolved into routine methodology for preparation of optic-based sensors [170–172]. In most cases, a chromophore is entrapped within the inorganic matrix as an optical probe. Many examples, covering practically all domains of analytical chemistry, have been published in recent years, making this sub-area of organically doped sol-gel materials one of the largest.

In the general area of sol-gel optical sensors, much attention has been devoted to the sensing of acidity [25, 173–181]. However, high interest has also been focused on the development of gas [182–188] and solvent sensors [189–193]. In all cases, the main strategy has been to incorporate within the sol-gel matrices dyes with optical properties sensitive to the presence of protons, gas or solvents. Following this approach, dye-doped sol-gel sensor materials have also been produced for detection of water (air moisture) [194], traces of hydrogen peroxide [195], hydrazine etc.

Most sol-gel optical sensors use emission and adsorption spectroscopy. Any intensity variation and/or wavelength shift observed on the spectra allow probing the dopant micro-environment. As an example, fluorescence quenching in the presence of O_2 is commonly used for O_2 sensing. However, the recent use of phosphorescence in O_2 sensing showed that this optic effect has longer lifetimes than fluorescence and thus the resulting sensor is more sensitive [186].

Different dyes can be incorporated within the same sol-gel host matrix to produce a multi-analyte detection system [176, 181].

One important task in chemical sensing is the development of suitable immobilizing matrices for the dopants. Many studies have showed that sol-gel processing parameters such as water/silane ratio, humidity and catalyst [178, 196] control the cage surface properties and thus affect the performance of the sensing dopant. In some case, sol-gel immobilization effectively enhanced the operating range of the probe compared to solution [177].

Dyes can also be co-entrapped with other organic moieties to modify their properties. It was recently reported that the co-entrapment of the surfactant cetyltrimethylammonium bromide (CTAB, the modifier) with an extensive series of pH indicators greatly modifies the indicating performance of the pH probe [25]. In these systems, the co-entrapment dopant/surfactant not only leads to much larger pK_i shifts, but also greatly improves the leaching profiles; half lives range from several months to several years. Moreover, the CTAB was found to stabilize the microscopic structure of the material upon heat-drying and thus to provide a potential solution to the problem of continuing structural changes inherent to sol-gel materials long after the completion of their synthesis.

The co-entrapment of the dye with liposomes [174] or dendrimers [197] was also reported in order to prevent the leaching of the sensing reagent.

In some cases, cracking of the sol-gel bulk material is a problem. In order to overcome this problem, organic-inorganic hybrid materials have been used as host matrices. Organically modified sol-gel processes give coatings with properties anywhere between sol-gels and silicones. They not only allow an easy incorporation of probes but also an easy tailoring of properties such as mechanical, hydrophobicity and refractive index. As an example, for some sensor applications such as dissolved oxygen sensing, sensor response is greatly enhanced by the use of the organically modified precursor methyltriethoxysilane compared to that of tetraethoxysilane [182, 183]. In some applications, silanisation of the dye (see Fig. 5.23), rather than the glass substrate, provides the advantage of a more homogeneous and well-defined distribution of the dye in the glass film [175, 184]. Moreover, the covalent coupling of the dye to the substrate allows prevention of any dye leaching and thus improves the sensor performance. Immobilization by chemical bonding is more appropriate for small indicators and sterically hindered analyte molecules. On the other hand, entrapment by steric hindrance is more adapted for large indicator molecules and small fast-diffusing analytes [196].

The organic modification of inorganic sol-gel materials to produce ormosils is one of the main strategies for controlling material polarity. Rottman *et al.* [189, 193] have reported the preparation of various dye-doped ormosils that are solvatochromic toward organic solvents such as alcohols and toluene and thus can be used as solvent sensors.

The sol-gel method offers the advantages of tailor-made porosity that is of paramount importance in sensing chemicals because it may control the analyte diffusion and the leachability of the optical probe.

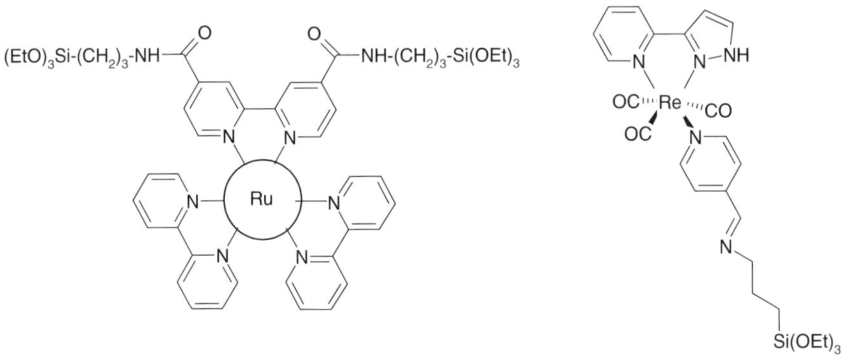

I   II

Fig. 5.23 Molecular structures of ruthenium polypyridyl complex (I) [17] and rhenium tricarbonyl-derived complex (II) [8], both incorporating silane pendant moieties

The preparation of hybrid organic-inorganic sol-gel systems which display the required porosity and rigidity has recently been reported for use in the discrimination of benzene and toluene contaminants emitted into the atmosphere [192]. Methytrimethoxysilane was co-condensed with tetramethoxysilane to tailor the polarity of the host matrix cavities and thus to trap non-polar molecules such as benzene and toluene. Discrimination of benzene and toluene can be easily made from UV-Vis absorption spectrophotometry. A higher sensitivity for aromatic hydrocarbons has been observed in porous silica modified by phenyl moieties compared to a pure porous silica sol-gel matrix [190].

The choice of optical probe is also relevant for the performance of the sensor. The co-immobilization of a fluorescent dye and the Saltzman reagent allows production of sensors sensitive to very low concentrations of NO_2 but irreversible. On the other hand, the use of a ruthenium complex in the same sol-gel system led to a sensor less sensitive to NO_2 but reversible [185]. The incorporation of lanthanide complexes was recently reported in order to produce pH sensors that appeared resistant to complex leaching and photodegradation [179, 180].

In the optical sensor area, the sol-gel method also has the advantage of easy shape-modeling. This process may be used to fabricate thin glass films that have well-defined optical and physical characteristics, such as thickness, porosity and refractive index. Moreover, the sol-gel method has the advantage of being compatible with fiber optic techniques [173, 185, 186, 191] that allow design of remote sensing systems suitable for analysis in toxic or corrosive environments.

5.6 Integrated Optics Based on Hybrid Material

Sol-gel hybrid, organic-inorganic, thin films with optical and good mechanical properties find interesting low cost applications in the field of integrated optics. Films are easily patterned by conventional sol-gel deposition techniques. As a result, high quality optical components including planar, ridge or buried waveguides can be obtained. Furthermore, more complex structures can be imagined such as structures with several guiding layers isolated from each other by inert layers. Depending on the composition of the guiding layer, passive circuits as well as active circuits were designed.

One way to prepare passive waveguides, initially developed by Krug *et al.* in 1992 and studied in depth by P. Etienne *et al.*, is to combine the inorganic sol-gel approach with polymer networking [198–201]. Hybrid systems were made by mixing two organic-inorganic precursors: methacryloxypropyltrimethoxysilane and the reaction product between zirconium-*n*-propoxide and methacrylic acid. Imbricate networks result from, on one hand, the hydrolysis and the condensation of alkoxy groups, and on the other hand the polymerization of double bonds (from methacrylate functions) under UV action as illustrated in Fig. 5.24. From a practical point of view, the starting solution is deposited on a substrate by dip or spin coating and dried at a temperature that is high enough to confer mechanical stability through the inorganic network (Zr-O-Zr and Si-O-Si), but as low as possible to have an optimal UV polymerization. As shown in Fig. 5.24 the dried film (2 μm thick) is then exposed to UV light through a mask defining the targetted pattern. Patterning of these photosensitive gel films by a UV beam interference method as well as by direct laser writing was successfully achieved [202, 203]. In fact, the UV irradiation induces both a refractive index change ($\Delta n \approx 0.025$) and a good stability to alcohol washing. An ultimate thermal treatment (T >120 °C) increases the crosslinking of the inorganic network, thereby leading to better system stability. The great versatility of this technique leads to the opportunity of tailoring a large variety of single or multilayer structures [201, 204, 205]. High quality basic planar waveguides for buried waveguides including ridges, multi-level guides and gratings were prepared for integrated optical circuit applications (splitter, coupler etc.). In terms of performance the propagation losses are not only defined by the intrinsic absorption of the material but are also conditioned by the structure of the guide and the fabrication process. One of the best performances was obtained for a buried waveguide. The propagation losses and coupling losses are around 0.1 dB cm^{-1} and 0.5 dB cm^{-1}, respectively, at 1.3 μm wavelength [206]. Efficient connections between optical fiber and organo-mineral waveguides have also been demonstrated. Splitters using Y-branch, couplers and more complex devices were also fabricated. Evolutions of this technology include the doping of the hybrid material with rare-earth ions leading to sol-gel material amplifiers [207].

Another technique based on UV photobleaching of organic chromophores incorporated in hybrid materials was used to fabricate channel waveguides. Hybrid thin films incorporating both push-pull chromophores (Disperse Red 1 molecules) and carbazole units covalently attached to the rigid backbone of the silica-based matrix were prepared by a sol-gel process [208]. A good confinement of the light was observed after 3 hours exposure through a mask using a 20 mW cm^{-2} broadband UV

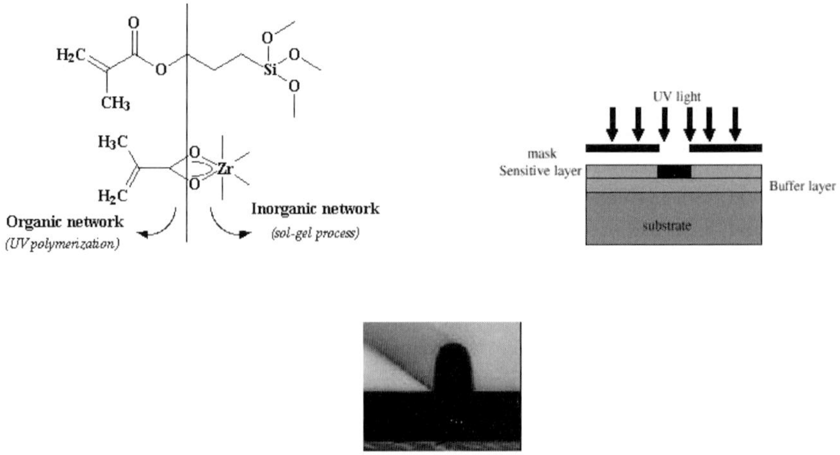

Fig. 5.24 Preparation of optical waveguides proposed by P. Etienne et al. (Groupe matériaux, Laboratoire de Verres, Université de Montpellier II)

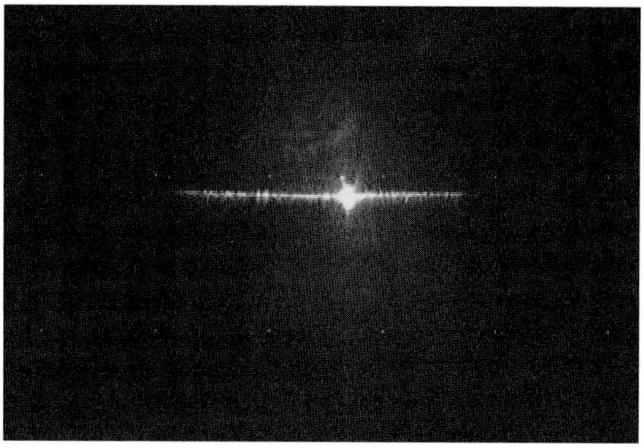

Fig. 5.25 Out-coupled beam of photobleached channel waveguide (layer thickness: 2.5 μm)

lamp centered on around 420 nm. A picture of a 1064 nm beam coupled out of such a waveguide, 2 μm × 5 μm, is shown in Fig. 5.25. Attenuation coefficients at 1064 nm were found to be of the order of 1 cm^{-1}. With the prospective use of chromophores with enhanced photostability, plasma-etched ridge structures were also fabricated (Fig. 5.26).

Concerning active waveguides, sol-gel mesostructured silica materials have recently attracted attention as potential optical materials [44, 114, 209–213]. These materials are synthesized via the polymerization of inorganic species (e.g. SiO_2, TiO_2, Ta_2O_5) around a surfactant micelle liquid crystal, which is then preserved in the mesoscopically ordered pore structure or eliminated by thermal treatment according to the properties required. Depending on the structure and porosity, the refractive index of mesoporous silica is low ($n = 1.15–1.4$) making these materials promising candidates for low-refractive-index supports. Specific properties can be conferred on silica-surfactant nanocomposites by doping the organic parts. A wide range of organic species including semiconducting polymers, organometallic complexes and laser dyes has already been incorporated [44, 114, 209, 211, 214–216]. In this case the material preparation involves the co-assembly of the inorganic and

Fig. 5.26 Scanning electron micrograph of ridge sol-gel waveguide (sample was etched on 1.4 μm out of 2.5 μm thick layers)

organic parts in the presence of the guest molecules. Recently dye-doped mesostructured materials have demonstrated their utility as potential laser materials by displaying amplified spontaneous emission. The active layer consists of a dye-doped silica/block copolymer nanocomposite. This approach allows the reaching of high doping concentration by simultaneously ensuring high dye dispersion and reduced dye aggregation. Deposited on low refractive index thin mesoporous supporting film and patterned by soft lithography, the active layer exhibits low-threshold amplified spontaneous emission (around 10 kW cm^{-2}) when optically pumped. Spontaneously emitted light is amplified by stimulated emission as it propagates along the waveguides.

5.7 Hierarchically Organized Hybrid Materials for Optical Applications

Since the recent development of organic templating growth of materials, new types of hybrid photonic materials whose structure and function are organized hierarchically are emerging. Ordered periodic micro-, meso- and macroporous materials allow the construction of composites with many guest types, e.g. organic molecules, inorganic ions, semiconductor clusters or polymers. These guest/host materials combine high stability of the inorganic host system, new structure forming mechanisms due to the confinement of guests in well-defined pores, and a modular composition. This could lead to applications of new optical materials in, for example, switches, nonlinear optics and lasers. Exploiting the periodic arrangement of the pore systems of mesoporous solids such as the MCM-41 type could generate new optical devices.

One of the first reports of dye-doped mesostructured materials was by H. S. Zhou and Honma [212] who described a new approach for incorporating dyes into mesoporous materials with applications in the fields of optics and photonics. The photosensitive and photosynthetic molecules phthalocyanine (Pc) and chlorophyll (Chl) have been used to dope oxide mesostructure materials such as SiO_2, V_2O_5, WO_3 and MoO_3, the material being directly synthesized by a self-organized co-assembly method involving mixing Pc or Chl with hexadecyltrimethylammonium chloride micelles. Chl and Pc were chosen as the functional molecules because of potential applications in solar cells, NO_x sensors, photocatalysis and nonlinear optical devices. Hoppe et al. [217] investigated doping rhodamine-B into MCM-41 synthesis. Spectroscopic studies of rhodamine-B doped mesostructured silica composites revealed a high monomer concentration relative to dimmer composites, confirming the power of the co-assembly process.

C. E. Fowler and co-workers first reported the direct synthesis of an ordered mesoporous silica containing covalently linked fluorescent organic chromophore functionalities [218]. These studies suggested that it should be possible to incorporate a range of organic chromophores into surfactant-silica mesophases and corresponding mesoporous replicas without extensive modification of the optical properties. An evaporation-induced self-assembly procedure [219] was used to process these mesostructured dye-functionalized silica systems as transparent thin films and monoliths [220].

The covalent attachment of dye molecules into ordered or semi-ordered porous MCM-41 type materials could be of general importance in a number of areas. In principle, the organic moieties can be dispersed and isolated from each other, which for some systems, for example with rhodamine derivatives, should minimize intermolecular quenching of fluorescence properties. Furthermore, the combination of mesoporosity and optical properties should give rise to interesting materials, particularly when host-guest interactions are being examined. Similarly, covalently linked chromophores could be used as sensors in separation technologies to detect molecules within the channel-like pores of MCM-41 phases fabricated in the form of thin membranes. Finally, as the optical properties of the chromophore can be highly sensitive to the local environment, such moieties could be used to probe the internal structure and dynamics of mesostructured silica in general.

Stucky and co-workers [210] reported the fabrication of a laser based on rhodamine 6G-doped mesoporous silica fibers. They showed the amplification of a guided mode in the fiber and the resulting directional, gain-narrowed emission and thus demonstrated for the first time that mesostructured systems can be applied as advanced optical materials. The processing of doped mesostructured materials was quickly extended to more easily processed block-copolymer templated film solutions that were then patterned into waveguiding structures by soft lithography [44]. Compared to dye-doped sol-gel glasses, superior lasing properties were observed for Rhodamine 6G-doped mesostructured materials. Indeed, dimerization of the dye molecule is considerably reduced in mesostructured materials.

A number of dyes such as rhodamine, azobenzene, coumarine derivatives have been integrated in mesoporous solids. M. Ganschow and co-workers [221] grafted a photochromic azo dye or a fluorescent laser dye (sulforhodamine B) in MCM-41.

G. Wirnsberger and co-workers [114] have recently shown that mesostructured materials are excellent hosts for photochromic dyes. Both dyes investigated, a spiropyran and a spirooxazine derivative, show direct photochromism, becoming colored upon UV illumination and bleaching thermally back to their colorless closed forms. The response times of SO-doped materials are very fast, being in the range of the best values reported so far for solid-state composites. The materials show long-term stability with no obvious competition between direct and reverse photochromism with increasing time.

Kinski and co-workers [222] investigated the nonlinear optical properties and polarization of para-nitroaniline (p-NA) entrapped in MCM-41 via the vapor phase. After the gas phase, loaded samples were aged in air for several weeks, and an increasing SHG was observed, indicating polar alignment of p-NA molecules within channels.

A fluoresceine derivative was occluded within the channels of a block-copolymer templated silica matrix to produce an optical pH sensor [223]. The fast response time observed was attributed to the high porosity of the dye-carrying mesoporous thin film, since the open pores enable a fast diffusion of the solution toward the dye molecules.

The organized pore systems were also used as hosts for growing or depositing inorganic materials for generating optical mesostructured materials. Specifically, the idea is to use the well-defined mesopores to generate ordered arrays of semiconductor nanocrystals, quantum dots. Following this approach, GaAs [224], InP [225], GaN [226], CdSe [227] and CdS [228] nanocrystals were grown inside mesoporous silica. However, an undesired crystal growth on the surface of the mesostructures was often observed [224, 225].

Mesoporous materials were also doped with semiconducting polymers such as poly[2-methoxy-5(2′-ethoxyhexloxy)-1,4-phenylenevinylene] (MEH-PPV) [229]. The polymer and the mesopore size was chosen to ensure that only a single chain could fit within the pores in order to study the polymer's photophysics and energy transfer through the polymer [230]. This work clearly demonstrated the utility of aligned mesopores by showing that energy transfer is conformation driven in MEH-PPV.

Already in these early studies several proposals were made to apply such mesoporous systems as optical materials and highly specific chemical sensors. These guest/host materials combine high stability of the inorganic host systems, new guest-structuring mechanisms produced by confinement in well defined pores and a modular composition. This could lead to new nonlinear optical, optical switching or conducting materials. The self-assembled host, especially, allows appreciable control of mesoscale dimensions, symmetry and orientational ordering of guest-host structures.

These composites can be structured hierarchically, leading to optical materials that have desired optical properties due to their electronic states and molecular arrangements, and that have the desired light-propagation of these complex structures of mesoporous materials.

Mesostructured silica functionalised with the dye moiety, (2,4-dinitrophenylamine), has been prepared with bimodal pore size distribution [231]. Dual templating methods are employed to control the hierarchical pore system. The macropores are formed by using crystalline arrays of monodisperse polystyrene (PS) spheres as macrotemplates for the *in situ* evaporation-induced deposition of organic dye-functionalised mesostructured silica (see Fig. 5.27). At the mesoscale level, the porosity of the wall is controlled by supramolecular templating using a reaction mixture containing tetraethylorthosilicate (TEOS), 3-(2,4-dinitrophenylaminopropyl)triethoxysilane (DNPTES) and cetyltrimethylammonium bromide (CTAB). Materials with bimodal porosity are of considerable interest. Indeed, for example, while micro- and mesopores may provide size- or shape-selectivity for guest molecules, the presence of additional macropores can offer easier transport and access to the active sites that should improve reaction efficiencies and minimize channel blocking.

H. Fan and co-workers [232], using a self-assembling "ink", combined silica-surfactant self-assembly with three rapid printing procedures – pen lithography, ink-jet printing, and dip-coating of patterned self-assembled monolayers – to form functional, hierarchically organized structures in seconds. Their approach combines molecular-scale evaporation-induced self-assembly (EISA) of organically modified mesophases with macroscopic, evaporative printing procedures. The resulting hierarchic structures exhibited form and function on multiple length scales and at mul-

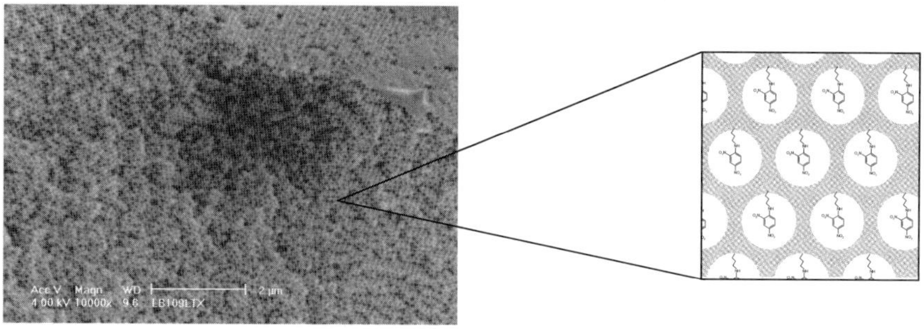

Fig. 5.27 Scanning electron micrograph of macroporous hybrid organically-modified silica reverse opal. Walls schematically represented in the insert are made of dye-functionalized mesostructured silica [231]

tiple locations. At the molecular scale, functional organic moieties are positioned on pore surfaces; on the mesoscale, mono-sized pores are organized into one-, two- or three-dimensional networks, providing size-selective accessibility from the gas or liquid phase; and on the macroscale, two-dimensional arrays and fluidic or photonic systems are defined. The ability to form arbitrary functional designs on arbitrary surfaces would be of practical importance for directly writing sensor arrays and fluidic or photonic systems.

The first reports of the optical properties of a hybrid material whose structure and function are organized hierarchically showed that these solids are promising candidates for optical applications, including lasers, light filters, sensors, solar cells, pigments, optical data storage, photocatalysis, and frequency doubling devices [233, 234].

5.8 Conclusions and Perspectives

Although most of the sol-gel based optical materials reviewed in this chapter are developed on the laboratory scale the research in this field has been exploding for the last two decades. These optically active hybrid solids are the fruit of multidisciplinary research in the areas of chemistry and physics. In particular, the progress in this field largely depends on the core competence of inorganic, polymer and organic chemists, and illustrates the central role of chemistry in the development of advanced materials with unprecedented performance.

After the paint industry, photonic materials are perhaps one of the first systems to take advantage of the hybrid materials approach. The demonstration that hybrid materials with specific optical properties could be synthesized was rapidly recognized. However, the uses of the hybrid approach to obtain vastly improved properties with opportunities for applications are only now being realized.

Significant improvements in the properties of photonic materials may be attributed to three factors:

- matrix improvements (mechanical integrity, transparency, thermal conductivity, reduced optical loss);
- improvements in dye characteristics (chemical, thermal and photochemical stability and high optical efficiencies);
- improved control of dye-matrix interactions.

Solid-state tunable lasers based on sol-gel methods represent an excellent example of the benefits of the hybrid materials approach. The development of dense hybrid materials, as compared to the nanoporous matrices obtained using traditional inorganic sol-gel derived matrices, was critical to this effort because it led to reduced scattering losses caused by pores and improved thermal conductivity and enabled optical quality polishing. The prospects for these material are extremely promising as hybrid solid-state dye lasers exhibit lower threshold powers for laser action (a few mJ), laser slope efficiency close to theoretical values, long laser lifetimes (10^6 pulses) and operation at high repetition rates (10 Hz).

The new hybrid nanocomposite materials developed for the fields of NLO, photorefractivity and PHB demonstrate the importance of covalent bonding of the optical guest molecule to the host backbone. This grafting allows one to effectively increase the dye content, achieve good control of dye dispersion and reduce dye mobility. Moreover, covalent bonding of the dopant offers an opportunity to tune the host-guest chemical interactions and to reduce electron-phonon coupling. The latter features are of paramount importance for improving PHB efficiency.

Photochromic hybrid materials possess attractive properties for such applications as optical switches and optical data storage media. The hybrid approach enables one to control the dye/matrix interaction and to improve matrix properties that reduce dye photofatigue. Diarylethene-doped hybrid matrices exhibit good quantum yields, photofatigue resistance, good color stability and a large refractive index change, which is of interest for wave-guide applications. Photochromic systems based on spirooxazine doped hybrids exhibit very fast thermal bleaching. Despite promising results in the photochromic field, it is evident that improved photochemical stability of these dyes is still needed if applications are to be realised.

The hybrid approach is also applicable to photonic materials whose properties are generated by an inorganic component. Indeed, vanadium (V) phosphorescence was reported from hybrid materials made from polydimethylsiloxane species crosslinked at the molecular level by oxovanadates [235]. The doping of trivalent lanthanides in traditional sol-gel matrices has been well studied, but it is only by using the hybrid approach that sufficiently low OH contents are achieved and good emission properties are observed. Some recent studies demonstrate the opportunity of using the hybrid approach to carefully control the *in situ* formation of lower valence states (Eu(II), Ce(III)) and may be quite important in designing new phosphors. A related topic, which has also received relatively little attention, is that of energy transfer between inorganic chromophores and organic dyes. Such energy transfer systems have been developed as probes and labels for aqueous chemical and biochemical applications. The hybrid approach offers the opportunity to bring these systems into the solid state.

There is no question that the development of hybrid photonic materials will continue in the years ahead because of both technological and scientific factors. The applications for dye-doped systems are to some degree limited by photochemical stability, and improvements are clearly necessary if these materials are to emerge beyond the demonstration level. In some instances, optical design can compensate for dye degradation; for example, rotating the gain medium or increasing sample thickness have both been used to improve laser lifetime. It is, however, the inherent nature of hybrid materials which ensures continued interest in this approach. Many devices or waveguides using integrated hybrid structure are now appearing.

The extraordinary amount of research that has appeared in the last 20 years in the field of hybrid materials indicates the growing interest of chemists, physicists and materials researchers to fully exploit this technical opportunity for creating materials and devices benefiting from the best of the three realms: inorganic, organic and biologic.

However, it is important to remember that major advances in the field of functional hybrids for photonics have been performed on the basis of more fundamental pioneer researches. It is now accepted that functional hybrids allow the construction of efficient optical materials and devices; however, optical properties can also be used as efficient probes for testing materials. Indeed, since the beginning of this field of research, numerous optical methods have been developed to investigate the internal structure, the evolution during processing, the texture of sol-gel derived hybrid systems and more generally organic-inorganic interactions and interfaces. Among these methods based on the used of optical molecular probes are emission spectroscopy of rhodamine or pyren as polarity probes [3, 236], fluorescein as pH probe [237], pyranine to evaluate the water/alcohol ratio [238], $ReCl(CO)_3$,2,2′ bipy as rigidochromic probe [239], Electrical Field Induced Second Harmonic measurements (EFISH) of methylnitroaniline as viscosity probe [126b, 240], fluorescence polarisation of rhodamine 6G (R6G) as a local viscosity probe [241], energy transfer study based on fluorescence life-time between green malachite (donor) and R6G (acceptor) as pore shape and size probes [242], and forced Rayleigh diffusion of molecules with cis-trans isomerization as viscosity probe [243]. Investigation of local structure and dynamics in hybrid sol-gel materials [14, 244, 245], and a study of dye-matrix interactions [246, 247], by using optical probing techniques or photochromic dyes as chemical probes are still in progress.

The basic knowledge of organic-inorganic interaction and interfaces will play a major role for designing hierarchically ordered hybrid structures. These advanced materials which are now just appearing satisfy the requirements for a variety of applications and, in particular, optical communications and chemical sensing. However, their future will depend on the major advances made in the tailoring of new classes of mesoscopic-macroscopic hybrid structures and on the improvements made in the different strategies used for their construction [5]:

- These new hybrids can be engineered by reacting molecular precursors or Nano Building Blocks (NBBs) in the presence of templates working at the nanometric (amphiphilic molecules or macromolecules) and at the micron scale (latex or silica spheres or PDMS (polydimethylsiloxane) stamps etc.).

- Another strategy consists of the use of the same precursors (molecular or NBBs) assembled through the tailoring of mesoscopic templated growth processes and phase separation phenomena.

The confinement of highly dispersed NBBs in the form of clusters or nanoparticles in mesoporous hybrid matrices carrying functional organic groups, or the organization of NBBs on textured substrates, are strategies also described in the recent literature. They could provide larger concentrations of active dots, better defined systems, and avoid coalescence into larger ill-defined aggregates. Numerous recent reports already have emphasized the specific magnetic, optical, electrochemical, chemical, and catalytic properties of nanostructured hybrid materials built from molecular clusters or colloids [5].

References

1 H. Schmidt, A. Kaiser, H. Patzelt, H. Sholze, *J. Phys.* **1982**, *12*, 275–278.

2 J. Livage, M. Henry, C. Sanchez, *Prog. Solid State Chem.* **1988**, *18*, 259–341.

3 D. Avnir, D. Levy, R. Reisfeld, *J. Phys. Chem.* **1984**, *88*, 5956–5959.

4 C. J. Brinker, G. W. Scherrer, *Sol-Gel Science, The Physics and Chemistry of Sol-Gel Processing*, Academic Press, San Diego, 1990.

5 C. Sanchez, G. J. de A. A. Soler-Illia, F. Ribot, T. Lalot, C. R. Mayer, V. Cabuil, *Chem. Mater.* **2001**, *13*, 3061–3083.

6 a) Y.Chujo, T. Saegusa, *Adv. Polym. Sci.* **1992**, *100*, 11–29 ; b) B. M. Novak, *Adv. Mater.* **1993**, *5*, 422–433 ; c) C. Sanchez, F. Ribot, *New J. Chem.* **1994**, *18*, 1007–1047 ; d) U. Schubert, N. Hüsing, A. Lorenz, *Chem. Mater.* **1995**, *7*, 2010–2027 ; e) D. A. Loy, K. J. Shea, *Chem. Rev.* **1995**, *95*, 1431–1442 ; f) P. Judenstein, C. Sanchez, *J. Mater. Chem.* **1996**, *6*, 511–525; g) R. J. P. Corriu, *C.R. Acad. Sci.* **1998**, *1*, 83–89; h) F. Ribot, C. Sanchez, *Comments Inorg. Chem.* **1999**, *20*, 327–371 ; i) C. Sanchez, F. Ribot, B. Lebeau, *J. Mater. Chem.* **1999**, *9*, 35–44.

7 a) *Better Ceramics Through Chemistry VII : Organic/Inorganic Hybrid Materials*, ed B. K. Coltrain, C. Sanchez, D. W. Schaefer, G. L. Wilkes, Materials Research Society, Pittsburgh, 1996, Vol. 435; (b) *Hybrid Materials*, ed R. Laine, C. Sanchez, C. J. Brinker, E. Gianellis, Materials Research Society, Pittsburgh, 1998, Vol. 519; c) *Hybrid Materials*, ed L. C. Klein, L. F. Francis, M. R. De Guire, J. E. Mark, Materials Research Society, Pittsburgh, 1999, Vol. 576; d) *Hybrid Materials*, ed R. Laine, C. Sanchez, C. J. Brinker, Materials Research Society, Pittsburgh, 2000, Vol. 628.

8 a) *Sol-Gel Optics I*, ed J. D. Mackenzie, D. R. Ulrich, SPIE, Washington, 1990, Vol. 1328; b) *Sol-Gel Optics II*, ed J. D. Mackenzie, SPIE, Washington, 1992, Vol. 1758; c) *Sol-Gel Optics III*, ed J. D. Mackenzie, SPIE, Washington, 1994, Vol. 2288; d) *Sol-Gel Optics IV*, ed J. D. Mackenzie, SPIE, Washington, 1997, Vol. 3136; e) *Organic-Inorganic hybrids for Photonics*, ed L. Hubert, S. I. Najafi, SPIE, Washington, 1998, Vol. 3469; f) *Sol-Gel Optics V*, ed B. S. Dunn, E. J. A. Pope, H. K. Schmidt, M. Yamane, SPIE, Washington, 2000, Vol. 3943; g) *Hybrid Organic-Inorganic Composites*, ed J. E. Mark, C. Y. C. Lee, P. A. Bianconi, American Chemical Society, Washington, 1995; i) B. Dunn, J. I. Zink, *J. Mater. Chem.* **1991**, *1*, 903–913.

9 B. Lebeau, C. Sanchez, *Curr. Opin. Solid State Mater. Sci.* **1999**, *4*, 11–23.

10 J.-P. Boilot, F. Chaput , T. Gacoin, L. Malier , M. Canva , A. Brun, Y. Lévy, J.-P. Galaup, *C. R Acad. Sci.* **1996**, *322*, 27–43.

11 D. Avnir, S. Braun, O. Lev, D. Levy, M. Ottolenghi in *Sol-Gel Optics, Processing*

and Applications, ed L.C. Klein, Kluwer Academic Publishers, Dordrecht, 1994, 539–582.

12 a) D. Levy, D. Avnir, *J. Phys. Chem.* **1988**, *9*, 4734; b) D. Levy, S. Einhorn, D. Avnir, *J. Non-Cryst. Solids* **1989**, *113*, 137–145.

13 C. Sanchez, B. Lebeau, *Mat. Res. Bull.* **2001**, *26*, 377–387.

14 T Keeling-Tusker, J. D. Brennan, *Chem. Mater.* **2001**, *13*, 3331–3350.

15 T. Itou, H. Matsuda, *Key Eng. Mater.* **1998**, *67*, 150.

16 G. Schottner, J. Kron, K. Deichmann, *J. Sol-Gel Sci. Technol.* **1998**, *13*, 183–187.

17 G. Schottner, *Chem. Mater.* **2001**, *13*, 3422–3435.

18 a) Aldrich, Fluka, Sigma, Supelco, Chiral Products, Catalog **1997**, p. 250; b) M. Reetz, A. Zonta, J. Simpelkamp, *Angew. Chem. Int. Ed.* **1995**, *34*, 301–303.

19 P. Belleville, C.Bonnin, J. J Priotton, *J. Sol-Gel Sci. Technol.* **2000**, *19*, 223–226.

20 E. Bescher, F. Piqué, D. Stulik, J. D. Mackenzie, *J. Sol-Gel Sci. Technol.* **2000**, *19*, 215–218.

21 a) T. Minami, *J. Sol-Gel Sci. Technol.* **2000**, *18*, 290–291, and references therein; b) A. Shirakura, *Research Laboratory for Packaging*, presented at the 10th International Workshop on Glass and Ceramics, Hybrids and Nanocomposites from Gels, Kirin, Yokohama, **1999**.

22 M. Faloss, M. Canva, P. Georges, A. Brun, F. Chaput, J.-P. Boilot, *Appl. Opt.* **1997**, *36*, 6760–6764.

23 B. Schaudel, C. Guermeur, C. Sanchez, K. Nakatani, J. Delaire, *J. Mater. Chem.* **1997**, *7*, 61–65.

24 B. Lebeau, C. Sanchez, S. Brasselet, J. Zyss, *Chem. Mater.* **1997**, *9*, 1012–1020.

25 C. Rottman, G. Grader, Y DeHazan, S. Melchior, D. Avnir, *J. Am. Chem. Soc.* **1999**, *121*, 8533–8543.

26 T. Dantas de Morais, F. Chaput, J.-P. Boilot, K. Lahlil, B. Darracq, Y. Levy, *Adv. Mater.* **1999**, *11*, 107–112.

27 a) D. Levy, J. M. S. Pena, C. J. Serna, J. M. Oton, L. Esquivias, *J. Non-Cryst. Solids* **1992**, *147–148*, 646–651; b) J. M. Oton, J. M. S. Pena, A. Serrano, D. Levy, *Appl. Phys. Lett.* **1995**, *66*, 929–931; c) D. Levy, M. N. Armenise, *Materials and devices for photonic circuits*, ed L. G. Hubert-Pfalzgraf, S. I. Najafi, SPIE, Washington, **1999**, *3803*, 12–17; d) D. Levy, *Molec. Cryst. Liquid Cryst.* **2000**, *354*, 747–753.

28 D. Avnir, *Acc. Chem. Res.* **1995**, *28*, 328–334.

29 Y. Kobayashi, Y. Kurokawa, Y. Imai, S. Muto, *J. Non-Cryst. Solids* **1988**, *105*, 198–200.

30 R. Reisfeld, D. Brusilovsky, M. Eyal, E. Miron, Z. Burstein, J. Ivri, *Chem. Phys. Lett.* **1989**, *160*, 43–44.

31 F. Salin, G. Le Saux, P. Georges, A. Brun, C. Bagnall, J. Zarzycki, *Opt. Lett.* **1989**, *14*, 785–787.

32 E. T. Knobb, B. Dunn, F. Fuqua, F. Nishida, J. I. Zink in *Ultrastructural processing of advanced materials*, eds. D. R. Uhlmann, D. R. Ullrich, Wiley, N.Y., **1992**, 519–529.

33 J. I. Zink, B. Dunn, in *Sol-Gel Optics, Processing and Applications*, ed L. Klein, Kluwer Academic Publishers, Dordrecht, 1994, 303–328.

34 S. Diré, F. Babonneau, C. Sanchez, J. Livage, *J. Mater. Chem.* **1992**, *2*, 239–244.

35 R. Reisfeld, G. Seybold, *Chimia* **1990**, *44*, 295–297.

36 R. E. Hermes, T. H. Allik, S. Chandra, J. A. Hutchinson, *Appl. Phys. Lett.* **1993**, *63*, 877–879.

37 M. D. Rhan, T. A. King, C. A. Capozzi, A. B. Seddon, in *Sol-Gel Optics III*, ed J. D. Mackenzie, SPIE, Washington, **1994**, *2288*, 364–371.

38 M. Canva, P. Georges, J.-F. Perelgritz, A. Brun, F. Chaput, J.-P. Boilot, *Appl. Opt.* **1995**, *34*, 428–431.

39 B. Lebeau, N. Herlet, J. Livage, C. Sanchez, *Chem. Phys. Lett.* **1993**, *206*, 15–20.

40 B. Dunn, F. Nishida, A.Toda , J. I. Zink, T. Allik, S. Chandra, J. Hutchinson, *Mater. Res. Soc. Symp. Proc.* **1994**, *329*, 267–277.

41 F. Del Monte, D. Levy, *Chem. Mater.* **1995**, *7*, 292–298.

42 M. Casalboni, F. De Matteis, P. Proposito, *Appl. Phys. Lett.* **1999**, *75*, 2172–2174.

43 A. E. Siegmann, *Lasers*, University Science Books, Mill Valley, USA, 1986.

44 P. D. Yang, G. Wirnsberger, H. C. Huang, S. R. Cordero, M. D. McGehee, B. Scott, T. Deng, G. M. Whitesides, B. F. Chmelka, S. K. Buratto, G. D. Stucky, *Science* **2000**, *287*, 465–467.

45 A. Donval, D. Josse, G. Kranzelbinder, R. Hierle, E. Toussaere, J. Zyss, G. Perpelitsa, O. Levi, D. Davidov, I. Bar-Nahum, R. Neumann, *Synthetic Metals* **2001**, *124*, 59–61.
46 C. W. Tang, S. A. Vanslyke, *Appl. Phys. Lett.* **1987**, *51*, 913–915.
47 T. Dantas de Morais, F. Chaput, J.-P. Boilot, K. Lahlil, B. Darracq, Y. Levy, C. R. Acad. Sci. **2000**, 4, 479–491.
48 T. Dantas de Morais, F. Chaput, J.-P. Boilot, K. Lahlil, B. Darracq, Y. Levy, *Adv. Mater. Opt. Electron.* **2000**, *10*, 69–79.
49 C.H. Chen, C.W. Tang in *Chem. Funct. Dyes Proc. Int. Symp.*, ed Z. Yoshida, Y. Shirota, Mita Press, Tokyo, 1993, p. 536.
50 A. H. Kärkkäinen, O. E. O. Hormi, J. T. Rantala, in *Sol-Gel Optics V*, ed B. S. Dunn, E. J. A. Pope, H. K. Schmidt, M. Yamane, SPIE, Washington, **2000**, *3943*, 194–209.
51 *Luminescent Materials*, ed G. Blasse, B. C. Grabmaier, Springer, Berlin, 1994.
52 C. W. E. van Eijk, J. Andriessen, P. Dorenbos, R. Visser, *Nucl. Instr. Nad. Meth. in Phys. Res. A*, **1994**, *348*, 546–550.
53 W. X. Que, Y. Zhou, Y. L. Lam, Y. C. Chan, C. H. Kam, L. H. Gan, G. R. Deen, *J. Electron. Mater.* **2001**, *30*, 6–10.
54 W. X. Que, C. H. Kam, Y. Zhou, Y. L. Lam, Y. C. Chan, *J. Appl. Phys.* **2001**, *90*, 4865–4867.
55 W. X. Que, Y. Zhou, Y. L. Lam, Y. C. Chan, C. H. Kam, *Thin Solid Films* **2000**, *358*, 16–21.
56 A. Huignard, T. Gacoin, F. Chaput, J.-P. Boilot, P. Aschehoug, B. Viana, *Mat. Res. Symp. Proc.* **2001**, *667*, G4.5.1.-G4.5.6.
57 N. I. Koslova, B. Viana, C. Sanchez, *J. Mater. Chem.* **1993**, *3*, 111–112.
58 B. Viana, N. Koslova, P. Aschehoug, C. Sanchez, *J. Mater. Chem.* **1995**, *5*, 719–724.
59 B. Viana, E. Cordoncillo, C. Philippe, C. Sanchez, F. J. Guaita, P. Escribano, in *Sol-Gel Optics V*, ed B. S. Dunn, E. J. A. Pope, H. K. Schmidt, M. Yamane, SPIE, Washington, **2000**, *3943*, 128–138.
60 L. D. Carlos, R. A. Ferreira, V. Zea Bermudez, C. Molina, L. A. Bueno, S. L. Ribeiro, *Phys. Rev. B* **1999**, *60*, 10042–10053.
61 B. Julian, E. Cordoncillo, P. Ecribano, B. Viana, C. Sanchez, *J. Sol-Gel Sci. Technol.* **2002**, in press and the same authors' private communication.
62 a) K. Dahmouche, L. D. Carlos, V. Zea Bermudez, , R.A. Ferreira , C. V. Santilli, S. L. Ribeiro, *J. Mater. Chem.* **2001**, *11*, 3249–3257; b) B. Yan, H. Zhang, *J. Adv. Mater.* **2001**, *33*, 39–43.
63 P. Etienne, P. Coudray, J. Porque, Y. Moreau, *Opt. Comm.* **2000**, *174*, 413–418.
64 M. Isakawi, J. Kuraki, S. Ito, *J. Sol-Gel Sci. Technol.* **1998**, *13*, 587–591.
65 E. Cordoncillo, B. Viana, P. Escribano, C. Sanchez, *J. Mater. Chem.* **1998**, *8*, 507–509.
66 E. Cordoncillo, J. Carda, H. Beltran, F. J. Guaita, A. Barrio, P. Escribano, B. Viana, C. Sanchez, *Bol. Soc. Esp. Ceram. Vidrio* **2000**, *39*, 95.
67 E. Cordoncillo, F. J. Guaita, P. Escribano, C. Philippe, B. Viana, C. Sanchez, *Opt. Mater.* **2001**, *18*, 309–320.
68 V. Bekiari, P. Lianos, U. L. Stangar, B. Orel, P. Judeinstein, *Chem. Mater.* 2000, *12*, 3095–3099.
69 V. Bekiari, E. Sthatos, P. Lianos, U. L. Stangar, B. Orel, P. Judeinstein, *Monarsh. Chem.* **2001**, *132*, 97–102.
70 R. B. Lessard, K. A. Berglund, D. G. Nocera, *Mat. Res. Soc. Symp. Proc.* **1989**, *155*, 119.
71 A. M. Klonkowski, S. Lis, Z. Hnatejko, K. Czarnobaj, M. Pietraszkiewicz, M. Elbanowski, *J. Alloys Compounds* **2000**, *300*, 55–60.
72 X. H. Chuai, H. J. Zhang, F. S. Li; S. B. Wang, G. Z. Zhou, *Mater. Lett.* **2000**, 46, 244–247.
73 L. S. Fu; Q. G. Meng, H. J. Zhang, S. B. Wang, K. Y. Yang, J. Z. Ni, *J. Phys. Chem. Solids* **2000**, *61*, 1877–1881.
74 H. H. Li, S. Inoue; D. Ueda, K. Machida, G. Adachi, *Bull. Chem. Soc. Jpn.* **2000**, *73*, 251–258.
75 P. A. Tanner, B. Yanand, H. J. Zhang, *J. Mater. Sci.* **2000**, *35*, 4325–4328.
76 H. Li, S. Inoue, K. Machida, G. Adachi, *J. Luminescence* **2000**, *87*, 1069–1072.
77 G. D. Qian, M. Q. Wang, *J. Am. Ceram. Soc.* **2000**, *83*, 703–708.
78 M. A. Garcia, A. Campero, *J. Non-Cryst. Solids* **2001**, *296*, 50–56.

79 X. L. Ji, B. Li, S. Jiang, D. Dong, H. J. Zhang, X. B. Jing, B. Z. Jiang. *J. Non-Cryst. Solids* **2000**, *275*, 52–58.
80 M. Bredol, T.H. Jüstel, S. Gutzov, *Opt. Mater.* **2001**, *18*, 337–341.
81 K. Driesen, C. Görller-Walrand, K. Binnemans, *Mater. Sci. Eng. C* **2001**, *15*, 255–258.
82 A. C. Franville, D. Zambon, R. Mahiou, S. Chou, Y. Troin, J. C. Cousseins, *J. Alloys Compounds*, **1998**, *275–277*, 831–834.
83 A. C. Franville, R. Mahiou, D. Zambon, J. C. Cousseins, *Solid State Sci.* **2001**, *3*, 211–222.
84 A. C. Franville, D. Zambon, R. Mahiou, Y. Troin, *Chem. Mater.* **2000**, *12*, 428–435.
85 D. W. Dong, S. C. Jiang, Y. F. Men, X. L. Ji, B. Z. Jiang, *Adv. Mater.* **2000**, *12*, 646–649.
86 a) R. J. P. Corriu, F. Embert, Y. Guari, A. Mehdi, C. Reyé, *Chem. Commun.* **2001**, 1116–1117; b) F. Embert, A. Mehdi, C. Reyé, R. J. P. Corriu, *Chem. Mater.* **2001**, *13*, 4542–4549.
87 H. R. Li, H. J. Zhang, H. C. Li, L. S. Fu, Q. G. Meng, *Chem. Commun.* **2001**, 1212–1213.
88 S. K. Yuh, E. P. Bescher, J. D. Mackenzie, in *Sol-Gel Optics III*, ed J. D. Mackenzie, SPIE, Washington, **1994**, *2288*, 248–254.
89 A. Vioux, D. Leclerc, *Heterog. Chem. Rev.* **1996**, *3*, 65–73.
90 M. Genet, V. Brandel, M. P. Lahalle, E. Simoni, *C. R. Acad. Sci.* **1990**, *311*, 1321–1325.
91 M. Genet, V. Brandel, M. P. Lahalle, E. Simoni, *Sol-Gel Optics I*, ed J. D. Makkenzie, D. R. Ulrich, SPIE, Washington, **1990**, *1328*, 194–196.
92 W. Nie, B. Dunn, C. Sanchez, P. Griesmar, *Mater. Res. Soc. Symp. Proc.* **1992**, *271*, 639–644.
93 J. C. Crano, R. J. Guglielmetti, in *Organic Photochromic and Thermochromic Compounds*, Plenum, New York, 1999, Vol. 1.
94 J. C. Crano, R. J. Guglielmetti, in *Organic Photochromic and Thermochromic Compounds*, Plenum, New York, 1999, Vol. 2.
95 *Applied Photochromic Polymer Systems*, ed McArdle, CB, New York, 1991.
96 J. Biteau, F. Chaput, Y. Yokoyama, J.-P. Boilot, *Chem. Lett.* **1998**, *4*, 359–360.
97 C. Sanchez, A. Lafuma, L. Rozes, K. Nakatani, J. A. Delaire, E. Cordoncillo, B. Viana ,P. Escribano, in *Organic-Inorganic Hybrids for Photonics*, ed L. Hubert, S. I. Najafi, SPIE, Washington, **1998**, *3469*, 192–200.
98 F. Chaput , J. Biteau, K. Lahlil, J.-P. Boilot, B. Darracq, Y. Lévy, J. Peretti, V. I. Savarov, G. Parent, A. Fernandez-Acebes, J.-M. Lehn, *Mol. Liq. Cryst.* **2000**, *344*, 77–82.
99 J. Biteau, F. Chaput, K. Lahlil, J.-P. Boilot, G. M. Tsivgoulis, J.-M. Lehn, B. Darracq, C. Marois, Y. Levy, *Chem. Mater.* **1998**, *10*, 1945–1950.
100 M. Irie, O. Miyatake, K. Uchida, T. Eriguchi, *J. Am. Chem. Soc.* **1994**, *116*, 9894–9900.
101 F. Matsui, H. Taniguchi, Y. Yokoyama, K. Sugiyama, Y. Kurita, *Chem. Lett.* **1994**, 1869–1872.
102 J. Peretti, J. Biteau, J.-P. Boilot, F. Chaput , V. I. Savarov, J.-M. Lehn, A. Fernandez-Acebes, *Appl. Phys. Lett.* **1999**, *74*, 1657–1659.
103 G. Kumar, D. Neckers, *Chem. Rev.* **1989**, *89*, 1915–1925.
104 B. Darracq, F. Chaput, K. Lahlil, Y. Levy, J-P. Boilot, *Adv. Mater.* **1998**, *10*, 1133–1136.
105 N. Landraud, J. Peretti, F. Chaput, G. Lampel, J.-P. Boilot, K. Lahlili, V. I. Safarov, *Appl. Phys. Lett.* **2001**, *79*, 4562–4564.
106 D. Levy, *Chem. Mater.* **1997**, *9*, 2666–2670.
107 J. Biteau, F. Chaput, J.-P. Boilot, *J. Phys. Chem.* **1996**, *100*, 9024–9031.
108 Y. Atassi, PhD dissertation, Ecole Nationale Supérieure de Cachan, France, 1996.
109 J. B. Flannery, *J. Chem. Soc. A* **1968**, 5660.
110 H. Nakazumi, R. Nagashiro, S. Matsumoto, K. Isagawa, in *Sol-Gel Optics III*, ed J. D. Mackenzie, SPIE, Washington, **1994**, *2288*, 402–409.
111 A. Léaustic, A. Dupont, P. Yu, R. Clément, *New. J. Chem.* **2001**, *25*, 1297–1301.
112 L. Hou, B. Hoffmann, M. Mennig, H. Schmidt, *J. Sol-Gel Sci. Technol.* **1994**, *2*, 635.
113 L. Hou, M. Mennig, H. Schmidt, in Proc. Eurogel 92 **1992**, 173–179.

114 G. Winrsberger, B. J. Scott, B. F. Chmelka, G. D. Stucky, *Adv. Mater.* **2000**, *12*, 1450–1454.
115 M. Mennig, K. Fies, H. Schmidt, *Mater. Res. Soc. Symp. Proc.* **1999**, *576*, 409–414.
116 B. Sahut, M. Rene, C. Sanchez, *French Patent* n°9907762, **1999** -C 90K 9/02.
117 C. Eychenne-Barron, A. Lafuma, F. Ribot, C. Sanchez, *Adv. Mater.* **2002**, *14*, 1496–1499.
118 Y. Imai, K. Adachi, K. Naka, Y. Chujo, *Polymer Bull.* **2000**, *44*, 9–15.
119 R. Nakao, N. Ueda, Y.S. Abe, T. Horii, H. Inoue, *Polym. Adv. Technol.* **1996**, *7*, 863–866.
120 L. Hou, H. Schmidt, *J. Mater. Sci.* **1996**, *31*, 3427–3434.
121 A. Zelichenok, F. Buchholtz, S. Yitzchaik, J. Ratner, M. Safro, V. Krongauz, *Macromolecules* **1992**, *25*, 3179–3183.
122 W. S. Kwak, J.C. Crano, *PPG Technology Journal*, **1996**, *2*, 45–59.
123 a)L. Hou, M. Mennig, H. Schmidt, in *Sol-Gel Optics* III, ed J. D. Mackenzie, SPIE, Washington, **1994**, *2288*, 328–339; b) L. Hou, B. Hoffmann, H. Schmidt, M. Mennig, *J. Sol-Gel Sci. Technol.* **1997**, *8*, 923–926; c) L. Hou, H. Schmidt, B. Hoffmann, M. Mennig, *J. Sol-Gel Sci. Technol.* **1997**, *8*, 927–929.
124 S. Benard, A. Léaustic, E. Rivière, P. Yu, R. Clément, *Chem. Mater.* **2001**, *13*, 3709–3716.
125 R. Sasay, H. Itoh, I. Shindachi, T. Shichi, K. Takagi, *Chem. Mater.* **2001**, *13*, 2012–2016.
126 a) E. Toussaere, J. Zyss, P. Griesmar, C. Sanchez, *Nonlinear Optics* **1991**, *1*, 349–354; b) P. Griesmar, C. Sanchez, G. Pucetti, I. Ledoux, J. Zyss, *Molec. Eng.* **1991**, *1*, 205–220.
127 L. Kador, R. Fischer, D. Haarer, R. Kasermann, S. Brück, H. Schmidt, H. Dürr, *Adv. Mater.* **1993**, *5*, 270–273.
128 R. J. Jeng, Y. M. Chen, A. K. Jain, S. K. Tripathy, J. Kumar, *Opt. Commun.* **1992**, *89*, 212–216.
129 R. J. Jeng, Y. M. Chen, A. K. Jain, J. Kumar, S. K Tripathy, *Chem. Mater.* **1992**, *4*, 972–975.
130 J. Kim, J. L. Plawski, E. Van Wagenen, G. M. Korenowski, *Chem. Mater.* **1993**, *5*, 1118–1125.
131 B. Lebeau, J. Maquet, C. Sanchez, E. Toussaere, R. Hierle, J. Zyss, *J. Mater. Chem.* **1994**, *4*, 1855–1860.
132 J. Kim, J. L. Plawsky, R. LaPeruta, G. M. Korenowski, *Chem. Mater.* **1992**, *4*, 249–252.
133 R. J. Jeng, Y. M. Chen, A. K. Jain, J. I. Chen, J. Kumar, S. K. Tripathy, *Chem. Mater.* **1992**, *4*, 1141–1144.
134 D. Riehl, F. Chaput, Y. Lévy, J.-P. Boilot, F. Kajzar, P. A. Chollet, *Chem. Phys. Lett.* **1995**, *245*, 36–40.
135 Z. Yang, C. Xu, B. Wu, L. R. Dalton, S. Kalluri, W. H. Steier, Y. Shi, J. H. Bechtel, *Chem. Mater.* **1994**, *6*, 1899–1901.
136 S. Kalluri, Y. Shi, W. H. Steier, Z. Yang, C. Xu, B. Wu, L. R. Dalton, *Appl. Phys. Lett.* **1994**, *65*, 2651–2653.
137 S. Maturunkakul, J. I. Chen, R. J. Jeng, S. Sengupta, J. Kumar, S. K. Tripathy, *Chem. Mater.* **1993**, *5*, 743–746.
138 a) R. Kasermann, S. Brück, H. Schmidt, L. Kador, in *Sol-Gel Optics III*, ed J. D. MacKenzie, SPIE, Washington, **1994**, *2288*, 321–327; b) H. W. Oviatt Jr, K. J. Shea, S. Kalluri, Y. Shi, W. H. Steier, L. R. Dalton, *Chem. Mater.* **1995**, *7*, 493–498; c) C. Claude, B. Garetz, Y. Okamota, S. Tripathy, *Mater. Lett.* **1992**, *14*, 336–342; d) J. R. Caldwell, R. W. Cruse, K. J. Drost, V. P. Rao, K. Y. Jeng, K. Y. Wong, Y. M. Cali, R. M. Mininni, J. Kenney, R. Binkley, *Mater. Res. Soc. Symp. Proc.* **1994**, *328*, 535; e) R. J. Jeng, Y. M. Chen, J. I. Chen, J. Kumar, S. K. Tripathy, *Polym. Prep.* **1993**, *34*, 292–293.
139 B. Lebeau, C. Sanchez, S. Brasselet, J. Zyss, G. Froc, M. Dumont, *New J. Chem.* **1996**, *20*, 13–18.
140 D. Blanc, P. Peyrot, C. Sanchez, C. Gonnet, *Opt. Eng.* **1998**, *37*, 1203–1207.
141 H. K. Kim, S.-J. Kang, S.-K Choi, Y.-H. Min, C.-S. Yoon, *Chem. Mater.* **1999**, *11*, 779–788.
142 D. Imaizumi, T. Hayakawa, T. Kasuga, M. Nogami, *J. Sol-Gel Sci. Technol.* **2000**, *19*, 383–386.
143 D. H. Choi, J. H. Park, J. H. Lee, S. D. Lee, *Thin Solid Films* **2000**, *360*, 213–221.
144 H. X. Xi, Z. Li, Z. X. Liang, *Chin. J. Polym. Sci.* **2001**, *19*, 421–427.

145 F. Chaumel, H. Jiang, A. Kakkar, *Chem. Mater.* **2001**, *13*, 3389–3395.

146 M. Blanchard-Desce, M. Barzoukas, F. Chaput, B. Darracq, M. Mladenova, L. Ventelon, K. Lahlil, J. Reyes, J.-P. Boilot, Y. Levy, in *Multiphonon and Light Driven Multielectron Processes in Organics: New Phenomena, Materials and Applications*, ed F. Kajzar, M. V. Agranovich, Kluwer Academic Publishers, Dordrecht, 2000, p. 183–197.

147 J. Reyes-Esqueda, B. Darracq, J. Garcia-Macedo, M. Canva, M. Blanchard-Desce, F. Chaput, K. Lahlil, J.-P. Boilot, A. Brun, Y. Levy, *Optics Commun.* **2001**, *198*, 207–215.

148 Y. H. Min, K. S. Lee, C. S. Yoon, L. M. Do, *J. Mater. Chem.* **1998**, *8*, 1225–1232.

149 F. Chaput, D. Riehl, J.-P. Boilot, K. Gargnelli, M. Canva, Y. Levy, A. Brun, *Chem. Mater.* **1996**, *8*, 312–314.

150 B. Darracq, M. Canva, F. Chaput, J.-P. Boilot, D. Riehl, Y. Levy, A. Brun, *Appl. Phys. Lett.* **1997**, *70*, 292–294.

151 a) Y. Zhang, T. Wada, H. Sasabe, *J. Mater. Chem.* **1998**, *8*, 809–828; b) P. Cheben, F. del Monte, D. J. Worsfold, D. J. Carlsson, C. P. Grover, J. D. Mackenzie, *Nature* **2000**,*408*,64–67.

152 S. Kulikov, J. P. Galaup, F. Chaput, J.-P. Boilot, *Spectral Hole-Burning Conference Proc.* Ascona, **1992**.

153 J.-P. Galaup, A.-V. Veret-Lemarinier, S. Kulikov, S. Arabei, J.-P. Boilot, F. Chaput, *Spectral Hole-burning and related spectroscopies Proc.* Tokyo, **1994**, *15*, 358–360.

154 S. G. Kulikov, A.-V. Veret-Lemarinier, J.-P. Galaup, F. Chaput, J.-P. Boilot, *Chem. Phys.* **1997**, *216*, 147–161.

155 S. M. Arabei, S. G. Kulikov, A.-V. Veret-Lemarinier, J.-P. Galaup, *Chem. Phys.* **1997**, *216*, 163–177.

156 M. Uo, M. Yamana, K. Soga, H. Inoue, A. Makishima, in *Sol-Gel Optics IV*, ed J. D. Mackenzie, SPIE, Washington, **1997**, *3136*, 2–9.

157 A.-V. Veret-Lemarinier, J.-P. Galaup, A. Ranger, F. Chaput, J.-P. Boilot, *J. Luminescence* **1995**, *64*, 223–229.

158 F. Bentivegna, M. Canva, P. Georges, A. Brun, F. Chaput, L. Malier, J.-P. Boilot, *Appl. Phys. Lett.* **1993**, *62*, 1721–1723.

159 B. Lebeau, C. Guermeur, C. Sanchez, *Mater. Res. Soc. Symp. Proc.* **1994**, *346*, 315–322.

160 P. Innocenzi, G. Brusatin, M. Guglielmi, R. Signorini, M. Meneghetti, R. Bozio, M. Maggini, G. Scorrano, M. Prato, *J. Sol-Gel Sci. Technol.* **2000**, *19*, 263–266.

161 P. Innocenzi, G. Brusatin, *Chem. Mater.* **2001**, *13*, 3126–3139.

162 D. W. McBranch, B. R. Mattes, A. Koskelo, J. M. Robinson, S. P. Love, in *Fullerenes and Photonics*, ed Z. H. Kafafi, SPIE, Washington, **1994**, *2284*, 15–20.

163 J. C. Hummelen, B. Knight, F. Lepec, F. Wudl, *J. Org. Chem.* **1995**, *60*, 532–538.

164 D. McBranch, V. Klimov, L. Smilowitz, M. Grigorova, J. M. Robinson, A. Koskelo, B. R. Mattes, in *Fullerenes and Photonics III*, ed Z. H. Kafafi, SPIE, Washington, **1996**, *2854*, 140–150.

165 C. Zhu, H. Xia, F. Gan, *Proc. Int. Cong. Glass XVII* **1995**, *4*, 204.

166 P. Innocenzi, G. Brusatin, M. Guglielmi, R. Signorini, R. Bozio, M. Maggini, *J. Non-Cryst. Solids* **2000**, *265*, 68–74.

167 R. Signorini, M. Meneghetti, R. Bozio, M. Maggini, G. Scorrano, M. Prato, G. Brusatin, P. Innocenzi, M. Guglielmi, *Carbon* **2000**, *38*, 1653–1662.

168 G. Brusatin, P. Innocenzi, M. Guglielmi, R. Signorini, R. Bozio, *Nonlinear Opt.* **2001**, *28*, 259.

169 D. Avnir, L. C. Klein, D. Levy, U. Schubert, A. B. Wojcik, in *The Chemistry of Organosilicon Compounds*, ed Z. Rappoport, Y. Apeloig, John Wiley & Sons Ltd, Chichester, 1998, Vol. 2, Chapter 48.

170 R. Zusman, C. Rottman, M. Ottolenghi, D. Avnir, *J. Non-Cryst. Solids* **1990**, *122*, 107–109.

171 O. Lev, M. Tsionsky, L. Rabinovich, V. Glezer, S. Sampath, I. Pankratov, J. Gun, *Anal. Chem.* **1995**, *67*, 22A-30A.

172 M. M. Collinson, *Mikrochim. Acta.* **1998**, *129*, 149–165.

173 O. Ben-David, E. Shafir, I. Gilath, Y. Prior, D. Avnir, *Chem. Mater.* **1997**, *9*, 2255–2257.

174 T. Nguyen, K. P. McNamara, Z. Rosenzweig, *Anal. Chim. Acta* **1999**, *400*, 45–54.

175 M. W. H. Lam, D. Y. K. Lee, K. W. Man, C. S. W. Lau, *J. Mater. Chem.* **2000**, *10*, 1825–1828.
176 J. Lin, D. Liu, *Anal. Chim. Acta* **2000**, *408*, 49–55.
177 C. Malins, H. G. Glever, T. E. Keyes, J. G. Vos, W. J. Dressick, B. D. MacGraith, *Sens. Actuators B* **2000**, *67*, 89–95.
178 C. Rottman, D. Avnir, in *Sol-Gel Optics V*, ed B. S. Dunn, A. J. A. Pope, H. K. Schmidt, M. Yamane, SPIE, Washington, **2000**, *3943*, 154–162.
179 A. Lobnik, N. Majcen, K. Niederreiter, G. Uray, *Sens. Actuators B* **2001**, *74*, 200–206.
180 S. Blair, M. P. Lowe, C. E. Mathieu, D. Parker, P. K. Senanayake, R. Kataky, *Inorg. Chem.* **2001**, *40*, 5860–5867.
181 W. J. Jin, J. M. Costa-Fernandez, A. Sanz-Medel, *Anal. Chim. Acta* **2001**, *431*, 1–9.
182 C. M. McDonagh, A. M. Shields, A. K. McEvoy, B. D. MacCraith, J. F. Gouin, *J. Sol-Gel Sci. technol.* **1998**, *13*, 207–211.
183 P. Lavin, C. M. McDonagh, B. D. MacCraith, *J. Sol-Gel Sci. technol.* **1998**, *13*, 641–645.
184 C. Malins, S. Fanni, H. G. Glever, J. G. Vos, B. D. MacCraith, *Anal. Commun.* **1999**, *36*, 3–4.
185 S. A. Grant, J. H. Satcher Jr, K. Bettencourt, *Sens. Actuators B* **2000**, *69*, 132–137.
186 M. A. Chan, J. L. Lawless, S. K. Lam, D. Lo, *Anal. Chim. Acta* **2000**, *408*, 33–37.
187 O. Worsfold, C. D. Dooling, T. H. Richardson, M. O. Vysotsky, R. Tregonning, C. A. Hunter, C. Malins, *J. Mater. Chem.* **2001**, *11*, 399–403.
188 S.-K. Lee, Y. B. Shin, H.-B. Pyo, S. H. Park, *Chem. Lett.* **2001**, 310–311.
189 C. Rottman, G. S. Grader, Y. De Hazan, D. Avnir, *Langmuir* **1996**, *12*, 5505–5508.
190 F. Abdelmalek, J. M. Chovelon, M. Lacroix, N. Jaffrezic-Renault, V. Matejec, *Sens. Actuators B* **1999**, *56*, 234–242.
191 F. L. Dickert, U. Geiger, P. Lieberzeit, U. Reutner, *Sens. Actuators B* **2000**, *70*, 263–269.
192 T.-H. Tran-Thi, M.-L. Calvo-Muñoz, J.-P. Bourgoin, C. Roux, A. Ayal, A. El-Mansouri, *PRA Proceedings Organic-Inorganic Hybrids* **2000**, *paper 8*, 1–14.
193 C. Rottman, G. Grader, D. Avnir, *Chem. Mater.* **2001**, *13*, 3631–3634.
194 J. M. Costa-Fernandez, A. S . Medel, *Anal. Chim. Acta* **2000**, *407*, 61–69.
195 A. Lobnik, M. Cajlakovic, *Sens. Actuators B* **2001**, *74*, 194–199.
196 C. Gojon, B. Duréault, N. Hovnanian, C. Guizard, *J. Sol-Gel Sci. Technol.* **1999**, *14*, 163–173.
197 M. D. Senarath-Yapa, S. S. Saavedra, *Anal. Chim. Acta* **2001**, *432*, 89–94.
198 H. Krug, F. Tiefensee, P. W. Oliveira, H. Schmidt, in *Sol-Gel Optics II*, ed J. D. MacKenzie, SPIE, Washington, **1992**, *1758*, 448–455.
199 S. I. Najafi, C. Y. Li, M. Andrews, J. Chisham, P. Lefebvre, J. D. Mackenzie, N. Peyghambarian, in *Functional Photonic Integrated Circuits*, ed M. N. Armenise, K.-K. Wong , SPIE, Washington, **1995**, *2401*, 110–115.
200 *Organic-Inorganic Hybrid Materials for Photonics*, ed L. G. Hubert-Pfalzgraf, S. I. Najafi, SPIE, Washington, **1998**, *3469*.
201 P. Etienne, P. Coudray, Y. Moreau, J. Porque, *J. Sol-Gel Sci. Techol.* **1998**, *13*, 523–527.
202 Y. Moreau, P. Arguel, P. Coudray, P. Etienne, J. Porque, P. Signoret, *Opt. Eng.* **1998**, *37*, 1130–1135.
203 R. B. Charters, B. Luther-Davies, F. Ladouceur, *23rd Australian Conference on Optical Fibre Technology*, Melbourne, Australia, **1998**, 37–40.
204 J. Porque, P. Coudray, Y. Moreau, P. Etienne, *Opt. Eng.* **1998**, *37*, 1105–1110.
205 M. A. Fardad, T. Touam, P. Meshkinfam, R. Sara, X. M. Du, M. P. Andrews, S. I. Najafi, *Elec. Lett.* **1997**, *33*, 1069–1070.
206 P. Coudray, P. Etienne, Y. Moreau, J. Porque, S. I. Najafi, *Opt. Commun.* **1997**, *143*, 199–202.
207 P. Etienne, P. Coudray, J. N. Piliez, J. Porque, Y. Moreau, in *Materials and Devices for Photonics Circuits*, ed M. N. Armenise, W. Percorella, L. G. Hubert-Pfalzgraf, S. I. Najafi, SPIE, Washington, **1999**, *3803*, 2–11.
208 A. C. Le Duff, M. Canva, T. Pliska, F. Chaput, E. Toussaere, G. I. Stegeman, J.-P. Boilot, Y. Levy, A. Brun, in *Linear, Nonlinear, and Power-Limiting Organics*, ed M. Eich, M. G. Kuzyk, C. M. Lawson, R. A. Norwood, SPIE, Washington, **2000**, *4106*, 21–30.

209 B. J. Scott, G. Wirnsberger, M. D. McGehee, B. F. Chmelka, G. D. Stucky, *Adv. Mater.* **2001**, *13*, 1231–1234.
210 F. Marlow, M. D. McGehee, D. Zhao, B. F. Chmelka, G. D. Stucky, *Adv. Mater.* **1999**, *11*, 632–634.
211 G. Wirnsberger, P. Yang, H. C. Huang, B. Scott, T. Deng, G. M. Whitesides, B. F. Chmelka, G. D. Stucky, *J. Phys. Chem. B* **2001**, *105*, 6307–6313.
212 H. S. Zhou, I. Honma, *Adv. Mater.* **1999**, *11*, 683–685.
213 S. Besson, T. Gacoin, C. Jacquiod, C. Ricolleau, J.-P. Boilot, *Nano-letters* (to be published, March 2002).
214 T. Q. Nguyen, J. J. Wu, V. Doan, B. J. Schwartz, S. H. Tolbert, *Science* **2000**, *288*, 652–656.
215 J. J. Wu, A. F. Gross, S. H. Tolbert, *J. Phys. Chem. B* **1999**, *103*, 2374–2384.
216 M. Ogawa, T. Nakamura, J. Mori, K. Kuroda, *J. Phys. Chem. B* **2000**, *104*, 8554–8556.
217 R. Hoppe, A. Ortlam, J. Rathousky, G. Schulz-Eckloff, A. Zukal, *Microporous Mater.* **1997**, *8*, 267–273.
218 C. E. Fowler, B. Lebeau, S. Mann, *Chem. Commun.* **1998**, 1825–1826.
219 C. J. Brinker, Y. Lu, A. Sellinger, H. Fan, *Adv. Mater.* **1999**, *11*, 579–585.
220 B. Lebeau, C. E. Fowler, S. R. Hall, S. Mann, *J. Mater. Chem.* **1999**, *9*, 2279–2280.
221 M. Ganschow, M. Wark, D. Whörle, G. Schulz-Ekloff, *Angew. Chem. Int. Ed.* **2000**, *39*, 161–163.
222 I. Kinski, H. Gies, F. Marlow, *Zeolites* **1997**, *19*, 375–381.
223 G. Wirnsberger, B. J. Scott, G. D. Stucky, *Chem. Commun.* **2001**, 119–120.
224 V. I. Srdanov, I. Alxneit, G. D. Stucky, C. M. Reaves, S. P. BenBaars, *J. Phys. Chem. B* **1998**, *102*, 3341–3344.
225 J. R. Agger, M. W. Anderson, M. E. Pemble, O. Terasaki, Y. Nozue, *J. Phys. Chem. B* **1998**, *102*, 3345–3353.
226 H. Winkler, A. Birkner, V. Hagen, I. Wolf, R. Schmechel, H. von Seggern, R. A. Fischer, *Adv. Mater.* **1999**, *11*, 1444–1448.
227 H. Parala, H. Winkle, M. Kolbe, A Wohlfart, R. A. Fischer, R. Schmechel, H. von Seggern, *Adv. Mater.* **2000**, *12*, 1050–1055.
228 [228]T. Hirai, H. Okubo, I. Komasawa, *J. Phys. Chem. B* **1999**, *103*, 4228–4230.
229 S. H. Tolbert, A. Firouzi, G. D. Stucky, B. F. Chmelka, *Science* **1997**, *278*, 264–268.
230 T. Q. Nguyen, J. J. Wu, V. Doan, B. J. Schwartz, S. H. Tolbert, *Science* **2000**, *288*, 652–656.
231 B. Lebeau, C. E. Fowler, S. Mann, C. Farcet, B. Charleux, C. Sanchez, *J. Mater. Chem.* **2000**, *10*, 2105–2108.
232 H. Fan, Y. Lu, A. Stump, S. T. Reed, T. Baer, R. Schunk, V. Perez-Luna, G. P. Lopez, C. J. Brinker, *Nature* **2000**, *405*, 56–60.
233 A. Stein, B. J. Melde, R. C. Schroden, *Adv. Mater.* **2000**, *12*, 1403–1419.
234 G. Wirnsberger, G. D. Stucky, *Chem. Mater.* **2001**, *13*, 3140–3150.
235 B. Alonso, J. Maquet, B. Viana, C. Sanchez, *New J. Chem.* **1998**, *22*, 935–939.
236 V. R. Kaufman, D. Avnir, *Langmuir*, **1986**, *2*, 717.
237 L. M. Shamanskya, M. Yanga, M. Olteanub, E. L. Chronister, *Mater. Lett.* **1993**, *26*, 113–120.
238 J.-C. Pouxviel, B. Dunn, J. I. Zink, *J. Phys. Chem.* **1989**, *93*, 2134–2139.
239 J. McKiernan, J.-C. Pouxviel, B. Dunn, J. I. Zink, *J. Phys. Chem.* **1989**, *93*, 2129–2133.
240 C. Sanchez, P. Griesmar, G. Puccetti, E. Toussaere, I. Ledoux, J. Zyss, *Nonlinear Optics* **1992**, *4*, 245–250.
241 R. Winter, D. W. Hua, X. Song, W. Mantulin, J. Jonas, *J. Phys. Chem.* **1990**, *94*, 2706–2713.
242 P. Levitz, J. M. Drake, J. Klafter, *J. Chem. Phys.* **1988**, *89*, 5224–5236.
243 D. W. Dozier, J. M. Drake, J. Klafter, *Phys. Rev. Lett.* **1986**, *56*, 197–200.
244 B. Dunn, J. I. Zink, *Chem. Mater.* **1997**, *9*, 2280–2291.
245 F. del Monte, M. L. Ferrer, D. Levy, *J. Mater. Chem.* **2001**, *11*, 1745–1751.
246 F. del Monte, J. D. MacKenzie, D. Levy, *Langmuir* **2000**, *16*, 7377–7382.
247 F. del Monte, M. L. Ferrer, D. Levy, *Langmuir* **2001**, *17*, 4812–4817.

6
Electrochemistry of Sol-Gel Derived Hybrid Materials

Pierre Audebert and Alain Walcarius

6.1
Introduction

Sol-gel derived hybrid materials have attracted substantial attention because they present special challenges and opportunities with respect to potential applications in various fields, including optical and electronic materials, solid electrolytes, protective coating technology, sensor devices, catalysis, separation science, biology, and electrochemistry [1, 2]. Hybrid materials lie at the interface of the organic and inorganic realms. By combining organic and inorganic components into a single composite, the versatility of sol-gel processes provides a rather straightforward way to produce a wide range of organic-inorganic hybrid materials with numerous promising applications [2]. Several well-documented reviews are available, dealing with their use in connection with electrochemistry [3–7], their application for electroanalytical purposes [7–11], and their exploitation in analytical chemistry [12–14].

Organic-inorganic hybrids are often classified into two main categories: (i) "class I materials", in which the organic and inorganic components are weakly linked through hydrogen bonding, van der Waals contacts, or electrostatic forces, and (ii) "class II materials", in which the constituents interact strongly through ionic or covalent bond formation [1]. Four different routes are mainly used to prepare organic-inorganic hybrid materials [2]:

- Impregnation of a porous inorganic matrix (most often silicon dioxide) by organic components, which can even be polymerized *in situ*, displaying a particular affinity for the host structure (mainly class I hybrids);
- Dispersion or solvation of the organic compound in a sol-gel mixture, which is often referred as sol-gel doping, where the organic component is physically entrapped in the three-dimensional inorganic structure during its formation (class I hybrids);
- Use of adducts with at least one direct, non-hydrolyzable heteroatom-carbon bond, that can be either fixed by post-synthesis grafting onto an as-synthesized inorganic support, or alternatively, be introduced as an organofunctional precursor in the starting sol and polycondensed with the other sol-gel precursor(s),

Gómez-Romero: Organic-Inorganic Materials. Pierre Audebert

ISBN: 3-527-30484-3

or even prepared by combining these two approaches (class II hybrids). Both amorphous and mesoporous crystalline materials can be obtained in to these ways. These processes can even be associated with impregnation or sol-gel doping to produce hybrids of classes I & II in a single material;
- Intercalation compounds and interpenetrating organic-inorganic polymers can be obtained by combining layered materials with organic polymer chains (class I and class II hybrids).

In principle, these approaches that exploit most often the versatility of sol-gel processing are offering exceptional opportunities to create a wide range of new products displaying tailor-made composition, structure, and properties, which can be tuned to fit the desired target application. These materials are not only combining the distinct properties of organic and inorganic components within a single composite, but new or enhanced phenomena as well as novel truly unique properties may arise as a result of the interface between the organic and inorganic worlds. The full control of all the experimental parameters affecting the synthetic pathways, however, will still require much work in the future, even though considerable advances have been performed recently [2].

This emerging field of materials science has generated considerable and increasing interest from the electrochemical community during the past decade [3–11]. On the one hand, electrochemistry was applied as a characterization technique to describe the basic behavior of sol-gel processed materials and to characterize mass transfer reactions in xerogels and, on the other hand, the intrinsic properties of organic-inorganic hybrids (mainly those based on silica) were exploited in various electroanalytical and sensor applications as well as in power source technology as solid electrolytes.

Although gelled solvents have been known for years (Napalm is a regrettable example of this), electrochemical investigations into gelled systems have been very slow to start, especially molecular electrochemistry, the main problem being the volume of the solution/gel to deal with, as well as the difficulty of gelling polymer-solvent mixtures in confined vessels of unspecified shape. This has been overcome with the use of hydrophobic Nafion gels [15]. Performing electro-chemistry in such gels requires dropping an aliquot of the hydrophobic gel on a millimeter size electrode and allowing it to dry, and using the resulting device as a working electrode in an aqueous electrolyte [16, 17]. Such systems have been shown to operate well, including in a few analytical applications [17]. In such systems, it was shown for the first time that, contrary to classical Nafion modified electrodes with ionically bounded redox sites [18], the included redox compounds were transported with a diffusion coefficient characteristic of the solvent used to prepare the gel. Another original exploitation of electro-chemistry was the determination of the glass transition temperature of a concentrated acrylate gel through the time dependence of the diffusion coefficient of included electroactive species [19]. This latter work has opened the way to fundamental investigations in oxide gels, and therefore the use of electrochemical techniques as a "spectroscopy of gelled matter". This aspect will be presented in the first part of this chapter.

motion was extremely stable and through which the gel could be dried up to one third of its initial weight (Fig. 6.1). Recently, these results were confirmed and extended with further details by Collinson et al., by using exclusively the newly developed experimental technique of ultramicroelectrodes (Fig. 6.2), although their studies were centered on the final drying period, disregarding the first steps of the sol-gel process [26]. A striking feature of their results is that the variation of the electroactive probe diffusion coefficient is strongly dependent upon the probe charge. While diffusion of cationic and, to a much lesser extent, neutral species was found to decrease during the drying step, there was no noticeable variation of the behavior of the anionic ones. This result can probably be explained by the existence of much stronger interaction between the silica pore walls with cations than with the neutral species, or above all with the anions. Up to now, the only published works focused on pure oxide gels, and no report is available to demonstrate if this is a special feature of silica, or on the contrary a more general behavior extendable to the wider family of oxide gels. In a different approach, Cox et al. have used cyclic voltammetry to demonstrate that diproportionation of uranium(V) in the gels is very sensitive to the local environment and could be a tool for testing the presence of anionic sites in the pores [27]. Finally, a recent study describes the electrochemically assisted deposition of a hybrid silica film, from the electrooxidation of a tetrafunctionalized phenothiazine [28]. The sol-gel polymerization in this case arises from the electrochemically-induced precipitation of the cation radical on the electrode surface. The triethoxysilane endgroups can then condense because of their high local concentration on the electrode, in the presence of the electrolyte solution.

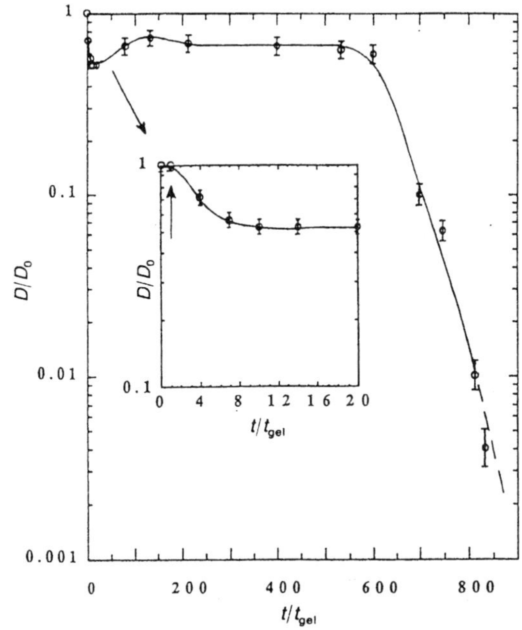

Fig. 6.1 Dependence of reduced diffusion coefficient of ferrocene in the gel on reduced gelation time for silica gel. *From [25, first], reproduced by permission of The Royal Society of Chemistry*

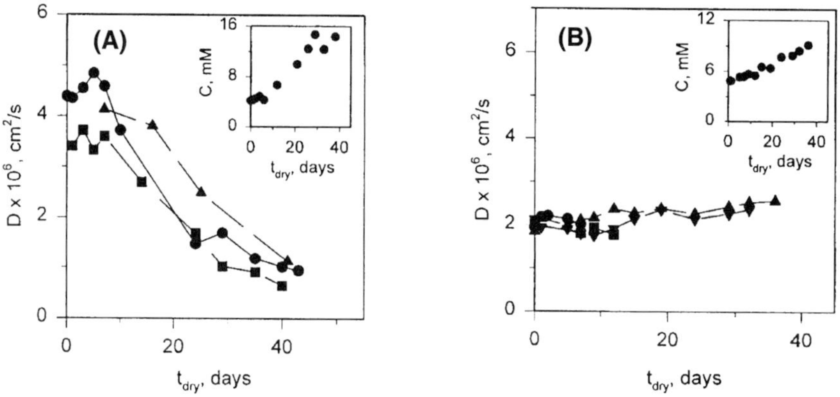

Fig. 6.2 Variation in the diffusion coefficient of gel-encapsulated ferrocene methanol ($FcCH_2OH$) (A), or $Fe(CN)_6^{3-}$ (B), with gel drying time. Single measurements on three different monoliths are shown. Inset: variation in the concentration of gel-encapsulated electroactive species with drying time. *Reprinted with permission from [26, first]. Copyright (1999) American Chemical Society*

In parallel to the efforts directed to the examination of the behavior of "free" molecules embedded in a gel, it was of still greater interest to look at the motion of redox probes anchored within the sol-gel polymer, both in the course of their formation/growing, as well as in the final state of the gel. Actually, the only reported work was aimed at investigations of the polymerization of zirconium and (to a lesser extent) titanium propoxide, in the presence of various chelating agents [29]. It was shown that the degree of polycondensation that can be attained was relatively low (depending however mainly on the chelate content). More importantly, it was confirmed that only weak bonds remained in the wet gel, in the bulk oxopolymer, so that the gel may be considered in fact as a kind of "dynamic solid", which is of relatively rare recognized occurrence. First results show that there is a similar possibility in the field of silica WOGs, and further work is in progress [29].

6.2.1.2 **Conducting Polymers – Sol-gel Composites**

The polymerization of conducting polymers into WOG structures is not an easy task, because the cation-radical precursors of conducting polymers are very often sensitive to weakly nucleophilic functions present in WOGs. An early partially successful attempt was made seven years ago by Sanchez et al. who polymerized pyrrole into a hybrid gel [30]. However, beyond the observable electrochemical response of the polymers, the other properties were poor or unreported. In fact, an easier approach was followed by other groups who proceeded by performing sol-gel polymerization and either adding a soluble conducting polymer to the starting sol (which is feasible in the case of polyaniline and some substituted polythiophenes), or, more often, performing the chemical polymerization of the heterocycle in the presence of a chemical oxidant, simultaneously to the sol-gel process [31].

Polymerization of pyrrole and aniline into sols was also described [30], but the properties of the resulting polymer films were not substantially different from those of classical electrochemically-grown liquid electrolyte films, although the authors reported that the films had incorporated some silica during the growth process. In a recent and noteworthy work, the same group claimed that pyrrole functionalization was achieved by *in situ* nucleophilic attack of the pyrrole cation-radical by silanols, leading to an electrogenerated hybrid film (Fig. 6.3) which in turn exhibited much better mechanical properties [32]. However, the electrochemical growth of materials in sols or WOGs appears less attractive than in xerogels and has been therefore less investigated.

6.2.2
Electrochemical Behavior of Xerogels and Sol-gel-prepared Oxide Layers

It is not obvious how to separate the field of xerogels (i.e. dried gels from which the gelation solvent has been extracted) from the field of oxides, since a whole family of materials exists in between. A xerogel is normally obtained upon simple drying, requiring sometimes a mild thermal treatment, while an oxide layer requires always a thermal treatment at a relatively elevated temperature. This latter depends upon the nature of the precursor and the type of xerogel initially prepared. However, if the curing step is performed at a lower temperature, or is not long enough, an intermediate state (oxopolymers) is obtained. This is often the case for electrochemical experiments, since pure oxide layers are sometimes too compact and hinder diffusion of the reactants or the electrolyte species through the film, while as-prepared xerogels are sometimes characterized by a lack of mechanical stability and strip off the electrode.

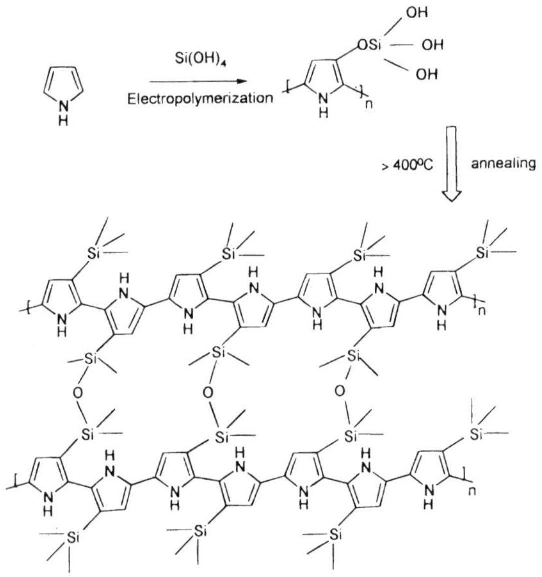

Fig. 6.3 Schematic diagram of the interpenetrating network system formed from annealing of the silanol-functionalized polypyrrole. *Reprinted with permission from [43]. Copyright (2000) American Chemical Society*

Beyond the fundamental work, a recent challenge is to use the versatility of sol-gel chemistry to realize new electrochemical electroluminescent devices. Electrochemical luminescence is light generation from *in situ* electrogenerated species. This phenomenon can also be used for electroanalytical purposes.

Again the different concerns exposed previously can be found in this research area. First, fundamental studies have been reported on xerogels that have been turned electroactive by insertion/functionalization of redox systems. Second, composites made by combining a ceramic component and conducting polymers have been produced and applied mainly as polyelectrolytes. Here, however, a third important category is devoted to composite study, essentially with DuPont's ionomer Nafion®, and applications emerge mainly in the field of electroluminescent and corrosion protection devices.

6.2.2.1 **Fundamental Studies**

Electrochemical activity in WOGs is mainly due to the transfer of electroactive species (physical diffusion), with negligible participation of electron exchange. In the xerogel state, either both phenomena are competitive, or the only occurring process is electron exchange between adjacent redox sites (electron pseudo-diffusion). Early work by Murray [24] has shown that the electrochemical response of freely diffusing molecules in organic polymers implies competition of the two aforementioned processes. On the other hand, we and others have shown that in all rigid backbone functionalized polymers, only electron exchange occurred. This latter point has been the focus of pertinent theoretical work, as described below.

Fundamental approach and theoretical background

There is some interesting theoretical background to the examination of electron exchange in rigid polymers, of which functionalized xerogels are a particularly representative and versatile family. About 20 years ago, Andrieux and Savéant [33] as well as Laviron [34] demonstrated for the first time *that in the case of a homogeneous distribution of redox sites* the charge transfer is equivalent to a diffusion process where the diffusion coefficient D would be equal to $kC^{\circ}\delta^2$, where k, C°, and δ are respectively the exchange rate constant *in the solid*, the concentration of electroactive sites and the average distance between these sites. The theory was later refined by Savéant and Blauch [35], who mainly discussed the influence of k. However, a further view, of particular interest concerning the application to xerogels and oxides, was the very recent approach of Andrieux and Audebert [36], who calculated the response of a layer with a fractal distribution of the redox sites. In particular, they have shown that not only did the electrochemical response become quite unclassical, with e.g. peak currents depending now on *fractional exponents of the scan rate* in cyclic voltammetry, but also that a monotonic dependence of these exponents does exist with the fractal dimension (provided the spectral dimension was known and constant, a reasonable assumption). In the same article, they applied their analysis to the electrochemical response of a previously published functionalized silica xerogel [37].

Experimental work

A sustained interest in the study of electroactive xerogels has appeared, not only for performing structural investigations, but also in view of possible further applications, such as development of new sensors, corrosion protection, production of solar cells etc. First of all, the behavior of non-bounded electroactive species included in the xerogels was examined. Many redox couples were investigated, including ferrocyanide, Fe^{3+}-phenanthroline, UO_2^{2+} [27], ferrocene methanol or Co^{3+}-phenanthroline [38], a cobalt porphyrin [39] and a cobalt phthalocyanine [40]. The silica xerogels were mainly obtained form tetramethoxysilane (TMOS) or tetraethoxysilane (TEOS) precursors, and apparently remained "fixed" on the electrode surface, despite one recent work showing that they may strip off after long standing in the sol, according to the drying process and the electrode nature [38]. The xerogels were porous enough to allow electrolyte wetting, and the conclusions were that the electrochemical behavior of some couples could be affected by the silica matrix, e.g. transport regime or the disproportionation rate of U(V); this latter for example was much faster than in aqueous solution [27].

All the systems dealt with attached redox centers and can therefore be considered as hybrid modified electrodes. Since ferrocene is one of the most common stable electroactive compounds, displaying in addition a not too complicated organic chemistry, it has been widely used as a functional moiety. The interesting point in the sol-gel building of modified electrodes is that the concentration of the electroactive species can be tailored at will, and this constitutes a quite pertinent parameter for the analysis of the gel structure through its electrochemical response. It has been shown that the response of silica xerogels was characteristic of a non-isotropic distribution of the ferrocenes in hybrid xerogels, whatever the starting monomer [37, 41], and could be explained in most cases by the repartition of the redox sites on internal pore surfaces, while in one special case a fractal distribution of the ferrocenes had to be invoked [36]. By contrast, zirconium and titanium oxopolymers did exhibit a behavior quite in accordance with an isotropic repartition of the sites, which was not unexpected in the case of gels structured on the basis of weak bonds [42].

Except in the case of the uranium species, little or no influence of the xerogel reactivity was observed on the chemistry or the stability of the electrogenerated species. On the contrary, the analysis of voltammetric curves resulted in a better understanding of the gel structure (and, in the most striking case, to the estimation of the fractal dimension of the network).

6.2.2.2 Composite Syntheses and Applications

Xerogels-conducting polymer composites

Similarly to what is found for WOGs, it is often easy to grow a conducting polymer into a xerogel coating, or even to create more complex composites, and examine their electrochemical behavior. Xerogels have the advantage of resulting in denser and thinner coatings, which enables the elaboration of thin composite films. However, there is a relatively scarce amount of works describing, either the synthesis, or the electrochemical properties of xerogel-based composites, compared to the huge amount of works devoted to composite materials issued from colloidal solutions of various oxides, or to coatings obtained from the deposition of analogous colloids (followed or not by a thermal treatment). These works have been compiled in a recent noteworthy review article [43].

A few years ago, polyaniline (PANI) was electrochemically grown into a TMOS xerogel cast on indium-tin oxide (ITO) electrodes [44]. Despite their poorer electrochemical properties and lower conductivities with respect to the conventional films, it could be shown that PANI was organized in these films in the form of tubules inside the xerogel matrix, which opens the way to new nanostructured materials. More recently, Roux et al. [45] have polymerized pyrrole through hybrid xerogel layers, either zirconium oxopolymers or silicon-zirconium oxopolymer composites. As usual, the films were prepared by casting a sol, followed by a mild thermal treatment, but the originality was that the pyrrole rings were attached to the inorganic polymers through an alkyl chain and a good complexant of the transition metal such as an acac (acetylacetone) or a salicylate was added in the medium. Polypyrrole can also be grown in the presence of additional pyrrole, and the growth kinetics associated with the formation of the conducting polymer have been determined: they are much slower than in solution. However, the conductivity of the films was still poor. Another remaining problem was that the xerogel layer is often relatively thick, and that the polypyrrole layer could never be grown throughout, leaving some unreacted xerogel at the xerogel-electrolyte interface.

A general disadvantage of the electrochemical syntheses appears to be that the electrode substrate has to be a conducting oxide (usually ITO). Other substrates, especially those among metals, exhibited insufficient adhesion of the xerogel, which leads to delamination of the composite in the early stages of the synthesis because of mechanical stress.

Layered V_2O_5 materials were also used as the inorganic host for either polyaniline or polypyrrole, and the resulting composites displayed good electrochemical properties as exemplified through reversible lithium ion insertion [46]. Enhanced performance was attained by synthesizing hybrid aerogels of conducting polymers-V_2O_5, which circumvents the need to add carbon to the electrode structures because of high conductivity (due to homogeneous distribution of the conducting polymer in the inorganic network) and good access to the vanadium oxide phase [47]. Such nanocomposite materials can be prepared by simultaneous polymerization of the interpenetrating organic and inorganic networks or, alternatively, by a sequential polymerization of each component. The polypyrrole/MoO_3 system constitutes another example of an organic-inorganic aerogel showing similar enhanced conductivity with respect to the corresponding xerogel [48]. These examples are part and parcel of the present increasingly growing interest in electrically conductive oxide aerogels in electrochemistry [49].

Ionomer-xerogel composites

An interesting challenge is to build new conducting materials associating the conductivity properties of ionomers and the mechanical and chemical stability of inorganic or hybrid polymers. Beyond the wide scope of the polyelectrolytes, some materials are also able to host electroactive and/or photoactive materials for emerging applications. The very first sol-gel polyelectrolyte composites described were made from polyacrylic acid and poly(sodiumstyrenesulfonate) [50], and the aim was to enhance the chemical stability of these well-known polyelectrolytes. In further approaches, Nafion® has been almost exclusively chosen for preparing ionomer-xerogel composites, most probably because of its chemical stability, its high ionic conductivity, and the ease of processing it from solution. In addition, the transparency of this polymer is an attractive feature. The first reported works were by Mauritz et al. [51] and by Seliskar and Heineman's group [52], and made by mixing a classical TEOS sol and a commercial Nafion solution, followed by aging and drying. Further works were describing impedance studies on analogous composites [53]. The next investigations were with applications, like sensing of Re(I) complexes [52] or electrogenerated chemiluminescence [54]. Apparently, diffusion into the composite was three times faster than in pure Nafion, indicating a quite interesting cooperativity in the composite, which was, in addition, completely transparent and therefore suitable for optical measurements.

Electrochemiluminescence

Electrochemiluminescence is the generation of light from excited species, which in turn are formed from the reaction between a sensitizer and an electrogenerated species of the initial compound. The widest example of such a compound is the complex ruthenium-tris(bipyridine) ($Ru(bpy)_3$) which emits light at 610 nm [55]. It can be used for detection of analytes by luminescence quenching. Collinson et al. [54] have shown that luminescence of $Ru(bpy)_3$ was detectable in Nafion/silica composites, and, again, that the response of the system was enhanced as compared to pure Nafion.

Corrosion protection

The application of sol-gel coatings remains a restricted area in corrosion protection, and only aims at forming a physical barrier between the corrosion agents and the coatings; nevertheless, this approach is somewhat limited, since previous examples show that in many cases the xerogel layer still allows the electrochemical reactions to occur and therefore displays a weak protective effect against corrosion. In some cases, however, the mechanical and thermal resistance of the xerogel provides a clear advantage compared to traditional coatings [56]. A recent improvement consisted in preparing a protective polypyrrole layer, which acts as the active protective barrier due to its high redox potential, recovered by an hybrid xerogel which insures the physical barrier against corrosion agents [57]. Interestingly, it appears again that a synergy takes place between the conducting polymer and the sol-gel components, since the achieved protection is better than that which would have been expected from a simple addition of the effects of both individual components. Another interesting approach is based on the system TiO_2-WO_3 which may act as a UV senzitizer , coupled with an electron pool [58]. The effect of the bilayer system acts as a significantly increased protection against corrosion.

6.2.3
Solid Polymer Electrolytes

Several materials prepared by sol-gel chemistry exhibit either electronic or ionic conductivity or both. They have found numerous applications as solid electrolytes in the development of batteries and fuel cell electrodes, as well as in the design of electrochromic devices [59]. Of course, this wide field of research is not the focus of the present chapter, but some aspects of these works which relate to the use of organic-inorganic hybrids are briefly mentioned hereafter. A more detailed description of the electrochemistry of sol-gel based electrolytes and sol-gel processing of electrochromic materials is given elsewhere [5].

6.2.3.1 Power Sources

Much of the interest in preparing solid electrolytes for applications in the construction of batteries or fuel cells involves protonic, sodium- and lithium-based ionic conductors, and electroactive metal oxides (V, Ce, Ru, or Ti oxides) [4, 5]. Beside the widely explored dispersion of ceramic into a polymer matrix (e.g. polyethylene oxide, Nafion) [60], most often doped with a lithium salt, efforts to produce class II hybrids were made, mainly by the groups of Popall and Poinsignon [61] or Dahmouche and Judeinstein [62, 63]. Protonic conductors were based on ammonium or sulfonic acid functions while lithium conductivity was introduced by neutralizing the acid groups with LiOH. Hybrids made of covalent binding between a silica network and poly(ethylene, propylene glycol) were also produced and served as rechargeable cathodes when doped with an appropriate lithium salt [63]. In general, interpenetrating organic-inorganic polymers gave rise to observed conductivity values higher than those obtained with covalently linked organic-inorganic electrolytes, presumably as a result of higher polymer chain mobility contributing to the overall ionic motion in the first case. Intercalation of organic compounds (mainly lithium conducting or electroactive polymers) into the layered structure of vanadium pentoxide was also exploited to combine the favorable features of each constituent and gaining therefore in increased ionic and electronic conductivity and enhanced mechanical integrity [64].

6.2.3.2 Electrochromic Devices

Electrochromism relies on the production of a persistent but reversible optical change during an electrochemical transformation. With respect to the production of all-solid-state smart windows and switchable displays, sol-gel technology has gained a clear-cut edge over other methods for preparing electrochromic electrode/electrolyte systems, often in thin-film configuration, and mainly based on metal oxides such as WO_3, V_2O_5, Nb_2O_5, TiO_2, CeO_2, or combinations of them, among others. Recent advances in this wide field are reported in several reviews [65]. Among them, some studies were devoted to the synthesis and application of organic-inorganic hybrids, but they are not widespread [66]. They are based either on the dispersion of organic dyes within a ceramic matrix (to extend the spectral color change) or on ceramic-organic polymer composites in which the organic polymer can be the electrochemically sensitive component (e.g. conducting polymer) or acts as a structure-tuning agent. Such electrochromic materials were reported to be useful for the preparation of digital papers [67].

6.3
Electroanalysis with Sol-gel Derived Hybrid Materials

Actually, the great majority of electrochemical applications involving sol-gel derived hybrid materials was in the field of electroanalysis, as reported in extensive reviews either devoted in part to [3–7, 12–14] or dealing entirely with [8–11] the intersection between electrochemistry and materials sciences. They include voltammetric detection of chemicals after selective preconcentration or permeation in a sol-gel hybrid, chemical sensors based on electrocatalysis, gas sensors, potentiometric sensors, flow-through amperometric detectors, and a wide range of biosensors [7]. The only exceptions are: (i) the electropolymerization and formation of interpenetrating ceramic networks and conducting polymers, and long-range charge transfer in redox polymers based on organically modified silicates, (ii) solid polymer electrolytes used for applications in batteries and fuel cells (ormolytes, organically modified electrolytes, and ormocers, organically modified ceramics), (iii) electrochemiluminescence, and (iv) corrosion protection. All these aspects are closely related to the basic electrochemistry of xerogels and sol-gel derived composites, which are treated in Sect. 6.2.2. of this chapter.

6.3.1
Design of Modified Electrodes

In spite of their attractive properties for advanced applications in electroanalysis, most sol-gel derived organic-inorganic hybrids are DC insulators, and very often not electroactive by themselves, so that their use in connection with electrochemistry requires a close contact between them and an electrode surface. Two main strategies were applied to this purpose: (i) the formation of so-called ceramic-carbon composite electrodes, corresponding to a dispersion of carbon particles within a ceramic-based material or, alternatively, to the incorporation of as-synthesized sol-gel derived hybrids into a carbon-based composite material, and (ii) the coating of sol-gel films in mono- or multilayer configuration onto the surface of conventional solid electrodes. Some other approaches were sparingly described. Each new electrode device was often characterized by cyclic voltammetry, either to assess its intrinsic electro-chemical activity, or to evaluate its performance with respect to solution-phase redox probes, or even to study both charge and mass transfer reactions within the material.

6.3.1.1 Bulk Ceramic-carbon Composite Electrodes (CCEs)

The concept of sol-gel derived composite ceramic-carbon electrodes was introduced and then largely developed by Lev's group [11, 68]. CCEs are obtained by performing the sol-gel process in the presence of a carbon powder dispersion, leading during gelation to the formation of a three-dimensional ceramic network (most often silicate) entrapping the carbon particles. A solid composite is formed after drying. It displays good mechanical stability due to the solid silicate backbone, good conductivity attributed to electron percolation through the interconnected carbon particles, and characteristic physico-chemical properties of the matrix which can be tuned by appropriate choice of the monomers constituting the starting solution of precursor(s). Both the nature, the particle size, and the weight percent of carbon in the material can influence strongly the performance of CCEs. In particular, the carbon content must be high enough to ensure conductivity: the percolation threshold is typically 4–5% for carbon black while 10–30% graphite is required to get appreciable volume conductivity (around 5 $(\Omega\ cm)^{-1}$) [68]. On the other hand, too high a carbon content induces loss of rigidity and even disintegration of the CCE structure. Porosity of CCEs is also affected by the water-to-silicon ratio in the starting sol [69]. CCE surfaces are easily renewed by mechanical smoothing.

The inherent ability of the sol-gel process to entrap selected components, in addition to carbon particles, within a ceramic matrix, simply by adding them in the starting sol has led to a wide range of chemicaly modified CCEs. As pioneered by Lev's group, CCEs were doped with metals [70], small organic species (e.g. mediators) [71], organometallic moieties [39], and enzymes, either alone or together with mediators [72]. Later on, this approach was extended by other groups for various application purposes (see Sect. 6.3.2.). Beside the straightforward "sol doping" route, an elegant way to produce chemically-modified CCEs is to incorporate one (or more) organo-trialkoxysilane precursor(s) in the starting sol and to proceed via the hydrolysis-co-condensation pathway to form a hybrid organic-inorganic structure acting as a binder for carbon particles [11]. Organically modified silicates are attractive because they can be manufactured in order to control the hydrophobicity of the composite electrode or to bring suitable chemical functions to the CCE surface. For example, the incorporation of hydrophobic organofunctional groups leads to restricted access of aqueous electrolyte solutions to the electrode surface and contributes therefore to lowering the background currents. The formation of nanocomposite structures containing specific organo-functional groups covalently attached to the ceramic backbone was only sparingly studied [73], but this approach is thought to attract increasing interest in the near future because it allows the durable linkage of a desired reactant at the electrode surface, by contrast with the doping processes that often suffer from leaching of the modifier into the external solution. Such organic function can also serve as an intermediate to bind additional electrode modifiers, as exemplified by the anchoring of carboxyl-bearing reagents (such as enzymes or mediators) by immobilized amine ligands [73]. An example of chemically integrated system in CCE is illustrated in Fig. 6.4 for a sol-gel derived, ferrocenyl-, methyl-, and amine-modified silicate-graphite composite electrode, where the silicate backbone ensures mechanical stability of the electrode, the amine groups participate in the durable immobilization of both the enzyme and the mediator, carbon allows percolation of the electrons and methyl groups impart hydrophobicity to the electrode surface. This figure illustrates the heterogeneous nature of the CCE matrix at various scales and gives a simple idea of the general arrangement of each component in the composite. In addition to CCEs, metal-ceramic composite electrodes, where carbon particles are replaced by gold grains, have been described and exploited for their catalytic properties [74].

The inherent versatility of the sol-gel molding technology has led to production of CCEs in a wide range of configurations, including monolithic rods, bulky cylinders, flat plates, or even thin films [68, 75]. A special case, developed by Wang's group, involves the combination of sol-gel chemistry with thick film technology to produce screen-printed silicate containing electrodes [76]. This is achieved by mixing a sol with a carbon ink (carbon particles + cellulose acetate + solvents), which is then coated by using a screen-printer onto ceramic plates and allowed to react, leading to a composite film after drying. This approach is also suitable for the preparation of single-use sensors incorporating organofunctional ligands, mediators, catalysts, or biomolecules, for specific applications [77]. A recent study has compared the screen-printed and sol-gel formats of cyclodextrin-modified mediator-containing biosensors [78].

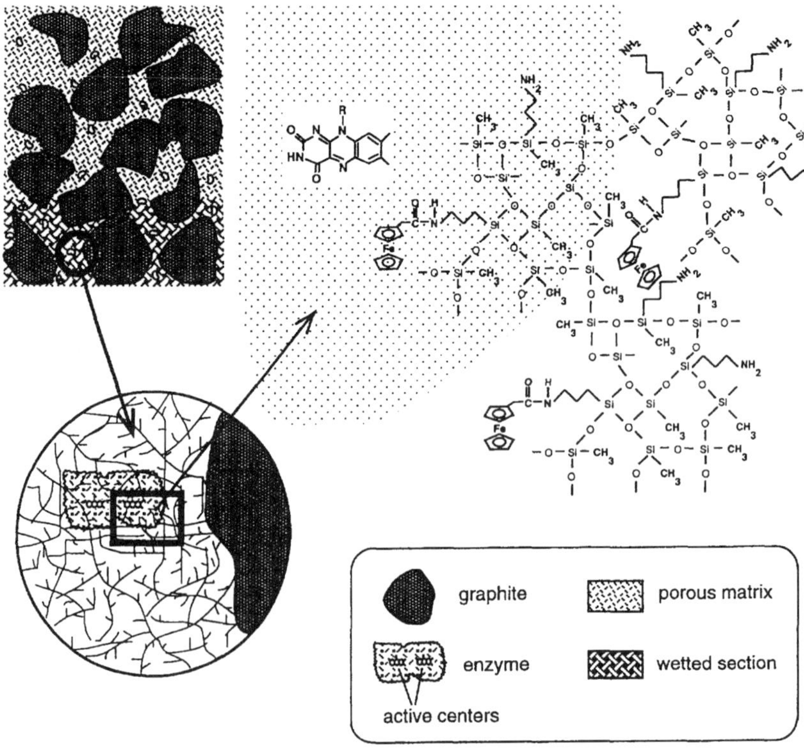

Fig. 6.4 A scheme for glucose oxidase doped and amine-, methyl-, and redox-modified silicate-graphite composite electrodes. *Reprinted from [73, first], Copyright (1996), with permission from Elsevier Science*

A unique property of CCEs is the ability to control their active section by tuning the wettability of their surface in aqueous solutions. This is achieved by monitoring the hydrophobic/hydrophilic balance of the composite material by an appropriate choice of its components [11]. The wetted section of CCEs can be controlled either by an appropriate selection of carbon powder (ranging from the hydrophobic carbon black to less hydrophobic graphite) [70], either by using a mixture of hydrophobic and hydrophilic organofunctional precursors (e.g. methyl- or phenyl-silicate are very hydrophobic) [11], or by incorporating hydrophilic or hydrophobic additives in the starting sol, which are then dispersed into the ceramic matrix upon gelation [79]. A similar effect was described with zeolite modified carbon paste electrodes for which a prolonged stay in aqueous electrolyte resulted in the progressive entrance of the external solution into the bulk paste due to the hydrophilic character of the aluminosilicate [80], while the active section of conventional non porous carbon paste electrodes is usually confined to their outermost surface. CCEs are more robust than carbon paste electrodes and their high mechanical stability allows prolonged operation in flowing streams, as required in flow detection analysis [40].

6.3.1.2 **Film-based Sol-gel Electrodes**

The simplest course to bring into contact a sol-gel derived material with an electrode is to exploit the fluidic character of the sol to cast it on the surface of a solid electrode, allowing gelation, and drying to get a xerogel film [4–10]. Deposition of the sol can be performed either by dropping a sol aliquot directly onto the solid substrate, or by spin-coating, or by dip-coating. These relatively simple processes are often sufficient to yield high quality films [81], and this strategy of film formation of electrode surfaces was the most widely used approach to combine sol-gel chemistry and electrochemical science [4–10].

The versatility of sol-gel technology for providing a wide range of material compositions, including the formation of nanocomposite organic-inorganic hybrids (via covalent binding, interpenetrating networks, doping of reactants etc.), has been exploited to construct several film-based sol-gel electrodes with various compositions and structures, in both mono- and multilayer configurations (Fig. 6.5). Hybrid coatings of both class I and class II have been reported [5, 7]. They were deposited essentially on glassy carbon (GC), indium-tin oxide (ITO), or platinum electrodes. The most widely used precursors are silicon alkoxides (TMOS, TEOS) and organofunctional silicon alkoxides (see e.g. many examples in recent reviews [5, 7]), while some other metal alkoxides (metal = Al, Ti, V, Zr, Ce, Nb etc.) were sparingly employed to form organic-inorganic hybrid films on an electrode surface [82, 83]. Of course, sol-gel processed films based on various metal alkoxides have found numerous applications in electrochemistry (electrochromic displays, photoelectronics, solid electrolyte devices, gas sensors etc. [84]), but most of them were not organically modified, so that they are not treated in this chapter. Class I hybrid films are made of ceramic oxo-polymers, either interpenetrated with organic macromolecules (ion exchange, redox, or conducting polymers) [43–45, 50–53], or doped with an organic reactant (e.g. electroactive species, charge transfer mediator, complexing ligand) [85, 86], or entrapping a biomolecule (mainly enzymes), alone or together with a mediator [9, 87, 88]. Class II hybrids offer the advantage of durable fixation of the organic reactants (redox active moieties [37, 42, 89], ion exchange sites [90, 91], complexation centers [92]) to the solid inorganic component of the film. Physical entrapment of biomolecules and/or mediators into these class II films was also performed [93, 94], which often exploited the particular affinity of the organic component for the reactive dopant; the resulting composites are called doped ormocers (organically modified ceramics) or doped ormosils (organically modified silicates). As also illustrated in Fig. 6.5, multilayer configurations can be obtained by the successive deposition of selected components. This was especially applied in the field of biosensors [9], where an enzyme layer was confined between the electrode surface and a silicate layer (preventing leaching of the enzymes into the external solution). An additional layer containing an appropriate mediator can even be deposited first onto the electrode surface (to act as an electron shuttle between the enzyme layer and the feeder electrode).

Even if film formation via the sol-gel route is intrinsically simple, obtaining electrode systems displaying good performance with respect to the target analytical application is not trivial. On the one hand, the film must adhere strongly to the electrode surface to prevent leaching (long-term stability) and, on the other hand, it must be porous enough to impart easy mass transfer of reactants to the electrode surface (high sensitivity). Controlling the film porosity can be achieved by modulation of its permeation properties, by tailoring the internal structure or incorporating charge-selective groups [90, 95, 96]. In case of restricted mass transport, an alternative to the latter structure modification is the use of charge-transfer mediators or conducting polymers, which promote electron exchanges between the electrode surface and the electroactive species located in the solution [37, 43, 88, 89, 94]. Another important point is related to the efforts directed to avoid the forma-

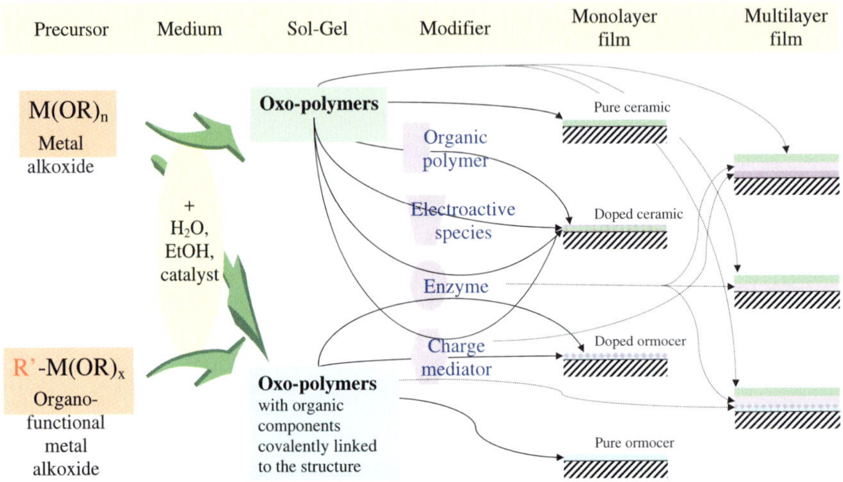

Fig. 6.5 Scheme illustrating the various ways to get organic-inorganic hybrid films on electrode surfaces

tion of cracks in the film, due to the shrinkage effect occuring during the gel-to-xerogel transformation, in order to get homogeneous coatings. This can be achieved by doping the starting sol with an appropriate organic polymer: examples of these additives that have been exploited in electrochemistry are poly(vinyl)alcohol-grafted-poly(vinyl)pyridine, poly(ethylene)glycol, or *N*-methyl-*2*-pyrrolidone [97, 98]. Finally, the organic components occluded within or attached to the inorganic backbone should be durably confined to the electrode/solution interface, in order to maintain long-life activity of the electrochemical device. This is efficiently achieved when using class II hybrids, for which the non-hydrolyzable covalent bonding prevents leaching of the organic moieties [86, 90–92] (note, leaching could occur via the loss of hydrolyzed organofunctional metal oxides), or when exploiting the particular affinity of the organic reactants for the ormosil structure [86, 87, 94], or when using a nanoglue such as glutaraldehyde [99]. Sometimes, the simple physical encapsulation can result in efficient and durable immobilization of the reactant in the ceramic host structure, especially when they are voluminous, such as biomolecules or large organometallic compounds [86, 90, 100].

6.3.1.3 **Other Electrode Systems**

In addition to bulk composites and sol-gel films, some other strategies were applied to fabricate electrochemical devices integrating organic-inorganic hybrid materials.

Sol-gel chemistry was applied to the production of silica membranes encapsulating large ionophores, such as crown ethers [101] or long-chain tetraalkylammonium [102]. Most often, they were used for potentiometric determinations in a similar way to that in which the glass electrode is used for monitoring proton activities. Such membranes can be used as synthesized [102] or as coated-on ISFET (ion-sensitive field-effect transistor) devices [101]. The ionophores are either physically entrapped within or covalently bonded to the silicate matrix [101, 103].

As an alternative to ceramic carbon electrodes, other carbon-based composites were used as a host matrix for as-synthesized sol-gel hybrid materials. Among them, the most common is carbon paste, a homogeneous mixture of graphite particles compacted with an organic binder, in which several organic-inorganic hybrids have been dispersed, such as polysiloxane immobilized amine ligands [104, 105] or enzyme-silica microreactors [106]. This approach was also largely applied to characterize the electrochemical response of silica gels coated with metal oxides and inorganic catalysts, as well as that of silica samples grafted with various organic groups [7, 8, 10]. It is especially attractive for investigating the electrochemistry of new nanostructured mesoporous materials (e.g. the MCM-41 family [107]) that are formed via a surfactant-assisted template route, which is not compatible with the production of CCE and for which the formation of homogeneous films remains difficult.

Finally, a new avenue for the production of sol-gel derived ormosil films is the initiation of hydrolysis-condensation reactions by electrochemical generation of the catalyst. This has been exemplified for the electrodeposition of a MTMOS sol-gel film on ITO electrodes by generating cathodically the OH^- catalyst [108].

6.3.2
Analytical Applications

6.3.2.1 Analysis of Chemicals

Two recurrent requirements in designing efficient electrochemical (and other) sensing devices are high selectivity and good sensitivity. These two key parameters can be adjusted and optimized when operating with chemically modified electrodes, including those based on sol-gel organic-inorganic hybrids, by an appropriate choice of the modifier composition, structure and properties. Improving sensitivity in trace analysis can be achieved by the chemical accumulation of the target analyte at the electrode surface (e.g. by reacting with the modifier). Preconcentration efficiency is governed by the particular affinity of the modifier for the analyte (expressed by the distribution coefficient), and by the speed of mass transfer processes. Also, a suitable choice of modifier will contribute to improving selectivity of detection by way of preferential binding of the target analyte over other species and/or by discriminating against interference by preventing some species from diffusion to the electrode surface. Enhanced selectivity can be obtained by immobilizing a catalyst at/onto the electrode surface, which, owing to the detection of the target analyte at a potential value close to its thermodynamic formal potential, requires the application of lower overvoltages (electro-catalysis), making the system less sensitive to electroactive interferent species. The use of electrocatalysts also contributes to improving the detection sensitivity by increasing the signal-to-background current ratio.

Electrocatalysis: detection of electroactive species and gas sensing

The principle of mediated electrocatalysis is described by Eqs. 1–3 in the case of the reduction of a substrate S_{Ox} into a product P_{Red}. Basically, if electrochemical reduction of S_{Ox} (Eq. 1) occurs at high overpotentials ($E_S^{\circ\prime} \ll E_S^{\circ}$), the use of a mediator couple (M_{Ox}/M_{Red}) allows lowering of the overpotential barrier by electrogenerating M_{Red} at $E_M^{\circ\prime} = E_M^{\circ}$ (Eq. 3), which then reacts chemically with S_{Ox} (Eq. 2). The overall process results therefore in the electrochemical transformation of S_{Ox} at a much lower overpotential than the initial substrate in the absence of mediator ($E_M^{\circ\prime} \gg E_S^{\circ\prime}$).

$$S_{Ox} + ne^- \rightarrow P_{Red} \qquad (E_S^{\circ\prime} \ll E_S^{\circ}) \tag{1}$$

$$S_{Ox} + M_{Red} \rightarrow P_{Red} + M_{Ox} \qquad (\text{occurring if } E_S^{\circ} > E_M^{\circ}) \tag{2}$$

$$M_{Ox} + ne^- \rightarrow M_{Red} \qquad (E_M^{\circ\prime} \cong E_M^{\circ}) \tag{3}$$

In this respect, resorting to electrodes modified with sol-gel derived organic-inorganic hybrids may take the advantage of immobilizing organic mediators onto/within a ceramic matrix. Several examples exploiting this approach in the electroanalysis of chemicals are described hereafter.

Electrocatalytic CCEs were prepared by dispersing various catalysts and graphite particles in a sol-gel derived organically modified network. An example from Wang's group involves phthalocyanine-dispersed CCEs, which exploit both the catalytic activity of the organometallic modifier and the mechanical stability of the composite for the flow injection analysis of H_2O_2, N_2H_4, oxalic acid, thiourea and cysteine [40]. On the other hand, the same group has demonstrated that nonsilicate metal alkoxide precursors offer a substantial decrease in the overvoltage of the NADH oxidation reaction without requiring any additional mediator [76], similarly to metal-ceramic composite electrodes that displayed electrocatalytic activity due to small gold particles [74]. Direct unmediated electrocatalysis by metal-doped CCEs (metal = platinum, rhodium, palladium) was also reported by Lev's group [70, 109]. Other examples by Cox and co-workers are based on rhodium polyoxometalates [110] and ruthenium metallodendrimers [111] entrapped within a CCE matrix and applied to the amperometric detection of proteins, peptides, and pollutants such as arsenic or cyanide. CCEs modified with encapsulated silico-, phospho-, germano- and isopolymolybdic compounds or manganous hexacyanoferrate were successfully employed for the amperometric sensing of either NO_2^-, BrO_3^-, ascorbic acid, or cysteine [112]. Most other CCEs doped with small organic mediators were directed to biosensing applications (see Sect. 6.3.2.2). Examples of electrocatalysis with film-based sol-gel electrodes are not widespread [113], even if organically modified silicate films with mediators covalently attached to the material [37, 42, 89, 114] are interesting candidates for this purpose. Considerable work was also conducted by the groups of Kubota and Gushikem [115] on electrocatalysts supported on silica gels coated with metal oxides (TiO_2, ZrO_2, Nb_2O_5) or titanium phosphate layers, applied to the voltammetric sensing of various solution-phase analytes, but these investigations are not considering sol-gel processes so they are not detailed here. The analytical performance (linear range, detection limit) of such electrocatalytic systems is reported elsewhere [8, 10].

The electrocatalytic properties of doped CCEs were extended to the fabrication of gas sensing devices. This field was developed by Rabinovich, Lev and co-workers [39, 70, 109, 116, 117]. Two major attractive properties of CCEs, with respect to application as gas sensors, are their hydrophobic structure permeable to gases and the ability to control the wetted section at the electrode/solution interface [109]. In such a way, the target gas can be introduced at the top of the composite electrode and can react after having diffused through the CCE to reach the wetted section of the electrode. CCE-based gas sensors exhibit high current densities, and efficient detection was ensured by selected catalysts, such as cobalt porphyrin, phthalocyanines [39, 116], or inert metals [70, 109]. In all these configurations, the catalyst is dispersed within the composite electrode matrix and the gaseous reactant is transformed upon an electrochemical excitation. A new pathway was recently proposed, on the basis of CCE containing naphthoquinone groups covalently bonded to the ceramic backbone, which was applied as an oxygen gas permeable electrode [117]. In this configuration (Fig. 6.6), oxygen diffuses from the back of the electrode to the wetted section at the composite/solution interface where it reacts with the hydroquinone moieties (QH_2 in Fig. 6.6); then charge transport is carried out by a self-exchange mechanism which leads to the regeneration of the mediator by electrochemical reduction of the quinone (Q in Fig. 6.6) on a carbon surface. Mediated oxidation and determination of gaseous monomethyl hydrazine and ammonia in a solid-state voltammetric cell employing a sol-gel electrolyte was also reported [118].

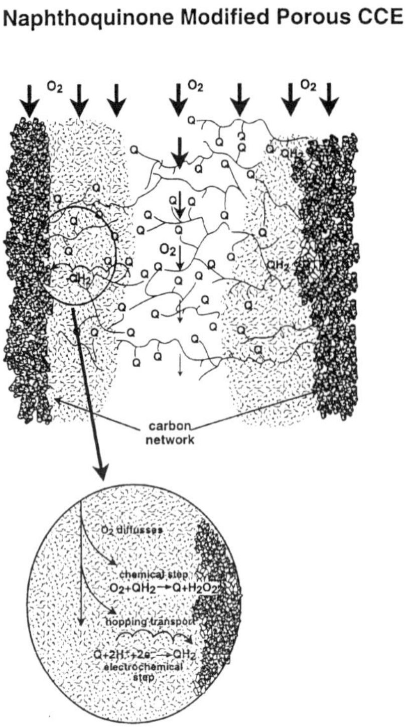

Fig. 6.6 Oxygen reduction on redox polymer-modified carbon ceramic gas electrodes. *Reprinted from [117], Copyright (2001), with permission from Elsevier Science*

Selective recognition – discrimination based on nature, charge, or size of the analyte
Preconcentration analysis is a well-established procedure applied to increase the sensitivity of chemically modified electrodes [2]. It involves generally two successive steps: (i) accumulation at open circuit by (as selective as possible) binding to active sites in/on the modifier, followed by (ii) the transfer to a pure electrolyte solution where the voltammetric quantification of the preconcentrated analyte is carried out. This analytical concept was utilized in sol-gel electrochemistry by focussing on the selection of appropriate organic functions liable to bind preferentially a target analyte, as well as in producing open ceramic frameworks enabling high diffusion rates for the reactants. Efforts were also directed to produce permselective coatings that are able to screen the analyte out from a mixture containing interferences.

Accumulation by adsorption on ormosils or complexation with immobilized ligands was performed on both class I and class II hybrids. Ni^{II} and Fe^{II} were determined by voltammetry after accumulation by complex formation with dimethylglyoxime [119] or dimethylphenanthroline [120] respectively, dispersed within a CCE. Another family of class I hybrids applied to preconcentration of anionic or cationic electroactive analytes is constituted by interpenetrated ion exchange polymers and ceramic binders [95, 121–123]. Accumulation in this case is governed by ion exchange, and significant improvements of mass transfer, in terms of enhanced diffusion processes, were reported for these hybrids in comparison to the pure organic polymers [95]. Both cation exchangers (e.g. polyethylene- or polystyrene-sulfonate and Nafion [95, 121]) and cation exchange polymers (e.g. poly(dimethyldiallyl-ammonium) or poly(vinylbenzyltrimethylammonium) [122, 123]) were used for this purpose. They were applied as thin composite films for the detection of cationic analytes (such as methylviologen, $Ru(NH_3)_6^{3+}$, $Ru(bpy)_3^{2+}$, $[Re^I(DMPE)_3]^+$ with DMPE = 1,2-bis-(dimethyl-phosphino)ethane), or anions ($Fe(CN)_6{}^{3-/4-}$).

Class II hybrids have been utilized in order to circumvent the activity loss of the sensor due to the progressive leaching of ligands that were simply encapsulated in the sol-gel matrix [124]. They involve the use of organofunctional metal alkoxides (e.g. Scheme 6.1) to form a covalent bond between the complexation centers and the inorganic matrix owing to durable retention capabilities. They were applied to the preconcentration and subsequent voltammetric detection of heavy metals such as copper and mercury [105, 125–128]. In particular, the preparation method of the hybrid material was found to affect greatly the performance of the analytical device, through both the accessibility and diffusion rates within the material, for which the more open structures gave rise to higher preconcentration efficiencies and therefore better sensitivity of the detection step [128]. The crystalline mesoporous organic-inorganic hybrids that combine highly ordered silicas with controlled texture and exceptionally high surface area with large amounts of accessible chemical groups are very attractive for this purpose [129]. This is illustrated in Fig. 6.7, where three silica samples grafted with thiol groups are compared with respect to the speed at which Hg^{II} analytes are reaching the complexing centers inside the mate-

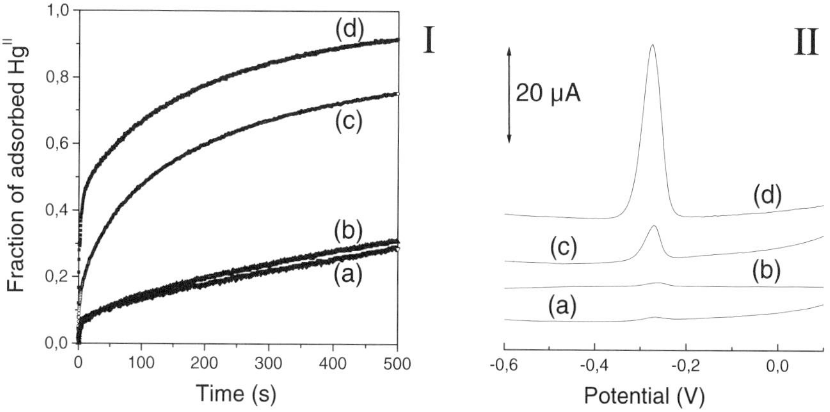

Fig. 6.7 (**I**) Uptake of Hg^{II}, as a function of time, by (a) MPS-K100, (b) MPS-K60, (c) MPS-MCM30, and (d) MPS-MCM60 from a 0.1 mM $Hg(NO_3)_2$ solution (in 0.5 M HCl); Hg(II) was in excess with respect to the –SH groups; Q_0 is the maximum amount of accessible –SH groups. Abbreviations: MPS: mercaptopropyl-grafted silica; K100: silica gel with 100 Å pore size (Kieselgel 100); K60: silica gel with 60 Å pore size (Kieselgel 60); MCM30: mesoporous silica of the MCM-41 type with 30 Å pore size; MCM60: mesoporous silica of the MCM-41 type with 60 Å pore size. (**II**) Electrochemical curves obtained with the corresponding materials incorporated in carbon paste electrodes, after 2 min accumulation in 0.001 mM Hg^{II}, recorded after transfer to an analyte-free electrolyte solution (5% thiourea in 0.1 M HCl) by applying anodic stripping voltammetry. Higher accumulation efficiencies are obtained for the ordered mesoporous structures and within that group for larger pore materials

Scheme 1

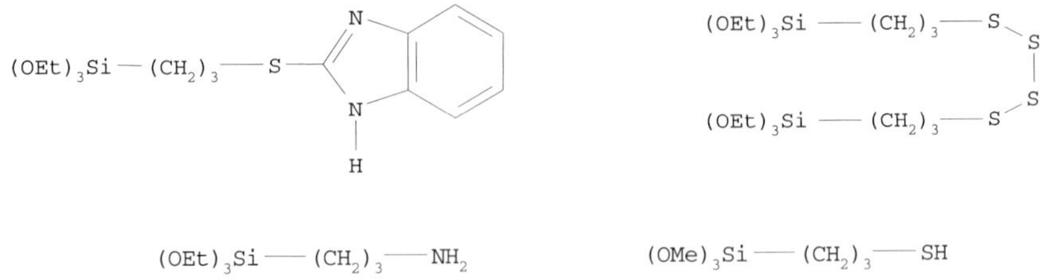

rial and to their relative performance in preconcentration electrochemical analysis. Higher accumulation efficiencies, and hence better voltammetric sensitivities, were obtained for the more organized materials and, to a lesser extent, for hybrids displaying the larger pore size [129].

Film-based class II hybrids containing permanent charges (e.g. $-NH_3^+$ or $-COO^-$ groups) have been coated on solid electrodes as permselective barriers to ionic electroactive species [90, 96]. Hence, large voltammetric currents were obtained for the positively charged methylviologen or $Ru(NH_3)_6^{3+}$ species at the $-COO^-$-containing films, while analytes such as $Fe(CN)_6^{3-}$ did not exhibit any significant signal. Using $-NH_3^+$-containing coatings prepared from 3-aminopropyltriethoxysilane (APTES) led to inversion of the charge selectivity [90]; as shown in Fig. 6.8, high partitioning of $Fe(CN)_6^{3-}$ into the film resulted in large voltammetric signals that increased with the soaking time of the electrode in the solution, while $Ru(NH_3)_6^{3+}$ species did not reach the electrode surface and gave rise to negligible currents. Similar selectivity was observed with glassy carbon or platinum electrodes covered by a delaminated Laponite® clay modified with octakis(3-amino-propylsilsesquioxane) by the sol-gel method [130]. Even more impressive discrimination was achieved by Makote and

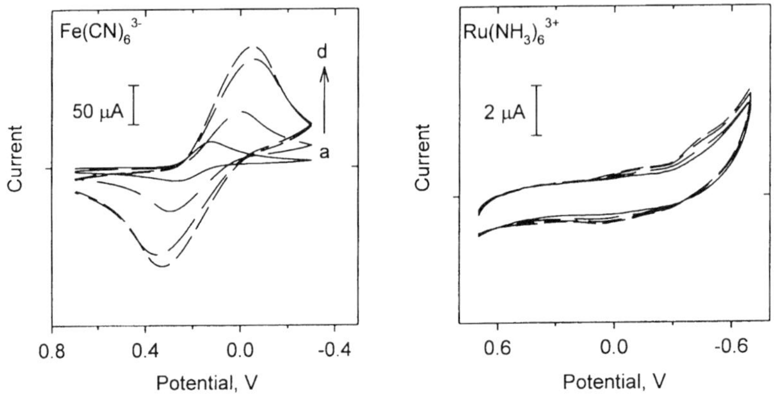

Fig. 6.8 Cyclic voltammograms of 1 mM $Fe(CN)_6^{3-}$ or 1 mM $Ru(NH_3)_6^{3+}$ in 0.05 M, pH 4.0 KHP buffer at a 1:1 MTMOS:APTES modified glassy carbon electrode. Curves were acquired (a) ca. 0 min, (b) 10 min, (c) 60 min, and (d) 180 min after the electrode was placed in solution. Scan rate: 100 mV s^{-1}. The $-NH_3^+$ containing coating leads to a marked selectivity for anions. *Reprinted from [96, second], Copyright (1999), with permission from Elsevier Science*

Collinson who have performed the synthesis of a phenyl-, methyl-silicate structure using dopamine as a templating agent [131]; the resulting material, when spin-coated on glassy carbon after extraction of the template, was found to recognize selectively the dopamine analyte over structurally related drugs [96]. This molecular imprinting approach combined with electrochemistry led also to the highly selective determination of dopamine in the presence of ascorbic acid, one of its major interferences.

Spectroelectrochemical sensing

Spectroelectrochemical sensing devices have typically two modes of selectivity; one is based on potential selection (amperometric or voltammetric detection) and the other arises from appropriate wavelength selection (optical sensing). Heineman, Seliskar and co-workers have highlighted the attractiveness of sol-gel processing of hybrid materials to design a new sensor concept that had a third level of selectivity [132]. It consists of an optically transparent electrode (most often ITO) coated with a thin permselective film based on interpenetrated silicate and ion exchange polymers. It operates in an attenuated total reflection (ATR) spectroscopic mode for detection of species within the film, so that only those species being significantly partitioned into the ormosil material are liable to be detected. The analyte partitioned into the film can undergo electrochemical modulation, and the resulting oxidation or reduction product is optically monitored as the device signal. This constitutes a new family of sol-gel-based sensors designed to improve selectivity by combining electrochemistry, spectroscopy, and selective partitioning through a coated thin film (three selectivity levels).

Feasibility of the concept was first evidenced for the model analyte $Fe(CN)_6^{4-}$ which was detected by the transmittance change of the ATR beam resulting from its electrochemical oxidation, after it had partitioned into an anion exchange composite formulated by entrapping poly(dimethyldiallylammonium chloride) in a cross-linked silica network [132]. The application of suitable potential steps or waveforms results in the generation of optically detectable $Fe(CN)_6^{3-}$ species, the intensity of the ATR signal being related to the concentration of the starting $Fe(CN)_6^{4-}$ analyte (Fig. 6.9). In the presence of a competing electroactive anion (e.g. $Ru(CN)_6^{4-}$), which also displays significant permeation into the film, selectivity for $Fe(CN)_6^{4-}$ was obtained by restricting the electrolysis potential window to the sole $Fe(CN)_6^{4-}$ oxidation. Alternatively, an appropriate selection of the wavelength for recording the ATR signal can lead to selectivity for $Ru(CN)_6^{4-}$ species [132]. When using a silicate/polymer composite characterized by cation exchange properties (e.g. ion exchange polymer = Nafion), selective partitioning of cations was observed while excluding anions, as exemplified for the selective detection of $Ru(bpy)_3^{2+}$ over $Fe(CN)_6^{4-}$ [132]. Good sensitivity in the μM to mM concentration range was achieved. Averaging the optical response upon continuously cycling potential resulted in significant lowering of the detection limit [133]; a simulation program was applied to optimize the excitation potential waveform [134]. A prototype waveguide spectroelectrochemical sensor was also developed to replace the multiple internal reflection optic consisting of a simple modified ITO electrode/glass substrate bilayer

[135]. Nonabsorbing analytes such as ascorbic acid or ascorbate can be detected by the spectroelectrochemical sensor if an appropriate mediator (e.g. $Ru(bpy)_3^{2+}$) is incorporated in the device [136]. Extension of the approach to a poly(vinyl alcohol) (PVA) – polyelectrolyte blend, where ceramic binder was replaced by PVA, was also described [137].

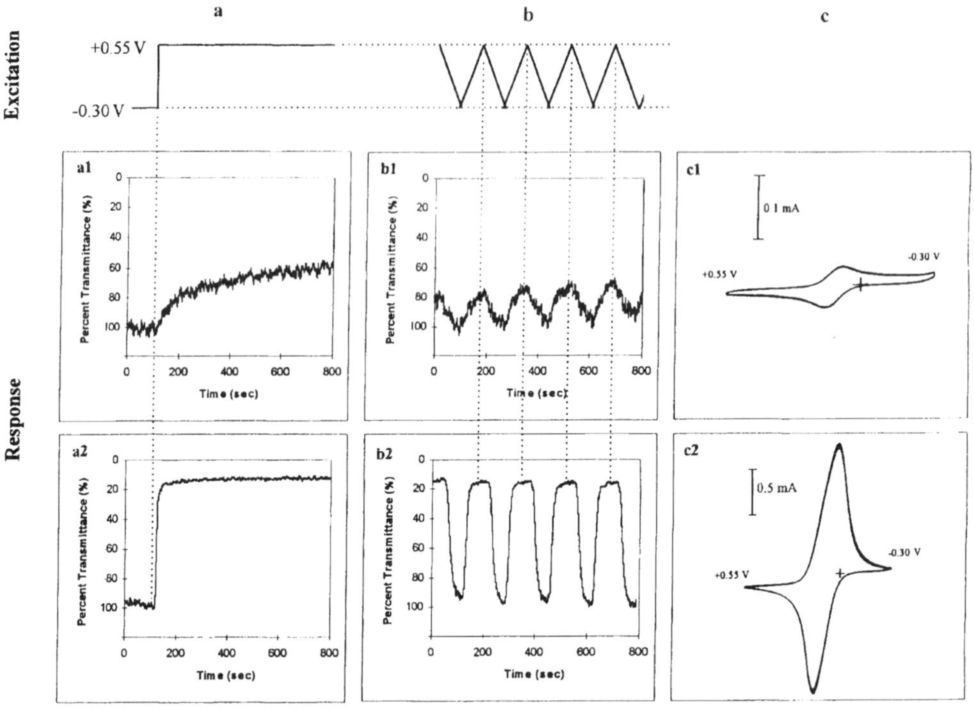

Fig. 6.9 A composite figure consisting of vertical columns and horizontal panels indicating spectroelectrochemical signals for a sensor made of ITO coated with a sol-gel derived anion-selective film. The horizontal panels depict results for the sensor exposed to two different concentrations of $Fe(CN)_6^{4-}$. The concentrations are as follows: panel 1,0.025 mM; panel 2:2.5 mM. The electrical potentials applied to the sensor are illustrated above the panels and are appropriate for a potential step (column a) and a repeating triangular potential modulation (column b). Column a depicts the changes in transmittance of the sensor induced by a step potential from −0.30 to +0.55 V for the 0.025 mM (a1) and 2.5 mM (a2) $Fe(CN)_6^{4-}$ solution exposures. Column b depicts the changes in transmittance of the same sensor induced by cyclic potential modulation, between the limits of +0.55 and −0.30 V for the 0.025 mM (b1) and 2.5 mM (b2) solution exposures. Column c depicts the cyclic voltammograms recorded simultaneously with the transmittance changes shown in column b, with c1 and c2 corresponding to the sensor exposures to 0.025 and 2.5 mM $Fe(CN)_6^{4-}$ solutions, respectively. Cyclic voltammograms were recorded about three cycles after changing the oxidizing potential as shown in a1 and a2. Cyclic voltammetry and optical parameters were as follows: $\nu = 10$ mV s^{-1}, *E/V* vs Ag/AgCl; $\lambda = 420$ nm. *Reprinted with permission from [132, first]. Copyright (1997) American Chemical Society*

Multiple or continuous analyses

The issue of routine analysis is central to the operation of any sensing device. To circumvent the often intricate and fastidious manipulations of conventional electrode instrumentation, screen-printing technology was used for the mass production of single-use disposable thick film electrodes. This method is compatible with the fabrication of sensing devices based on chemically modified electrodes [2], and Wang's group has adapted the process to the case of sol-gel derived ceramic carbon composites [69, 76]. This is achieved by dispersing carbon particles and hydroxylpropyl-cellulose (alone or together with an additional modifying agent) into a sol solution, which is then printed on a ceramic substrate and allowed to gelify to form the composite strip after curing. A strict control of the water content during acid-catalyzed hydrolysis enables tailoring of the microstructure of such sol-gel-derived strips, for which the resulting tunable macroporosity can be used to impart mass transport limitations or to increase diffusion rates [69]. This new kind of disposable electrode was especially exploited in the biosensing field, but its compatibility with the incorporation of catalysts or complexing ligand makes the methodology applicable to voltammetric analysis after preconcentration and electrocatalysis [77].

The mechanical stability of CCEs (ceramic-carbon composite electrodes) makes them promising for use as electrochemical flow detectors. Several examples are available where a CCE is operating in a wall-jet configuration for the flow injection analysis of chemicals [40, 112, 138]. For example, higher sensitivity and reproducibility was achieved with CCEs for the detection of neurotransmitters as compared to corresponding carbon paste electrodes, when evaluated as amperometric detectors in liquid chromatography [138]. A similar system was applied as an amperometric detector in capillary electrophoresis [139]. Detection of cyanides and peptides after chromatographic separation was also performed using CCEs doped with an appropriate catalyst, such as a rhodium polyoxometalate or ruthenium metallodendrimer [110, 111]. CCEs modified with cuprous oxide were suitable for the detection of carbohydrates after separation in capillary electrophoresis [140]. Finally, the electrochemiluminescence of $Ru(bpy)_3^{2+}$ immobilized in poly(styrenesulfonate)-silica-triton X-100 composite films was applied to the flow injection analysis of oxalate, tripropylamine and NADH [141].

Potentiometric determinations

Sol-gel-derived organic-inorganic hybrids have been manufactured in membrane forms which were then applied as ion-sensitive electrodes (ISE) or coated on ISFET-based sensors (ISFET = ion-sensitive field-effect transistor) for the potentiometric determination of non-electroactive cations and anions. The simplest approach consists in doping the starting sol with an appropriate ionophore (e.g. tetraalkyl-ammonium for the detection of Cl^- ions [102, 103], or other more complex systems [142]), but it implies that encapsulation is strong enough to avoid (or at least limit) leaching of the active material into the external solution (this can be achieved when bulky dopants are dispersed in a xerogel matrix). From this point of view, the use of organic functions linked covalently to the ceramic backbone is more attractive, but often requires synthesis efforts to produce the corresponding precursors [125,

143–145]. Examples of organofunctional silicon alkoxide precursors containing covalently attached ionophores, neutral carriers, permanent charge structures or electroactive groups sensitive to ionized species are depicted in Scheme 6.2.

Another route to potentiometric sensing is to exploit the pH-sensitive response of redox functions in ormosils. This was demonstrated for the naphthoquinone-silicate electrode, which shows a linear dependence of the voltammetric peak potentials against pH on the 2–7 range [117].

6.3.2.2 Biosensors

Electrochemical biosensors are modified electrode devices that rely on the intimate coupling of a biomolecule (most often an enzyme) and an electrode transducer that converts a specific biological recognition event into a measurable electrical signal. Successful operation of biosensors involves therefore effective immobilization of the biomolecule while retaining its activity over prolonged periods of time, as well as its affinity for and accessibility towards the target analyte. In addition, many amperometric biosensors, particularly the oxidase-based ones, require the use of charge transfer mediators or electrocatalysts. To be efficient, these latter should be immobilized in the close vicinity of the electrode surface while keeping mobile enough to facilitate communication between the active sites of the biomolecule and

Scheme 2

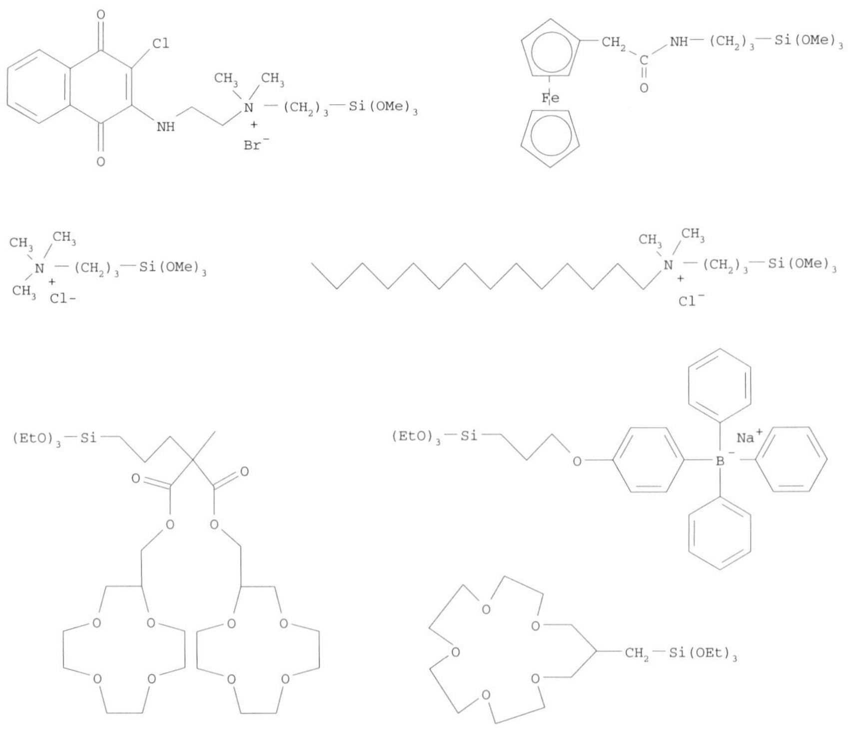

the electrode surface. For both these immobilization purposes, sol-gel materials were found to be attractive for constructing electrochemical biosensing devices, as reviewed by Wang in a recent report [9].

Sol-gel materials and immobilization of biomolecules

Basically, five methods were used to immobilize a biomolecule at the surface of or within an electrode material. They include adsorption, reticulation by using polyfunctional reagents (like glutaraldehyde), encapsulation or physical entrapment into a host matrix, covalent grafting, or molecular assembling. Because of its aqueous chemistry and the possibility of producing porous ceramics at room temperature, the sol-gel process is an obvious candidate for bioencapsulation. Considerable work was performed in this field during the last decade to entrap labile biological molecules with catalytic, recognition, and transduction functions, within porous ceramic networks [146–148]. Fig. 6.10 illustrates the generic route to prepare such composite materials, which simply involves the addition of the biological element to the sol mixture, allowing it to gelify, to age and to dry at room temperature, to give a doped siloxane network. Encapsulation is thus based on polymer lattice formation around the biomolecule which is progressively entrapped within the inorganic oxide. It was demonstrated in several cases that the biomolecules retain most of their solution activity upon encapsulation [146, 147]. A recent review covers the state-of-the-art of sol-gel bioencapsulation, including about 50 references on amperometric devices [149]. The versatility of sol-gel processes to produce a wide range of materials in various configurations (films, CCEs, monoliths etc.) guarantees these materials an important role in the realm of electrochemical biosensors [5, 7–11].

Unmediated biosensors (first generation)

Electrochemical biosensors of the first generation are based on the direct detection of the electroactive species generated (or consumed) during the biological event, without using any charge transfer co-factor. The feasibility of amperometric biosensing with using enzymes immobilized within a ceramic matrix was demonstrated for silica-based bioceramic films coated on solid electrodes, and applied to the

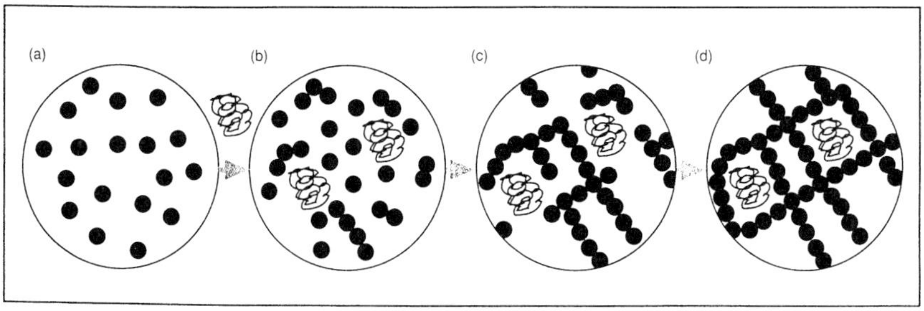

Fig. 6.10 Protein entrapment in a silicate matrix during sol-gel polymerization. *Reprinted with permission from [147, third]. Copyright (1994) American Chemical Society*

detection of glucose [87, 98, 150], hydrogen peroxide [100, 151], phenols [152], and urea [153], by involving glucose oxidase, peroxidase, tyrosinase, and urease, respectively. The composite films were obtained by casting the silica sol containing the enzyme by dropping or dip-coating on either glassy carbon or platinum substrates. The addition of an organic polymer into the starting sol, such as poly(vinyl alcohol)-g-poly(4-vinylpyridine) or poly(ethylene oxide), was very useful to avoid damage due to shrinkage and to produce crack-free films [98, 100, 151–153]. Mixing the sol in a cellulose solution and adding carbon particles in suspension enables the process to be compatible with thick-film technology, leading to the production of screen-printed disposable strip bioelectrodes [76, 77]. The microstructure of such films can be tuned by controlling the water content during the acid-catalyzed hydrolysis step, which leads to monitoring mass transfer reactions to the active centers and, hence, the sensitivity of the amperometric signal [76]. Adding gold nanoparticles within the biocomposites was found to improve charge transfer efficiency and to allow quantification of hydrogen peroxide at lower overpotentials [74, 154]. In addition to the single layer configuration, the use of a sol-gel silica layer overcoated on an enzyme deposit on solid electrodes, in which the silica film acts as a physical barrier to prevent the enzyme from leaching into the external solution, was also reported [97]. Sol-gel derived bioelectrodes were also produced by immobilizing enzymes in ceramic structures other than silica, such as aluminium-, vanadium-, or titanium oxides [82, 83, 155]. Recently, silica-based hybrid coatings on electrodes applied to glucose biosensing have been described to be biocompatible, which opens the door to future implantable biosensors [156]. Due to the usually high overpotentials required by these systems, however, efforts were directed to the production of second-generation biosensors by exploiting the ability of the sol-gel process to immobilize charge mediators within the ceramic matrix or to produce organic-inorganic hybrids containing the mediator covalently attached to the ceramic structure.

Mediated biosensors (second generation)

Amperometric biosensors of the second generation are utilizing redox mediators or electrocatalysts, which are interacting with species involved in the biological event; the former act as shuttles for the electrons between the electroactive species and the electrode surface. Organically modified silicates are attractive candidates to form integrated chemical systems containing spatially arranged biomolecule-mediator couples on (or within) electrode devices for biosensing applications [5, 7–11].

A simple strategy is to cover the electrode surface with a mediator layer, then an enzyme layer and, finally, a sol-gel overcoat to ensure mechanical stability [157]. The mediator can be directly entrapped in the bulk of a single-layer silicate film [87, 88, 94] and immobilization was more durable as the mediator was bigger and the xerogel structure was less open (species more strongly confined in the porous network). This approach was also applied in multilayer configurations of typically: silicate-mediator film/enzyme layer/protective silicate overcoat [158]. Mediators based on ferrocene, viologen, or tetrathia-fulvalene derivatives have been incorporated in ceramic-carbon bioelectrodes, but slow leaching in solution was observed because of the rather weak linkage between the composite material and the small mediator species [68, 72], while better stability was obtained when covalently attaching the mediator center to the enzyme [72]. Another way to improve mediator immobilization is to fix it onto the surface of carbon particles prior to their dispersion into the starting composite sol, e.g. by self-assembling the mediator onto gold-covered carbon powder pretreated with glutaraldehyde [159]. An even more general approach consists in the formation of redox-modified silicates containing the organic mediator covalently attached to the polysiloxane network, as a nanocomposite organic-inorganic hybrid, which is formed upon co-condensation of organo-functional silicon alkoxides and tetraalkoxysilanes. This was exemplified with ferrocene-based mediators in both CCEs (Fig. 6.4) [73] and thin film coatings [160]. Performance of mediated biosensors is linked to a compromise between proper entrapment of both mediator and biomolecule (long-term stability) and accessibility to the target analyte (sensitivity) concomitant to permanent activity of the electron shuttle (selectivity).

New directions

Even if preparation and some successful uses of sol-gel bioelectrodes have been achieved, little information is actually available on the factors affecting the structure of the composite materials at the microscale and even nanoscale levels, and little attention was given to the structure-activity relationship in such biosensing devices. Progress in rational design of (bio)organically modified sol-gel materials with tailor-made properties in a wide range would help in developing biosensors with improved performance. Of course, controlling all the parameters influencing the preparation of such multicomponent and heterogeneous systems is not an easy task. The resort to spatially organized structures such as the surfactant-templated mesoporous materials could be a promising route, especially because these rather new structured materials can host large molecules like enzymes or proteins [161]. A pioneering work by Cosnier et al. [162] illustrates the possibility of immobilizing glucose oxidase into mesoporous titanium dioxide and to apply the resulting system coupled to polypyrrole as an amperometric biosensor for glucose, with ensuring low overpotential for detection of the enzymatic product (hydrogen peroxide) due to the catalytic effect of titanium dioxide.

In addition to amperometric biosensors based on biocatalytic reactions, Wang et al. have introduced the first disposable immunosensor exploiting sol-gel technology for immobilizing selected antigens or antibodies [163]. It is based on rabbit immunoglobulin G and was applied to the electrochemical detection of phenolic compounds.

Another promising avenue is the formation of supramolecular assemblies organized in the space by adsorption on/within mechanically stable ceramic porous structures. An example of such a device is illustrated in Fig. 6.11 and involves the immobilization of bilayer vesicles (prepared from ferrocenic diacetylene lipid and the cell surface receptor ganglioside GM1) on a sol-gel thin film electrode. This latter offers a way to molecular recognition for which the corresponding amperometric signal is affected by the presence of enterotoxin [164]. Binding of the enterotoxin hinders the electrochemical communication between the incorporated redox lipids and the electrode surface, which can be exploited for toxin quantification.

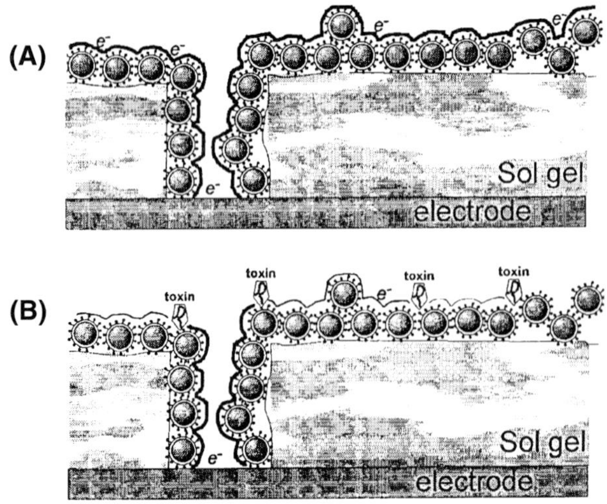

Fig. 6.11 (A) Schematic diagram illustrating lateral electron transport on vesicles adsorbed on the sol-gel thin-film electrodes. (B) Schematic diagram showing current decline as a result of toxin binding that blocks the electron transport path. *Reprinted with permission from [164]. Copyright (2000) American Chemical Society*

6.4 Conclusions

This chapter has highlighted the abundance of investigations performed at the intersection between the chemistry of sol-gel derived hybrid materials and electrochemical science. From the materials science point of view, electrochemistry applied in gelled systems or oxide xerogels, respectively doped with electroactive moieties, is very promising for pereforming a "spectroscopic" characterization of microstructures and internal reactivity. With respect to analytical chemistry, the attractive properties of (bio)organic-inorganic nano-composites can be exploited in various fields of electroanalysis, including preconcentration analysis and permselectivity, electrocatalysis, potentiometry, spectroelectrochemistry, and biosensors. The emergence of crystalline mesoporous materials is expected to bring a new dimension to these fields in the near future.

References

1 H. Eckert, M. Ward eds, Special Issue of *Chem. Mater.* **2001**, *13*.

2 C. Sanchez, F. Ribot, *New J. Chem.* **1994**, *18*, 1007; D. A. Loy, K. J. Shea, *Chem. Rev.* **1995**, *95*, 1431; D. Avnir, *Acc. Chem. Res.* **1995**, *28*, 328; U. Schubert, N. Hüsing, A. Lorentz, *Chem. Mater.* **1995**, *7*, 2010; J. E. Mark, C. Y.-C. Lee, P. A. Bianconi, *Hybrid Organic-Inorganic Composites, ACS Symp. Ser.* **1995**, Vol. *585*; T. Saegusa, *Macromol. Symp.* **1995**, *98*, 719; J. D. Mackenzie, *ACS Symp. Ser.* **1995**, *585*, 226; R. J. P. Corriu, D. Leclercq, *Angew. Chem. Int. Ed.* **1996**, *35*, 1420; J. Wen, G. L. Wilkes, *Chem. Mater.* **1996**, *8*, 1667; P. Judeinstein, C. Sanchez, *J. Mater. Chem.* **1996**, *6*, 511; A. Vioux, D. Leclercq, *Heterog. Chem. Rev.* **1996**, *3*, 65; M. P. Andrews, *Proc. SPIE-Int. Soc. Opt. Eng.* **1997**, *2997*, 48; J. Livage, *Curr. Opinion Solid State Mater. Sci.* **1997**, *2*, 132; R. J. P. Corriu, D. Leclercq, *Comments Inorg. Chem.* **1997**, *19*, 245;

S. Mann, S. L. Burkett, S. A. Davis, C. E. Fowler, N. H. Mendelson, S. D. Sims, D. Walsh, N. T. Whilton, *Chem. Mater.* **1997**, *9*, 2300; A. N. Seddon, *Crit. Rev. Opt. Sci. Technol.* **1997**, *CR68*, 143; N. Hüsing, U. Schubert, *Angew. Chem. Int. Ed.* **1998**, *37*, 22; G. Cerveau, R. J. P. Corriu, *Coord. Chem. Rev.* **1998**, *178–180*, 1051; E. Bescher, J. D. Mackenzie, *Mater. Sci. Eng.* **1998**, *C6*, 145; J. J. E. Moreau, M. Wong Chi Man, *Coord. Chem. Rev.* **1998**, *178–180*, 1073; K. Moller, T. Bein, *Chem. Mater.* **1998**, *10*, 2950; J. Livage, *Bull. Mater. Sci.* **1999**, *22*, 201; E. H. Lan, B. C. Dave, J. M. Fukuto, B. Dunn, J. I. Zink, J. S. Valentine, *J. Mater. Chem.* **1999**, *9*, 45; A. Stein, B. J. Melde, R. C. Schroden, *Adv. Mater.* **2000**, *12*, 1403; I. Gill, A. Ballesteros, *Trends Biotechnol.* **2000**, *18*, 282; A. D. Pomogailo, *Russian Chem. Rev.* **2000**, *69*, 53; C. Sanchez, B. Lebeau , F. Ribot, M. In, *J. Sol-Gel Sci. Technol.* **2000**, *19*, 31; E. Ruiz-Hitzki, B. Casal, P. Aranda, J. C. Galvan, *Rev. Inorg. Chem.* **2001**, *21*, 125; P. Gomez-Romero, *Adv. Mater.* **2001**, *13*, 163; C. Sanchez, G. J. de A. A. Soler-Illia, F. Ribot, T. Talot, C. R. Mayer, V. Cabuil, *Chem. Mater.* **2001**, *13*, 3061.

3 B. C. Dave, B. Dunn, J. I. Zink, in *Access Nanoporous Mater. [Proc. Symp.]* (Eds.: T. J. Pinnavaia, M. F. Thorpe), Plenum, New York, USA, **1995**, 141–159.

4 K. S. Alber, J. A. Cox, *Mikrochim. Acta* **1997**, *127*, 131.

5 O. Lev, Z. Wu, S. Bharathi, V. Glezer, A. Modestov, J. Gun, L. Rabinovich, S. Sampath, *Chem. Mater.* **1997**, *9*, 2354.

6 O. Lev, S. Bharathi, S. Sampath, J. Gun, L. Rabinovich, in *Book of Abstracts of the 217th ACS National Meeting (Anaheim, Calif., March 21–25, 1999)*, **1999**, PHYS-072.

7 A. Walcarius, *Chem. Mater.* **2001**, *13*, 3351.

8 A. Walcarius, *Electroanalysis* **1998**, *10*, 1217.

9 J. Wang, *Anal. Chim. Acta* **1999**, *399*, 21.

10 A. Walcarius, *Electroanalysis* **2001**, *13*, 701.

11 L. Rabinovich, O. Lev, *Electroanalysis* **2001**, *13*, 265.

12 J. Lin, C. W. Brown, *Trends Anal. Chem.* **1997**, *16*, 200.

13 M. M. Collinson, *Mikrochim. Acta* **1998**, *129*, 149.

14 M. M. Collinson, *Crit. Rev. Anal. Chem.* **1999**, *29*, 289.

15 P. Aldebert, M. Pineri, French Patent N° 86 05 792 (**1986**).

16 P. Audebert, P. Aldebert, B. Divisia-Blohorn, J. M. Kern, *J. Chem. Soc., Chem. Commun.* **1989**, 939; P. Audebert, P. Aldebert, B. Divisia-Blohorn, F. Michalak, *J. Electroanal. Chem.* **1990**, *296*, 117; C. P. Andrieux, P. Audebert, P. Aldebert, P. Hapiot, B. Divisia-Blohorn, *J. Electroanal. Chem.* **1990**, *296*, 129.

17 P. Audebert, P. Aldebert, B. Divisia-Blohorn, F. Michalak, *J. Electroanal. Chem.* **1992**, *322*, 301.

18 J. Leddy, A. J. Bard, *J. Electroanal. Chem.* **1985**, *189*, 85. See also parts of reviews on chemically modified electrodes, e. g. A. Merz, *Top. Curr. Chem.* **1990**, *152*, 49; H. D. Abruña, *Coord. Chem. Rev.*, **1988**, *86*, 135, or P. Audebert, *Trends Electrochem.* **1994**, *3*, 459.

19 P. Audebert, B. Divisia-Blohorn, J. P. Cohen-Adad, *Polymer Bull.* **1997**, *39*, 225.

20 M. W. Espenscheid, A. R. Ghatak-Roy, R. B. Moore III, R. M. Penner, M. N. Szentirmay, C. R. Martin, *J. Chem. Soc., Faraday Trans. I* **1986**, *82*, 1051; J. Wang, in *Electroanalytical Chemistry*, Vol. 16 (Ed: A. J. Bard), Marcel Dekker, New York, **1989**; K. Kalcher, *Electroanalysis* **1990**, *2*, 419; J. A. Cox, R. K. Jaworski, P. J. Kulesza, *Electroanalysis* **1991**, *3*, 869; S. A. Wring, J. P. Hart, *Analyst* **1992**, *117*, 1215; K. Kalcher, J.-M. Kauffmann, J. Wang, I. Svancara, K. Vytras, C. Neuhold, Z. Yang, *Electroanalysis* **1995**, *7*, 5; M. A. T. Gilmartin, J. P. Hart, *Analyst* **1995**, *120*, 1029; J. A. Cox, M. E. Tess, T. E. Cummings, *Rev. Anal. Chem.* **1996**, *15*, 173; S. Alegret, *Analyst* **1996**, *121*, 1751; D. Leech, in *Electroactive Polymer Electrochemistry, Part II Methods and Applications*, (Ed: M. E. G. Lyons), Plenum Press, New York, **1996**, 269–296; G. G. Wallace, M. Smyth, H. Zhao, *Trends Anal. Chem.* **1999**, *18*, 245; A. Walcarius, *Anal. Chim. Acta* **1999**, *384*, 1; J. Labuda, M. Vanickova, M. Buckova, E. Korgova, *Chem. Papers* **2000**, *54*, 95.

21 A. J. Bard, L. R. Faulkner, *Electrochemical Methods. Fundamentals and Applications*, 2nd ed., Wiley, New York, **2001**.

22 S. M. Jones, S. E. Friberg, *J. Mater. Sci. Lett.* **1996**, *15*, 1172; M. A. Aegerter, C. R. Avellaneda, A. G. Pawicka, M. Atik, *J. Sol-Gel Sci. Technol.* **1997**, *8*, 689; D. Camino, D. Deroo, D. Salardenne, *Sol. Energy Mater. Sol. Cells* **1995**, *39*, 349.

23 This chapter.

24 R. H. Terrill, J. E Hutchinson, R. W. Murray, *J. Phys. Chem. B* **1997**, *101*, 1535; M. E. Williams, H. Masui, J. W. Long, J. Malik, R. W. Murray, *J. Am. Chem. Soc.* **1997**, *119*, 1997; P. Audebert, B. Divisia-Blohorn, J. P. Cohen-Adad, *Polymer Bull.* **1997**, *39*, 225.

25 P. Audebert, P. Griesmar, C. Sanchez, *J. Mater. Chem.* **1991**, *1*, 699; P. Audebert, P. Hapiot, C. Sanchez, P. Griesmar, *J. Mater. Chem.* **1992**, *2*, 12.

26 M. M. Collinson, P. J. Zambrano, H. Wang, J. S. Taussig, *Langmuir* **1999**, *15*, 662; A. R. Howells, P. J. Zambrano, M. M. Collinson, *Anal. Chem.* **2000**, *72*, 5275.

27 M. E. Tess, J. A. Cox, *J. Electroanal. Chem.* **1998**, *457*, 163.

28 N. Leventis, M. Chen, *Chem. Mater.* **1997**, *9*, 2621.

29 P. Audebert, H. Cattey, C. Sanchez, P. Hapiot, *J. Phys. Chem.* **1998**, *102*, 1193; P. Audebert, S. Sallard, S. Sadki, submitted for publication.

30 B. Alonso, P. Audebert, F. Ribot, B. Chapusot, C. Sanchez, *J. Sol-Gel Sci. Technol.* **1994**, *2*, 809.

31 M. Onoda, T. Moritake, T. Matsuda, S. Nakamura, *Synth. Met.* **1995**, *71*, 2225.

32 S. Komaba, F. Fujihana, T. Osaka, S. Aiki, S. Nakamura, *J. Electrochem. Soc.* **1998**, *145*, 1126.

33 C. P. Andrieux, J. M. Savéant, *J. Electroanal. Chem.* **1980**, *111*, 337.

34 E. Laviron, *J. Electroanal. Chem.* **1980**, *112*, 143.

35 D. Blauch, J. M. Savéant, *J. Am. Chem. Soc.* **1992**, *114*, 3323.

36 C. P. Andrieux, P. Audebert, *J. Phys. Chem.* **2001**, *105*, 444.

37 P. Audebert, R. J. P. Corriu, G. Cerveau, N. Costa, *J. Electroanal. Chem.* **1995**, *413*, 89.

38 M. M. Collinson, H. Wang, R. Makote, A. Khramov, *J. Electroanal. Chem.* **2002**, *519*, 65.

39 M. Tsionsky, O. Lev, *Anal. Chem.* **1995**, *67*, 2409.

40 J. Wang, P. V. A. Pamidi, C. Parrado, D. S. Park, J. Pingarron, *Electroanalysis* **1997**, *9*, 908.

41 P. Audebert, C. Sanchez, H. Cattey, *New J. Chem.* **1996**, *20*, 1023.

42 P. Audebert, P. Calas, G. Cerveau, N. Costa, R. J. P. Corriu, *J. Electroanal. Chem.* **1995**, *372*, 275; S. O'Brien, J. M. Keates, S. Barlow, M. J. Drewitt, B. R. Payne, D. O'Hare, *Chem. Mater.* **1998**, *10*, 4088.

43 R. Gangopadhyay, A. De, *Chem. Mater.* **2000**, *12*, 608.

44 M. M. Verghese, K. Ramanathan, S. M. Ashraf, M. N. Kamalasanan, B. D. Malhotra, *Chem. Mater.* **1996**, *8*, 822.

45 S. Roux, P. Audebert, J. Pagetti, M. Roche, *New J. Chem.* **2000**, *24*, 885; ibid., *New J. Chem.* **2002**, in press.

46 F. Leroux, G. Goward, W. P. Power, L. F. Nazar, *J. Electrochem. Soc.* **1997**, *144*, 3886.

47 B. C. Dave, B. Dunn, F. Leroux, L. F. Nazar, H. P. Wong, *Mater. Res. Soc. Symp. Proc.* **1996**, *435*, 611.

48 W. Dong, B. Dunn, *Mater. Res. Soc. Symp. Proc.* **1999**, *576*, 269.

49 D. R. Rolison, B. Dunn, *Analyst* **2001**, *11*, 963.

50 K. Nakanishi, N. Soga, *J. Non-Cryst. Solids* **1992**, *139*, 1; ibid., *J. Am. Ceram. Soc.* **1991**, *74*, 2518; Y. Shi, C. R. Seliskar, *Chem. Mater.* **1997**, *9*, 821.

51 K. A. Mauritz, P. L. Shao, R. B. Moore, *Chem. Mater.* **1995**, *7*, 192; K. A. Mauritz, I. D. Stephanitis, S. V. Scheetz, R. W. Pope, R. K. Milkes, H.-H. Huang, *J. Appl. Polym. Sci.* **1995**, *55*, 181.

52 Z. Hu, A. F. Slaterbeck, C. J. Seliskar, T. H. Ridgeway, W. H. Heineman, *Langmuir* **1999**, *15*, 767.

53 R. A. Zoppi, V. P. Yoshida, S. P. Nunes, *Polymer* **1997**, *39*, 1309; R. A. Zoppi, S. P. Nunes, *J. Electroanal. Chem.* **1998**, *445*, 39.

54 A. N. Khramov, M. M. Collinson, *Anal. Chem.* **2000**, *72*, 2943.

55 A. W. Knight, *Trends Anal. Chem.* **1999**, *18*, 47, and references cited therein; R. D. Gerardi, N. W. Barnett, S. W. Lewis, *Anal. Chim. Acta* **1999**, *378*, 1.

56 T. P. Chou, C. Chandrasekaran, S. J. Limmer, S. Seraji, Y. Wu, M. J. Forbess, C. Nguyen, G. Z. Cao, *J. Non-Cryst. Solids* **2001**, *290*, 153; M. Guglielmi, *J. Sol-Gel Sci. Technol.* **1997**, *8*, 443; P. De Lima Neto, M. Atik, L. A. Avaca, M. A. Aegerter, *J. Sol-Gel Sci. Technol.* **1994**, *2*, 529.

57 S. Roux, P. Audebert, J. Pageti, M. Roche, *J. Mater. Chem.* **2001**, *11*, 1.

58 T. Tatsuma, S. Saitoh, Y. Ohko, A. Fujishima, *Chem. Mater.* **2001**, *13*, 2838.

59 J. Livage, *Solid State Ionics* **1992**, *50*, 307; B. Dunn, G. C. Farrington, B. Katz, *Solid State Ionics* **1994**, *70/71*, 3.

60 S. Skaarup, K. West, B. Zachau-Christiansen, M. Popall, J. Kappel, J. Kron, G. Eichinger, G. Semrau, *Electrochim. Acta* **1998**, *43*, 1589; B. Baradie, J. P. Dodelet, D. Guay, *J. Electroanal. Chem.* **2000**, *489*, 101; H. J. Walls, J. Zhou, J. A. Yerian, P. S. Fedkiw, S. A. Khan, M. K. Stowe, G. L. Baker, *J. Power Sources* **2000**, *89*, 156.

61 Y. Charbouillot, D. Ravaine, M. B. Armand, C. Poinsignon, *J. Non-Cryst. Solids* **1988**, *103*, 325; M. Popall, X.-M. Du, *Electrochim. Acta* **1995**, *40*, 2305; L. Depre, J. Kappel, M. Popall, *Electrochim. Acta* **1998**, *43*, 1301; L. Depre, M. Ingram, C. Poinsignon, M. Popall, *Electrochim. Acta* **2000**, *45*, 1377.

62 K. Dahmouche, M. Atik, N. C. Mello, T. J. Bonagamba, H. Panepucci, M. A. Aegerter, P. Judeinstein, *Mater. Res. Soc. Symp. Proc.* **1996**, *435*, 363; ibid., *J. Sol-Gel Sci. Technol.* **1997**, *8*, 711; ibid., *Solar Energy Mater. Solar Cells* **1998**, *54*, 1; K. Dahmouche, P. H. de Souza, T. J. Bonagamba, H. Panepucci, P. Judeinstein, H. Pulcinelli, C. V. Santilli, *J. Sol-Gel Sci. Technol.* **1998**, *13*, 909.

63 P. Judeinstein, M. E. Brik, J. P. Bayle, J. Courtieu, J. Rault, *Mater. Res. Soc. Symp. Proc.* **1994**, *346*, 937; M. E. Brik, J. J. Titman, J. P. Bayle, P. Judeinstein, *J. Polym. Sci., Part B: Polym. Phys.* **1996**, *34*, 2533.

64 M. G. Kanatzidis, C.-G. Wu, H. O. Marcy, D. C. DeGroot, C. R. Kannerwurf, *Chem. Mater.* **1990**, *2*, 222; E. Ruiz-Hitzky, A. Aranda, B. Casal, J. C. Galvan, *Adv. Mater.* **1995**, *7*, 180.

65 A. Agrawal, J. P. Cronin, R. Zhang, *Sol. Energy Mater. Sol. Cells* **1993**, *31*, 9; M. A. Aegerter, C. O. Avellaneda, A. Pawlicka, M. Atik, *J. Sol-Gel Sci. Technol.* **1997**, *8*, 689; B. Orel, A. Surca, U. O. Krasovec, *Acta Chim. Slov.* **1998**, *45*, 487; J. Livage, D. Ganguli, *Sol. Energy Mater. Sol. Cells* **2001**, *68*, 365.

66 N. Hagfeld, N. Vlachopouls, S. Gilbert, M. Grätzel, *Proc. SPIE* **1994**, *2255*, 297; G.-W. Jang, C.-C. Chen, R. W. Gumbs, Y. Wei, J.-M. Yeh, *J. Electrochem. Soc.* **1996**, *143*, 2591; I. D. Brotherston, D. S. K. Mudigonda, J. M. Osborn, J. Belk, J. Chen, D. C. Loveday, J. L. Boehme, J. P. Ferraris, D. L. Meeker, *Electrochim. Acta* **1999**, *44*, 2993; J.-H. Choy, Y.-I. Kim, B.-W. Kim, N.-G. Park, G. Campet, J.-C. Grenier, *Chem. Mater.* **2000**, *12*, 2950; H. P. Oliveira, C. F. O. Graeff, C. A. Brunello, E. M. Guerra, *J. Non-Cryst. Solids* **2000**, *273*, 193.

67 N. Kobayashi, *Nippon Gazo Gakkaishi* **1999**, *38*, 122.

68 M. Tsionsky, G. Gun, V. Glezer, O. Lev, *Anal. Chem.* **1994**, *66*, 1747; G. Gun, M. Tsionsky, O. Lev, *Mat. Res. Soc. Symp. Proc.* **1994**, *346*, 1011; ibid., *Anal. Chim. Acta* **1994**, *294*, 261.

69 J. Wang, D. S. Park, P. V. A. Pamidi, *J. Electroanal. Chem.* **1997**, *434*, 185.

70 J. Gun, M. Tsionsky, L. Rabinovich, Y. Golan, I. Rubinstein, O. Lev, *J. Electroanal. Chem.* **1995**, *395*, 57.

71 S. Sampath, O. Lev, *J. Electroanal. Chem.* **1998**, *446*, 57.

72 I. Pankratov, O. Lev, *J. Electroanal. Chem.* **1995**, *393*, 35; S. Sampath, I. Pankratov, J. Gun, O. Lev, *J. Sol-Gel Sci. Technol.* **1996**, *7*, 123; S. Sampath, O. Lev, *Electroanalysis* **1996**, *8*, 1112.

73 J. Gun, O. Lev, *Anal. Chim. Acta* **1996**, *336*, 95; ibid., *Anal. Lett.* **1996**, *29*, 1933.

74 J. Wang, P. V. A. Pamidi, *Anal. Chem.* **1997**, *69*, 4490; S. Bharathi, O. Lev, *Chem. Commun.* **1997**, 2303; S. Bharathi, O. Lev, *Anal. Commun.* **1998**, *35*, 29; S. Bharathi, N. Fishelson, O. Lev, *Langmuir* **1999**, *15*, 1929.

75 L. Rabinovich, J. Gun, O. Lev, D. Aurbach, B. Markovsky, M. D. Levi, *Adv. Mater.* **1998**, *10*, 577; D. Aurbach, M. D. Levi, O. Lev, J. Gun, L. Rabinovich, *J. Appl. Electrochem.* **1998**, *28*, 1051.

76 J. Wang, P. V. A. Pamidi, D. S. Park, *Anal. Chem.* **1996**, *68*, 2705; J. Wang, P. V. A. Pamidi, M. Jiang, *Anal. Chim. Acta* **1998**, *360*, 171.

77 Y. Guo, A. R. Guadalupe, *Sensors & Actuators B* **1998**, *46*, 213.

78 R. Kataky, R. Dell, P. K. Senanayake, *Analyst* **2001**, *126*, 2015.

79 S. Sampath, O. Lev, *Anal. Chem.* **1996**, *68*, 2015.

80 A. Walcarius, L. Lamberts, E. G. Derouane, *Electrochim. Acta* **1993**, *38*, 2257; ibid. 2267.

81 D. B. Mitzi, *Chem. Mater.* **2001**, *13*, 3283.

82 Z. Liu, B. Liu, M. Zhang, J. Kong, J. Deng, *Anal. Chim. Acta* **1999**, *392*, 135; Z. Liu, J. Deng, D. Li, *Anal. Chim. Acta* **2000**, *407*, 87; Z. Liu, B. Liu, J. Kong, J. Deng, *Anal. Chem.* **2000**, *72*, 4707; D. Chen, B. Liu, Z. Liu, J. Kong, *Anal. Lett.* **2001**, *34*, 687.

83 P. V. A. Pamidi, D. S. Park, J. Wang, *Polym. Mater. Sci. Eng.* **1997**, *76*, 513; G. Sbervelieri, M. Z. Atashbar, W. Wlodarski, G. Faglia, E. Comini, *Chem. Sens., Tech. Dig. Int. Meet., 7th* **1998**, 657; A. M. Castellani, Y. Gushikem, *J. Colloid Interface Sci.* **2000**, *230*, 195; A. M. Castellani, J. E. Goncalves, Y. Gushikem, *Electroanalysis* **2001**, *13*, 1165; X. Chen, G. Cheng, S. Dong, *Analyst* **2001**, *126*, 1728.

84 J. P. Zheng, P. J. Cygan, T. R. Jow, *J. Electrochem. Soc.* **1995**, *142*, 2699; J. M. Bell, J. P. Matthews, I. L. Skryabin, J. Wang, B. G. Monsma, *Renewable Energy* **1998**, *15*, 312; A. Yasumori, H. Shinoda, Y. Kameshima, S. Hayashi, K. Okada, *J. Mater. Chem.* **2001**, *11*, 1253; J. W. Long, L. R. Qadir, R. M. Stroud, D. R. Rolison, *J. Phys. Chem. B* **2001**, *105*, 8712; J. Sakamoto, W. Dong, B. Dunn, *Proc. Electrochem. Soc.* **2001**, *2000–36*, 197; Y. Li, K. Galatsis, W. Wlodarski, M. Ghantasala, S. Russo, J. Gorman, S. Santucci, M. Passacantando, *J. Vac. Sci. Technol. A* **2001**, *19*, 904; H. Wang, M. Yan, Z. Jiang, *Thin Solid Films* **2001**, *401*, 211.

85 M. M. Collinson, C. G. Rausch, A. Voigt, *Langmuir* **1997**, *13*, 7245.

86 M. D. Petit-Dominguez, H. Shen, W. R. Heineman, C. J. Seliskar, *Anal. Chem.* **1997**, *69*, 703.

87 U. Künzelmann, H. Böttcher, *Sensors & Actuators B* **1997**, *38–39*, 222; K. Ogura, K. Nakaoka, M. Nakayama, M. Kobayashi, A. Fujii, *Anal. Chim. Acta* **1999**, *384*, 219.

88 B. Wang, S. Dong, *Talanta* **2000**, *51*, 565.

89 J. Wang, M. M. Collinson, *J. Electroanal. Chem.* **1998**, *455*, 127.

90 C. C. Hsueh, M. M. Collinson, *J. Electroanal. Chem.* **1997**, *420*, 243.

91 L. Coche-Guérente, V. Desprez, P. Labbé, *J. Electroanal. Chem.* **1998**, *458*, 73.

92 Y. Guo, A. R. Guadalupe, *J. Pharm. Biomed. Anal.* **1999**, *19*, 175.

93 P. C. Pandey, S. Upadhyay, H. C. Pathak, *Sensors & Actuators B* **1999**, *60*, 83; P. C. Pandey, S. Upadhyay, H. C. Pathak, I. Tiwari, V. S. Tripathi, *Electroanalysis* **1999**, *11*, 1251.

94 B. Wang, B. Li, Z. Wang, G. Cheng, S. Dong, *Anal. Chim. Acta* **1999**, *388*, 71.

95 B. Barroso-Fernandez, M. T. Lee-Alvarez, C. J. Seliskar, W. R. Heineman, *Anal. Chim. Acta* **1998**, *370*, 221.

96 R. Makote, M. M. Collinson, *Chem. Mater.* **1998**, *10*, 2440; H. Wei, M. M. Collinson, *Anal. Chim. Acta* **1999**, *397*, 113.

97 J. Li, S. N. Tan, H. Ge, *Anal. Chim. Acta* **1996**, *335*, 137; J. Li, S. N. Tan, J. T. Oh, *J. Electroanal. Chem.* **1998**, *448*, 69; S. H. Jang, M. G. Han, S. S. Im, *Synth. Met.* **2000**, *110*, 17.

98 B. Wang, B. Li, Q. Deng, S. Dong, *Anal. Chem.* **1998**, *70*, 3170.

99 E. I. Iwuoha, S. Kane, C. Ovin Ania, M. R. Smyth, P. R. Ortiz de Montellano, U. Fuhr, *Electroanalysis* **2000**, *12*, 980.

100 B. Wang, B. Li, Z. Wang, G. Xu, Q. Wang, S. Dong, *Anal. Chem.* **1999**, *71*, 1935.

101 K. Kimura, T. Sunagawa, M. Yokoyama, *Chem. Commun.* **1996**, 745; K. Kimura, T. Sunagawa, S. Yajima, S. Miyake, M. Yokoyama, *Anal. Chem.* **1998**, *70*, 4309.

102 W. Kim, S. Chung, S. B. Park, S. C. Lee, C. Kim, S. D. Sung, *Anal. Chem.* **1997**, *69*, 95; W. Kim, D. D. Sung, G. S. Cha, S. B. Park, *Analyst* **1998**, *123*, 379.

103 K. Kimura, H. Takase, S. Yajima, M. Yokoyama, *Analyst* **1999**, *124*, 517; K. Kimura, S. Yajima, H. Takase, M. Yokoyama, Y. Sakurai, *Anal. Chem.* **2001**, *73*, 1605.

104 C. A. Borgo, R. T. Ferrari, L. M. S. Colpini, C. M. M. Costa, M. L. Baesso, A. C. Bento, *Anal. Chim. Acta* **1999**, *385*, 103.

105 A. Walcarius, N. Lüthi, J.-L. Blin, B.-L. Su, L. Lamberts, *Electrochim. Acta* **1999**, *44*, 4601.

106 J. Kulys, *Biosens. Bioelectron.* **1999**, *14*, 473.

107 A. Walcarius, C. Despas, P. Trens, M. J. Hudson, J. Bessière, *J. Electroanal. Chem.* **1998**, *453*, 249; A. Walcarius, J. Bessière, *Chem. Mater.* **1999**, *11*, 3009.

108 R. Shacham, D. Avnir, D. Mandler, *Adv. Mater.* **1999**, *11*, 384.

109 L. Rabinovich, J. Gun, M. Tionsky, O. Lev, *J. Sol-Gel Sci. Technol.* **1997**, *8*, 1077; L. Rabinovich, O. Lev, G. A. Tsirlina, *J. Electroanal. Chem.* **1999**, *466*, 45.

110 M. E. Tess, J. A. Cox, *Electroanalysis* **1998**, *10*, 1237; J. A. Cox, S. D. Holmstrom, M. E. Tess, *Talanta* **2000**, *52*, 1081.

111 S. D. Holmstrom, J. A. Cox, *Anal. Chem.* **2000**, *72*, 3191; J. Seneviratne, S. D. Holmstrom, J. A. Cox, *Talanta* **2000**, *52*, 1025.

112 P. Wang, X. Wang, Y. Yuan, G. Zhu, *J. Non-Cryst. Solids* **2000**, *277*, 22; P. Wang, X. Wang, L. Bi, G. Zhu, *Analyst* **2000**, *125*, 1291; P. Wang, X. Wang, X. Jing, G. Zhu, *Anal. Chim. Acta* **2000**, *424*, 51; P. Wang, X. Wang, G. Zhu, *Electrochim. Acta* **2000**, *46*, 637; ibid. *Electroanalysis* **2000**, *12*, 1493; ibid. *New J. Chem.* **2000**, *24*, 481; P. Wang, X. Jing, W. Zhang, G. Zhu, *J. Solid State Electrochem.* **2001**, *5*, 369; P. Wang, Y. Yuan, Z. Han, G. Zhu, *J. Mater. Chem.* **2001**, *11*, 549.

113 J. A. Cox, K. S. Alber, M. E. Tess, T. E. Cummings, W. Gorski, *J. Electroanal. Chem.* **1995**, *396*, 485; J. B. Laughlin, K. Miecznikowski, P. J. Kulesza, J. A. Cox, *Electrochem. Solid-State Lett.* **1999**, *2*, 574.

114 T. Inagaki, H. S. Lee, T. A. Skotheim, Y. Okamoto, *J. Chem. Soc., Chem. Comun.* **1989**, 1181; S. Ikeda, N. Oyama, *Anal. Chem.* **1993**, *65*, 1910.

115 L. T. Kubota, Y. Gushikem, J. Perez, A. A. Tanaka, *Langmuir* **1995**, *11*, 1009; C. R. M. Peixoto, L. T. Kubota, Y. Gushikem, *Anal. Proc.* **1995**, *32*, 503; L. T. Kubota, F. Gouvea, A. N. Andrade, B. G. Milagres, G. de Oliveira Neto, *Electrochim. Acta* **1996**, *41*, 1465; E. F. Perez, G. de Oliveira Neto, A. A. Tanaka, L. T. Kubota, *Electroanalysis* **1998**, *10*, 111; C. A. Pessôa, Y. Gushikem, *J. Electroanal. Chem.* **1999**, *477*, 158.

116 M. Tsionsky, O. Lev, *J. Electrochem. Soc.* **1995**, *142*, 2154.

117 L. Rabinovich, V. Glezer, Z. Wu, O. Lev, *J. Electroanal. Chem.* **2001**, *504*, 146.

118 J. A. Cox, K. S. Alber, *J. Electrochem. Soc.* **1996**, *143*, L126; S. D. Holmstrom, Z. D. Sandlin, W. H. Steinecker, J. A. Cox, *Electroanalysis* **2000**, *12*, 262.

119 J. Wang, P. V. A. Pamidi, V. B. Nascimento, L. Angnes, *Electroanalysis* **1997**, *9*, 689.

120 Z. Ji, A. R. Guadalupe, *Electroanalysis* **1999**, *11*, 167.

121 Z. Hu, A. F. Slaterbeck, C. J. Seliskar, T. H. Ridgway, W. R. Heineman, *Langmuir* **1999**, *15*, 767; Z. Hu, W. R. Heineman, *Anal. Chem.* **2000**, *72*, 2395.

122 Y. Shi, C. J. Seliskar, *Chem. Mater.* **1997**, *9*, 821.

123 L. Gao, C. J. Seliskar, W. R. Heineman, *Anal. Chem.* **1999**, *71*, 4061.

124 D. Levy, B. K. Iozefzon, I. Gigozin, I. Zamir, D. Avnir, M. Ottolenghi, O. Lev, *Sep. Sci. Technol.* **1992**, *27*, 589.

125 L. M. Aleixo, M. B. Souza, O. E. S. Godinho, G. de Oliveira Neto, Y. Gushikem, J. C. Moreira, *Anal. Chim. Acta* **1993**, *271*, 143.

126 Y. Guo, A. R. Guadalupe, O. Resto, L. F. Fonseca, S. Z. Weisz, *Chem. Mater.* **1999**, *11*, 135.

127 M. Etienne, J. Bessière, A. Walcarius, *Sensors & Actuators B*, **2001**, *76*, 531.

128 S. Sayen, M. Etienne, J. Bessière, A. Walcarius, submitted for publication.

129 A. Walcarius, M. Etienne, personal communication.

130 L. Coche-Guérente, S. Cosnier, V. Desprez, P. Labbé, D. Petridis, *J. Electroanal. Chem.* **1996**, *401*, 253.

131 R. Makote, M. M. Collinson, *Chem. Commun.* **1998**, 425.

132 Y. Shi, A. F. Slaterbeck, C. J. Seliskar, W. R. Heineman, *Anal. Chem.* **1997**, *69*, 3679; Y. Shi, C. J. Seliskar, W. R. Heineman, *Anal. Chem.* **1997**, *69*, 4819.

133 A. F. Slaterbeck, T. H. Ridgway, C. J. Seliskar, W. R. Heineman, *Anal. Chem.*, **1999**, *71*, 1196.

134 A. F. Slaterbeck, M. G. Stegemiller, C. J. Seliskar, T. H. Ridgway, W. R. Heineman, *Anal. Chem.* **2000**, *72*, 5567.

135 S. E. Ross, C. J. Seliskar, W. R. Heineman, *Anal. Chem.* **2000**, *72*, 5549.

136 J. M. DiVirgilio-Thomas, W. R. Heineman, C. J. Seliskar, *Anal. Chem.* **2000**, *72*, 3461.

137 M. Maizels, C. J. Seliskar, W. R. Heineman, *Electroanalysis* **2000**, *12*, 1356.

138 P. V. A. Pamidi, C. Parrado, S. A. Kane, J. Wang, M. R. Smyth, J. Pingarron, *Talanta* **1997**, *44*, 1929.

139 L. Hua, S. N. Tan, *Anal. Chim. Acta* **2000**, *403*, 179; ibid., *Fresenius J. Anal. Chem.* **2000**, *367*, 697.

140 L. Hua, S. N. Tan, *Anal. Chem.* **2000**, *72*, 4821; L. Hua, L. S. Chia, N. K. Goh, S. N. Tan, *Electroanalysis* **2000**, *12*, 287.

141 H. Wang, G. Xu, S. Dong, *Analyst* **2001**, *126*, 1095.

142 W. Wroblewski, M. Chudy, A. Dybko, Z. Brzozka, *Anal. Chim. Acta* **1999**, *401*, 105; M. Ben Ali, R. Kalfat, H. Sfihi, J. M. Chovelon, H. Ben Ouada, N. Jaffrezic-Renault, *Sensors & Actuators B* **2000**, *62*, 233; J. Liu, X. Wu, Z. Zhang, S. Wakida, K. Higashi, *Sensors Actuators B* **2000**, *66*, 216.

143 J. R. Fernandes, L. T. Kubota, Y. Gushikem, G. de Oliveira Neto, *Anal. Lett.* **1993**, *26*, 2555; L. L. Lorencetti, Y. Gushikem, L. T. Kubota, G. de Oliveira Neto, J. R. Fernandes, *Mikrochim. Acta* **1995**, *117*, 239; R. V. S. Alfaya, A. A. S. Alfaya, Y. Gushikem, S. Rath, F. G. R. Reyes, *Anal. Lett.* **2000**, *33*, 2859.

144 P. C. Pandey, S. Upadhyay, H. C. Pathak, C. M. D. Pandey, *Electroanalysis* **1999**, *11*, 950.

145 P. Tien, L.-K. Chau, Y.-Y. Shieh, W.-C. Lin, G.-T. Wei, *Chem. Mater.* **2001**, *13*, 1124.

146 S. Braun, S. Rappoport, R. Zusman, D. Avnir, M. Ottolenghi, *Mater. Lett.* **1990**, *10*, 1; S. Braun, S. Shtelzer, S. Rappoport, D. Avnir, M. Ottolenghi, *J. Non-Crystal. Solids* **1992**, *147/148*, 739; D. Avnir, S. Braun, O. Lev, M. Ottolenghi, *Chem. Mater.* **1994**, 6, 1605.

147 L. M. Ellerby, C. R. Nishida, F. Nishida, S. A. Yamanaka, B. Dunn, J. S. Valentine, J. I. Zink, *Science* **1992**, *255*, 1113; S. Wu, L. M. Ellerby, J. S. Cohen, B. Dunn, M. A. El-Sayed, J. S. Valentine, J. I. Zink, *Chem. Mater.* **1993**, *5*, 115; B. C. Dave, B. Dunn, J. S. Valentine, J. I. Zink, *Anal. Chem.* **1994**, *66*, 1120A.

148 R. Wang, U. Narang, P. N. Prasad, F. V. Bright, *Anal. Chem.* **1993**, *65*, 2671; U. Narang, M. H. Rahman, J. H. Wang, P. N. Prasad, F. V. Bright, *Anal. Chem.* **1995**, *67*, 1935; I. Gill, A. Ballesteros, *J. Am. Chem. Soc.* **1998**, *120*, 8587.

149 I. Gill, *Chem. Mater.* **2001**, *13*, 3404.

150 S. Yang, Y. Lu, P. Atanossov, E. Wilkins, X. Long, *Talanta* **1998**, *47*, 735.

151 B. Wang, J. Zhang, G. Cheng, S. Dong, *Anal. Chim. Acta* **2000**, *407*, 111.

152 B. Wang, S. Dong, *Talanta* **2000**, *51*, 565; ibid. *J. Electroanal. Chem.* **2000**, *487*, 45.

153 W.-Y. Lee, K. S. Lee, T.-H. Kim, M.-C. Shin, J.-K. Park, *Electroanalysis* **2000**, *12*, 78; W.-Y. Lee, S.-R. Kim, T.-H. Kim, K. S. Lee, M.-C. Shin, J.-K. Park, *Anal. Chim. Acta* **2000**, *404*, 195.

154 S. Bharati, M. Nogami, *Analyst* **2001**, *126*, 1919.

155 V. Glezer, O. Lev, *J. Am. Chem. Soc.* **1993**, *115*, 2533.

156 A. Kros, M. Gerritsen, V. S. I. Sprakel, N. A. J. M. Sommerdijk, J. A. Jansen, R. J. M. Nolte, *Sensors & Actuators B* **2001**, *81*, 68.

157 T.-M. Park, E. I. Iwuoha, M. R. Smyth, B. D. MacCraith, *Anal. Commun.* **1996**, *33*, 271; T.-M. Park, E. I. Iwuoha, M. R. Smyth, R. Freaney, A. J. McShane, *Talanta* **1997**, *44*, 973; T.-M. Park, E. I. Iwuoha, M. R. Smyth, *Electroanalysis* **1997**, *9*, 1120.

158 U. Narang, P. N. Prasad, F. V. Bright, K. Ramanathan, N. D. Kumar, B. D. Malhotra, M. N. Kamalasanan, S. Chandra, *Anal. Chem.* **1994**, *66*, 3139; S. L. Chut, J. Li, S. N. Tan, *Analyst* **1997**, *122*, 1431; J. Li, S. Chia, N. K. Goh, S. N. Tan, H. Ge, *Sensors & Actuators B* **1997**, *40*, 135; P. C. Pandey, S. Upadhyay, H. C. Pathak, *Electroanalysis* **1999**, *11*, 59; P. C. Pandey, S. Upadhyay, *Sensors & Actuators B* **2001**, *76*, 193; P. C. Pandey, S. Upadhyay, I. Tiwari, S. Sharma, *Electroanalysis* **2001**, *13*, 1519.

159 S. Sampath, O. Lev, *Adv. Mater.* **1996**, *9*, 410.

160 P. Audebert, C. Sanchez, *J. Sol-Gel Sci. Technol.* **1994**, *2*, 809.

161 M. E. Gimon-Kinsel, V. L. Jimenez, L. Washmon, J. Balkus, Jr., *Stud. Surf. Sci. Catal.* **1998**, *117*, 373; Y.-J. Han, G. D. Stucky, A. Butler, *J. Am. Chem. Soc.* **1999**, *121*, 9897; J. He, X. Li, D. G. Evans, X. Duan, C. Li, *J. Mol. Catal. B: Enzym.* **2000**, *11*, 45; H. Takahashi, B. Li, T. Sasaki, C. Miyazaki, T. Kajino, S. Inagaki, *Chem. Mater.* **2000**, *12*, 3301; Q. Feng, J. Xu, M. Lin, H. Dong, Y. Wei, *Polym. Mater. Sci. Eng.* **2000**, *83*, 502; H. Dong, J. Xu, Q. Feng, Y. Wei, *Polym. Mater. Sci. Eng.* **2000**, *83*, 504; Y. Wei, J. Xu, Q. Feng, H. Dong, M. Lin, *Mater. Lett.* **2000**, *44*, 6; Y. Wei, J. Xu, Q. Feng, M. Lin, H. Dong, W.-J. Zhang, C. Wang, *J. Nanosci. Nanotechnol.* **2001**, *1*, 83; H. Takahashi, B. Li, T. Sasaki, C. Miyazaki, T. Kajino, S. Inagaki, *Microporous Mesoporous Mater.* **2001**, *44–45*, 755; H. H. P. Yiu, P. A. Wright, N. P. Botting, *Microporous Mesoporous Mater.* **2001**, *44–45*, 763; ibid., *J. Mol. Catal. B: Enzym.* **2001**, *15*, 81.

162 S. Cosnier, C. Gondran, A. Senillou, M. Graetzel, N. Vlachopoulos, *Electroanalysis* **1997**, *9*, 1387; S. Cosnier, A. Senillou, M. Graetzel, P. Comte, N.Vlachopoulos, N. Jaffrezic Renault, C. Martelet, *J. Electroanal. Chem.* **1999**, *469*, 176.

163 J. Wang, P. V. A. Pamidi, K. Rogers, *Anal. Chem.* **1998**, *70*, 1171.

164 T. Peng, Q. Cheng, R. C. Stevens, *Anal. Chem.* **2000**, *72*, 1611.

7
Multifunctional Hybrid Materials Based on Conducting Organic Polymers. Nanocomposite Systems with Photo-Electro-Ionic Properties and Applications

Monica Lira-Cantú and Pedro Gómez-Romero

7.1
Introduction

The design of hybrid systems in which organic and inorganic components add up to form new and improved materials is mostly restrained by our limited imagination. Indeed, the synthesis of nanocomposite materials by combination of dissimilar components at the molecular level is a continuing source of surprises, uncovering unexpected synergies in which the performance of the final system goes beyond the sum of those of their components. And the overwhelming variety of materials that can be found in each of the organic and inorganic realms is the best guaranty of an endless world of hybrid organic-inorganic combinations.

Thus, although the mainstream in hybrid materials has rested on inorganic frameworks built from different types of silicon compounds – from silicates to polysiloxanes to silica, as shown in previous chapters – many other types of fascinating hybrid materials based on radically different approaches are presently being developed and many others remain yet to be exploited. In this respect, this and the following chapters will provide different examples of some new uncharted territories opened in different lines of the general field of hybrid materials.

One such field is that of hybrid materials based on conducting organic polymers (COPs) [1]. The unique combination of useful properties found in these materials includes electronic conductivity (e^- or h^+), ionic transport, reversible electroactivity, electrooptical properties typical of semiconductors as well as electrochromic, pH- and composition-dependent properties, all of them to add to their polymeric nature. That is an excellent basis for the design of hybrid materials in which *either* of these properties *or their combinations* work to enhance or combine with those of a myriad inorganic phases with electronic, magnetic, photochemical, electrochemical, optical or catalytic properties. A large variety of functional hybrid materials can thus be designed and fabricated in which *multifunctionality* can be easily built to address specific technological needs.

This chapter will present an overview of the field of hybrid materials in which the common factor is the presence of conducting organic polymers (COPs). The topic is broad in itself and includes molecular and supramolecular materials as well

Gómez-Romero: Organic-Inorganic Materials. Monica Lira-Cantú

ISBN: 3-527-30484-3

as extended, cluster and nanocomposite solids, all of which span a large variety of applications as functional materials. Our discussion will include a short historical overview, rooted in the chemistry of conducting polymers themselves, an analysis of the different types of hybrid materials from a chemical point of view and a final review of applications.

Modern interest in what we now know as intrinsically conductive organic polymers began before their electrical (conducting) properties were discovered. Thus, the first polyacetylene (PAc) films were obtained by 1971 [2]. On the other hand, the first conducting polymer studied in detail as such was an inorganic compound poly(sulfur nitride) $(SN)_x$ and it was the experience and interest in this material [3] that led to the investigation and discovery of conducting properties in trans-polyacetylene in 1977 [4] and soon after in other conjugated polymers [1].

Back then, the electrical properties of these materials were a revolutionary discovery, with maximum conductivities around 50 S cm^{-1} [5] that have increased today up to 10^5 S cm^{-1} (remarkable in comparison with those of copper, one of the best metallic conductors, with conductivities of 10^6 S cm^{-1} at room temperature). Materials scientists had at hand inexpensive, processable, lightweight "electrically conductive plastics". But it was not just their conductivity. In the last thirty years research on conducting polymers has unearthed a wealth of useful properties and diverse lines of application, an explosive growth that led to the award of the Nobel Prize in Chemistry in 2000 to Shirakawa, MacDiarmid and Heeger.

It is precisely their characteristic combination of useful functional properties together with their light weight and low cost that make COPs excellent candidates to form functional hybrid materials. Furthermore, several approaches can be followed to integrate all sorts of active inorganic species into COPs, but also to incorporate COPs into extended mineral phases, natural or synthetic, thus acting as *inserting* or *inserted* components, respectively. This versatility is directly related to the intermediate *size* scale and *dimensionality* of polymers in between those of molecular or cluster species and extended solid-state phases (Fig. 7.1).

In hybrid materials based on COPs, the ability of the conjugated polymer to carry charge (electrons or holes) is normally at the heart of most applications. Furthermore, the relation between COPs' electrochemistry and their electronic conductivity, directly related to the broad spectrum of possible doping levels of these polymers, is crucial in the development of different types of functional materials.

Yet, the introduction of distinct additional species allows for specific properties to be added to the hybrid material. These appended species can be of any nature. They are most frequently inorganic compounds providing radically different or complementary behavior to that of COPs, but they can also be dissimilar organic species that combine to form what we could label as Organic-Organic hybrids. From a structural point of view there is also a good variety of possibilities and viable inorganic components can range from zero-dimensional molecular or cluster species, to one-, two- or three-dimensional extended structures, polymers or nanosized inorganic particles. In every case, the hybrid approach bestows the possibility to play with the characteristics of dissimilar or complementary species and the opportunity to find novel combinations with improved or synergic properties.

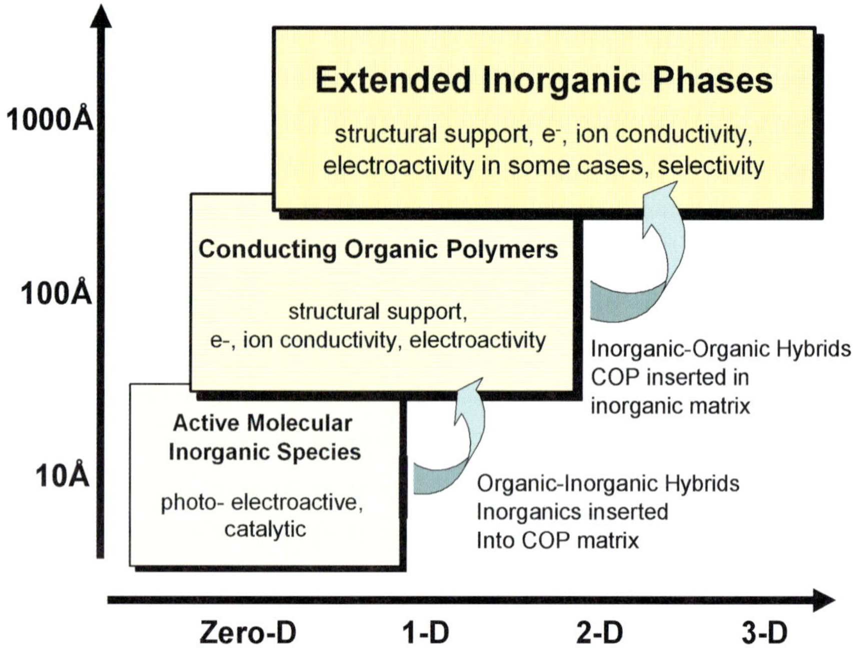

Fig. 7.1 The intermediate size and dimensionality of Conducting Organic Polymers (COPs) makes them ambivalent elements for the design of hybrid materials. Thus COPs can constitute polymer matrices where molecular or cluster inorganic species are inserted or they can in turn be the inserted species in extended mineral inorganic phases

As we will discover in this chapter, the inorganic components combined with COPs, whether added as doping species, as extended phases or as covalently linked units, can be used as electroactive, photoactive or catalytic agents providing the basis for the application of the resulting hybrid materials in different types of devices. These include insertion electrodes in rechargeable lithium batteries or supercapacitors, electrochromic or photochromic components, display devices, LEDs, photovoltaics, or novel energy-conversion systems, as well as electrocatalysts, proton-pump electrodes, sensors, or chemiresistive detectors familiarly known as artificial "noses" or "tongues". But before going into the chemistry and applications of the most surprising combinations let us take a look at the background related to the conjugated polymers that constitute the basis for these remarkable hybrid materials.

7.2 Conducting Organic Polymers (COPs): from Discovery to Commercialization

The report in 1977 of an organic polymer able to conduct electricity represented a remarkable breakthrough and the starting point for dreams of many new technological prospects and the corresponding follow-up applications: lighter electrical components, thinner batteries, flexible light-emitting diodes, etc. This meant many new research lines for scientists and, given the sturdiness and low cost of the prospective materials, many potential market opportunities for the industry. From the initial euphoria in the eighties followed the recognition in the nineties of several problems to be solved before COPs could make it to the marketplace, not the least of which was the poor solubility and unmoldability that characterize these polymers in their doped, conductive state.

Yet, COPs like polyaniline (PAni) can be dissolved in *concentrated* acids (ca. 97%) and precipitated in methanol or water, retaining their conductivity [6, 7], but these procedures are of little practical use for industry. Several lines of work emerged in order to overcome the problem of solubility and processability of COPs without damaging their electroactivity and conductivity characteristics [8]. First of all, numerous polymers with side chains, from substituted monomer derivatives, were synthesized without any remarkably positive results. Although these derivatives provide a certain level of control of the physical properties and solubility, the electronic conductivity of the resulting polymers was generally poor in comparison to that of their original counterparts, and none of these derivatives went commercial [9].

Recently many research groups have realized that for applications, solubility and processability of conducting polymers are related to their dispersability [10]. In other words, what we call "polymer solutions" are nothing but stable dispersions with ultrafine particle sizes of polymers dispersed in certain solvents. The unusually high surface tension that characterizes these dispersions leads to agglomeration and self-assembly of the particles. This self-organization and the fact that at a certain critical concentration the particles stick together (like a collar of pearls) touching one another gives rise to electron transfer and mobility in the resulting material. This seems to be the best approach towards COP's processability while retaining their stability and electrically conducting properties [9].

With the hurdle of processability surmounted, the promise of commercialization is finally being realized for conducting organic polymers. Interestingly, the first of these materials that made it to the marketplace were COPs with other distinct organic species as doping agents. This type of material, which is briefly described below, could be suitably labeled as organic-organic hybrids (O–O).

Thus, it was in 1992 when the first processable, electrically conductive organic polymer was reported [11]. It was polyaniline doped with an anionic surfactant, namely dodecylbenzenesulfonic acid (PAni-DBSA). Currently this material is patented and commercialized by UNIAX Corporation, Neste Oy (Finland) (see Table 7.1). In 1993 the company Baytron P, Bayer AG Co reported on poly(3,4-ethylenedioxythiophene (PEDOT) doped with polystyrene sulphonate, (PSS) [10], with good processability and applications in photography, antistatic coatings or as an electrode for capacitors. Since then, several materials based on conducting and processable organic polymers have been commercialized. A product sold under the name of PANIPOL, from Panipol Ltd, is a solution of PAni that can be blended with other polymers such as Polyethylene (PE) or poly(vinylchloride) (PVC). The resulting materials combine the electrical and optical properties of the conducting polymer with the good mechanical characteristics of the other bulk polymers. As a last example we can mention LIGNO-PANI™ from GeoTech Chemical Co. Ltd., formed by grafting polyaniline chains into lignin. It is a redox-active, highly dispersible, cross-linkable material. Table 7.1 describes some of these commercialized COPs.

Pron et al. have recently reviewed the processability and related applications of COPs [12]. In that paper, the authors discuss synthetic routes to some of the main conjugated polymers such as poly(acetylene), polyheterocyclic polymers, poly(p-phenylene vinylene)s, aromatic poly(azomethine)s and poly(aniline) with special emphasis on the preparation of solution and/or processible polyconjugated systems. The authors cover the area of dopant engineering and discuss the mechanism for increasing COP's electronic conductivity to obtain the desired properties of the doped polymer system [12].

Twenty-five years after the discovery of their remarkable conducting properties and only ten after controlling their processability, development and application of conducting organic polymers keeps growing in many different areas, beginning with antistatic films or coatings, conductive fabrics, anticorrosion paints, printed circuit boards, etc. All the above applications are based on the intrinsic properties of COPs, but we should not forget the modulation of properties and the synergies provided by the integration of other *dissimilar organic* compounds such as some of the doping agents mentioned above. It is in this respect that we can talk about Organic-Organic hybrid systems, a specific subarea presently under strong development. Table 7.2 shows several representative examples of materials currently under study where conducting organic polymers are doped with organic compounds (such as DTMBP-PO_4H, PSS, PCBM, etc.) as well as their applications.

These materials based on COPs and organic compounds are part of a new generation of lightweight, processable, flexible materials that are breaking new ground especially in the fields of sensors [13], photovoltaic devices and solar cells [10, 14–17], LEDs [18], actuators [19], lithium batteries and capacitors [20, 21], etc.

7.3
Organics and Inorganics in Hybrid Materials

The doping of COPs with organic species shows the importance of added components to modulate the properties of conducting polymers. Yet, in general, those organic dopants are *not* active species. In stark contrast, the chemistry of most transition metal compounds, whether molecular or solids, is full of examples of all sorts of properties potentially useful in the design of multifunctional materials.

Yet, historically, the origins of hybrid organic-inorganic materials are to be found in the search for improved structural materials and centered on different forms of

Tab. 7.1 Selected examples of commercial Conducting Organic Polymers (COPs). Most of them are presented as "solutions" (actually dispersions). A brief list of their possible applications, characteristics and properties is also given.

Product Name	Company	Composition	Application	Characteristics
Baytron P®	Bayer AG	PEDOT-PSS	Photographic film Antistatic coating Electrode for capacitors Corrosion resistant coating	Multiton production Water solution(1.3% weight) Conductivity: 1–10 S cm^{-1}
Panipol®	Panipol Ltd Neste Oyj	Panipol® EB (Emerald. Base); Panipol® EB + sulfonic acid dop. Panipol® Toluene Solutions	Injection moulded, Antistatic Printed circuit boards. Electrochromic "smart" windows Electrochromic automobile rear vision systems Conductive fabrics Paint primers	Conductive polymer materials and blends that can be processed from melt and/or solution.
Ormecon®	Ormecon-Lacquer-systems	Polyaniline	Antistatic coating Anticorrosion paints (Corrpassiv®) Nanotechnology Sensors, etc.	Dispersion in solvents with binders Surface resistance 10^{1-}–10^{9} ohm
PLED	Uniax Co. DuPont Displays	PAni-DBS: PAni · Dodecylbenzene sulfonic acid	PLED- Polymer light Emitting Diode. Liquid CrystalDisplays	More than 14,500 h working (75% finally)
Ligno-Pani™	Catize (GeoTech Chemical Co)	Polyaniline and Lignine	Paint additive, Corrosion resistant Anti-static fabrics, Packing Conductive Inks, Printed Circuit Boards, Sensors, etc.	Dark green color, 2–3 Conductivity up to 10 S cm^{-1}, MW 5–15000 Stable up to 300 °C Soluble
PPy-Latex	DSM	Polypyrrole and silica latex	Paint additive	

Tab. 7.2 Selected examples of conducting organic polymers (COP) doped with organic molecules (O-O Materials) and their applications

	COP + Organic Compound[a]	Application	Ref.
1	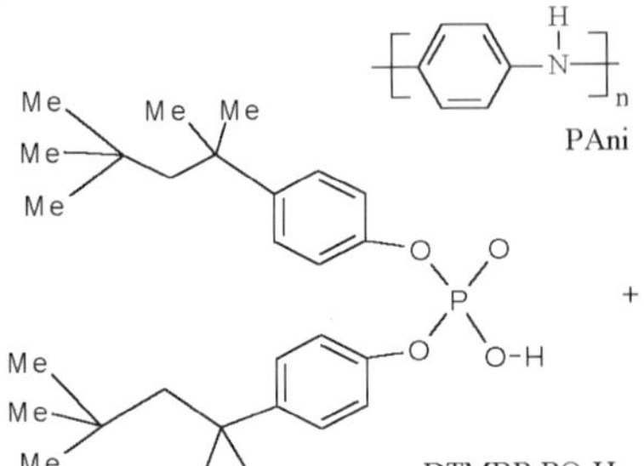 PAni + DTMBP-PO_4H	Sensor: Ca-selective electrode	13
2	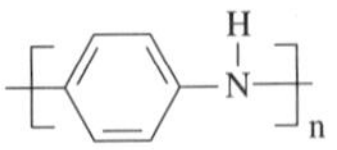 CF_3COOH + $LiFCH_3SO_4$ $K_4Fe(CN)_6$† PAni PMT Ppy	Electronic Nose	363
3	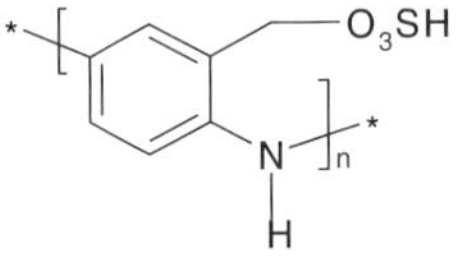 SPAni + PI	Actuators	19
4	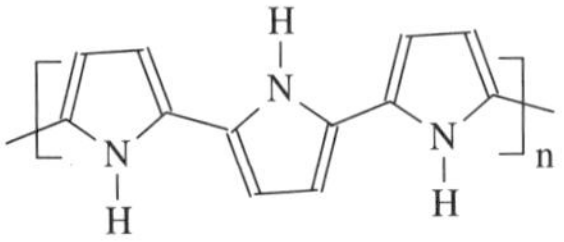 Ppy + DBS-Na+	Microrobots Micromuscles	389
5	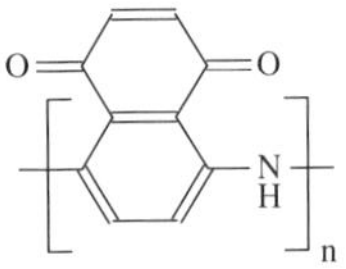 PANQ	Supercapacitors and Lithium batteries	20
6	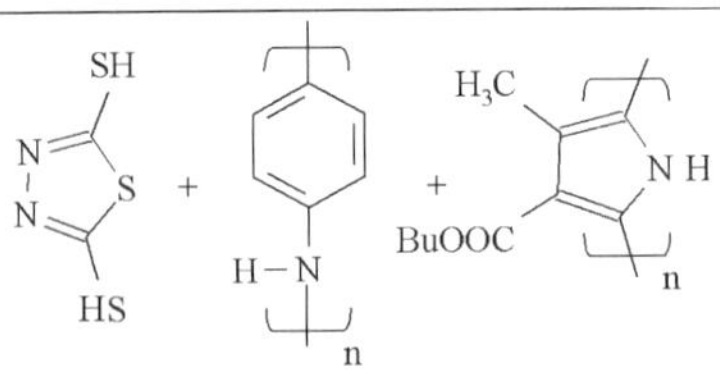 DMcT + PAni + PBMPPy	Cathodes for Lithium Batteries	337

Tab. 7.2 Fortsetzung

	COP + Organic Compound[a]	Application	Ref.
7	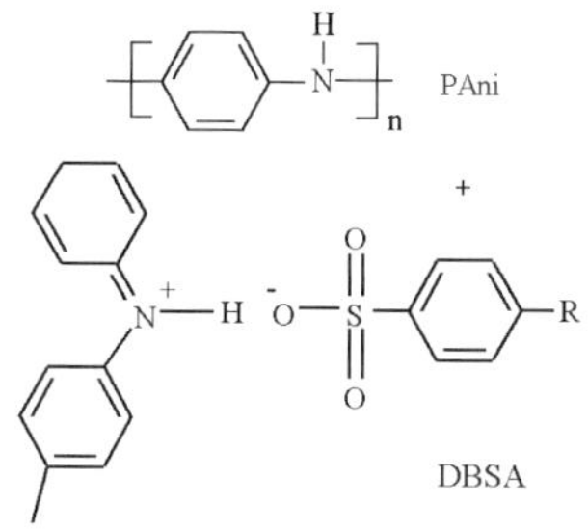 PAni + DBSA	LED Processable PAni	11
8	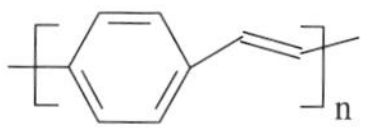 PPV 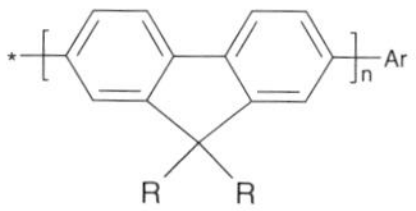 + Polyfluorene	LED	18
9	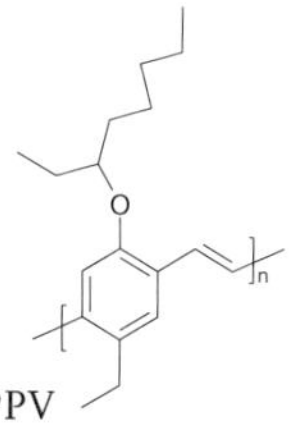 MEH-PPV	Lasers	390, 391
10	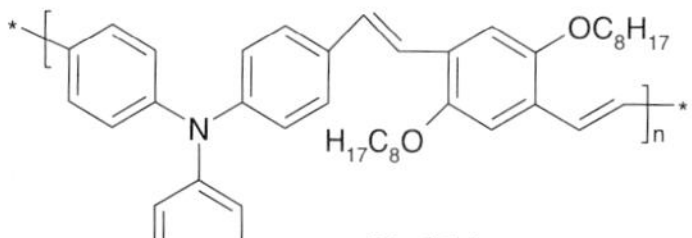 PA-PPV	Photovoltaic Devices	16
11	BAYTRON P® 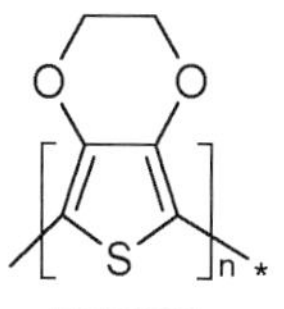 PEDOT 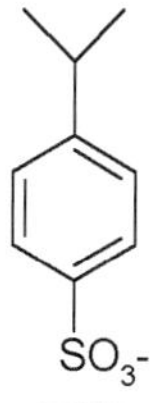 PSS⁻	Photovoltaic Devices	10, 15
12	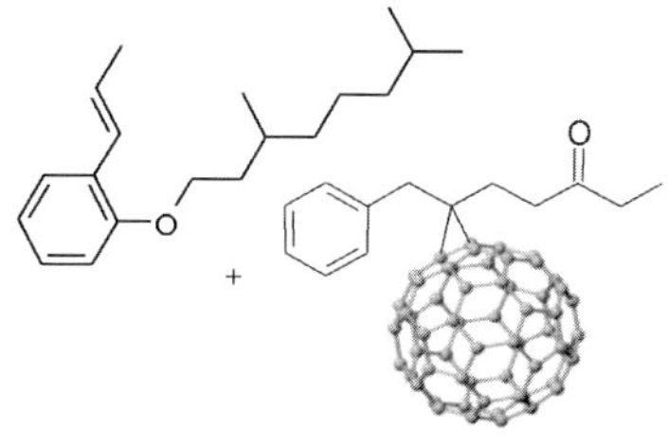 MDMO-PPV + PCPM	Photovoltaic Devices	392

Tab. 7.2 Fortsetzung

	COP + Organic Compound[a]	Application	Ref.
13	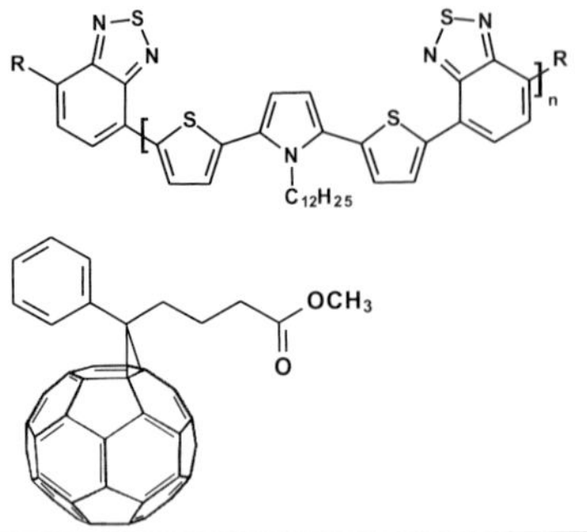	Photovoltaic Devices Heterojunctions in Energy conversion Devices	17
14	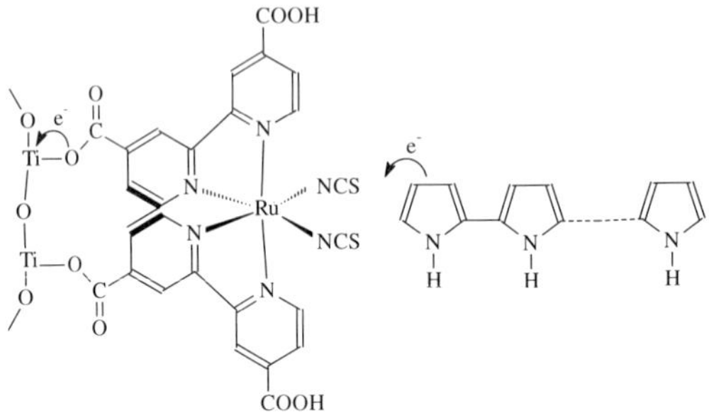 TiO_2 + $RuL_2(NCS)_2$ + PPy	Photovoltaic Devices Solar Cell	14

a) Names:
1) DTMBP-PO_4H: Bis[4-(1,1,3,3-tetramethylbutyl) phenyl] phosphoric acid;
2) PMT: Poly(3-methylthiophene);
3) SPAni: Sulfonated polyaniline and PI: polyimide;
4) DBS-Na+: Dodecyl benzene sulfonate;
5) PANQ: Poly(5-amino-1,4-naphthoquinone);
6) DMcT: 2,5-Dimercapto-1,3,4-thiodiazole and PBMPPy: Poly(3-butylcarboxylate-4-methylpyrrole)
7) DBSA: Dodecylbenzene sulfonic acid;
8) PPV: Poly(p-phenylenevinylene);
9) MEH-PPV: Poly(2-methoxy-5-(2′-ethylhexyloxy)-1,4-phenylene vinylene);
10) PA-PPV: Phenyl amino-p-phenylenevinylene;
11) PEDOT: Poly(3,4-ethylenedioxythiophene) and PSS: Polystyrene sulfonate;
12) MDMO-PPV: Poly[2-methoxy-5-(3′,7′-dimethyl-octyloxy)1,4-phenylenevinylene];
13) PTPTB-I (R=H, Br, n = 1–4): Poly-(N-dodecyl-2,5-bis(2′-thienyl)pyrrole-(2,1,3-benzothiadiazole)) and PCBM: Phenyl-C_{61}-butyric acid methylester (1-(3-methoxycarbonyl)-propyl-1-phenyl-[6,6]C_{61});
14) $RuL_2(NCS)_2$ dye: cis-di(thiocyanate)-N,N′-bis(2,2′-bipyridyl-4,4′-dicarboxylic acid) ruthenium (II) complex;

silicates or siloxane derivatives. Thus, the design of early organic-inorganic nanocomposites strove to get hybrid polymers [22] and materials based on mixed silicon-carbon networks prepared by sol-gel methods [23–28] which could also entrap additional active species [29]. In these initial stages scientists aimed at obtaining structural materials in the twilight between inorganic glasses and organic polymers [23]. But nowadays expectations go far beyond mechanical strength and thermal and chemical stability. As we have seen in previous chapters, novel hybrid materials are also sought for improved optical [30–33], and electrical [34, 35] properties, luminescence [27, 36–41], ionic conductivity [42–44], and selectivity [45–48], as well as chemical [49–51] or biochemical [52–54] activity.

Chemical activity is precisely at the core of functional materials. Sensors, selective membranes, all sorts of electrochemical devices from actuators to batteries or supercapacitors, supported catalysts or photoelectrochemical energy conversion cells, are some important devices based on functional materials. In this type of material mechanical properties are secondary (though certainly not unimportant) and the emphasis is on reactivity, reaction rates, reversibility or specificity. Furthermore, in a wider sense, functional materials also seek to exploit all sorts of physical properties from magnetic to electronic to optical. The hybrid approach is also most rewarding in this multifunctional context by combining organic and inorganic species with complementary chemical activities and physical properties to yield functional and multifunctional materials.

7.3.1 Classifications

This review will center on a particularly fruitful and wide group of hybrid functional materials: namely, those containing conducting organic polymers (COPs) as components. We will find examples in all types of conducting organic polymers (see for example reference [55] for a general review of types and formulae), although most research work has focused on hybrids with polypyrrole (PPy), polyaniline (PAni) or polythiophene (PT) and their derivatives. This category of hybrids based on COPs alone accounts for a large number of materials and applications, as will be apparent below.

As has been mentioned before, the intermediate size and dimensionality of these polymers allows their use as inserting or inserted species. As proposed earlier [1], this ambivalence offers a way to classify and create some order in the study of this particular type of hybrid material. Namely, it allows us to consider two distinct classes of materials: i) those in which the organic polymers are the matrix where inorganic species are inserted or integrated (we will call this type O–I, organic-inorganic hybrids) and ii) materials in which the organic polymer itself is inserted into more extended solid mineral matrices (called I–O, inorganic-organic materials). As is frequently the case, we will find examples of hybrids that challenge that classification: for instance, hybrid materials where neither organic nor inorganic dominates as a clear matrix, materials where both are dispersed at the nanometric level. We will generically refer to these types of materials as *hybrid nanocomposite (NC) materials.*

In addition to this ad hoc classification we will also follow the one presented in Chapter I to distinguish weakly bound hybrids (class I, where the organic and inorganic components only interact through ionic, H-bonds, etc.) from covalently linked hybrids (class II).

A schematic representation of hybrid nanocomposite material (NC), Inorganic-Organic Materials (IO) and Organic-Inorganic Materials (OI) is shown in Fig. 7.2. This figure illustrates the nature of the different types of hybrids and relates the dispersion level between the organic and the inorganic compounds and the molecular dimension of the resulting hybrid materials. For hybrid nanocomposite materials (NC) dimensions can be very variable depending on the particular case, but in principle they can reach the 10^2 nm (1000 Å) crystallite or crystal grain dimension. On the other hand, IO Materials, with COPs incorporated into inorganic hosts, will be structured around the 10^1 nm (100 Å) polymer chain dimension. Finally we distinguish the group of organic-inorganic materials (OI) in which the minimal chemical unit usually falls into the 10^0 nm, 10 Å dimension.

7.4
Synergy at the Molecular Level: Organic-Inorganic (OI) Hybrid Materials

In this section we will describe and discuss hybrid functional materials, with conducting organic polymers as the host or major component, that incorporate inorganic species to form hybrids with the added functionality of the latter.

One method to introduce molecular inorganic species into a COP network relies upon the introduction of suitable coordination sites, groups or molecules which could lead to the formation of metal complexes. Those ligands or molecules can get incorporated to the polymer chain either through functionalization, by means of covalent bonds or by copolymerization with the usual monomers in a general approach that has been reviewed before [1, 56, 57]. These methods would lead to

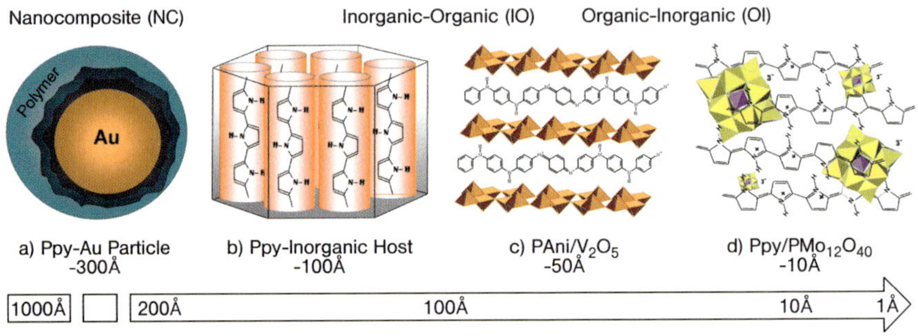

Fig. 7.2 From nanocomposites to molecular-size hybrid organic-inorganic materials: Examples of different hybrids with various dispersion levels and dimensionalities a) gold nanoparticle encapsulated polypyrrole [229, 269], b) Polypyrrole intercalated in a zeolite-type host (based on [200]), c) Polyaniline intercalated into V_2O_5 [165, 166], d) Polypyrrole-$PMo_{12}O_{40}$ heteropolyanion [105]

strongly bonded class II hybrids. On the other hand, the introduction of molecular inorganic species into a COP network can also be accomplished by taking advantage of the doping process of the polymer leading to the incorporation of charge-balancing species into the structure, in which case class I hybrids are normally obtained.

The latter is the case for hybrid organic-inorganic materials with COPs and tetrachloroferrate [58–63], hexacyanoferrate [64–72] and other cyano-metalate complexes [73–75], coordination compounds such as oxalatometalates [76, 77] or EDTA complexes [78, 79], tetrathiomolybdate [80–87], carboranes [88] and macrocyclic compounds like metal phthalocyanine [79, 89–100] or porphyrine [101, 102] complexes. Among polynuclear metal complexes polyoxometalates have been by far the most widely studied (see [1, 57] and Table 7.3). Different examples of these oxide clusters have been used to modify or get incorporated into polyacetylene, polypyrrole, poly(N-methylpyrrole), polyaniline, and polythiophenes. Fig. 7.3 shows a couple of relevant examples of O–I hybrids made with conducting polymers and cluster species. Table 7.3 collects some representative examples of hybrid O–I materials reported in the literature.

Given the weak nature of O–I bonds in class I hybrids and the possible mobility of the inorganic molecules within the polymers, there are two major points concerning the nature of insertion adducts formed between COPs and active doping anions. One is the *spatial distribution* of the inorganic doping molecules and their concentration within the polymer matrix, the second relates to the *reversible nature* of the doping process.

The concentration and distribution of active inorganic species in OI hybrids can vary widely depending on the particular conditions of preparation. In principle, the amount of active anions incorporated into the hybrid is limited by the doping level attained for each particular polymer. But in addition, the presence of other (inactive) anions in the reaction or doping medium can decrease even further the amount of active species in the hybrid. On the other hand, it must be stressed that different applications will impose different requirements on the design of a particular material, its form and composition. For example, a catalytic material should be designed with the inorganic active species concentrated at the surface of the polymer, whereas energy-storage applications would require a bulk distribution [1].

A second important point concerning class I (loosely bound) OI hybrids, which is related to the *reversible nature of the doping process*, is the possibility of suffering the deinsertion of active species upon reduction of the conjugated polymer. This would simply correspond to the inverse process taking place during oxidative doping with insertion of anions. This possibility is most apparent in applications that require redox cycles of the hybrid materials. These include most notably electrodes for rechargeable batteries and ultracapacitors but also electrocatalysts or sensors.

p-Doped COPs containing simple conventional anions such as ClO_4^- do actually suffer the deintercalation of those anions upon reduction. Nonetheless, the processes of doping/undoping and charge compensation are not as straightforward as they might seem. It has been shown [104] that even for COPs doped with simple

Tab. 7.3 Selected examples of Hybrid Organic-Inorganic (OI) materials found in the literature

COP	Inorganic part	Applications	Ref.
MDMO-PPV[a]	PCBM	General studies	392
		Solar cells	103
PPy	Heteropolys	Electrocatalysis	393
		General studies	106, 374, 394
		Catalysis	395
		General studies	396, 397
		Batteries	68, 105, 107, 113, 398
		General studies	99
		Batteries	1
PAni	Heteropolys	Electrocatalysis	399
		Catalysis	400
		General studies	394, 396, 401, 402
		Catalysis	403, 404
		General studies	405, 406
		Batteries	107, 111, 113, 398, 407
		General studies	110
		Batteries	68, 118
		General studies	408–410
		Batteries	1
PAc	Heteropoly	General studies and catalysis	411–414
PAni	$[PMo_{12}O_{40}]^{3-}$	Cathode for lithium batteries	1, 57, 68, 107, 113, 118, 398, 407, 415
PPy	$[SiCr(H_2O)W_{11}O_{39}]_5$	General studies	416
PAni	Heteropoly/M-Cresol	General studies	410
PPy	Fe_2O_3	General studies	417
PPy	$[Fe(CN)_6]^{+-}_{3}$	General studies	114–117, 124, 129
		Batteries	1, 57, 67–69, 407
PAni	$[Fe(CN)_6]^{+-}_{3}$	Batteries	68
COPs	$FeCl_4^-$	Photocatalysis	62
		General studies	58, 59, 61
		Magnetism	60
		Photocatalysis	63
COPs	MoS_4^-	General studies	80, 81, 83–85
Ppy	Tris(oxolato) metalate	General studies	77
PPy	$Au(CN)_2$	General studies	75
PPy	$Ni(CN)_4$	General studies	74
PPy	Co porphyrin	General studies	78, 79, 95–97, 101
PAni	Prussian Blue	Electrochromism	125
		Catalysis	379, 418
		Electrochromism	126, 419
PPy	Prussian Blue	Batteries	420
		General studies	123, 127
		Solar cells	128
PPy	M-EDTA complexes	General studies	78, 79

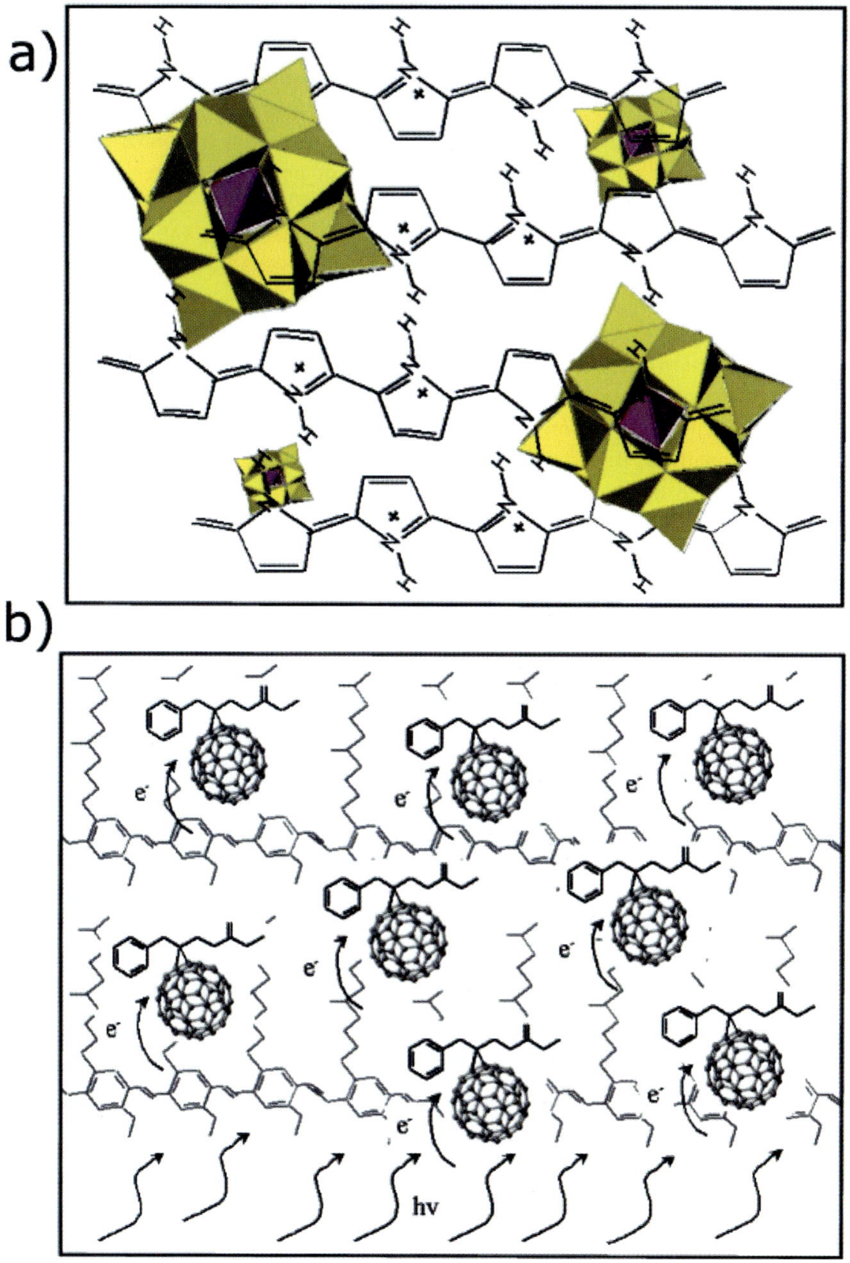

Fig. 7.3 Examples of hybrid Organic-Inorganic (OI) materials with molecular species integrated into conducting polymer matrices: a) electroactive hybrid formed by the phosphomolybdate heteropolyanion and polypyrrole [1, 57], b) photoactive hybrid formed by a C_{60} derivative and MDMO-PPV [103]

anions, charge balancing during reduction can take place also by insertion of cations under certain conditions.

In OI hybrids, though, the challenge is to avoid the deinsertion of the active anion incorporated during the oxidation of the polymer. Large anions with high negative charge will have lower diffusion coefficients within the polymeric matrix and will be more likely to stay. This is the case for polyoxometalates, which have been shown to get effectively anchored within PAni, PPy, PT or even PAc matrices. The evidence for this anchoring is varied and has included cyclic voltammetry of the hybrids [1, 57, 105–107] as well as quartz-crystal microbalance studies [108–110].

The retention of large active anions in these hybrid materials has important consequences regarding their redox insertion mechanism. The permanence of the anions in the polymer matrix upon reduction forces the insertion of cations for the necessary charge balance (Fig. 7.4), with the inverse process of cation expulsion taking place upon reoxidation, thus converting p-doped polymers into cation-inserting redox materials. In addition to the above mentioned studies, direct evidence of this cation-based mechanism was obtained from analysis of lithium after discharge of a hybrid PAni-PMo12 electrode showing the incorporation of lithium ions corresponding well to the amount of charge involved in the redox reaction [111]. The possibility of converting p-doped COPs into cation-inserting polymers opens new possibilities for their application, for example as selective membranes or sensors, or for their integration in energy storage devices such as electrochemical supercapacitors or rechargeable lithium batteries, the latter being most commonly based on cation-producing anodes and a source-sink mechanism, which prevented the efficient use of conventional COPs due to their anion-insertion mechanism [1, 21, 111–113].

Whereas bulky molecular clusters of the polyoxometalate type are easily anchored within COPs the question remains as to whether smaller active anions would also be. The electroactive hexacyanoferrate anion $[Fe(CN)_6]^{3-/4-}$ (HCF) and its hybrid with PPy or PAni provide interesting examples in this respect. Some experimental evidence suggested that the permanence or loss of HCF anions in PPy was potential dependent [114]; furthermore, cation incorporation into PPy-HCF hybrids has been reported to take place only when the reduction of the HCF anion alone was involved, whereas the electrochemical reduction of the polymer (all in aqueous KNO_3) led to the deinsertion of hexacyanoferrate ions [115, 116]. The nature of counterionic cations also plays a role in the redox behavior of HCF doped into Ppy [115, 117], and finally the nature of the electrolyte has been shown to play a role, with non-aqueous electrolytes used in lithium batteries preventing the loss of the HCF anion from its hybrids [67, 69–71, 118]. All these factors, which seem to complicate the redox chemistry and insertion/deinsertion properties of these hybrid materials, can nevertheless be turned into useful behavior, as in the latter example of Li battery applications. In another example of electrochemical control, films of PPy-HCF hybrids were used to demonstrate the feasibility of switching between a strongly bound state for HCF in the polymer to the release of controlled amounts of the anion by cathodic current pulses in what could be considered a model for drug-release chemical systems [115].

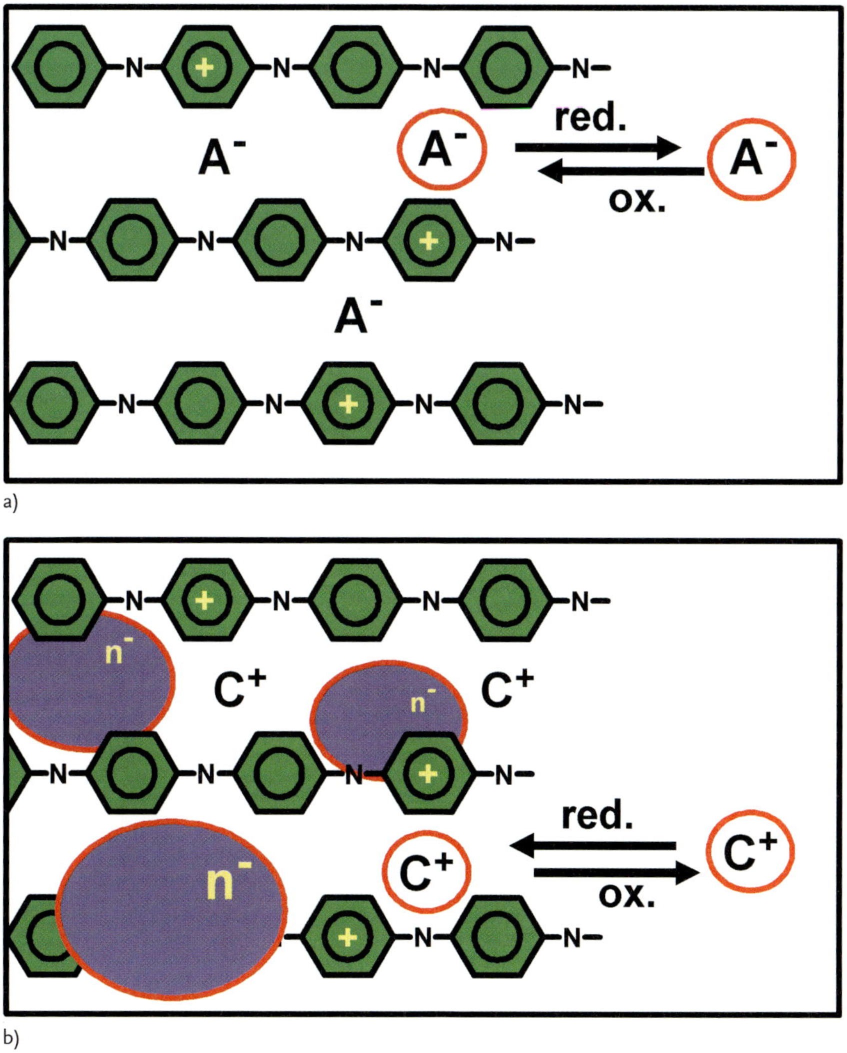

Fig. 7.4 Schematic representation of the anchoring of large anionic dopants in OI hybrid materials, leading to a change in the insertion mechanism from a) anion to b) cation insertion

In principle the "active-anion" doping approach seems obviously limited to anionic molecules when dealing with the most common p-doped COPs. However, there are several ways to overcome this expected limitation. A suitable substitution of neutral ligands with anionic functional groups, for example, can turn neutral or even cationic complexes into anions. Thus, sulfonated ligands were used for instance for the integration of the metal phthalocyanine complexes mentioned above or iron trisphenantroline sulfonic acid complexes [119] into COPs. In addition to the doping species, the polymers themselves can also be derivatized by sulfonation [120] or substitution with other anionic groups such as alkylcarboxylic acids [121]. This approach, sometimes called self-doping, also leads to cation-inserting materials. In the above examples, intended applications were the development of cathodes for lithium batteries [120] and proton-pump electrodes [121] respectively. Another way to introduce cationic species into p-doped COPs is the previous doping of the COPs with anionic organic polyelectrolytes which – similarly to the inorganic polyoxometalets described above – have a hindered diffusion out of the matrix [119].

Since we have just discussed OI hybrids with the hexacyanoferrate anion as inorganic doping component, it might be pertinent to mention here a related large group of hybrids formed by COPs and inorganic polymers of the Prussian blue family ($M[Fe(CN)_6]$) [64, 65, 72, 114–117, 122–131, 132 Ni-PB], where M is Fe in Prussian Blue but can be a number of other transition metals. Despite their common $[Fe(CN)_6]$ moiety, these are inorganic polymers with little relation to molecular ferricyanide, and form with COPs materials which could be considered as true hybrid interpenetrating polymers, although in that case the structure of the hybrid and its intercalate nature are less well established. In this respect these hybrids could be considered a type of material in between the OI hybrids discussed above and the IO systems that will be described next.

7.5
COPs Intercalated into Inorganic Hosts: Inorganic-Organic (IO) Materials

The integration of conducting organic polymers into larger, extended inorganic hosts, whether natural or synthetic, takes us to a very different field of hybrid materials. In these so-called IO systems the inorganic host normally contributes a solid crystal structure where the structure of the polymer is to be constrained. As we will see below though, there are some cases where the polymeric quasi-amorphous nature of the inorganic components themselves leads to systems better described as hybrid polymer blends.

In general, the integration of a polymer into an extended inorganic lattice can be best performed by polymerization of the corresponding monomers after intercalation into the inorganic host. Nevertheless, it is possible sometimes to form simultaneously both the inorganic matrix and the conducting polymer chains to form IO hybrid materials in "one-pot" syntheses [1]. It should be understood that once formed inside the lattice the polymer is fixed and the process irreversible. In this sense it is not correct to talk about an inorganic "host" for the intercalation of the polymers. Yet, despite the irreversible character of the process, the inclusion of COPs within the structure of extended solid phases leads in many cases to remarkable hybrid materials with synergic properties.

In general IO hybrid materials based on COPs were initially studied as intercalation derivatives of conducting polymers in inorganic phases [133–136] and have since evolved towards the development of materials with well-defined microstructures controllable at the nanometer level, as in hybrids based on mesoporous structures [137–139]. These materials can possess novel electrical and mechanical properties and the structural constraint of COPs grown in mesoporous templates allows for the study of these polymers in restrained configurations, as in a recent study on the energy transport properties of hybrids formed by MEH-PPV in mesoporous silica [139].

Examples of a large variety of IO materials are described in Table 7.4 and can be found in the literature as early as 1975 with the intercalation of aromatic compounds in silicates (smectites like montmorillonite and hectorite) [136] as well as conjugated polymers such as polyaniline [140] or polypyrrole [141]. Also by 1979 the reaction of pyrrole and FeOCl was reported to give $(Ppy)_{0.34}FeOCl$ [142]. The intercalation or encapsulation of polymers into MoS_2 [135, 143, 144] or into MoO_3 [145–151] constituted also intensive topics of research in the early nineties.

One of the most extensively studied and better characterized IO systems is the family of hybrids based on different forms of V_2O_5 (from crystalline to xerogel) as the inorganic phase. We can analyze it in more detail as a representative example of IO hybrid materials. Indeed, there have been a few major lines of work dealing with several related hybrids: V_2O_5 with PAni or its derivatives [118, 152–169], PPy-V_2O_5 [153, 156, 162, 170–177], as well as hybrids formed between PT derivatives and V_2O_5 [153, 154, 175, 178–180] such as PEDOT-V_2O_5 [180]. For a given hybrid, the inorganic starting matrix could be either crystalline V_2O_5 or the quasi-amorphous V_2O_5 xerogel material obtained by drying the gel resulting from the acidic hydrolysis of vanadate [181, 182]. Most studies have been based on the latter, aiming at the production of hybrids interacting at the molecular level with reduced kinetic barriers for their formation, although hybrids based on crystalline V_2O_5 have also been reported [167]. A related inorganic phase not so extensively studied but also of interest for the insertion of conducting polymers into its layered structure is vanadyl phosphate α-$VOPO_4$, [159, 183]. Furthermore, in addition to binary hybrids of V_2O_5 and conducting polymers, other more complex combinations have been prepared and studied, such as the material integrating poly(2,5-dimercapto-1,3,4-thiadiazole) (PDMcT) and polyaniline (PANI) with V_2O_5 [184]

Among hybrids made of V_2O_5 and related compounds we find that the main target application is as electrodes for lithium batteries [107, 118, 156, 163, 165, 166, 173, 175, 177, 179, 180, 184, 185].

These COPs/V_2O_5 hybrid materials can be formed by diffusion of monomers into the inorganic network followed by their polymerization, or else by simultaneous formation of the organic and inorganic networks. The crystal structure, insertion processes and the degree of order in these materials can be followed by means of powder X-ray diffraction. Fig. 7.5 shows powder diffraction patterns for V_2O_5 grown as oriented films (a) and as homogeneous powder (b), as well as for several PAni/V_2O_5 hybrids (c1, c2, d). For V_2O_5 the effect of material growing conditions on order and structure is already evident. The presence of preferential orientation in films (a) is evidenced by the increased intensity of *00l* diffraction peaks (especially *001*) as compared with powder samples (b). The corresponding hybrid materials maintain these structural differences. The insertion of conducting polymers (PAni in the case of these data) within V_2O_5 leads to an expansion of the interlayer

Tab. 7.4 Selected examples of Hybrid Inorganic-Organic (IO) materials found in the literature

COP	Inorganic	Applications	Ref.
COPs	MoS_2	General studies	135, 186, 187
COPs	$RuCl_3$	General studies	188
PAni	UO_2PO_4	General studies	197
COPs	$Zr(HPO_4)_2$	General studies	421, 422
PAni	α-$Sn(HPO_4)_2$–H_2O	General studies	423
PPy	α-$Sn(HPO_4)_2 \cdot H_2O$	General studies	424
Ppy, PAni	H_3SbO_4	General studies	425
COPs	HUO_2PO_4	General studies	197, 421, 422
COPs	Layered Acidic Zr-Co Phosphate	General studies	426
PAni	V_2O_5	Batteries	148, 156, 160, 427
		General studies	149, 338
		Batteries, general studies	107, 162, 165, 167, 173–175, 185
		Batteries	428
		General studies	118, 163
		Batteries	429, 166, 177
PEDOT	V_2O_5	Batteries	180
PT-based	V_2O_5	General studies	153, 154, 178
		Batteries	175, 179, 180
PPy	V_2O_5	General studies	153, 170, 171, 172
		Batteries	107, 156, 162, 173, 175, 177
		General studies	174, 176, 339, 430–433
PAni	α-$VOPO_4$	General studies	159, 183
PAni	MoO_3	General studies	147, 150
PPV	MoO_3	Batteries	145, 146, 151
COP	$CdPS_3$	General studies	150
COP	$NiPS_3$	General studies	150
MEH-PPV[a]	Mesoporous Silica	General studies	139
PAni	Silica	General studies	434
COP	Mica-type silicates	General	134, 435
COPs	Silica Host	General studies	212
PAni	MCM-41	General studies	137, 138, 207, 211, 212
COP	Zeolite	General studies	136, 198–201, 203–205, 207, 211, 436
		Nanostructures	204
		Photoluminescence	437

a MEH-PPV: Poly[2-methoxy-5-(2′-ethyl-hexyloxy)-1,4-phenlylenevinylene]

Tab. 7.4 Selected examples of Hybrid Inorganic-Organic (IO) materials found in the literature

COP	Inorganic	Applications	Ref.
PPV	Mesoporous Silica MCM-41	Photoluminescence	437
COP	Clays	General studies	140, 438–442
PAc	Al_2O_3	General studies	136, 212, 443–445
PAni	FeOCl	General studies	192–196
PPy	FeOCl	General studies	189–191
PF	FeOCl	General studies	133
Ppy	γ-Fe_2O_3	Batteries	265
PAni	Layered $HNbMoO_6$	General studies	446
PAni	Layered MnO_2	Li-ion transport	244
PAni	Layered Silicate	Synthesis	447
PAni	Tin(IV) phosphate host	General studies	448
COPs		As templates (intermediate hybrid material)	51, 229, 449–454
PAni	Montmorillonite	General studies	361, 442, 455–457
Pani	Graphite Oxide	General studies Anode for batteries	343, 458–461 462

spacing. The shift of the intense 001 peak in the X-ray diffraction pattern (Fig. 7.5c1) indicates a change in the c axis from 11.5 Å for V_2O_5 to ca. 14 Å for PAni/V_2O_5, which is consistent with the substitution of the xerogel water molecules for PAni chains. Furthermore, a systematic study of aniline-rich reaction mixtures has shown that it is possible to identify two intercalate phases among polycrystalline PAni-V_2O_5 hybrids, corresponding to a single and a double layer of PAni inserted in the inter-layer space [164]. Traces c1 and c2 in Fig. 7.5 correspond to these single and double intercalates which show spacings of ca. 14 Å and 22 Å respectively. The inset figure shows schematic drawings and spacings corresponding to these materials and the hydrated xerogel.

In addition to the expansion of the unit cell, insertion of polyaniline into V_2O_5 results in loss of crystallinity as shown by the broader, less intense 001 peak, as well as by the broad diffuse scattering feature centered at ca. 25° in 2θ. Indeed, for long reaction times, the diffraction patterns of aged materials shift from truly IO hybrids with a V_2O_5 dominated structure to materials in which the structure of the organic chains seems to take over [1]. This is shown by the final loss of crystallinity evidenced in Fig. 7.5d, an X-ray diffraction pattern where peaks characteristic of V_2O_5 have disappeared and only diffuse scattering features from the polymer are left. This final material could therefore be considered a borderline case between I–O and O–I hybrids and could be rightfully labeled as a hybrid polymeric material.

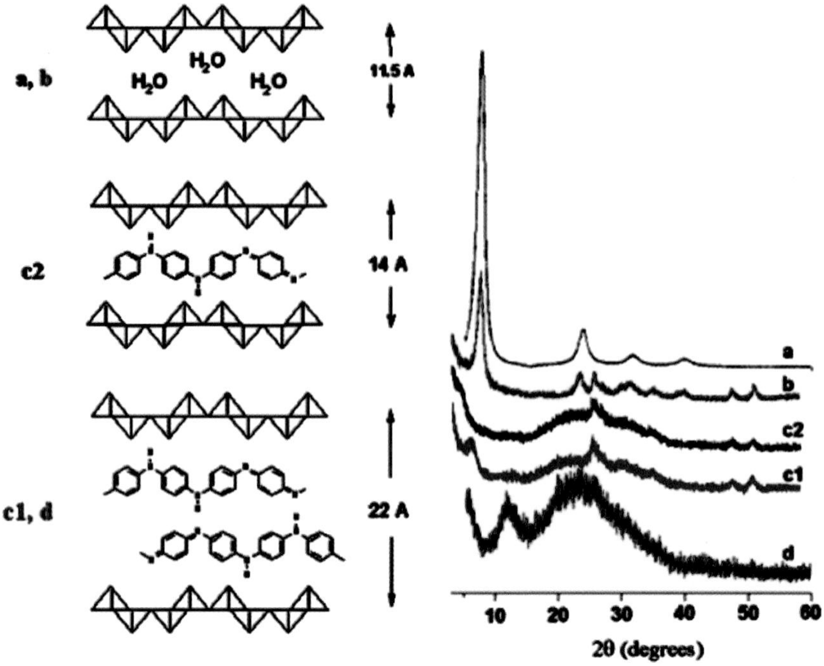

Fig. 7.5 X-ray diffraction patterns for V_2O_5 and related hybrid materials: a) Films of V_2O_5 showing preferential orientation (enhanced intensity for *00l* peaks), b) V_2O_5 xerogel powder. c1) PAni-V_2O_5 hybrid with a monolayer of PAni inserted within the oxide slabs (001 spacing of 14 Å), c2) PAni-V_2O_5 hybrid with a double layer of PAni inserted within the oxide slabs (001 spacing of 22 Å). d) PAni-V_2O_5 hybrid prepared for an extended period of 194 hours (as compared with one hour for samples c1 and c2) showing the collapse of V_2O_5 structure (from Lira-Cantu [1, 57, 164])

Layered transition metal sulfides like MoS_2 [135, 186, 187], chlorides like $RuCl_3$ [188], and oxychlorides like FeOCl [133, 189–196], silicates and phosphates [136], UO2PO4 [197], $CdPS_3$ [150] and $NiPS_3$ [150] are other examples of hybrid inorganic-organic I-O materials.

7.5.1
Mesoporous Host or Zeolitic-type Materials (silicates inclusive)

We can also find very interesting research work on the incorporation of conducting organic polymers into a mesoporous channel host. Some initial researches were reported back in 1989 by Bein and coworkers on the incorporation of polypyrrole in zeolite-type inorganic host structures [198–205] followed by the incorporation of other COPs like polyaniline [137, 196, 206] or the formation of wires or filaments of COPs in this kind of inorganic material [137, 138, 205, 207]. This type of synthesis provided a restrained environment for the ordered growth of single-chain polyaniline molecules akin to molecular wires [137, 138, 205, 207].

Initially the idea of incorporating organic molecules or polymers into inorganic hosts was used for the formation of purely inorganic mesoporous materials by removal of the organic structure-forming templates, as initially obtained by researchers of Mobil Corporation back in 1992 [208, 209]. But true hybrid materials have also been produced. For instance, taking advantage of pre-formed mesoporous MCMs (Mobil Crystalline Materials) [208, 209], all sorts of organic components have been inserted, from phthalocyanines [210] to conducting organic polymers [137, 138, 204, 207, 211], for the application of membranes [210, 212], as catalyst, or production of molecular wires [138, 205], etc.

As proposed by Bein et al. [138], initial work on the incorporation of conducting organic polymers into inorganic host materials was intended as a first step towards the design of electronic devices based on molecules instead of on bulk semiconductors. And more recently a paper published by Schwartz et al. presented a study about the energy transfer in hybrid materials made of COP intercalated in ordered mesoporous silica glass [139]. This work studies the energy transfer and its quantification in COPs and the importance of the architecture of the hybrid material that forces the energy transport to take place along the polymer chain. These studies allow researchers to understand the conduction mechanism of COPs in order to optimize polymer-based optoelectronic devices [139, 213]. A schematic representation of the hybrid material under study is shown in Fig. 7.6.

Fig. 7.6 Schematic representation of COP intercalated in the 22 Å wide channels of an ordered mesoporous silica glass [139]. Graphic courtesy of Daniel Schwartz of D.I.S.C. Co.

7.6
Emerging Nanotechnology: Toward Hybrid Nanocomposite Materials (NC)

In a general way, all hybrids described in this chapter are nanocomposite materials since they are formed by well-defined distinct components dispersed at the nanometer level. But we will deal in this section with a particular type of hybrid nanocomposite material formed by nanoparticles of inorganic compounds (not molecular) dispersed in conducting organic polymers. In these cases, neither the COP nor the inorganic species can be considered as a clearly dominating matrix in the final material. We could therefore consider them as true composite materials where each component still retains its own extended structure but with an increased interfacial interaction derived from the nanometric particle size. These hybrid nanocomposites, which have been recently reviewed [214], constitute an entire area which could comprise a full chapter in this book. We will try however to cover the topic in this section by giving a general description of nanocomposite hybrids based on conducting organic polymers. We will describe the evolution of these materials from earlier research work, when *composite* materials were made of inorganic particles at the micrometer range level. Finally we will describe true nanocomposite materials where the organic and inorganic components interact with each other almost at the molecular level.

Nanocomposite materials made of inorganic species and electrically conducting polymers were first reported only a few years after the discovery of the conductivity of COPs back in the 1980s. The sought-for synergy between organic and inorganic materials was initially investigated at the macroscopic level by the study of interactions between thin film interfaces. An illustrative example is the preparation of an electrochromic element composed of bilayers [216]. In this case, an oxidative layer was composed of COPs (PAni, PPy or PT) and a reductive layer was made of oxides such as WO_3, MoO_3 or TiO_2. It is obvious that this "composite" material was not a true organic-inorganic nanocomposite, but showed a way to combine organic and inorganic active components in an ensemble where their complementary behavior and the mutual reinforcement of their functionalities was used for better performance. Since then, many other examples of true hybrid nanocomposite mate-

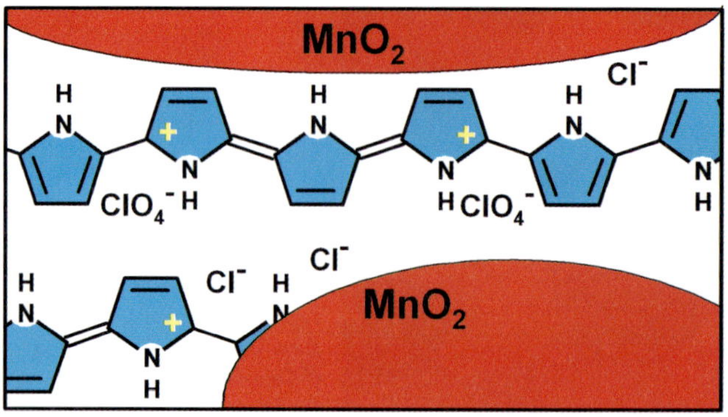

Fig. 7.7 Example of a Hybrid Nanocomposite Material based on MnO_2 particles dispersed in polypyrrol [215]

rials began to appear, either with metals or with oxides as in the case of polyaniline with MnO_2 [217] (See also Fig. 7.7), and even examples of unexpected combinations like those produced by the incorporation of COPs into customized inorganic materials like porous glass [218].

Until the end of the 80's one could find in the literature composite materials made of inorganic materials coated with COPs as antistatic films [219], or COPs in textile materials as coatings [220], or the use of COPs to coat inorganic mineral particles of talc, mica, $CaCO_3$, clay, hydroxyapatite, etc. [221]. Some of these hybrid nanocomposite materials are listed in Table 7.5.

But it was not until the decade of the 1990s when a mainstream line of research on hybrid nanocomposites was consolidated. Two main groups of hybrids could be distinguished, those containing metal nanoparticles and those incorporating inorganic compounds with small particles sizes dispersed in COPs. The first group of materials is formed by nanosized metal particles dispersed in COPs [222–228] where, frequently, the metals used are precious metals from group 10, Pd and Pt. These hybrids are designed mainly for catalytic purposes, like proton and oxygen reduction [222, 225], hydrogen oxidation [222] as well as hydrogenation reactions [223].

On the other hand, gold (Fig. 7.8) and silver metal nanoparticles have also been used as templates for the synthesis of hollow nanometer-sized conductive polymer capsules [229, 230]. Although in that particular line of work, by the group of Feldheim, the final purpose was the use of well-arranged metal nanoparticles to obtain ordered COP superstructures, their approach shows a way to the production of COP-coated metal nanoparticles that will likely find many other uses and applications.

Although metals are the most common elements formed in nanodispersions within COPs, there have also been interesting studies on dispersions of other active elements such as carbon [227, 228, 231–236] or sulfur [237] for use in energy-storage devices [231, 232, 237], as electrocatalysts for environmental remediation in Cr(VI) reduction [228, 238], or as sensors and "artificial noses" [235, 236].

In the second group of materials (with nanodispersed compounds) we can mention several hybrid materials based on colloidal particles of transition metal oxides such as γ-Fe_2O_3, [238, 239], MnO_2, [215, 240–245], CuO [246], TiO_2 [14, 16, 247–250] and WO_3 [126, 241, 251–253], as well as tin oxide [254–256] and silica [34, 255, 256]. In addition to the more extensively studied work on oxides, recent reports have extended this family of hybrid composites to other colloidal semiconducting particles such as CdS [248, 257–264], CdSe [259, 260, 263, 264] and copper or silver halides [258]. All these nanocomposite materials are summarized in Table 7.3.

A detailed analysis of the work published on this category of hybrid nanocomposites shows that particle sizes of the inorganic components have been systematically decreasing with time. Before 1990, only reports on hybrid composite materials in the micrometer range were reported. An example is the synthesis of nanocomposite materials of COPs coating inorganic particles of dimensions between 10 and 300 μm [219, 221]. At the beginning of the 1990s smaller particle sizes were used and currently one can easily find composite materials in the nanometer size

Tab. 7.5 Selected examples of Hybrid Nanocomposite (NC) materials found in the literature

COP	Inorganic part	Applications	Ref.
COP	Au	Nanocomposite	463
Ppy	Au		449, 464, 465
Pac	Au	Nanoparticle assembly	269
PPE-PAni	Ag	Composite fibers	466
COP	Nanoparticles: Pt, RuO_2, Ag, Pd, Ni, MnO_2, Cd, Co, Mo, etc.		281, 467
Ppy	Ferrite		281
COP	Au wire	Electronic circuits	468
PMeT(Y)	CdS_2	Solar cells	257
COP	$CdS_2 + SiW_{12}O_{40}$	Photovoltaics	469
PAni	CdS		247, 261
COP	CdS	Films	469
COP	CdSe	Solar cell	330
COP	Perovskites		8
COP	Cations		8
PAni	80% TiO_2	Solar cells	270
PPy	TiO_2		247, 255
Ppy	Dye	Solar cells	329
Pani	WO_3	Films	241
COP	WO_3, MoO_3 or TiO_2	Layers	216
COP'	Alkali metal particles	Anode for battery	470
COP	Silica, g-Fe_2O_3, WO_3, Pd, Pt		471
COP	Cu-Fluorhectorite		472
COP	Fe-polymer		473
PAni+PSS			
Pani	Silica		110, 136–138, 211, 214, 255, 256, 267, 434, 474, 475
Ppy	Silica		34, 134, 214, 255, 256
COP	SiO/Si		206
COP	Alumina		476
PPy-based	Zirconium		477
Ppy	Glass	Films Multilayer	478
Ppy, PAni	Porous Vycor Glass	Sensor	479, 480
COP	Silica Cu-Ni	Dispersion of substances	481

Tab. 7.5 Fortsetzung

COP	Inorganic part	Applications	Ref.
Pthyophene	Silane	Scratch resistant, transparent coatings	35
PAni	Al		482–484
COP	Graphite		458
Ppy	C		231–234
PAni Ppy	MnO_2 or b-MnO_2		215, 217, 240, 242, 243, 245, 485, 486
PPy	$LiMn_2O_4$		485, 486
COP	Hg-Ag		487
Ppy	Pt	Catalyst	227, 228, 488
Pani	Pd	Catalyst	266
COP	Pb, Sn, Ni, Co, Ti, In, Cd, Fe, Cr, Ga, Zn, Ti, Be, Mg, Na, K, Li		489
COP	Host polymer + Ce^{+3}, Mo^{+6}, W, Cr, Cu^{+2}, Ag^{+}, etc.		490
PAni	Cobalt	Nanowires	454
PAni or Ppy	C Nanotubes		491, 492
PAni	$BaTiO_3$		493
PPy	Bi_2S_3		334
COPS	$BaTiO_3$, Ag, Ni or Cu + Coated graphite fibers		494
PPy	Fe_2O_3		214, 239
PAni	$MoCl_6$		
	BF_3		495
COP	Gels	MEMS	496
COP	Talc, mica, $CaCO_3$, Clay, Hydroxyapatite		219, 221
Pani	Latex	Conductive layers	497
Pani	Rubber		498
PAni	Nafion		499
PAni	Nylon 6		499
COP	Textile fiber	Coating	500
Ppy	Fiber glass		501
Ppy	Fiber glass	Stabilization	502

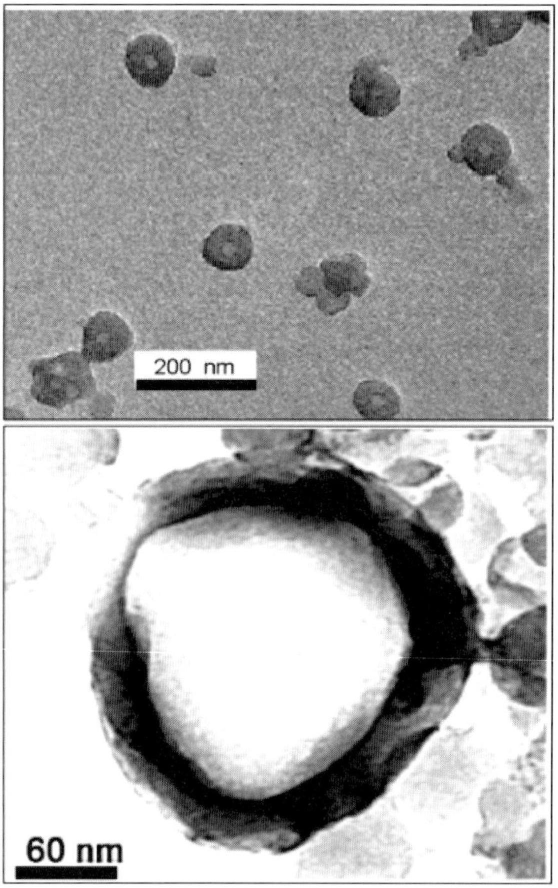

Fig. 7.8 Microphotograph showing the encapsulation of a gold nanoparticle (about 200 nm) in polypyrrol. With permission of D. Feldheim [229]

range. To illustrate this we will take as an example the composite materials formed by the dispersion of γ-Fe_2O_3 particles in COPs. In 1994 a hybrid with oxide particle size of about 100 Å was reported [239], whereas a recent paper describes a hybrid nanocomposite with the same inorganic material, γ-Fe_2O_3, where the particle size has been reduced to 80 Å [265]. Other examples published in 1998 and 1999 show the use of Pt metal particles dispersed in COPs with a 400 nm particle size [266] or particles of silica as small as 300 nm dispersed in Polyaniline [267] respectively. More recently in 2001, a report of a hybrid nanocomposite made of COP and CdS particles of about 1–2 nm was published by Godovsky et al. [268].

The effort to reduce the size of the inorganic particles down to the nanosize range is a common factor in the work of nanotechnologists. The properties of micrometer-long or larger metal particles are far from the rich optical, electronic or catalytic properties of metal nanoparticles. This trend towards smaller active units is reinforced by the trend towards miniaturization, which deals with the need to fabricate smaller, cheaper, more controllable, efficient and faster electronic devices. These attractive characteristics are leading researchers to look for hybrid nanocomposite materials where the organic and inorganic components are linked together at a molecular level and in controllable arrangements leading to improved properties (see Chapter 10).

An illustrative example showing the potential of hybrid materials in the development of nanostructured devices is described by Feldheim et al. [269]. They report the synthesis of a hybrid nanocomposite material by the assembly of metal nanoparticle arrays of gold or silver using conducting organic polymers as molecular bridges [269]. The metal nanoparticles of about 5 nm in diameter are linked together by phenylacetylene structures of different lengths. The distance between nanoparticles can be controlled by the length of the COP chain used. These hybrid composite materials could be useful for the fabrication of nanotransistors and nano-tunnel diodes [269].

7.7 Current Applications and Future Trends

We could sort out the applications of COP-hybrids into two major groups according to the main activity at work in each case. The first group could include materials with applications relying upon the *conductivity* of COPs while the second group would include those materials in which the *electroactivity* of COPs is used. In the first group for example, we could list their application as electrostatic materials, conducting adhesives, electromagnetic shielding, antistatic clothing, diodes, transistors, etc. Usually the polymers in this group are used because of their light weight, processability or low cost. In the second group we could consider applications where the polymer undergoes chemical modification as part of its use as a functional material, for example for application in sensors, rechargeable batteries, ion exchange membranes, capacitors, photovoltaics, electrochromics, molecular electronics, "smart" materials, switches, etc. In the following section we will describe some of these applications with special emphasis on those based on their photo-electro-ionic properties and will show examples where hybrid organic-inorganic functional and multifunctional materials have opened the way to new applications.

7.7.1 Electronic and Opto-electronic Applications

The first integrated circuit initiated in 1958 the era of microelectronics, computing and information processing, growing at an accelerated rate ever since, as the number of transistors per unit area doubled every 1.5 years, a trend that has come to be known as Moore's law.

Ultra-dense, ultra-fast and molecular-sized logic devices are required in order to keep pace in modern and future electronic equipment. To achieve this goal, traditional inorganic semiconductors are being replaced by organic molecules, polymeric or even biological materials [270].

Conducting organic polymers (COPs) constitute a very important part of this new field of "organic electronics" since they offer a combination of most of the characteristics required for microelectronics applications to be added to their low cost and plastic nature. Manipulation of their electronic conductivity by chemical or electrochemical means enables high charge carrier mobility in the polymer backbone. They can be used in electronics in two major applications: as interconnector materials or as electronic materials by themselves [271]. Several publications describing the multiple advantages of COPs in this area have appeared recently [12, 270–277].

Other related interesting applications can also be found in the literature, for example COPs have been applied as conducting resists (charge dissipators for electron beam lithography) by the incorporation of the polymer into a multilayer resist system, acting as a conducting interlayer between the imaging resist and a planarizing underlayer [271]. Fig. 7.9 shows an example of the application of COPs as conducting resist in which the electric circuit is based mainly on poly (3-octylthiophene) (P3OT) [278].

In spite of the relatively large number of recent reports making use of COPs as electronic components, similar applications for O–I hybrid materials are less abundant, although they have been around for some time. For instance, Crivello et al. used a polyaniline-osmium salt hybrid as a resist system. Exposure to an electron beam or ultraviolet radiation decomposed the osmium salt, producing protonic acids which in turn doped the conducting polymer, changing the properties of the material [279].

On the other hand, a whole line of thought has developed concerning I-O hybrids to set the basis for electronic applications. Thus, the incorporation of conducting organic polymers into highly ordered inorganic hosts, such as silica, has been

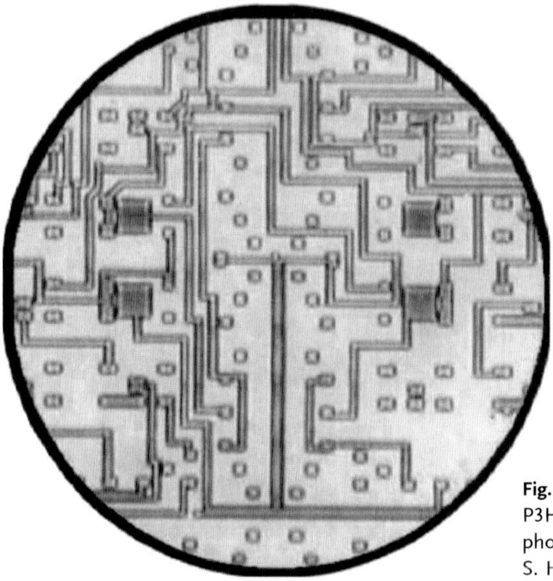

Fig. 7.9 Conductive circuit of P3HT formed by laser direct-write photolithography. With permission, S. Holdcroft [278]

reported as a useful way to obtain high-performance electronic devices. The hybrid material is applied in order to eliminate the disordered structure typically found in one-dimensional (film) COPs. This arrangement of the polymer chains enhances alignment, stability, electronic and optical properties of the corresponding COP [275].

Properties like magnetism, conductivity and nonlinear optical behavior have been reported by Lacroix [280] for different hybrid materials. In his review, materials based on chromophores integrated in different inorganic environments were analysed. The inorganic phases were, specifically, layered structures of $Mn_2P_2S_6$ and $[Mn^{II}Cr^{III}(ox)_3]^-$ and the relevance of centro- and noncentrosymmetric structures and their relation to conductivity and magnetism were discussed. Kim et al. reported a layer-by-layer assembly of COPs and magnetic nanoparticles [281]. The multifunctional films obtained have both electrical conductivity and ferrimagnetic properties. The ferrite nanoparticles were uniformly distributed within the film, making it a good candidate for its application in electronics, photonics, etc. [281].

An example of the formation and characterization of ultrathin films of controlled architecture and thickness has recently been reported. In this later case, bilayers and multilayer structured films made of monolayers of heteropolyanions ($[PMo_{12}O_{40}]^{3-}$) and polyaniline could find applications as sensors, luminiscent devices, LEDs or hybrid electronics [282].

Other kinds of hybrid materials that show magnetic properties are those based on alternating layers of magnetic polyoxometalate clusters and organic donor molecules [283]. In this example, the organic molecules are based on tetrathiafulvalene (TTF)-type donors which can introduce electronic delocalisation in the final hybrid material. These hybrid organic-inorganic materials are discussed in detail in Chapter 9 of this book and won't be detailed here.

On the other side of the application range but directly related to electronic materials are light emitting diodes (LEDs). Electroluminescence is well known in inorganic semiconductor materials. It is a process which requires the injection of electrons from one electrode and holes from the other, followed by the recombination of oppositely charged carriers and the radiative decay of the excited electron-hole state (exciton) produced by the recombination process [274]. For organic semiconductors, electroluminescence was first observed back in the 60s in anthracene single crystals [284]. By the 80s, researchers from the Kodak Co. developed electroluminescent organic thin films [285], and by 1990, Burroughes et al. reported the first case of electroluminescence in COPs [286, 287].

The discovery of electroluminescence in organic thin films by Tang and Van Slyke of the Kodak Co. was based on a two-layer molecular device made of an aromatic diamine as a hole-transporting layer and a hydroxyquinoline aluminum compound as an emissive layer. In the device, an indium tin oxide (ITO) was used as a hole-injection electrode and a magnesium-silver alloy as the electron-injection electrode [285]. These organic thin films are currently known as Organic Light Emitting Diodes or OLEDS, and are electronic devices made by placing a series of organic thin films between two conductors. When electrical current is applied, a bright light is emitted. OLEDs are lightweight, durable, power-efficient and ideal for portable applications. OLEDs have fewer and simpler manufacturing steps and also use both fewer and lower-cost materials than LCD displays. In principle, an OLED can replace the current LED technology in many applications. OLEDs are self-luminous and do not require backlighting, thus thinner and more compact displays can be fabricated. The low voltage (only 2 to 10 V) provides maximum efficiency and reduces heat and electric interference in electronic devices.

Two types of OLEDs have been proposed and described, the Small Molecule Organic Light Emitting Diode (SMOLED) and a second type made of large molecules or polymers, better known as Polymer Light Emitting Diode (PLED). The main difference between these two types is the kind of manufacturing technique employed for their production. SMOLEDs must be deposited as multilayers by chemical vapor deposition (CVD) techniques, which makes their production difficult and high-cost, whereas PLEDs require simpler coating techniques such as spin coating, although sometimes photolithography is employed.

A polymer light-emitting diode (PLED) consists of a thin layer of undoped conjugated polymer sandwiched between two electrodes on top of a glass substrate. The most studied COP for PLED applications is the polymer poly(phenylene vinylene) (PPV) and its derivatives. The conversion efficiencies for these COPs have been reported to be slightly higher than 1% photons/charge carrier. Usually, the PPV is spin-coated on top of a patterned indium-tin-oxide (ITO) bottom electrode which forms the anode. The cathode on top of the polymer consists of an evaporated metal layer (e.g. Ca). Under forward bias, electrons and holes are injected from the cathode and the anode, respectively, into the polymer. Driven by the applied electric field, the charge carriers move through the polymer over a certain distance until recombination takes place. Recent reviews on the application of conducting organic polymers (COPs) as light-emitting diodes have been published [277, 288–290].

Besides well-known COP properties such as electrical, electrochemical and optical properties, together with their easy fabrication and flexibility as uniform films, they exhibit other properties that make them suitable to be used as LEDs: their emission at wavelengths spanning the whole visible spectrum, and their high fluorescence efficiencies [291]. Unfortunately this wide emission in COPs (or small organic molecules) is a drawback when a pure emission color is required. This is a problem especially for pure red-emitting LEDs. Almost all red-emitting COPs actually emit some orange-red light which must be filtered in order to obtain pure red emission, with a consequent loss of efficiency. There have been some reports of red-emitting LEDs made, either of blue-emitting and red-emitting COPs, or by the application of COPs and porphyrins (see [291] and references cited therein). Other examples are the application of true organic-inorganic hybrid materials, for example by the incorporation of ruthenium or rhenium polypyridine complexes [291, 292] europium (Eu^{+3}) complexes [291] or Er-doped Si nanocrystals [293] into COPs. This new type of LED based on hybrid organic-inorganic nanocomposites, which we could call HOILEDs, has been the subject of recent studies that have analyzed the advantages of this nano-scale level hybridization on the properties and performances of the resulting materials [294].

7.7.2
Photovoltaic Solar Cells

Solar energy is one of the topics in energy conversion where both a huge potential and an urgent need for improvement are most evident. Thus, the power density of sunlight after filtering through the atmosphere is close to 1000 W m^{-2}; this means that four hours of daylight shining onto an area like that of a large car are equivalent to the energy stored in one gallon of gasoline [295]. Or in global terms, there would be enough energy to supply the present world energy needs if only 0.1% of the Earth's surface were covered by solar cells with an efficiency of 10% [296]. But it is obvious that improving the limited present efficiencies of state-of-the-art photovoltaic (PV) cells would not hurt.

To understand how conducting organic polymers can be applied as photoconversion materials we must understand the working mechanism of semiconductors in actual photovoltaic cells and their similarities with the physical properties of COPs.

The electronic structure of a semiconductor is described in terms of energy bands. Ideally, its electronic structure consists of a filled valence band and an empty conduction band, which are separated by a characteristic energy gap (E_g). The absorption of a photon of light with the right energy excites an electron from the valence band into the conduction band, leading to the formation of charge carriers, an electron (e^-) in the conduction band and a hole (h^+) in the valence band [295–297]. By avoiding recombination of the e^-/h^+ pair, or reaction of any of them with external species, this e^-/h^+ separation leads to the production of photocurrent in a process very much alike the inverse of the LED working principle.

For conducting organic polymers the situation is similar to semiconductors. These polymers are insulators in their neutral (undoped) state but they can be made conductive by oxidation (p-doping) and/or reduction (n-doping) of the polymer either by chemical or electrochemical means [55, 270, 288, 297, 298]. The characteristic semiconductivity of COPs results from their -conjugation formed during oxidation and/or reduction. The π-bonds are delocalized over the entire molecule and two groups of molecular orbitals (or pseudobands) are produced, corresponding to predominantly bonding (π) and antibonding (π*) combinations. The lower energy -orbitals correspond to the valence band, and the π*-orbitals to the conduction band. The magnitude of the energy gap (E_g) is a characteristic of a specific semiconducting material. A semiconductor with a wide bandgap (e.g. >5 eV) would require very high energy photons (UV) to excite electrons into the conduction band, a typical behavior of insulating materials. On the other hand, small bandgaps (e.g. <1 eV), like those found for many common semiconductors, allow excitation with lower energy (visible) light. The band gaps of most semiconducting materials are typically between 0.1 eV and 2.2 eV, whereas most semiconducting conjugated polymers have a band gap between 1.5–3 eV (Fig. 7.10). This range is comparable to the energies of photons within the visible spectrum [297]. A recent review of the synthetic principles for bandgap control in linear π-conjugated systems has been reported by Roncali [299].

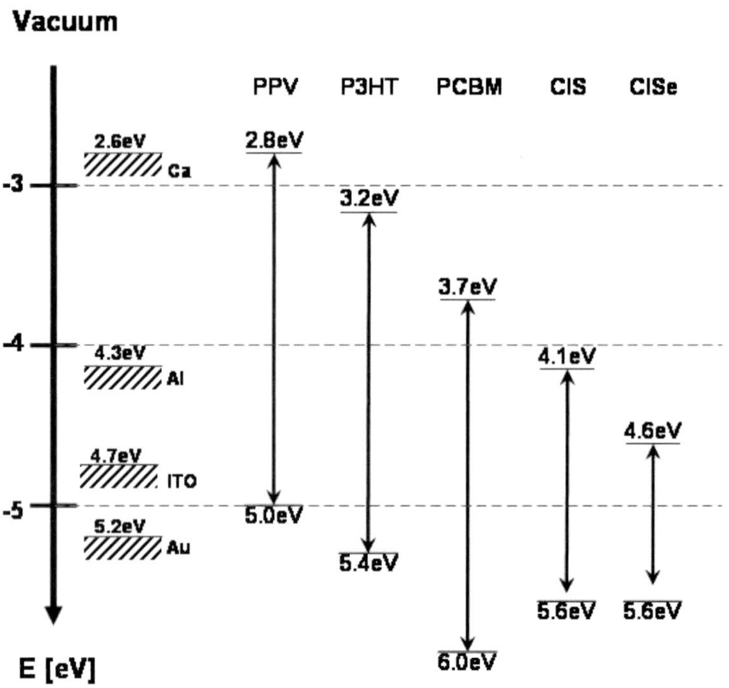

Fig. 7.10 Band and Fermi level energies of some conducting organic polymers used in plastics solar cells. PPV: poly(phenylene-vinylene); P3HT: poly(3-hexylthiophene); PCBM: 1-(3-methoxycarbonyl) propyl-1-phenyl; CIS: Copper Indium diSulfide ($CuInS_2$); CISe: Copper Indium diSelenide ($CuInSe_2$); MDMO-PPV: 3,7-dimethyloctyloxy methyloxy poly(phenylene-vinylene)

The key event in the conversion of light to energy in a photovoltaic cell is the electron-hole charge separation described before. If the electron and hole produced by an incident photon are kept apart from each other, a photovoltage builds up, electrons can be collected as photocurrents and electrical energy is produced. Otherwise, their recombination will mainly produce thermal energy. In order to assist this electron-hole charge separation an electric field is needed, and it is provided by an interface to another material exhibiting a different work function. This contact is called a "junction" and plays a crucial role in the performance and efficiency of the solar cell. There are different types of junctions in solar cells: (i) homojunctions, provided by the interface of two differently doped elements of the same semiconductor; (ii) heterojunctions, made of two dissimilar semiconductors; (iii) a Schottky junction, formed when a metal and a semiconductor are joined together, and (iv) a MIS, metal-insulator-semiconductor junction where a thin (3 nm) insulating layer is sandwiched between a metal and a semiconductor.

Current photovoltaic solar cells are made of inorganic semiconductors with photosensitivity in the solar spectrum region. Silicon, whether in the form of single crystals or as polycrystalline or amorphous deposits, is by far the predominating material, although the development of III–V (e.g. GaAs, InP) and II-VI (e.g. CdS, CuInSe) compound semiconductors has grown quickly and they are also used for particular applications. In general, high-purity materials, preferably with high crystallinity, are required for the fabrication of efficient photovoltaic devices, which makes for expensive devices.

An alternative to these conventional photovoltaic cells, developed during the last thirty years, makes use of a semiconductor-electrolyte interface. These so-called photoelectrochemical solar cells lie in between physics and chemistry, between bulk semiconductors and nanocomposite materials. After an initial thrust in the development of these cells (in the 1970's) certain problems were apparent. Namely, narrow band-gap materials tended to suffer photocorrosion, whereas semiconductors stable under illumination, typically oxides of metals like Ti or Nb, have a wide band gap (corresponding to the UV region) and consequently insensitivity to visible light [295, 296]. However, these oxide semiconductors can be used in an alternative design of solar cell technology (known as dye-sensitized solar cells) that separates the absorption process from charge-transport in the semiconductor [296]. This is possible by using a dye sensitizer that absorbs in the visible spectrum, leading to electron-hole separation, and is able at the same time to inject the electrons generated into the semiconducting oxide (most frequently TiO_2) and to transfer the holes to a shuttle redox species dissolved in the electrolyte [14, 300, 301]. This type of cell was originally introduced by Gerisher and Tributsch through their initial studies on the sensitization of semiconductor (ZnO single crystals)-electrolyte interfaces [302].

The high performance of dye-sensitized solar cells, which yield efficiencies of up to 10–11%, as reported by Grätzel and coworkers [303], is due to the microcrystalline TiO_2 nanoparticles used, which provide a high interface for electrode-dye-electrolyte interactions, and a large number of groups around the world are currently working on this type of cell [14, 301, 304–311]. The most commonly used dyes are complexes of Ru and bipyridines or terpyridines, although other organic dyes and transition metal complexes have also been used. A review of dyes, based on cyanines for example, has been reported by Mishra et al. [312].

7.7.2.1 **Nanocomposite and Hybrid Solar Cells**

Dye-sensitized systems, making use of microcrystalline TiO_2 nanoparticles with dyes adsorbed on them, could already be considered as early examples of nanocomposite designs for solar cells. Indeed, these so-called Grätzel cells have been sometimes included as one category among the type of systems generically referred to as "organic solar cells" (to distinguish them from inorganic semiconductor cells). Other arbitrary groups within this type have been labeled as Molecular Organic Solar Cells (MOSC) and Polymer Organic Solar Cells (POSC).

In the context of this book, however, it is quite clear that all those denominations share the strategy and characteristics of the hybrid approach, that is, dissimilar materials (frequently organic and inorganic), each with its own chemical character, are integrated and interact at a molecular level, leading to some type of cooperative or synergic behavior while keeping their own molecular nature.

In this section we will discuss nanocomposite hybrid materials and structures specifically designed for photoelectrochemical energy conversion. Grätzel cells, though related, are considered a category on their own and that is why they have been discussed above separately.

In the hybrid supramolecular architectures designed for solar energy conversion, organic-inorganic interfaces are taken to the molecular limit, forming what we could consider as molecular heterojunctions, as will be discussed below. This is the most recent group of materials under development in the field of photoconversion and therefore it is not surprising that their present conversion efficiencies are still low (1–2%). However, this is a quickly-moving field that will likely witness remarkable progress in the near future.

Some examples of molecular systems for photoconversion make use of dyes with high absorption coefficients and a good match of the solar spectrum, integrated in blends of different organic dye-semiconductors. For instance, a combination of a conjugated polymer and a soluble perylene-tetracarboxylic derivative [313], or copper-phthalocyanine in combination with perylene to form Donor-Acceptor architectures [314], has been reported.

The most exciting developments, nonetheless, are being made with hybrid organic-inorganic solar cells where an inorganic semiconductor or electron acceptor is combined with an organic compound (e.g. conducting polymer) to form polymer solar cells.

Conjugated polymers have received enormous attention in recent years in relation to their application in optoelectronics and photovoltaics and several reviews have been published [103, 270, 296, 297, 304]. The main working principle in these materials is the photoinduced charge transfer observed from the π-conjugated polymer to a suitable acceptor molecule or polymer [103, 296].

Buckminsterfullerene (C_{60}) and related derivatives (see review on C_{60}-based electroactive organofullerenes by Martin et al. [315]) have been some of the acceptors more widely used and studied in combination with COPs for this particular application [103, 316–321] (Fig. 7.11). A recent review of this growing area was recently made by Sariciftci et. al. [103]. Different reports where C_{60} is combined with poly(3-hexylthiophene), polymethyl methacrylate, polyvinyl carbazole, bicyanovinylpyridine, polyimide and 2-cyclooctylamino-5-nitropyridine compounds can be found in the literature [322]. In this context, we can find hybrids based on fullerenes with common insulating organic polymers [323], and also with electrical conducting organic polymers [324–326]. In both cases, the properties that characterize fullerenes are improved to finally apply in devices where photoluminescent materials and optical limiter (OL) materials are required [323].

The initial thrust in the development of polymer photoactive materials came from the possibility of manufacturing inexpensive plastic LEDs and photoconversion devices. Yet, for the specific application in solar cells, very low efficiencies soon turned out to be a major problem to add to the presumably low stability of organic polymers under intense illumination [103]. The efficiency problem was due to the low charge generation observed in the polymer layer at that time. The discovery of the photoinduced charge transfer from π-conjugated polymer to fullerene and the

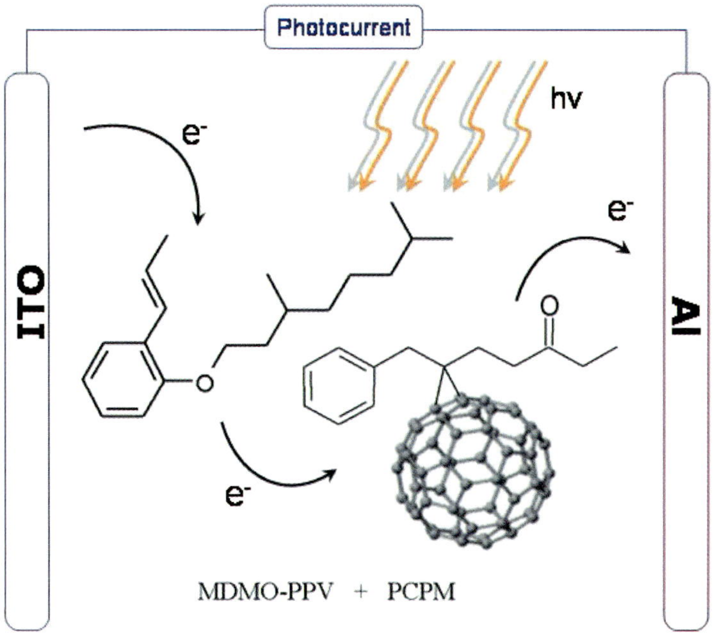

Fig. 7.11 Schematic representation of a polymer-based solar cell

related recent work of Sariciftci et al. have opened up the possibility of obtaining higher efficiency in plastic solar cells and photodiodes. Thus, the introduction of polymer/C_{60} bilayers moderately improved the efficiency of plastic solar cells, although only to low values in comparison with conventional semiconducting systems [103].

It was well-established then that the internal quantum efficiency (or absorbed photon to converted electron ratio) of photoinduced charge separation approaches unity for a donor-acceptor pair [327]. However, the power conversion efficiency was affected due to a variety of factors. For example the photoexcitations that are created far away from the heterojunction are more likely to recombine, and there is only a small fraction of the incident photons that are absorbed near the heterojunction [327].

To minimize these problems new solar cells applying the concept of bulk heterojunction materials were developed. In this design the photoactive layer consists of a class I (weakly bound) homogeneous hybrid COP/C_{60} film with a uniform bulk molecular composition.

Sariciftci et al. have demonstrated the efficiency improvement of solar cells by the application of these bulk heterojunctions (Fig. 7.12). The charge carriers photogenerated within the COP/C_{60} hybrid layer are being collected at electrodes via diffusion as well as field-induced migration. The application of these heterojunctions provides electrical power to an external circuit [103, 327]. However, for an efficient transport of holes (through the donor polymer matrix) and electrons (through the acceptor fullerene phase), a truly homogeneous distribution of the two components is required. This could be ensured through the design and synthesis of class II (covalently linked) hybrid organic-inorganic materials with intrinsic donor-acceptor properties and simultaneous electron-hole transport capabilities leading to structures of the type depicted in Fig. 7.12c. This is precisely the idea behind a new type of hybrid material called "double-cable" polymer that has been proposed very recently by Cravino et al. [327]. That design takes the p/n-heterojunction to the molecular level while forcing a regular arrangement of donors and acceptors in the form of a bicontinuous interpenetrating network. In Fig. 7.12 we can compare the three different types of junctions used for nanocomposite hybrid solar cells described above. In Fig. 7.12a, the bilayer junction initially used is shown, Fig. 7.12b shows a bulk molecular heterojunction and Fig. 7.12c corresponds to the "double-cable" or molecular heterojunction model.

In addition to exploiting molecular species like fullerenes, nanocomposite hybrid materials for application in solar cells have also been designed based on nanoparticles of otherwise conventional inorganic semiconducting materials. Among the different types of hybrids that have been applied in these devices, we can mention nanosized PbS [14, 16, 248–250, 253, 301, 308, 328, 329], CdSe [259, 260, 263, 264] or CdS [248, 257–264] dispersed in COPs. Recently, the group of Alivisatos reported a hybrid solar cell of a thiophene derivative and CdSe nanorods with a conver-

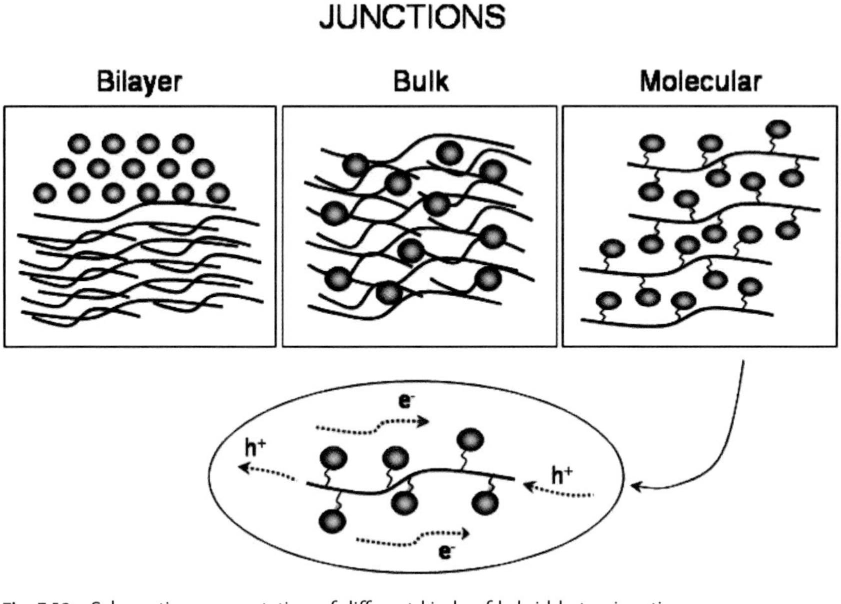

Fig. 7.12 Schematic representation of different kinds of hybrid heterojunctions: bilayer, bulk and molecular [327]

sion efficiency of almost 7% using monochromatic light and 1.7% using the full solar spectrum [330].

An advantage of using hybrid materials in solar cells is that the polymeric systems with colloids below 50 nm present low scattering effects. Also, the photoinduced charges can be effectively separated and the carrier traps can be reduced when reducing the grain boundary effect between the particles and the polymer [268]. In these cases, the charges generated by the inorganic nanoparticles are subsequently transported by the polymer [328]. The synergy offered by the interaction of organic and inorganic components of a hybrid material is patent in cases like a Poly(4-undecyl-2,2′-bithiophene) layer [309] on dye-sensitised TiO_2 [309, 329] where COPs provide the hole conduction complementary of electronic conduction in TiO_2 for an efficient photovoltage buildup.

Also the growth of organic semiconductors like quaterthiophene on n-doped GaAs and Si substrates has been reported as a possibility for heterojunctions in solar cell applications [331] and several references on this aspect can be found [14, 257, 270, 332, 333].

Another recent example of a photoactive hybrid nanocomposite material is the electropolymerization of pyrrole and its codeposition with Bi_2S_3 nanoparticles on chemically deposited bismuth sulfide substrates. The materials were designed to explore new approaches to improve light-collection efficiency in polymer photovoltaics. It seems that in these hybrid materials a photoinduced charge transfer reaction takes place that is compatible with the electron-accepting nature of bismuth sulfide nanoparticles [334].

Polyaniline and TiO_2 have also been combined in hybrid materials. In that case the hybrid material is used to improve photoconducting properties in both optoelectronic-based applications [14, 50, 248–250, 253, 261] and solar cells [14]. In these materials, TiO_2 can be present as a thin layer on a conducting organic polymer film [249, 250, 335] or it can be used as a true mixture of 80% TiO_2 in the conducting polymer [253, 270]. In the first case we could talk about the formation of a hybrid heterostructure, where the organic-inorganic interaction is critical to properties and applications but without the molecular-level dispersion found in nanocomposites like the latter example.

7.7.3
Energy Storage and Conversion Devices: Batteries, Fuel Cells and Supercapacitors

7.7.3.1 Rechargeable Batteries

Back in the 80s when COPs were a relatively novel discovery, their reversible electrochemistry added to their light weight and low cost made them very promising materials for application in thin, plastic energy storage devices like rechargeable lithium batteries. However, conducting polymers were not problem-free. Their effective capacity to store charge turned out to be relatively low and some of their characteristic properties turned into a handicap in certain cases. For instance, the well-known anion intercalation-deintercalation taking place during cycling of p-doped COPs (the type used for cathodes) (see Fig. 7.4) is detrimental to the energy density of a cell formed in combination with cation-generating lithium anodes (due to the electrolyte limiting the capacity of the whole cell) [1, 21, 112]. These limitations have been some of the major obstacles for the application of organic conducting polymers as lithium insertion cathodes, making them lose ground in favor of inorganic active materials in the race for achieving improved performance in rechargeable Li batteries.

To overcome this problem and revamp the use of COPs in certain energy storage applications, new hybrid organic-inorganic materials have been synthesized and studied. For the specific purpose of electrode development, one such combination was for example the use of a hybrid nanocomposite material made of polyaniline with organosulfur [20, 336], dimercaptan [337] or polysulfide [237] species for rechargeable lithium batteries [21]. Besides this type of battery material we can find others such as polypyrrole-amorphous molybdenum sulfide [80, 81], or polypyrrole and manganese dioxide [240, 242–244] with MnO_2 particles dispersed in the polymer matrix [240, 242, 243] or as a layered oxide [244].

At the same time, Inorganic-Organic (IO) hybrid materials were made with conducting organic polymers (polyaniline, polypyrrole, etc.) and transition metal oxides or chalcogenides (V_2O_5, TiS_2). Indeed, hybrid materials based on different forms of V_2O_5 have been some of the most widely studied for Li batteries, for example Polypyrrole/V_2O_5 [107, 118, 173, 175, 338, 339], Polyaniline/V_2O_5 [107, 118, 160, 162, 165, 167, 175, 340], sulfonated polyaniline V_2O_5 [340, 341] or poly(3,4-ethylenedioxythiophene)V_2O_5 [179]. We can also find a remarkable example of a polypyrrole-maghemite hybrid as a lithium insertion cathode for rechargeable batteries [265]. Other examples of hybrid materials that were targeted for battery applications were those based on COPs and heteropolyanions [107, 113, 118], $[Fe(CN)_6]^{+3-}$ [66, 68, 69, 71, 117, 124], although this particular type of hybrid has shown much better performance in combination with acidic electrolytes than in non-aqueous lithium electrolytes. This has resulted in a remarkable improvement in cyclability of POM-COPs in acidic electrolytes (up to 2000 cycles) and the consequent application of these hybrid materials as electrodes in electrochemical supercapacitors [342].

Fig. 7.13 presents two representative types of these kinds of hybrid materials based on conducting organic polymers that have been used as cathodes for rechargeable lithium batteries, extended systems such as Polyaniline/V_2O_5 [165] and molecular systems like Polyaniline/$[Fe(CN)_6]^{3-/4-}$ [69].

In Fig. 7.13 we observe the first cycle of the hybrid Polyaniline/V_2O_5 and its comparison with the corresponding V_2O_5 xerogel.

Fig. 7.13b shows a representative cycle (10th cycle) obtained for the other representative type of energy-storage hybrids, namely molecular hybrid materials. In that particular case data corresponded to the hybrid Polyaniline/$[Fe(CN)_6]^{3-/4-}$ (PAni-HCF) collected at a C/15 discharge rate between 3.9 and 2.7 V. The mean discharge voltage of 3.2(3) V and the specific charge of 137 Ah kg^{-1} yields a specific energy of 438 Wh kg^{-1} for the active cathode material. Furthermore, a very important and promising feature from this PAni-HCF hybrid is the high stability of the values of specific charge upon repeated cycling, with no sign of decrease after 40 cycles

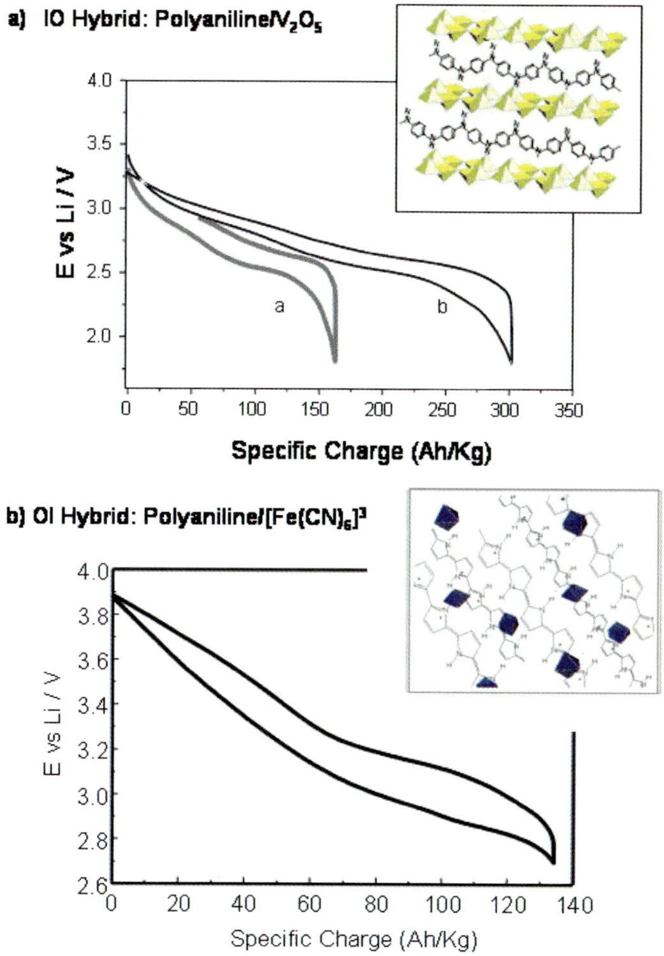

Fig. 7.13 Charge-discharge curves for two representative types of hybrid materials used as cathodes in rechargeable lithium batteries: a) Polyaniline/V_2O_5 [165], b) Polyaniline/$[Fe(CN)_6]^3$ [70, 71]

(not shown) both at C/15 and C/5 regimes [70, 71]. The experimental specific capacity corresponds very well to the expected value (140 Ah kg^{-1} corresponding to the cycling of 4 electrons per formula unit $\{(C_6H_5N)_6[Fe(CN)_6]\}_n$, i.e., one electron per hexacyanoferrate and 0.5 electrons per aniline ring). This serves as the perfect example of how the hybrid approach allows us to exploit the electroactivity of molecular species impossible to harness outside the polymer matrix.

In addition to cathode materials, fewer but also remarkable hybrids were designed as *anode* active materials, such as hybrids with carbon [232], graphite [343] or sulfur [237].

7.7.3.2 Fuel Cells and Electrocatalysis

The actual electrodes applied in commercial polymer membrane fuel cells (PEM FCs) are based on Pt dispersed on carbon. This electroactive composite is mixed with Nafion® (proton conductor) to facilitate the transport of protons and increase conversion rates. The main problem arises with the use of carbon, which, besides providing high electrical conductivity, limits the transport of oxygen, water and protons due to its impermeability. To avoid this problem COPs have been applied as proton and water permeable materials. Polypyrrole-Polystyrene sulfonated composites have been shown to possess proton and electron conductivity; when Pt is dispersed into its matrix this hybrid organic-inorganic material can be applied as supported catalyst in electrodes for polymer membrane fuel cells [344]. A similar hybrid material made of Nafion®, PAni and Pt particles has been investigated for the electrochemical reduction of oxygen [345].

For most electrode materials used in fuel cells, the electrochemical reaction is limited to the catalytic surface and the bulk material is not exploited in the catalytic process. One way to overcome this problem can be the dispersion at the molecular level of the electrocatalytic material within a conducting polymer matrix. In this way each catalytic center will be accessible to the reactive molecules, as described by Becerik et al. [346]. Platinum dispersed into poly-3-methylthiophene (P3MT) has been applied as electrocatalyst and in fuel cells in this way [347].

In addition to conventional fuel cells like PEMFC, fed with hydrogen, an exploratory research is the glucose oxygen fuel cell. It has been reported that this kind of fuel cell has been used in cardiac pacemakers [348]. In this research, D-glucose has been electrochemically oxidized in a phosphate-buffered medium and it has been observed that gluconic acid is the main product of the reaction. The formation of gluconic acid inhibits the further oxidation of glucose, thus a platinum-dispersed polypyrrole film in neutral media has been applied. This work also reports the increase of electrode activity and the reduction of poisoning effect when lead, tin, bismuth or ruthenium is added to the Pt-Ppy material [348].

Supercapacitors represent a different (and complementary) approach to batteries for the storage of charge, based on interfacial processes, and allow for higher power density but lower energy density than batteries. Originally supercapacitors harnessed double-layer charge storage on the interface between microporous carbon materials and suitable electrolytes in a purely electrophysical capacitive mechanism, but more recently a new type of device known as an electrochemical supercapacitors is leading the way towards higher specific energy systems while maintaining high values of specific power. These are based on redox active materials able to cycle repeatedly between two oxidation states and therefore represent an approximation between traditional supercapacitors and batteries.

Conducting organic polymers, which can be p- or n-doped polymers, have also been proposed and studied as active materials for electrochemical supercapacitors. The main feature of COPs is the possibility of energy storage due to a Faradaic process (not electrostatic as in double-layer carbon supercapacitors). The Faradaic process takes place at the surface of the electrode material (like in a battery) giving rise to what is known as pseudocapacitance [349]. As supercapacitors, COPs can deliver high specific power for a short period of time (10–100 s), although with values of specific energy significantly lower than those of the corresponding batteries [349].

The most desirable configuration for supercapacitors based on COPs makes use of an n-doped polymer as negative electrode and a p-doped polymer as the positive electrode. This configuration is the most promising in terms of energy and power [349]. Some reviews on application of COPs in supercapacitors have been recently published; among them are [20, 349–353].

A few recent papers on the application of hybrid organic-inorganic materials as supercapacitors can be found in the lliterature. Hu et al. reported the application of composite thin films composed of polyaniline and platinum microparticles [354]. Polypyrrole/carbon nanotubes [355], where multiwalled carbon nanotubes are coated with polypyrrole, and the chemical polymerization of pyrrole over the surface of porous graphite [343, 356] have also been reported. The modification of the electrochemical properties of polypyrrole with Pt or carbon black leading to enhanced properties relative to the neat polymer has been reported. These modified material were proposed as supercapacitors but also in other applications like pollution remediation and sensing [228]. Finally, another recent hybrid supercapacitor material is the one reported by Gomez-Romero et al., in which the reversible redox chemistry of polyoxometalates in acidic media is exploited in hybrid materials where heteropolyanions are incorporated into a conducting polymer matrix such as Polyaniline. In comparison with the use of similar materials for Li batteries, these materials cycle reversibility for thousands of cycles with only minor capacity losses during cycling [342].

7.7.4 Sensors

COPs have been widely used in different kind of sensors [357]. Some of the most attractive applications are as pH sensors. For example, conductive molecular composites prepared by chemical oxidation of polyanilines within the expandable wide channels of the 3D-coordination polymers $[(Me_3E)_3Fe(CN)_6]$ (where $E = Sn$ or Pb) have been reported. The dramatic shift in optical response of the molecular composite could be used as an optical pH sensor over a wide pH range [358].

Conducting organic polymers are also used as active templates in biological sensors. In this context, COPs have been used in order to immobilize biological species such as enzymes, antibodies, DNA, receptors, cells, etc. The basis for using COPs as biosensors lies in their electrochemical properties. Reviews on the field have been recently published by Sadik [359] and McQuade et al. [360]. Among all the examples of organic sensors that we can find in the literature, only a small amount can be properly considered as organic-inorganic materials. A good example is a sensor applied for the detection of ascorbic acid formed by polypyrrole polymerized within montmorillonite clay [361]. Sensors to analyze the aroma of virgin olive oil based on conducting organic polymers, polypyrrole, and tetrasulfonated nickel phthalocyanine (NiPcTs) have also been reported [362, 363]. Other examples include the application of sensors for the discrimination of odors for trim plastic materials used in automobiles. The sensors were based on conducting organic polymers and potassium ferrocyanide, $K_4Fe(CN)_6$ [364]. Hybrid materials have also been applied as biosensors [365]. Thus, an amperometric hydrogen peroxide biosensor was designed based on horseradish and tobacco peroxidase entrapped into a conducting redox-polymer immobilized on either glassy-carbon or platinum electrodes. In this case the hybrid material was made of polypyrrole functionalized with an Os-complex [366]. Casella et al. obtained a hybrid organic-inorganic material by dispersion of copper into polyaniline film. The final film was applied as amperomeric sensor and was capable of oxidizing sugars and amino acids [367]. A similar amperometric glucose biosensor has been recently developed by the immobilization of glucose oxidase on a COP matrix entrapping a Fe-heteropolyanion [368]. In this later case there is some evidence of the formation of a multilayer self-assembling structure. Yamamoto et al. reported the fabrication of a hybrid organic-inorganic material based on polypyrrole/sulfated poly(b-hydroxyethers) in which fine Pd particles were dispersed [365]. The material was applied as biosensor upon glucose oxidase immobilization.

A novel hybrid material made of 3-alkylpolypyrrole–tin(IV) oxide composites has been used as active sensing material for a number of classes of volatile organic compounds (VOCs) like butane and propane at concentrations well below their lower limit of explosion (LLE) [369]. Novel applications of hybrid materials as artificial tongues can be exemplified by the work of Riul et al. [370, 371], with multi-array sensors based on Langmuir-Blodgett films of the 16-mer polyaniline oligomer, polypyrrole, and a ruthenium complex and self-assembled films of an azobenzene-containing polymer [370]. These materials, in the form of ultrathin films, were used to fabricate a high-sensitivity artificial tongue composed of four sensors that were able to distinguish easily the four basic tastes (salty, sour, sweet, and bitter), in addition to discriminating between different brands of coconut water, or detecting inorganic contaminants in ultrapure water.

7.7.5 Catalysis

There are many inorganic compounds with catalytic activities which can be used for the development of hybrid electrodes in combination with conducting organic polymers. This combination of a catalytic and a conducting component is very well suited for the development of hybrid electrocatalysts [56, 372]. Electrocatalytic reactions studied with hybrid electrodes range from the reduction of O_2 [92, 373] and protons [225, 374] or the oxidation of H_2 [222]; oxidation [101, 375] or hydrogenation [376] of organic substrates or inorganic oxoanions [377, 378]; electrocatalytic reduction of CO_2 [379] or environmental pollution remediation by electroreduction of Cr(VI) [234].

A different kind of inorganic-organic hybrid has been designed and prepared by immobilization of a Dawson-type polyoxometalate (POM) of formula $[(CH_3)_4N]_6 P_2Mo_{18}O_{62} \cdot 9\,H_2O$ on the surface of a wax-impregnated graphite electrode (WIGE) by the sol-gel technique. The electrocatalytic behavior of the P_2Mo_{18}-WIGE toward the reduction of chlorate, hydrogen peroxide and nitrite in unimolar H_2SO_4 aqueous solution was reported [380].

Metal particles in between the micro and nanoparticle size have been dispersed into COP matrices and are still the subject of catalytic studies by several research groups. Metals include Pt, Ru, Ta-Pt, Pt-Pb, Pt-Sn, Cu, Ni.P, etc, CuO_2 or CuCl [381]. These hybrid nanocomposite materials are applied to many processes in the chemical industry and their most important feature lies in their dispersity and surface processes. A high degree of dispersion and high surface area are required for catalysis. COPs can provide the high surface area and high porosity required for the metal support. Due to their electrical conduction properties, COPs are also ideal supports because they can function as electron-transport materials for the metallic nanoparticles dispersed in them. A representative example is the dispersion of Pt particles into Ppy, PAni or 2,5-dimethoxyaniline. In the case of the 2,5-dimethoxyaniline, the final hybrid material was applied for the electroreduction of methanol in polymer membrane fuel cells [381, 382].

Oxidation of alcohols, polyhydroxyl compounds, formic acid, carbon monoxide, hydrogen, etc are some of the reactions catalyzed by these hybrid materials. They have also been applied in the detection of amino acids and polyoxoalcohols [381]. Bouzek et al. reported an example of a hybrid catalyst formed by metal nanoparticles dispersed in conducting polymer films. These materials were used as electrocatalysts for the anodic oxidation of hydrogen and provide an excellent example of different approaches that can be followed for the synthesis of this type of hybrid. The syntheses of the materials were carried out following three different procedures: (i) cathodic deposition of platinum from a hexachloro-platinate complex $[PtCl_6]^{2-}$ solution onto the pre-synthesised polymer film, (ii) incorporation of colloidal platinum particles into the polymer film during electropolymerisation and (iii) incorporation of a tetrachloro-platinate complex $[PtCl_4]^{2-}$ during the electropolymerisation as counterion and its subsequent cathodic reduction [383].

7.7.6 Membranes

The development of membranes for specific applications, such as selective membranes, separation membranes, etc., could be considered a new emerging area for the development of conducting organic polymers and related hybrid materials. Its beginnings could be traced back 15 to 20 years with the application of COPs for ion permeation [384]. Membranes based on COPs can work by size exclusion or by chemical affinity; based on these operation modes different types of membranes have been reported: (i) ion exchange, (ii) ion permeation, (iii) gas separation and (iv) drug or chemical permeation [8]. Fig. 7.14 shows a cross-section of a conducting organic polymer membrane made of polypyrrole nanotubules [385]. The authors report the synthesis of the polymer inside of a track-etch membrane by polymerization of monomers; the resulting hollow tubules have enhanced electrical properties compared to those of the bulk materials [385].

Concerning related hybrid materials, Michalska et al. reported a true organic-inorganic material based on COPs and applied as a permselectivity membrane [386]. The material is applied for cation selectivity and is made by the incorporation of $Fe(CN)_6^{3-/4-}$ anion into a polypyrrole matrix. They showed that the permselectivity of the membranes is dependent on the electrolyte solution, and both cation and anion exchange occurs. They carried out experiments for the modification of the polymer surface by applying a thin Nafion film and demonstrated improved performance as a cation-exchange membrane [386].

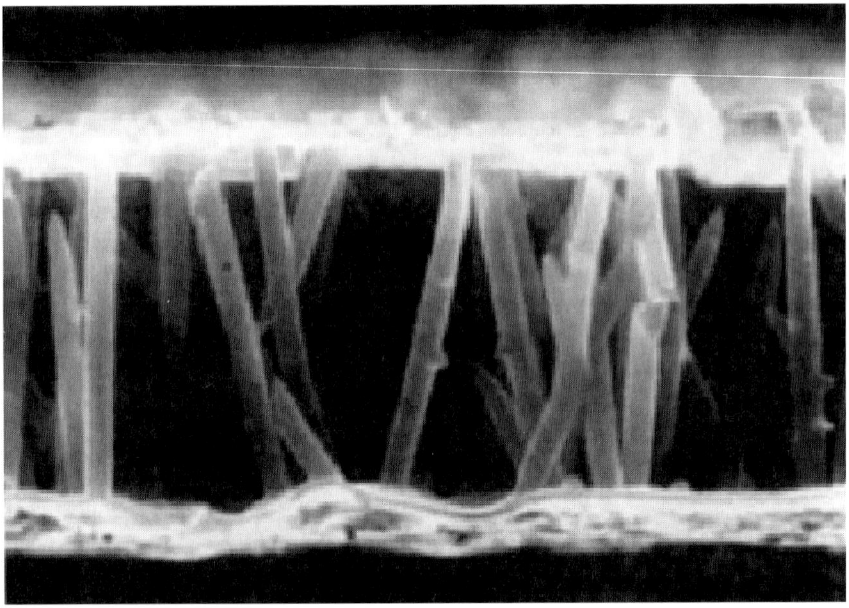

Fig. 7.14 Image of a membrane of conducting organic polymer synthesized inside the pores of a track-etch membrane. With permission of N. Somerdijk [385]

7.7.7
Biomaterials

Hybrid biomaterials will be the subject of the last chapter (Chapter 11) and therefore will not be discussed here in any detail. However, we have decided to include here a brief account of a handful of applications of novel biomaterials related to COPs just to give the reader a feeling for a field that is breaking barriers between disciplines and will probably yield amazing results in the near future.

The remarkable mechanical characteristics of naturally occurring, biological hybrid organic-inorganic materials, like the high specific strength and modulus and high toughness found in bones or ivory, make these natural hybrids fascinating topics of study begging to be mimicked or reproduced artificially by candidate materials. There are very different ways to approach the synthesis of biomaterials, and interestingly many different groups with wildly diverse backgrounds have discovered the benefits of self-assembly and the synergies of bio-hybrid materials.

Bioencapsulation of biological materials such as enzymes, antibodies, living microbial, plant and animal cells, DNA, etc within ceramic matrices is possible by the application of sol-gel techniques, as will be discussed in detailed in Chapter 11. In most cases the final applications is the synthesis of nanocomposites for sensors, catalysis or diagnostic [387]. There is also the appealing possibility of creating cheap and easy-to-fabricate devices with self-assembling and self-healing/repairing capabilities based on biomaterials. On the other hand biomolecular electronics (BME) and related nanotechnology applications constitute also a fast growing area where biomaterials will make surprising contributions. Thus, beyond conventional candidates like carbon nanotubes or molecular junctions, biomolecules are beginning to perform as state-of-the-art nanomaterials. As examples we can mention the synthesis of self-organized films of DNA which interconnect planar metallic nanopatterns. Another example is the application of ordered metalloproteins immobilized in a nanocircuit forming a hybrid covalently bound biologic-inorganic system, both cases to be applied in electronic devices [388].

7.8
Conclusions and Prospects

Conducting organic polymers constitute an excellent basis for the development of a wide variety of functional and multifunctional hybrids in combination with all sorts of inorganic species, from zero-dimensional, molecular or cluster inorganics to 3-D extended mineral phases. The approaches and methods for the synthesis of these hybrids vary widely depending on the particular nature of the material sought, on its components and applications, but in general we have seen how the main interest in these hybrids stems from a combination of the unique properties found in the polymers and those provided by the inorganic species. The hybrid approach always yields materials that combine both sets of properties, but it also allows for the generation of synergic combinations, materials and properties and even the isolation of altogether new materials. In certain cases, the hybrid approach allows harnessing of the chemical activity and properties of molecular species which could not be used or exploited outside the polymerframework. The use of polyoxometalate clusters or even smaller molecular species for energy storage applications is a remarkable example.

In this chapter we have concentrated our discussion on hybrid materials with photo- and electroionic activities, specific properties leading to a wide range of applications from energy storage and conversion in rechargeable batteries and supercapacitors, fuel cells, photoelectrochemical and hybrid solar cells and LEDs, to sensors and electrochromics, selective membranes or catalytic materials. In all these fields hybrids have shown great promise. There are however many other fields opened to the use of hybrids based on COPs, such as novel functional biomaterials, equally relevant and with plenty of exciting surprises predictably waiting ahead.

Another trend that can be extracted from the analysis of this chapter is the increasing importance of control at the nanometer dimension that has accompanied the historical evolution of hybrid materials in general and those based on COPs in particular. As discussed in this text, this trend will continue and will involve in the future further evolution from statistic control of molecules and materials towards local and stochastic control and design of nanostructured hybrids.

So far, conducting polymers have found commercial success in a relatively reduced number of specific niche applications. The work presented in this chapter shows how hybrid materials based on COPs expand upon their properties and in many cases even help to overcome their limitations. Therefore, as conducting organic polymers find new applications and technological niches for their commercialization it can be predicted that the parallel development of functional hybrids will continue to grow and flourish, providing the basis for further development and a wider range of properties and multifunctionalities in a growing number of applications.

Acknowledgments

M.L.-C. would like to express her gratitude to Dr. Limin Song, Dr. David Calabro and Dr. Jose Santiesteban for their support and attention during her years of work at ExxonMobil Research & Engineering (Annandale, N.J., USA). P.G.-R. is grateful for the partial financial support from the Spanish MCyT (grants MAT2001–1709-C04–01 and MAT2002–04529-C03) and the total dedication of his Ph.D. students, who made possible our research on hybrid materials and other exciting topics.

References

1 P. Gomez-Romero, *Adv. Mater.* **2001**, *13*, 163.
2 H. Shirakawa, S. Ikeda, *Polymer Journal* **1971**, *2*, 231.
3 C. M. Mikulski, P. J. Russo, M. S. Saran, A. G. MacDiarmid, A. F. Garito, A. J. Heeger, *J. Am. Chem. Soc.* **1975**, *97(22)*, 6358–6363.
4 H. Shirakawa, E. J. Louis, A. G. MacDiarmid, C. K. Chiang, A. J. Heeger, *J. Chem. Soc. Chem. Comm.* **1977**, 578–580.
5 H. Naarmann, in J. W. Sons (Ed.): *Polymers to the year 2000 and beyond. Chapter 4*, John Wiley & Sons 1993.
6 Y. Cao, P. Smith, J. Heeger, *Synth. Met.* **1989**, *32*, 263.
7 A. Andretta, Y. Cao, J. C. Chaing, A. J. Heeger, P. Smith, *Synth. Met.* **1988**, *26*, 383.
8 P. Chandrasekhar, in *Conducting Polymer Fundamentals and Applications: A Practical Approach*, Kluwer Academic Press, Dordrecht, 1999.
9 B. Wessling, *Chem. Innov.* **2001**, *Jan*, 35–40.
10 L. Groenendaal, F. Jonas, D. Freitag, H. Pielartzik, R. Reynolds, *Adv. Mater.* **2000**, *12*, 481–494.
11 Y. Cao, P. Smith,A. J. Heeger, *Synth. Met.* **1992**, *48*, 91.
12 A. Pron, P. Rannou, *Progress in Polymer Science* **2002**, *27*, 135–190.
13 T. Lindfors, A. Ivaska, *Analytica Chimica Acta* **2001**, *437*, 171–182.
14 K. Murakoshi, R. Kogure, Y. Wada, S. Yanagida, *Chem. Lett.* **1997**, *(5)*, 471–472.
15 L. S. Roman, O. Inganäs, T. Granlund, T. Nyberg, M. Sevensson, M. R. Andersson, J. C. Hummelen, *Advanced Materials* **2000**, *12*, 189–195.
16 A. C. Arango, L. R. Johnson, V. N. Bliznyuk, Z. Schlesinger, S. A. Carter, H. H. Hörhold, *Advanced Materials* **2000**, *12*, 1689–1692.
17 A. Dhanabalan, J. K. J. van Duren, P. A. van Hal, J. L. J. van Dongen, A. J. Janssen, *Advanced Functional Materials* **2001**, *11*, 255.
18 M. T. Bernius, M. Inbasekaran, J. O'Brien, W. Wu, *Advanced Materials* **2000**, *12*, 1737.
19 M. Kaneko, K. Kaneto, *Reactive and Functional Polymers* **1998**, *37*, 155–161.
20 K. Naoi, *J. Electrochem. Soc.* **1997**, *144*, L173.
21 P. Novak, K. Mueller, K. S. V. Santhanam, O. Haas, *Chem. Rev.* **1997**, *97*, 207–281.
22 H. R. Allcock, *Adv. Mater.* **1994**, *6*, 106–115.
23 B. M. Novak, *Adv. Mater.* **1993**, *5*, 422–433.
24 D. A. Loy, K. J. Shea, *Chem. Rev.* **1995**, *95*, 1431–1442.
25 P. Judeinstein, C. Sanchez, *J. Mater. Chem.* **1996**, *6(4)*, 511–525.
26 K. G. Sharp, *Adv. Mater.* **1998**, *10*, 1243–1248.
27 C. Sanchez, F. Ribot, B. Lebeau, *J. Mater. Chem.* **1999**, *9*, 35–44.
28 C. Sanchez, B. Lebeau, *Mater. Res. Bull.* **2001**, *26*, 377–387.
29 P. Judeinstein, H. Schmidt, *J. Sol-Gel Sci. Technol.* **1994**, *3(3)*, 189–197.
30 J. H. Harreld, B. Dunn, J. I. Zink, *J. Mater. Chem.* **1997**, *7*, 1511–1517.
31 C. Roscher, R. Buestrich, P. Dannberg, O. Rosch, M. Popall, *Mater. Res. Soc. Symp. Proc.* **1998**, *519*, 239–244.
32 H. K. Kim, S.-J. Kang, S.-K. Choi, Y.-H. Min, C.-S. Yoon, *Chem. Mater.* **1999**, *11*, 779–788.
33 J. Livage, C. Sanchez, *Mclc S&T Sect. B: Nonlinear Opt.* **1999**, *21*, 125–141.
34 S. P. Armes, S. Maeda, M. Gill, *Polym. Mater. Sci. Eng.* **1993**, *70*, 352–353.
35 M. Mager, F. Jonas, A. Eiling, U. Guntermann, *Patent Ger. Offen. DE 19650147 A1 10 Jun, 8 pp.* **1998**.
36 S. Dire, F. Babonneau, C. Sanchez, J. Livage, *J. Mater. Chem.* **1992**, *2*, 239–244.
37 N. I. Koslova, B. Viana, C. Sanchez, *J. Mater. Chem.* **1993**, *3*, 111–112.
38 S. M. Jones, L. Kotorman, S. E. Friberg, *J. Mater. Sci.* **1996**, *31*, 1475–1479.
39 E. Cordoncillo, B. Viana, P. Escribano, C. Sanchez, *J. Mater. Chem.* **1998**, *8(3)*, 507–509.
40 V. de Zea Bermudez, L. D. Carlos, M. C. Duarte, M. M. Silva, C. J. R. Silva, M. J. Smith, M. Assuncao, L. Alcacer, *J. Alloys Compd.* **1998**, *275–277*, 21–26.

41 T. Dantas de Morais, F. Chaput, K. Lahlil, J.-P. Boilot, *Adv. Mater.* **1999**, *11*, 107–112.

42 T. Fujinami, K. Sugie, K. Mori, M. A. Mehta, *Chem. Lett.* **1998**, *(7)*, 619–620.

43 L. Depre, J. Kappel, M. Popall, *Electrochim. Acta* **1998**, *43*, 1301–1306.

44 I. Honma, Y. Takeda, J. M. Bae, *Solid State Ionics* **1999**, *120*, 255–264.

45 P. Hernan, C. Del Pino, E. Ruiz-Hitzky, *Chem. Mater.* **1992**, *4*, 49–55.

46 P. Aranda, A. Jimenez-Morales, J. C. Galvan, B. Casal, E. Ruiz-Hitzky, *J. Mater. Chem.* **1995**, *5*, 817–825.

47 M. Barboiu, C. Luca, C. Guizard, N. Hovnanian, L. Cot, G. Popescu, *J. Membr. Sci.* **1997**, *129*, 197–207.

48 A. Jimenez-Morales, J. C. Galvan, P. Aranda, E. Ruiz-Hitzky, *Mater. Res. Soc. Symp. Proc.* **1998**, *519*, 211–216.

49 P. Battioni, E. Cardin, M. Louloudi, B. Schollhorn, G. A. Spyroulias, D. Mansuy, T. G. Traylor, *Chem. Commun. (Cambridge)* **1996**, *(17)*, 2037–2038.

50 M. Buechler-Skoda, R. Gill, D. Vu, C. Nguyen, G. Larsen, *Appl. Catal.* **1999**, *A*, 185(2), 301–310.

51 K. J. Ciuffi, H. C. Sacco, J. B. Valim, C. M. C. P. Manso, O. A. Serra, O. R. Nascimento, E. A. Vidoto, Y. Iamamoto, *J. Non-Cryst. Solids* **1999**, *247*, 146–152.

52 M. T. Reetz, *Adv. Mater.* **1997**, *9*, 943–954.

53 B. Wang, B. Li, Q. Deng, S. Dong, *Anal. Chem.* **1998**, *70*, 3170–3174.

54 J. D. Brennan, J. S. Hartman, E. I. Ilnicki, M. Rakic, *Chem. Mater.* **1999**, *11*, 1853–1864.

55 J. Heinze, *Topics in Current Chemistry*, Springer-Verlag, Berlin, 1990, 1–47, Vol. 152.

56 A. Deronzier, J.-C. Moutet, *Coord. Chem. Rev.* **1996**, *147*, 339–371.

57 P. Gomez-Romero, M. Lira-Cantu, *Hybrid Nanocomposite Materials.* Chapter for the *Kirk-Othmer Encyclopedia of Chemical Technology*, John Wiley & Sons, New York, 2002, *http://www.mrw.interscience.wiley.com/kirk/articles/hybdgome.a01/abstract.ht ml.*

58 J. Przyluski, M. Zagorska, A. Pron, Z. Kucharski, J. Suwalski, *J. Phys. Chem. Solids* **1987**, *48*, 635–640.

59 A. Pron, J. Suwalski, S. Lefrant, *Synth. Met.* **1987**, *18*, 25–30.

60 S. Flandrois, A. Boukhari, A. Pron, M. Zagorska, *Solid State Commun.* **1988**, *67*, 471–475.

61 I. Kulszewicz-Bajer, A. Pawlicka, J. Plenkiewicz, A. Pron, S. Lefrant, *Synth. Met.* **1989**, *30*, 335–339.

62 J. F. Rabek, J. Lucki, M. Zuber, B. J. Qu, W. F. Shi, *Polym. Prepr.* **1990**, *31*, 486–487.

63 M. S. A. Abdou, S. Holdcroft, *Chem. Mater.* **1994**, *6*, 962–968.

64 J. Han, S. Lee, W. K. Paik, *Bull. Korean Chem. Soc.* **1992**, *13*, 419–425.

65 S. Lee, J. Han, W. K. Paik, *Synth. Met.* **1993**, *55*, 1129–1134.

66 A. P. Chattaraj, I. N. Basumallick, *J. Power Sources* **1993**, *45*, 237–242.

67 G. Torres-Gomez, P. Gomez-Romero, *Synth. Met.* **1998**, *98*, 95–102.

68 G. Torres-Gomez, M. Lira-Cantu, P. Gomez-Romero, *J. New Mater. Electrochem. Syst.* **1999**, *2*, 145–150.

69 G. Torres-Gomez, K. West, S. Skaarup, P. Gomez-Romero, *J. Electrochem. Soc.* **2000**, *147*, 2513–2516.

70 P. Gomez-Romero, G. Torres-Gomez, *Adv. Mater.* **2000**, *12*, 1454–1456.

71 G. Torres-Gomez, E. M. Tejada-Rosales, P. Gomez-Romero, *Chem. Mater.* **2001**, *13*, 3693–3697.

72 P. J. Kulesza, K. Miecznikowski, M. A. Malik, M. Galkowski, M. Chojak, K. Caban, A. Wieckowski, *Electrochimica Acta* **2001**, *46*, 4065–4073.

73 R. Cameron, S. I. Yaniger, *US Patent US4855361 Aug 8*, **1989**.

74 R. Cervini, R. J. Fleming, K. S. Murray, *J. Mater. Chem.* **1992**, *2*, 1115–1121.

75 R. Cervini, R. J. Fleming, B. J. Kennedy, K. S. Murray, *J. Mater. Chem.* **1994**, *4*, 87–97.

76 M. Takakubo, *Synth. Met.* **1987**, *18*, 53–58.

77 N. S. Allen, K. S. Murray, R. J. Fleming, B. R. Saunders, *Synth. Met.* **1997**, *87*, 237–247.

78 B. R. Saunders, K. S. Murray, R. J. Fleming, D. G. McCulloch, L. J. Brown, J. D. Cashion, *Chem. Mater.* **1994**, *6*, 697–706.

79 B. R. Saunders, R. J. Fleming, K. S. Murray, *Chem. Mater.* **1995**, *7*, 1082–1094.

80 D. Belanger, G. Laperriere, L. Gravel, *J. Electrochem. Soc.* **1990**, *137*, 365–366.

81 D. Belanger, G. Laperriere, F. Girard, D. Guay, G. Tourillon, *Chem. Mater.* **1993**, *5*, 861–868.
82 F. Fusalba, D. Belanger, *J. Mater. Res.* **1999**, *14(5)*, 1805–1813.
83 B. Garcia, F. Roy, D. Belanger, *J. Electrochem. Soc.* **1999**, *146*, 226–231.
84 F. Girard, S. Ye, G. Laperriere, D. Belanger, *J. Electroanal. Chem.* **1992**, *334*, 35–55.
85 F. Girard, S. Ye, D. Belanger, *J. Electrochem. Soc.* **1995**, *142*, 2296–2301.
86 S. Lavallee, G. Laperriere, D. Belanger, *J. Electroanal. Chem.* **1997**, *431*, 219–226.
87 S. Ye, F. Girard, D. Belanger, *J. Phys. Chem.* **1993**, *97*, 12373–12378.
88 C. Masalles, S. Borrós, C. Viñas, F. Teixidor, *Adv. Mater.* **2000**, *12*, 1199–1202.
89 R. A. Bull, F.-R. Fan, A. J. Bard, *J. Electrochem. Soc.* **1983**, *130*, 1636.
90 A. Elzing, A. Van der Putten, W. Visscher, E. Barendrecht, *J. Electroanal. Chem. Interfacial Electrochem.* **1987**, *233*, 113–123.
91 S. Kuwabata, K. Okamoto, O. Ikeda, H. Yoneyama, *Synth. Met.* **1987**, *18*, 101–104.
92 R. Jiang, S. Dong, *J. Electroanal. Chem. Interfacial Electrochem.* **1988**, *246*, 101–117.
93 M. Kawashima, H. Tanaka, M. Sakaguchi, *J. Electroanal. Chem. Interfacial Electrochem.* **1990**, *286*, 123–131.
94 C. S. Choi, H. Tachikawa, *J. Am. Chem. Soc.* **1990**, *112*, 1757–1768.
95 B. R. Saunders, K. S. Murray, R. J. Fleming, *Synth. Met.* **1992**, *47*, 167–178.
96 B. R. Saunders, K. S. Murray, R. J. Fleming, Y. Korbatieh, *Chem. Mater.* **1993**, *5*, 809–819.
97 B. R. Saunders, K. S. Murray, R. J. Fleming, D. G. McColloch, *Synth. Met.* **1995**, *69*, 363–364.
98 V. C. Nguyen, K. Potje-Kamloth, *Thin Solid Films* **1999**, *338*, 142–148.
99 W. Paik, I.-H. Yeo, H. Suh, Y. Kim, E. Song, *Electrochimica Acta* **2000**, *45*, 3833–3840.
100 R. P. Mikalo, G. Appel, P. Hoffmann, D. Schmeisser, *Synthetic Metals* **2001**, *122*, 249–261.
101 I. De Gregori, M. Carrier, A. Deronzier, J. C. Moutet, F. Bedioui, J. Devynck, *J. Chem. Soc., Faraday Trans.* **1992**, *88*, 1567–1572.
102 F. Bedioui, Y. Bouchier, C. Sorel, J. Devynck, L. Coche-Guerrente, A. Deronzier, J. C. Moutet, *Electrochim. Acta* **1993**, *38*, 2485–2491.
103 C. J. Brabec, N. S. Sariciftci, J. C. Hummelen, *Adv. Func. Mater.* **2001**, *11*, 15–26.
104 J. Y. Lim, W.-K. Paik, I.-H. Yeo, *Synth. Met.* **1995**, *69*, 451–454.
105 P. Gomez-Romero, M. Lira-Cantu, *Adv. Mater.* **1997**, *9*, 144–147.
106 G. Bidan, E. Genies, M. Laprowski, *EP 323351*, **1989**.
107 M. Lira-Cantu, P. Gomez-Romero, *Recent Res. Dev. Phys. Chem.* **1997**, *1*, 379–401.
108 B. Keita, A. Mahmoud, L. Nadjo, *J. Electroanal. Chem.* **1995**, *386*, 245–251.
109 P. Wang, Y. F. Li, *J. Electroanal. Chem.* **1996**, *408*, 77–81.
110 A. Mahmoud, B. Keita, L. Nadjo, *J. Electroanal. Chem.* **1998**, *446*, 211–225.
111 M. Lira-Cantu, P. Gomez-Romero, *Chem. Mater.* **1998**, *10*, 698–704.
112 N. Furukawa, K. Nishio, in *Applications of Electroactive Polymers*, ed B. Scrosati, Chapman & Hall, London, 1993, 150–181.
113 M. Lira-Cantu, P. Gomez-Romero, *Ionics* **1997**, *3*, 194–200.
114 M. Zagorska, H. Wycislik, J. Przyluski, *Synth. Met.* **1987**, *20*, 259–268.
115 L. L. Miller, B. Zinger, Q. X. Zhou, *J. Am. Chem. Soc.* **1987**, *109*, 2267–2272.
116 G. Lian, S. Dong, *J. Electroanal. Chem. Interfacial Electrochem.* **1989**, *260*, 127–136.
117 W. Breen, J. F. Cassidy, M. E. G. Lyons, *J. Electroanal. Chem. Interfacial Electrochem.* **1991**, *297*, 445–460.
118 M. Lira-Cantu, G. Torres-Gomez, P. Gomez-Romero, *Mater. Res. Soc. Symp. Proc.* **1999**, *548*, 367–376.
119 T. Iyoda, A. Ohtani, T. Shimidzu, K. Honda, *Synth. Met.* **1987**, *18*, 725–730.
120 C. Barbero, M. C. Miras, B. Schnyder, O. Haas, R. Kotz, *J. Mater. Chem.* **1994**, *4*, 1775–1783.
121 H. K. Youssoufi, F. Garnier, A. Yassar, S. Baiteche, P. Srivastava, *Adv. Mater.* **1994**, *6*, 755–758.
122 J. Tietje-Girault, J. M. Anderson, I. MacInnes, M. Schroder, G. Tennant, H. H.

Girault, *J. Chem. Soc. Chem. Commun.* **1987**, *14*, 1095–1097.
123 O. Ikeda, H. Yoneyama, *J. Electroanal. Chem.* **1989**, *265*, 323–327.
124 S. Dong, G. Lian, *J. Electroanal. Chem. Interfacial Electrochem.* **1990**, *291*, 23–39.
125 N. Leventis, Y. C. Chung, *J. Electrochem. Soc.* **1990**, *137*, 3321–3322.
126 B. P. Jelle, G. Hagen, S. Noedland, *Electrochim. Acta* **1993**, *38*, 1497–1500.
127 A. A. Karyakin, M. F. Chaplin, *J. Electroanal. Chem.* **1994**, *370*, 301–303.
128 P. Somani, S. Radhakrishnan, *Materials Chemistry and Physics* **2001**, *70*, 150–155.
129 H. Sung, T. Lee,W.-K. Paik, *Synth. Met.* **1995**, *69*, 485–486.
130 S. D. Han, G. Campet, S. Y. Huang, M. C. R. Shastry, J. Portier, J. C. Lassegues, H. S. Dweik, *Act. Pass. Elect.* **1995**, *18*, 31–37.
131 M. Tolgyesi, A. Szucs, C. Visy, M. Novak, *Electrochemical Acta* **1995**, *40*, 1127–1133.
132 P. J. Kulesza, K. Miecznikowski, M. Chojak, M. A. Malik, S. Zamponi, R. Marassi, *Electrochimica Acta* **2001**, *46*, 4371–4378.
133 C. G. Wu, H. O. Marcy, D. C. DeGroot, J. L. Schindler, C. R. Kannewurf, W. Y. Leung, M. Benz, E. LeGoff, M. G. Kanatzidis, *Synth. Met.* **1991**, *41*, 797–803.
134 V. Mehrotra, E. P. Giannelis, *Solid St. Ionics* **1992**, *51*, 115–122.
135 M. G. Kanatzidis, R. Bissessur, D. C. DeGroot, J. L. Schindler, C. R. Kannewurf, *Chem. Mater.* **1993**, *5*, 595–596.
136 E. Ruiz-Hitzky, P. Aranda, *An. Quim. Int. Ed.* **1997**, *93*, 197–212.
137 C. G. Wu, T. Bein, *Science* **1994**, *264*, 1757–1759.
138 C.-G. Wu, T. Bein, *Chem. Mater.* **1994**, *6*, 1109–1112.
139 T.-Q. Nguyen, J. Wu, S. H. Tolbert, B. J. Schwartz, *Advanced Materials* **2001**, *13*, 609.
140 P. Cloos, A. Moreale, C. Braers, C. Badot, *Clay Min.* **1979**, *14*, 307.
141 D. Vande Poel, in *Adsorption de composés aromatiques sur le montmorillonite saturée en Cu (II)*, Catholic Unoiversity of Louvian, Leuven, 1975.
142 A. F. Diaz, K. K. Kanazawa, G. P. Gardini, *J. Chem. Soc., Chem. Commun.* **1979**, 635–636.
143 R. Bissessur, J. L. Schindler, C. R. Kannewurf, M. G. Kanatzidis, *Mol. Cryst. Liq. Cryst.* **1993**, *245*, 249–254.
144 R. Bissessur, M. G. Kanatzidis, J. L. Schindler, C. R. Kannewurf, *J. Chem. Soc. Chem. Commun.* **1993**, 1582–1585.
145 L. F. Nazar, X. T. Yin, D. Zinkweg, Z. Zhang, S. Liblong, *Mater. Res. Soc. Symp. Proc.* **1991**, *210*, 417–422.
146 L. F. Nazar, Z. Zhang, D. Zinkweg, *J. Am. Chem. Soc.* **1992**, *114*, 6239–6240.
147 P. G. Hill, P. J. S. Foot, R. Davis, *Mater. Sci. Forum* **1995**, *191*, 43–46.
148 L. F. Nazar, T. Kerr, B. Koene, in *Fundamental Materials Science Series*, eds. T. J. Pinnavia and M. F. Thorpe, Kluwer, Dordrecht, 1995, p. 405.
149 T. A. Kerr, H. Wu, L. F. Nazar, *Chem. Mater.* **1996**, *8*, 2005.
150 P. G. Hill, P. J. S. Foot, R. Davis, *Synth. Met.* **1996**, *76*, 289–292.
151 T. A. Kerr, F. Leroux, L. F. Nazar, *Chem. Mater.* **1998**, *10*, 2588–2591.
152 M. G. Kanatzidis, C. G. Wu, H. O. Marcy, C. R. Kannewurf, *J. Am. Chem. Soc.* **1989**, *111*, 4139–4141.
153 C. G. Wu, H. O. Marcy, D. C. DeGroot, C. R. Kannewurf, M. G. Kanatzidis, *Mater. Res. Soc. Symp. Proc.* **1990**, *173*, 317–322.
154 C. G. Wu, M. G. Kanatzidis, H. O. Marcy, D. C. DeGroot, C. R. Kannewurf, *NATO ASI Ser.* **1990**, *Ser. B*, *248*, 427–433.
155 M. G. Kanatzidis, C. G. Wu, Y. J. Liu, D. C. DeGroot, J. L. Schindler, H. O. Marcy, C. R. Kannewurf, *Mater. Res. Soc. Symp. Proc.* **1991**, *233*, 183–194.
156 V. D. Pokhodenko, V. G. Koshechko, V. A. Krylov, *J. Power Sources* **1993**, *45*, 1–5.
157 Y. J. Liu, D. C. DeGroot, J. L. Schindler, C. R. Kannewurf, M. G. Kanatzidis, *J. Chem. Soc., Chem. Commun.* **1993**, *7*, 593–596.
158 T. Fujii, N. Katagiri, O. Kimura, T. Kabata, Y. Kurosawa, Y. Hayashi, H. Iechi, T. Ohsawa, *Synthet. Metal.* **1995**, *71*, 2225–2226.
159 H. Nakajima, G.-E. Matsubayashi, *J. Mater. Chem.* **1995**, *5*, 105–108.

160 F. Leroux, B. E. Koene, F. L. Nazar, *J. Electrochem. Soc.* **1996**, *143*, L181–183.
161 C. G. Wu, D. C. DeGroot, H. O. Marcy, J. L. Schindler, C. R. Kannewurf, Y. J. Liu, W. Hirpo, M. G. Kanatzidis, *Chem. Mater.* **1996**, *8*, 1992–2004.
162 F. Leroux, G. Goward, W. P. Power, L. F. Nazar, *J. Electrochem. Soc.* **1997**, *144*, 3886–3895.
163 M. Lira-Cantu, P. Gomez-Romero, *Journal of New Materials for Electrochemical Systems* **1999**, *2*, 141–144.
164 M. Lira-Cantu, P. Gomez-Romero, *J. Solid State Chem.* **1999**, *147*, 601–608.
165 M. Lira-Cantu, P. Gomez-Romero, *J. Electrochem. Soc.* **1999**, *146*, 2029–2033.
166 M. Lira-Cantu, K. Cuentas-Gallegos, G. Torres-Gomez, P. Gomez-Romero, *Bol. Soc. Esp. Ceram. Vidrio* **2000**, *39*, 386–390.
167 S. Kuwabata, T. Idzu, C. R. Martin, H. Yoneyama, *J. Electrochem. Soc.* **1998**, *145*, 2707–2710.
168 C.-G. Wu, Y.-C. Liu, S.-S. Hsu, *Synthetic Metals* **1999**, *102*, 1268–1269.
169 P. R. Somani, R. Marimuthu, A. B. Mandale, *Polymer* **2001**, *42*, 2991–3001.
170 C. G. Wu, M. G. Kanatzidis, H. O. Marcy, D. C. DeGroot, C. R. Kannewurf, *Polym. Mater. Sci. Eng.* **1989**, *61*, 969–973.
171 D. C. DeGroot, J. L. Schindler, C. R. Kannewurf, Y. J. Liu, C. G. Wu, M. G. Kanatzidis, *Mater. Res. Soc. Symp. Proc.* **1992**, *274*, 133–138.
172 B. C. Dave, B. S. Dunn, F. Leroux, L. F. Nazar, H. P. Wong, *Mater. Res. Soc. Symp. Proc.* **1996**, *435*, 611–616.
173 H. P. Wong, B. C. Dave, F. Leroux, J. Harreld, B. Dunn, L. F. Nazar, *J. Mater. Chem.* **1998**, *8*, 1019–1027.
174 J. H. Harreld, B. Dunn, W. Cheng, F. Leroux, L. F. Nazar, *Mater. Res. Soc. Symp. Proc.* **1998**, *519*, 191–200.
175 G. R. Goward, F. Leroux, L. F. Nazar, *Electrochim. Acta* **1998**, *43*, 1307–1313.
176 G. J. F. Demets, F. J. Anaissi, H. E. Toma, *Electrochim. Acta* **2000**, *46*, 547–554.
177 S. Kuwabata, S. Masui, H. Tomiyori, H. Yoneyama, *Electrochim. Acta.* **2000**, *46*, 91–97.
178 M. G. Kanatzidis, C. G. Wu, H. O. Marcy, D. C. DeGroot, C. R. Kannewurf, *Chem. Mater.* **1990**, *2*, 222–224.
179 C.-W. Kwon, A. Vadivel Murugan, G. Campet, J. Portier, B. B. Kale, K. Vijaymohanan, J.-H. Choy, *Electrochemistry Communications* **2002**, *4*, 384–387.
180 A. V. Murugan, C.-W. Kwon, G. Campet, B. B. Kale, T. Maddanimath, K. Vijayamohanan, *Journal of Power Sources* **2002**, *105*, 1–5.
181 J. Livage, N. Gharbi, M. C. Leroy, M. Michaud, *Mater. Res. Bull.* **1978**, 1117.
182 J. Livage, *Chem. Mater.* **1991**, *3*, 578–593.
183 A. Destefanis, S. Foglia, A. A. G. Tomlinson, *J. Mater. Chem.* **1995**, *5*, 475–483.
184 N.-G. Park, K. S. Ryu, Y. J. Park, M. G. Kang, D.-K. Kim, S.-G. Kang, K. M. Kim, S.-H. Chang, *Journal of Power Sources* **2002**, *103*, 273–279.
185 M. Lira-Cantú, Ph.D. Thesis, Universita Autonoma de Barcelona, 1997.
186 R. Bissessur, J. L. Schindler, C. R. Kannewurf, M. Kanatzidis, *Mol. Cryst. Liq. Cryst. Sci. Technol.* **1994**, *Sect. A*, *245*, 249–254.
187 L. Wang, J. Schindler, J. A. Thomas, C. R. Kannewurf, M. G. Kanatzidis, *Chem. Mater.* **1995**, *7*, 1753–1755.
188 L. Wang, P. Brazis, M. Rocci, C. R. Kannewurf, M. G. Kanatzidis, *Mater. Res. Soc. Symp. Proc.* **1998**, *519*, 257–264.
189 M. G. Kanatzidis, L. M. Tonge, T. J. Marks, H. O. Marcy, C. R. Kannewurf, *J. Am. Chem. Soc.* **1987**, *109*, 3797–3799.
190 M. G. Kanatzidis, M. Hubbard, L. M. Tonge, T. J. Marks, H. O. Marcy, C. R. Kannewurf, *Synth. Met.* **1989**, *28*, C89-C95.
191 M. G. Kanatzidis, H. O. Marcy, W. J. McCarthy, C. R. Kannewurf, T. J. Marks, *Solid State Ionics* **1989**, *32–33*, 594–608.
192 M. G. Kanatzidis, C. G. Wu, H. O. Marcy, D. C. DeGroot, C. R. Kannewurf, A. Kostikas, V. Papaefthymiou, *Adv. Mater.* **1990**, *2*, 364–366.
193 M. G. Kanatzidis, C. G. Wu, H. O. Marcy, D. C. DeGroot, J. L. Schindler, C. R. Kannewurf, M. Benz, E. LeGoff, *ACS Symp. Ser.* **1992**, *499*, 194–219.

194 K. Prassides, C. J. Bell, A. J. Dianoux, C. G. Wu, M. G. Kanatzidis, *Physica B* **1992**, *180–181*, 668–670.

195 M. G. Kanatzidis, C. G. Wu, D. C. DeGroot, J. L. Schindler, M. Benz, E. LeGoff, C. R. Kannewurf, *NATO ASI Ser.* **1993**, *Ser. B*, *305*, 63–72.

196 C. G. Wu, D. C. DeGroot, H. O. Marcy, J. L. Schindler, C. R. Kannewurf, Y. Bakas, V. Papaefthymiou, W. Hirpo, J. P. Yesinowski, Y. J. Liu, M. G. Kanatzidis, *J. Am. Chem. Soc.* **1995**, *117*, 9229–9242.

197 Y. J. Liu, M. G. Kanatzidis, *Inorg. Chem.* **1993**, *32*, 2989–2991.

198 P. Enzel, T. Bein, *J. Physical. Chem.* **1989**, *93*, 6270.

199 P. Enzel, T. Bein, *J. Chem. Soc., Chem. Commun.* **1989**, 1362.

200 T. Bein, P. Enzel, *Angew. Chem. Int. Ed. Engl.* **1989**, *28*, 1692–1693.

201 T. Bein, P. Enzel, F. Beuneu, L. Zuppiroli, *Electron Transf. Biol. Solid St.* **1990**, 434.

202 P. Enzel, T. Bein, *Chem. Mater.* **1992**, *4*, 819–824.

203 P. Enzel, T. Bein, *Chem. Mater.* **1992**, *4 (4)*, 819–824.

204 T. Bein, P. Enzel, *NATO ASI Ser. E* **1993**, *246*, 51–60.

205 L. Zuppiroli, F. Beuneu, J. Mory, P. Enzel, T. Bein, *Synth. Met.* **1993**, *57*, 5081–5087.

206 C.-G. Wu, J. Y. Chen, *Chem. Mater.* **1997**, *9*, 399.

207 T. Bein, *Stud. Surf. Sci. Catal.* **1996**, *102*, 295–322.

208 J. S. Beck, J. C. Vartuli, W. J. Roth, M. E. Lenowicz, C. T. Kresge, K. D. Schmidtt, C. T.-W. Chu, D. H. Olson, E. W. Sheppard, S. B. McCunelli, J. B. Higgins, J. B. Schenker, *J. Am. Chem. Soc.* **1992**, *114*, 10834–10843.

209 C. T. Kresge, M. E. Leonowicz, W. J. Roth, J. C. Vartuli, J.S. Beck, *Nature* **1992**, *359*, 710.

210 A. Ortlam, M. Wark, G. Schulz-Ekloff, J. Rathousky, A. Zukal, *Stud. Surf. Sci. Catal.* **1998**, *117*, 357.

211 C. G. Wu, T. Bein, *Stud. Surf. Sci. Catal.* **1994**, *84*, 2269–2276.

212 V. Parvulescu, L. Buhoci, G. Roman, B. Albu, G. Popescu, *Separation and Purification Technology* **2001**, *25*, 25–32.

213 T.-Q. Nguyen, J. J. Wu, V. Doan, B. J. Schwartz, S. H. Tolbert, *Science* **2000**, *288*, 652.

214 R. Gangopadhyay, A. De, *Chem. Mater.* **2000**, *12*, 608–622.

215 A. H. Gemeay, H. Nishiyama, S. Kuwabata, H. Yoneyama, *J. Electrochem. Soc.* **1995**, *142*, 4190.

216 I. Toshiyasu, M. Taskaaki, M. Jun, O. Katsuaki, T. Shigeyuki, K. Mamoru, US Patent US4750816 A1, June 14, **1988**.

217 Y. Susumu, K. Shiego, *JP Patent JP63250482 A2 April 6,* **1988**.

218 U.S. Patent 4636430A1 (Jan. 13, 1987). H. Moehwald (to BASF Akiengesellschaft).

219 P. Kathirgamanathan, P. N. Adams, A. M. Marsh, D. Shah, Japanese Patent JP01060639 A2, March 7, **1989**.

220 Europ. Patent EP0302590A2 (Feb. 8, 1989). H. H. Kuhn, W. C. Kimbrell (to Milliken Res. Corp. – US).

221 P. Kathirgamanathan, P. N. Adams, A. M. Marsh, D. Shah, US Patent US4750816, **1989**.

222 C. C. Chen, C. S. C. Bose, K. Rajeshwar, *J. Electroanal. Chem.* **1993**, *350*, 161–176.

223 M. Hasik, A. Drelinkiewicz, M. Choczynski, S. Quillard, A. Pron, *Synth. Met.* **1997**, *84*, 93–94.

224 H.-S. Li, M. Josowicz, D. R. Baer, M. H. Engelhard, J. Janata, *J. Electrochem. Soc.* **1995**, *142*, 798–805.

225 C. S. C. Bose, K. Rajeshwar, *J. Electroanal. Chem.* **1992**, *333*, 235–256.

226 H.-S. Li, M. Josowicz, M. Engelhard, D. R. Baer, J. Janata, *Proc. Electrochem. Soc.* **1994**, *94–14*, 123–131.

227 K. Rajeshwar, C. S. C. Bose, U.S. Patent US 5334292 A 2 Aug, **1994**.

228 K. Rajeshwar, C. Wei, W. Wampler, C. S. C. Bose, S. Basak, S. German, D. Evans, V. Krishna, *Polym. Prepr. (Am. Chem. Soc., Div. Polym. Chem.)* **1994**, *35*, 234–235.

229 S. M. Marinakos, D. A. Shultz, D. L. Feldheim, *Adv. Mater.* **1999**, *11*, 34–38.

230 D. L. Feldheim, C. D. Keating, *Chem. Soc. Rev.* **1998**, *27*, 1.

231 W. A. Wampler, C. Wei, K. Rajeshwar, *J. Electrochem. Soc.* **1994**, *141*, L13-L15.

232 W. A. Wampler, C. Wei, K. Rajeshwar, *Chem. Mater.* **1995**, *7*, 585–592.

233 W. A. Wampler, K. Rajeshwar, R. G. Pethe, R. C. Hyer, S. C. Sharma, *J. Mater. Res.* **1995**, *10*, 1811–1822.
234 W. A. Wampler, S. Basak, K. Rajeshwar, *Carbon* **1996**, *34*, 747–755.
235 G. A. Sotzing, J. N. Phend, R. H. Grubbs, N. S. Lewis, *Chem. Mater.* **2000**, *12*, 593–595.
236 N. S. Lewis, C. Lewis, R. Grubbs, G. A. Sotzing, Patent Int. Appl. WO 2000020852 A1 13 *Apr.* **2000**.
237 E. Genies, French Patent FR 2591605 A1 19 Jun, **1987**.
238 M. Wan, W. Zhou, J. Li, *Synt. Met.* **1996**, *78*, 27–31.
239 G. Bidan, O. Jarjayes, J. M. Fruchart, E. Hannecart, *Adv. Mater.* **1994**, *6*, 152–155.
240 H. Yoneyama, A. Kishimoto, S. Kuwabata, *J. Chem. Soc., Chem. Commun.* **1991**, *15*, 986–987.
241 H. Yoneyama, S. Hirao, S. Kuwabata, *J. Electrochem. Soc.* **1992**, *139*, 3141–3146.
242 H. Yoneyama, Japanese Patent JP 04136195 A2 11 May, **1992.**
243 S. Kuwabata, A. Kisimoto, T. Tanaka, H. Yoneyama, *J. Electrochem. Soc.* **1994**, *141*, 10–15.
244 K. Gong, W. Zhang, *Mat. Res. Soc. Symp. Proc.* **1997**.
245 M. Biswas, S. Sinha Ray, Y. Liu, *Synthetic Metals* **1999**, *105*, 99–105.
246 C.-L. Huang, R. E. Partch, E. Matijevic, *J. Colloid Interface Sci.* **1995**, *170*, 275–283.
247 M. Matsumura, T. Ohno, S. Saito, M. Ochi, *Chem. Mater.* **1996**, *8*, 1370–1374.
248 M. Matsumura, T. Ohno, *Adv. Mater.* **1997**, *9*, 357–359.
249 A. C. Arango, S. A. Carter, P. J. Brock, *Appl. Phys. Lett.* **1999**, *74*, 1698–1700.
250 A. C. Arango, P. J. Brock, S. A. Carter, *Mater. Res. Soc. Symp. Proc.* **1999**, *561*, 149–153.
251 H. Yoneyama, Y. Shoji, *J. Electrochem. Soc.* **1990**, *137*, 3826–3830.
252 P. K. Shen, H. T. Huang, A. C. C. Tseung, *J. Electrochem. Soc.* **1992**, *139*, 1840–1845.
253 N. R. Avvaru, N. R. de Tacconi, K. Rajeshwar, *Analyst* **1998**, *123*, 113–116.
254 S. Maeda, S. P. Armes, *Chem. Mater.* **1995**, *7*, 171–178.
255 S. Maeda, S. P. Armes, *Synth. Met.* **1995**, *73*, 151–155.
256 M. D. Butterworth, R. Corradi, J. Johal, S. F. Sascelles, S. Maeda, S. P. Armes, *J. Colloid Interface Sci.* **1995**, *174*, 510–517.
257 C. H. Nguyen, M. Dieng, C. Sene, P. Chartier, *Solar Energy Materials and Solar Cells* **2000**, *63*, 23–35.
258 M. Lal, M. Joshi, D. N. Kumar, C. S. Friend, J. Winiarz, T. Asefa, K. Kim, P. N. Prasad, *Mater. Res. Soc. Symp. Proc.* **1998**, *519*, 217–225.
259 P. Chartier, H. Nguyen Cong, C. Sene, *Sol. Energy Mater. Sol. Cells* **1998**, *52*, 413–421.
260 D. E. Fogg, L. H. Radzilowski, B. O. Dabbousi, R. R. Schrock, E. L. Thomas, M. G. Bawendi, *Macromolecules* **1997**, *30(26)*, 8433–8439.
261 H. Yoneyama, M. Tokuda, S. Kuwabata, *Electrochim. Acta* **1994**, *39*, 1315–1320.
262 K. Jackowska, M. Skompska, *Pol. J. Chem.* **1990**, *64*, 657–663.
263 K. Jackowska, M. Skompska, *Mater. Symp. Electrochem. Sect. Pol. Chem. Soc.* **1988**, *112*, 13325.
264 K. Jackowska, H. T. Tien, *J. Appl. Electrochem.* **1988**, *18*, 357–362.
265 C.-W. Kwon, A. Poquet, S. Mornet, G. Campet, J. Portier, J.-H. Choy, *Electrochemistry Communications* **2002**, *4*, 197–200.
266 A. Drelinkiewicz, M. Hasik, M. Choczynski, *Ater. Res. Bull.* **1998**, *33*, 739–762.
267 V. Aboutanos, J. N. Barisci, L. A. P. Kane-Maguire, G. G. Wallace, *Synthetic Metals* **1999**, *106*, 89–95.
268 D. Y. Godovsky, A. E. Varfolomeev, D. F. Zaretsky, R. L. N. Chandrakanthi, A. Kündig, C. Weder, W. Caseri, *J. Mater. Chem.* **2001**, *11*, 2465–2469.
269 D. Feldheim, *Interfase, The Electrochem. Soc.* **2001**, *Fall 2001*, 22–24.
270 K. Gurunathan, A. V. Murugan, R. Marimuthu, U. P. Mulik, D. P. Amalnerkar, *Materials Chemistry and Physics* **1999**, *61*, 173–191.
271 M. Angelopoulos, *IBM. J. Res. & Dev.* **2001**, *45*, 57–62.
272 D. L. Pearson, L. Jones III, J. S. Schumm, J. M. Tour, *Synth. Metals* **1997**, *84*, 303–306.

273 S. R. Marder, B. Kippelen, A. K.-Y. Jen, N. Peyghambarian, *Nature* **1997**, *388*, 845–851.
274 R. H. Friend, R. W. Gymer, A. B. Holmes, J. H. Burroughes, R. N. Marks, C. Taliani, D. D. C. Bradley, D. A. Dos Santos, J. L. Brédas, M. Lögdlund, W. R. Salaneck, *Nature* **1999**, *397*, 121–128.
275 G. D. Stucky, *Nature* **2001**, *410*, 885–886.
276 T. Yamamoto, *Macromol. Rapid Commun.* **2002**, *23*, 583–606.
277 T. Swager, *Nature Mat.* **2002**, *1*, 151–152.
278 S. Holdcroft, *Adv. Mater.* **2001**, *13*, 1753–1765.
279 J. V. Crivello, J. H. W. Lam, *Macromol.* **1977**, *10*, 1307.
280 P. G. Lacroix, *Chem. Mater.* **2001**, *13*, 3495–3506.
281 H. S. Kim, B. H. Sohn, W. Lee, J.-K. Lee, S. J. Choi, S. J. Kwon, *Thin Solid Films* **2002**, *419*, 173–177.
282 P. J. Kulesza, M. Chojak, K. Miecznikowski, A. Lewera, M. A. Malik, A. Kuhn, *Electrochemistry Communications* **2002**, *4*, 510–515.
283 M. Clemente-León, E. Coronado, P. Delhaes, C. J. Gómez-García, C. Mingotaud, *Adv. Mater.* **2001**, *13*, 574–577.
284 M. Pope, H. Kallaman, P. Magnante, *J. Chem. Phys.* **1963**, *38*, 2042–2043.
285 C. W. Tang, S. A. Van Slyke, *Appl. Phys. Lett.* **1987**, *51*, 913–915.
286 R. H. Friend, J. H. Burroughes, D. D. Bradley, US Patent US5247190, **1993**.
287 J. H. Burroughes, D. D. C. Bradley, A. R. Brown, R. N. Marks, R. H. Friend, P. L. Burns, A. L. Holmes, *Nature* **1990**, *347*, 539–541.
288 D. Braun, *Mat. Today* **2002**, *June*, 32–39.
289 R. Friend, J. Burroughes, T. Shimoda, *Physics World* **1999**, *June*, 35–40.
290 M. Leadbeater, *Spie´s OEMagazine* **2002**, *June*, 14–17.
291 J. Pei, X.-L. Liu, W.-L. Yu, Y.-H. Lai, Y.-H. Niu, Y. Cao, *Macromolecules* **2002**, *35*, 7274–7280.
292 P. K. Ng, X. Gong, S. H. Chan, L. S. M. Lam, W. K. Chan, *Chem. Eur. J.* **2001**, *7*, 4358.
293 J. Ji, J. L. Coffer, *J. Phys. Chem. B* **2002**, *106*, 3860–3863.
294 N. Tessler, D. J. Pinner, P. K. H. Ho, *Optical Mater.* **2001**, *17*, 155–160.
295 C. F. Gay, *Kirk-Othmer Encyclopedia of Chemical Technology*, John Wiley & Sons, New York, **2000**, online posting date: December 4.
296 M. Grätzel, *Nature* **2001**, *414*, 338–344.
297 G. G. Wallace, P. C. Dastoor, D. L. Officer, C. O. Too, *Chem. Inov.* **2000**, *30*, 14–22.
298 A. O. Patil, A. J. Heeger, F. Wudl, *Chem. Rev.* **1988**, *88*, 183–200.
299 J. Roncali, *Chem. Rev.* **1997**, *97*, 173–205.
300 K. Murakoshi, R. Kogure, Y. Wada, S. Yanagida, *Solar Energy Mat. Solar Cells* **1998**, *55*, 113–125.
301 D. Gebeyehu, C. J. Brabec, N. S. Sariciftci, *Thin Solid Films* **2002**, *403–404*, 271–274.
302 H. Gerischer, H. Tributsch, *Ber. Bunsenges. Phys. Chem.* **1968**, *72*, 437–445.
303 D. O. Reagen, M. Grätzel, *Nature* **1991**, *353*, 737.
304 D. Wöhrle, D. Meissner, *Adv. Mater.* **1991**, *3*, 129138.
305 W. C. Sinke, M. M. Wienk, *Nature* **1998**, *395*, 544–545.
306 D. Meissner, *Photon* **1999**, *Marz/April*, 24–27.
307 A. Hagfeldt, M. Grätzel, *Acc. Chem. Res.* **2000**, *33*, 269–277.
308 D. Gebeyehu, C. J. Brabec, N. S. Sariciftci, D. Vangeneugden, R. Kiebooms, D. Vanderzande, F. Kienberger, H. Schindler, *Synthetic Metals* **2001**, *125*, 279–287.
309 S. Spiekermann, G. Smestad, J. Kowalik, L. M. Tolbert, M. Grätzel, *Synth. Met.* **2001**, *121*, 1603–1604.
310 C. J. Brabec, A. Cravino, D. Meissner, N. S. Sariciftci, M. T. Rispens, L. Sanchez, J. C. Hummelen, T. Fromherz, *Thin Solid Films* **2002**, *403–404*, 368–372.
311 Y. Saito, T. Kitamura, Y. Wada, S. Yanagida, *Synthetic Metals* **2002**, *131*, 185–187.
312 A. Mishra, R. K. Behera, P. K. Behera, B. K. Mishra, G. B. Behera, *Chem. Rev.* **2000**, *100*, 1973–2011.
313 J. J. Dittmer, R. Lazzaroni, P. Leclere, P. Moretti, M. Granstrom, K. Petritsch, E. A. Marseglia, R. H. Friend, J. L. Bredas, H. Rost, A. B. Holmes, *Solar Energy Materials and Solar Cells* **2000**, *61*, 53–61.

314 K. Petritsch, J. J. Dittmer, E. A. Marseglia, R. H. Friend, A. Lux, G. G. Rozenberg, S. C. Moratti, A. B. Holmes, *Solar Energy Materials and Solar Cells* **2000**, *61*, 63–72.

315 N. Martin, L. Sánchez, B. Illescas, I. Pérez, *Chem. Rev.* **1998**, *98*, 2527–2547.

316 G. Yu, J. Gao, J. C. Hummelen, F. Wudl, A. J. Heeger, *Science* **1995**, *270*, 1789–1791.

317 N. S. Sariciftci, *Progress in Quantum Electronics* **1995**, *19*, 131–159.

318 H. Neugebauer, C. Brabec, J. C. Hummelen, N. S. Sariciftci, *Solar Energy Materials and Solar Cells* **2000**, *61*, 35–42.

319 D. Gebeyehu, C. J. Brabec, F. Padinger, T. Fromherz, J. C. Hummelen, D. Badt, H. Schindler, N. S. Sariciftci, *Synthetic Metals* **2001**, *118*, 1–9.

320 M. F. Durstock, B. Taylor, R. J. Spry, L. Chiang, S. Reulbach, K. Heitfeld, J. W. Baur, *Synthetic Metals* **2001**, *116*, 373–377.

321 T. Munters, T. Martens, L. Goris, V. Vrindts, J. Manca, L. Lutsen, W. De Ceuninck, D. Vanderzande, L. De Schepper, J. Gelan, *Thin Solid Films* **2002**, *403–404*, 247–251.

322 N. V. Kamanina, *Synth. Met.* **2002**, *127*, 121–128.

323 P. Innocenzi, G. Brusatin, M. Guglielmi, R. Signorini, R. Bozio, M. Maggini, *Journal of Non-Crystalline Solids* **2000**, *265*, 68–74.

324 M. Biswas, S. Sinha Ray, *Synthetic Metals* **2001**, *123*, 135–139.

325 S. Bouchtalla, A. Deronzier, J. M. Janot, J. C. Moutet, P. Seta, *Synth. Met.* **1996**, *82*, 129–132.

326 K. Yoshino, Y. Kawagishi, S. Tatsuhara, H. Kajii, S. Lee, M. Ozaki, Z. V. Vardeny, A. A. Zakhidov, *Superlattices Microstruct.* **1999**, *25*, 325–341.

327 A. Cravino, N. S. Sariciftci, *J. Mater. Chem.* **2002**, *12*, 1931–1943.

328 N. P. Gaponik, D. V. Sviridov, *Ber. Bunsen-Ges. Phys. Chem.* **1997**, *101*, 1657.

329 D. Gebeyehu, C. J. Brabec, N. S. Sariciftci, D. Vangeneugden, R. Kiebooms, D. Vanderzande, F. Kienberg, H. Schindler, *Synth. Met.* **2002**, *125*, 279–287.

330 W. U. Huynh, J. J. Dittmer, A. P. Alivisatos, *Science* **2002**, *295*, 2425–2427.

331 J. Ackermann, C. Videlot, A. El Kassmi, *Thin Solid Films* **2002**, *403–404*, 157–161.

332 J. Gallardo, A. Duran, D. di Martino, R. M. Almeida, *Journal of Non-Crystalline Solids* **2002** (in press, uncorrected proof).

333 T. Mikayama, H. Matsuoka, M. Ara, K. Uehara, A. Sugimoto, K. Mizuno, *Solar Energy Materials and Solar Cells* **2001**, *65*, 133–139.

334 M. E. Rincón, H. G. Martínez, R. Suárez, J. G. Bañuelos, *Solar Energy Mater. Solar Cells* **2003**, *77*, 239–254.

335 P. Aranda, E. Ruiz-Hitzky, *Acta Polym.* **1994**, *45*, 59–67.

336 M. Maxfield, L. W. Shackelette, J. J. Wolf, S. M. Savner, US Patent US4472489 Sept 18. **1984**.

337 T. Tatsuma, T. Sotomura, T. Sato, D. A. Buttry, N. Oyama, *J. Electrochem. Soc.* **1995**, *142*, L183.

338 B. C. Dave, B. Bunn, F. Leroux, L. F. Nazar, H. P. Wong, *Mater. Res. Soc. Symp. Proc.* **1996**, *435*, 611–616.

339 J. H. Harreld, B. Dunn, L. F. Nazar, *International Journal of Inorganic Materials* **1999**, *1*, 135–146.

340 F. Huguenin, E. A. Ticianelli, R. M. Torresi, *Electrochimica Acta* **2002**, *47*, 3179–3186.

341 F. Huguenin, R. M. Torresi, D. A. Buttry, J. E. Pereira da Silva, S. I. Cordoba de Torresi, *Electrochim. Acta.* **2001**, *46*, 3555–3562.

342 P. Gomez-Romero, M. Chojak, K. Cuentas-Gallegos, J. A. Asensio, P. J. Kulesza, N. Casan-Pastor, M. Lira-Cantu, *Electrochem. Com.* **2003**, *5*, 149–153.

343 J. H. Park, J. M. Ko, O. O. Park, D.-W. Kim, *Journal of Power Sources* **2002**, *105*, 20–25.

344 Z. Qi, M. C. Lefebvre, P. G. Pickup, *Journal of Electroanalytical Chemistry* **1998**, *459*, 9–14.

345 E. K. W. Lai, P. D. Beattie, F. P. Orfino, E. Simon, S. Holdcroft, *Electrochimica Acta* **1999**, *44*, 2559–2569.

346 I. Becerik, S. Suzer, F. Kadirgan, *Journal of Electroanalytical Chemistry* **2001**, *502*, 118–125.

347 H. B. Mark, Jr., J. F. Rubinson, J. Krotine, W. Vaughn, M. Gold-

schmidt, *Electrochimica Acta* **2000**, *45*, 4309–4313.
348 I. Becerik, F. Kadirgan, *Synthetic Metals* **2001**, *124*, 379–384.
349 M. Mastragostino, C. Arbizzani, F. Soavi, *Journal of Power Sources* **2001**, *97–98*, 812–815.
350 A. Rudge, J. Davey, I. Raistrick, S. Gottesfeld, J. P. Ferraris, *Journal of Power Sources* **1994**, *47*, 89–107.
351 A. Rudge, I. Raistrick, S. Gottesfeld, J. P. Ferraris, *Electrochimica Acta* **1994**, *39*, 273–287.
352 M. Kalaji, P. J. Murphy, G. O. Williams, *Synthetic Metals* **1999**, *102*, 1360–1361.
353 M. Mastragostino, C. Arbizzani, F. Soavi, *Solid State Ionics* **2002**, *148*, 493–498.
354 C.-C. Hu, E. Chen, J.-Y. Lin, *Electrochimica Acta* **2002**, *47*, 2741–2749.
355 K. Jurewicz, S. Delpeux, V. Bertagna, F. Beguin, E. Frackowiak, *Chemical Physics Letters* **2001**, *347*, 36–40.
356 R. Pacios, D. D. C. Bradley, *Synthetic Metals* **2002**, *127*, 261–265.
357 J. Janata, M. Josowicz, *Nature Mat.* **2002**, *2*, 19–24.
358 A. M. A. Ibrahim, *J. Mater. Chem.*, **1998**, *8*, 841–846.
359 O. A. Sadik, *Electroanalysis* **1999**, *11*, 839–844.
360 D. T. McQuade, A. E. Pullen, T. M. Swager, *Chem. Rev.* **2000**, *100*, 2537–2574.
361 P. W. Faguy, W. Ma, J. A. Lowe, W.-P. Pan, T. Brown, *J. Mater. Chem.* **1994**, *4*, 771–772.
362 A. Guadarrama, M. L. Rodríguez-Méndez, J. A. de Saja, J. L. Ríos, J. M. Olías, *Sensors and Actuators B: Chemical* **2000**, *69*, 276–282.
363 A. Guadarrama, M. L. Rodriguez-Mendez, C. Sanz, J. L. Rios, J. A. de Saja, *Analytica Chimica Acta* **2001**, *432*, 283–292.
364 A. Guadarrama, M. L. Rodríguez-Méndez, J. A. de Saja, *Anal. Chim. Acta* **2002**, 455 (1), 41–47.
365 H. Yamato, T. Koshiba, M. Ohwa, W. Wernet, M. Matsumura, *Synth. Met.* **1997**, *87*, 231–236.
366 S. Gaspar, K. Habermuller, E. Csoregi, W. Schuhmann, *Sensors and Actuators B: Chemical* **2001**, *72*, 63–68.
367 I. G. Casella, T. R. I. Cataldi, A. Guerrieri, E. Desimoni, *Anal. Chim. Acta* **1996**, *335*, 217–22.
368 G. L. Turdean, A. Curulli, I. C. Popescu, C. Rosu, G. Palleschi, *Electroanal.* **2002**, *14*, 1550–1556.
369 N. Guernion, B. P. J. de Lacy Costello, N. M. Ratcliffe, *Synth. Met.* **2002**, *128*, 139–147.
370 A. Riul Jr, D.S. Dos Santos Jr, K. Wohnrath, R. Di Tommazo, A. C. P. L. F. Carvalho, F. J. Fonseca, O.N. Oliveira Jr, D. M. Taylor, L. H. C. Mattoso, *Langmuir* **2002**, 18, 239–245.
371 M. Ferreira, A. Riul Jr, K. Wohnrath, F. J. Fonseca, O.N. Oliveira Jr, L. H. C. Mattoso, *Anal. Chem.* **2003**, 75, 953–955.
372 D. Curran, J. Grimshaw, S. D. Perera, *Chem. Soc. Rev.* **1991**, *20*, 391–404.
373 S. Holdcroft, B. L. Funt, *J. Electroanal. Chem. Interfacial Electrochem.* **1988**, *240*, 89–103.
374 G. Bidan, E. M. Genies, M. Lapkowski, *J. Electroanal. Chem. Interfacial Electrochem.* **1988**, *251*, 297–306.
375 P. Moisy, F. Bedioui, Y. Robin, J. Devynck, *J. Electroanal. Chem. Interfacial Electrochem.* **1988**, *250*, 191–199.
376 A. Deronzier, J.-C. Moutet, *Platinum Met. Rev.* **1998**, *42*, 60–68.
377 B. Fabre, G. Bidan, *Electrochim. Acta* **1997**, *42*, 2587–2590.
378 B. Fabre, G. Bidan, *J. Chem. Soc., Faraday Trans.* **1997**, *93*, 591–601.
379 K. Ogura, N. Endo, M. Nakayama, H. Ootsuka, *J. Electrochem. Soc.* **1995**, *142*, 4026–4032.
380 X. Wang, Z. Kang, E. Wang, C. Hu, *Journal of Electroanalytical Chemistry* **2002**, *523*, 142–149.
381 A. Malinauskas, *Synthetic Metals* **1999**, *107*, 75–83.
382 M. Musiani, *Electrochimica Acta* **2000**, *45*, 3397–3402.
383 K. Bouzek, K.-M. Mangold, K. Juttner, *Electrochimica Acta* **2001**, *46*, 661–670.
384 P. Burgmayer, R. W. Murray, *J. Am. Chem. Soc.* **1982**, *104*, 6139.
385 A. Kros, R. J. M. Nolte, N. A. J. M. Sommerdijk, *Adv. Mater.* **2002**, *14*, 1779–1782.

386 A. J. Michalska, E. A. H. Hall, *Electroanalysis* **1999**, *11*, 756–762.

387 I. Gill, *Chem. Mater.* **2001**, *13*, 3404–3421.

388 R. Cingolani, R. Rinaldi, G. Maruccio, A. Biasco, *Physica E: Low-dimensional Systems and Nanostructures* **2002**, *13*, 1229–1235.

389 E. W. H. Jager, O. Inganänas, I. Lundström, *Advanced Materials* **2001**, *13*, 76.

390 M. D. McGehee, A. J. Heeger, *Advanced Materials* **2000**, *12*, 1655.

391 S.-C. Ng, H.-F. Lu, S. O. Chan, A. G. Fujii, T. Laga, K. Yoshino, *Advanced Materials* **2000**, *12*, 1122.

392 E. Smela, O. Inganäs, I. Lundström, *Science* **1995**, *268*, 1735.

393 M. Zaragoza, A. Pron, *Synt. Metals* **1987**, *18*, 123–125.

394 M. Lapkowski, G. Bidan, M. Fournier, *Synth. Met.* **1991**, *41*, 411–414.

395 J. Pozniczek, A. Bielanski, I. Kulszewiez-Bajer, M. Zargorska, A. Pron, *J. Molecular Catalysis* **1991**, *69*, 223–233.

396 M. Lapkowski, G. Bidan, M. Fournier, *Synth. Met.* **1991**, *41*, 407–410.

397 H. Sung, H. So, W. K. Paik, *Electrochim. Acta* **1994**, *39*, 645–650.

398 P. Gomez-Romero, N. Casan-Pastor, M. Lira-Cantu, *Solid State Ionics* **1997**, *101–103*, 875–880.

399 G. Bidan, E. M. Genies, M. Lapkowski, *J. Chem. Soc., Chem. Commun.* **1988**, *5*, 533–535.

400 B. Keita, D. Bouaziz, L. Nadjo, *J. Electroanal. Chem. Interfacial Electrochem* **1988**, *255*, 303–313.

401 M. Lapkowski, G. Bidan, M. Fournier, *Pol. J. Chem.* **1991**, *65*, 1547–1561.

402 A. Pron, *Synth. Metals* **1992**, *46*, 277–283.

403 M. Hasik, A. Pron, I. Kulszewiczbajer, J. Pozniczek, A. Bielanski, Z. Piwowarska, R. Dziembaj, *Synthet. Metal.* **1993**, *55*, 972–976.

404 M. Hasik, J. Pozniczek, Z. Piwowarska, R. Dziembaj, A. Bielanski, A. Pron, *J. Mol. Catal.* **1994**, *89*, 329–344.

405 V. W. Jones, M. Kalaji, G. Walker, C. Barbero, R. Kotz, *J. Chem. Soc. Faraday Trans.* **1994**, *90*, 2061–2064.

406 M. Barth, W. Turek, M. Lapkowski, *Pol. J. Appl. Chem.* **1996**, *39*, 271–275.

407 P. Gomez-Romero, M. Lira-Cantu, N. Casan-Pastor, Spanish Patent. ES 2120324 A1, 16 Oct **1998**.

408 M. Barth, M. Lapkowski, S. Lefrant, *Electrochim. Acta* **1999**, *44*, 2117–2123.

409 A. Mahmoud, B. Keita, L. Nadjo, O. Oung, R. Contant, S. Brown, Y. de Kouchkovsky, *J. Electroanal. Chem.* **1999**, *463*, 129–145.

410 V. D. Pokhodenko, Y. I. Kurys, O. Y. Posudievsky, *Synthetic Metals* **2000**, *113*, 199–201.

411 M. Zagorska, I. Kulszewicz-Bajer, E. Lukomska-Godzisz, A. Pron, I. Glowacki, J. Ulanski, S. Lefrant, *Synth. Met.* **1990**, *37*, 99–106.

412 I. Kulszewicz-Bajer, M. Zagorska, J. Pozniczek, A. Bielanski, K. Kruczala, K. Dyrek, A. Pron, *Synth. Met.* **1991**, *41*, 39–42.

413 I. Kulszewicz-Bajer, M. Zagorska, A. Pron, D. Billaud, J. J. Ehrhardt, *Mater. Res. Bull.* **1991**, *26*, 163–170.

414 J. Pozniczek, I. Kulszewicz-Bajer, M. Zagorska, K. Kruczala, K. Dyrek, A. Bielanski, A. Pron, *J. Catal.* **1991**, *132*, 311–18.

415 P. Gomez-Romero, M. Lira-Cantu, *Proc. Electrochem. Soc. (Lithium Batteries)* **1997**, *96–17*, 158–161.

416 S. Cheng, T. Fernandez-Otero, E. Coronado, C. J. Gomez-Garcia, E. Martinez-Ferrero, C. Gimenez-Saiz, *J. Phys. Chem. B* **2002**, *106*, 7585–7591.

417 R. Mathur, D. R. Sharma, S. R. Vadera, N. Kumar, *Acta Materialia* **2001**, *49*, 181–187.

418 K. Ogura, N. Endo, M. Nakayama, *J. Electrochem. Soc.* **1998**, *145*, 3801–3809.

419 M. Morita, *J. Appl. Polym. Sci.* **1994**, *52*, 711–719.

420 K. Otsuka, S. Osada, Japanese Patent JP 62290759 A2, **1987**.

421 K. J. Chao, T. C. Chang, S. Y. Ho, *J. Mater. Chem.* **1993**, *3*, 427.

422 Y. J. Liu, M. G. Kanatzidis, *Chem. Mater.* **1995**, *7*, 1525.

423 A. J. G. Zarbin, D. J. Maia, M.-A. De Paoli, O. L. Alves, *Synth. Met.* **1999**, *102*, 1277–1278.

424 A. B. Goncalves, A. S. Mangrich, A. J. G. Zarbin, *Synthetic Metals* **2000**, *114*, 119–124.

425 F. Beleze, J. G. Zarbin, *J. Braz. Chem. Soc.* **2001**, *12*, 542–547.
426 N. S. P. Bhuvanesh, J. Gopalakrishnan, *Inorg. Chem.* **1995**, *34*, 3760.
427 L. F. Nazar, T. Kerr, B. Koene, in *Proc. Symp. Access Nanoporous Mater.*, eds T. J. Pinnavaia and M. F. Thorpe, Plenum, New York, 1995, 405–427.
428 J. H. Harreld, B. Dunn, L. F. Nazar, *Int. J. Inorg. Mater.* **1999**, *1*, 135–146.
429 V. D. Pokhodenko, V. A. Krylov, Y. I. Kurys, O. Y. Posudievsky, *Phys. Chem. Chem. Phys.* **1999**, *1*, 905–908.
430 J. Harreld, H. P. Wong, B. C. Dave, B. Dunn, L. F. Nazar, *J. Non-Crystalline Solids* **1998**, *225*, 319–324.
431 F. Huguenin, E. M. Girotto, R. M. Torresi, D. A. Buttry, *J. Electroanalytical Chemistry* **2002**, *536*, 37–45.
432 G. J.-F. Demets, F. J. Anaissi, H. E. Toma, M. B. A. Fontes, *Materials Research Bulletin* **2002**, *37*, 683–695.
433 F. Huguenin, E. M. Girotto, G. Ruggeri, R. M. Torresi, *J. Power Sources* **2003**, *114*, 133–136.
434 Y. Wang, X. Wang, J. Li, Z. Mo, X. Zhao, X. Jing, F. Wang, *Adv. Mater.* **2001**, *13*, 1582.
435 E. P. Giannelis, V. Mehrotra, European Patent EP0444574 Sept 4, **1991**.
436 P. Enzel, T. Bein, *Chem. Commun.* **1989**, 1326–1327.
437 A. G. Pattantyus-Abraham, M. O. Wolf, *Mater. Res. Soc. Symp. Proc.* **1999**, *560*, 291–295.
438 A. Moreale, P. Cloos, C. Badot, *Clay Min.* **1985**, *20*, 29.
439 K. Chibwe, W. Jones, *J. Chem. Soc., Chem. Commun.* **1989**, *14*, 926–927.
440 A. Okada, A. Usuki, *Materials Science and Engineering C* **1995**, *3*, 109–115.
441 H. L. Frisch, B. Xi, Y. Qin, M. Rafailovich, N. L. Yang, X. Yan, *High Perf. Polym.* **2000**, *12*, 543–549.
442 W. Jia, E. Segal, D. Kornemandel, Y. Lamhot, M. Narkis, A. Siegmann, *Synthetic Metals* **2002**, *128*, 115–120.
443 W. Krawiec, L. G. Scanlon Jr., J. P. Fellner, R. A. Vaia, S. Vasudevan, E. P. U.-F. Giannelis, *J. Power Sources* **1995**, *54*, 310–315.
444 A. Chandra, *Fractals: Complex Geometry, Patterns, and Scaling in Nature and Society* **2001**, *9*, 171–175.
445 S. Sinha Ray, *Materials Research Bulletin* **2002**, *37*, 813–824.
446 H.-J. Nam, H. Kim, S. H. Chang, S.-G. Kang, S. H. Byeon, *Solid St. Ionics* **1999**, *120*, 189–195.
447 C. S. Liao, L. K. Lin, US Patent US6136909, **2000**.
448 D. J. Maia, O. L. Alves, M. A. DePaoli, *Synth. Met.* **1997**, *90*, 37–40.
449 S. M. Marinakos, L. C. Brousseau, A. Jones, D. L. Feldheim, *Chem. Mater.* **1998**, *10*, 1214–1219.
450 K. A. Mauritz, *Materials Science and Engineering: C* **1998**, *6*, 121–133.
451 M. Wan, J. Huang, Y. Shen, *Synthetic Metals* **1999**, *101*, 708–711.
452 K. A. Carrado, *Applied Clay Science* **2000**, *17*, 1–23.
453 M. Wan, J. Liu, H. Qiu, J. Li, S. Li, *Synthetic Metals* **2001**, *119*, 71–72.
454 H. Cao, Z. Xu, H. Sang, D. Sheng, C. Tie, *Adv. Mater.* **2001**, *13*, 121.
455 H. Inoue, H. Yoneyama, *J. Electroanal. Chem.* **1987**, *233*, 291.
456 C. O. Oriakhi, X. Zhang, M. M. Lerner, *Applied Clay Science* **1999**, *15*, 109–118.
457 H. Shioyama, *Synth. Metals* **2000**, *114*, 1–15.
458 B. Coffey, P. V. Madsen, T. O. Peohler, P. C. Searson, *J. Electrochem. Soc.* **1995**, *142*, 321.
459 S. Higashika, K. Kimura, Y. Matuso, Y. Sugie, *Carbon* **1999**, *37*, 354–356.
460 P. Liu, K. Gong, *Carbon* **1999**, *37*, 706.
461 H. Shioyama, *Synth. Met.* **2000**, *114*, 1–15.
462 B. Veeraraghavan, J. Paul, B. Haran, B. Popov, *J. Power Sources* **2002**, *109*, 377–387.
463 G. Skandan, A. Singhal, *Powder Metallurgy* **2000**, *43*, 313–317.
464 A. D. Pomogalio, *Russ. Chem. Rev.* **2000**, *69*, 53–80.
465 C. J. Brumilik, V. P. Menon, C. R. Martin, *J. Mater. Res.* **1994**, *9*, 1174.
466 C.-H. Hsu, US Patent US6001475 Dec 14, **1999**.
467 C. S. C. Bose, K. Rajeshwar, US Patent US5334292 Aug 2, **1994**.

468 S.-J. Choi, S.-M. Park, *Adv. Mater.* **2000**, *12*, 1547–1549.

469 C. Sene, H. N. Cong, M. Dieng, P. Chartier, *Materials Research Bulletin* **2000**, *35*, 1541–1553.

470 L. W. Shacklette, J. E. Toth, R. L. Elsenbaumer, US Patent US4695521 Sept. 22, **1987**.

471 A. De, R. Gangopadhyay, *2000 Photomicrographs* **2000**, 608–622.

472 V. Mehrotra, E. P. Giannelis, *Mat. Res. Soc. Symp. Proc.* **1990**.

473 P. Brandt, R. D. Fisher, E. S. Martinez, R. D. Calleja, *Angew. Chem. Int. Ed.* **1989**, *28*, 1265.

474 A. Walcarius, *Chem. Mater.* **2001**, *13*, 3351–3372.

475 S. H. Jang, M. G. Han, S. S. Im, *Synth. Met.* **2000**, *110*, 17–23.

476 C. R. Martin, *Acc. Chem. Res.* **1995**, *28*, 61.

477 S. Roux, P. Audebert , J. Pagetti, M. Roche, *New J. Chem* **2002**, *3*, 298–304.

478 H. Cattey, P. Audebert, *Synt. Metals* **1998**, *93*, 127–131.

479 A. J. G. Zarbin, M.-A. De Paoli, O. L. Alves, *Synthetic Metals* **1999**, *99*, 227–235.

480 P. T. Sotomayor, I. M. Raimundo Jr, A. J. G. Zarbin, J. J. R. Rohwedder, G. O. Neto, O. L. Alves, *Sensors & Actuators B* **2001**, *74*, 157–162.

481 S. Armes, *Mater. World* **2000**, *8*, 15–17.

482 O. Kimura, T. Kabata, T. Ohsawa, German Patent DE3841924 July 13, **1989**.

483 O. Kimura, T. Kabata, T. Ohsawa, US Patent US4886572 A1, Dec 12, **1989**.

484 K. L. Tan, S. L. Lim, E. T. Kang, *Synthetic Metals* **1998**, *92*, 213–222.

485 A. Du Pasquier, F. Orsini, A. S. Gozdz, J. M. Tarascon, *J. Power Sourc.* **1999**, *81–82*, 607–611.

486 M. Nishizawa, K. Mukai, S. Kuwabata, C. R. Martin, H. Yoneyama, *J. Electrochem. Soc.* **1997**, *144*, 1923.

487 J. Janata, J. Langmater, *Anal. Chem.* **1991**, *28*, 372.

488 R. Mazeikiene, A. Malinauskas, *Synth. Metals* **1997**, *89*, 77.

489 M. Makoto, U. Arimitsu, Japanese Patent JP62246926 A2 April 18, **1987**.

490 W. P. Roberts, L. A. Schulz, US Patent US4604427 Aug 5, **1986**.

491 A. Hassanien, M. Gao, M. Tokumoto, L. Dai, *Chemical Physics Letters* **2001**, *342*, 479–484.

492 G. Z. Chen, M. S. P. Shaffer, D. Coleby, G. Dixon, W. Zhou, D. J. Fray, A. H. Windle, *Advanced Materials* **2000**, *12*, 522–526.

493 R. C. Patil, S. Radhakrishnan, S. Pethkar, K. Vijayamohanan, *J. Mater. Res.* **2001**, *16*, 1982–1988.

494 S. Negi, K. Godon, S. M. Khan,I. Khan, in I.M. Khan and J.S. Harrison eds., *Field Responsive Polymers*. Oxford University Press, New York, **1998**.

495 H. Tsuneo, D. Toshio, I. Kaoru, Japanese Patent JP 62161830 A2, **1987**.

496 S. Yuasa, Japanese Patent JP1041673 A2, Feb 13, **1989**.

497 D. A. Upson, US Patent US4526706 7 Feb, **1985**.

498 M. Oda, Japanese Patent JP1069662 March 15,**1989**.

499 T.-H. Ahn, Y.-H. Park, Korean Patent KR9706708 April 29, **1997**.

500 H. H. Kuhn, W. C. Kimbrell, European Patent EP0302590 A2, Dec 13, **1989**.

501 P. R. Newman, L. F. Warren, E. F. Witucki, US Patent US4617228, Oct 14, **1986**.

502 E. F. Witucki, L. F. Warren, R. Newman, US Patent US4692225, Sept. 8, **1987**.

8
Layered Organic-Inorganic Materials: A Way Towards Controllable Magnetism

Pierre Rabu and Marc Drillon

8.1
Introduction

Today, seeking new materials for new devices concerns not only the solid-state chemists and physicists, but also the communities of molecular and organo-metallic chemistry, and those of biochemistry. This general interest is especially clear in the field of organic-inorganic multifunctional materials, whose design necessitates the investigation of new concepts and principles developed in these different disciplines.

Basically, the so-called hybrid organic-inorganic compounds involve different families of materials, and a complete description of their physical properties is beyond the scope of this review. We will focus here on the structure-magnetic property relationships in layered compounds, made of organic and inorganic subunits, by ruling out materials such as amorphous or glassy solids, and coated or core-shell nanoparticles whose properties are described elesewhere in this book.

Today, the molecular approach combining two distinct subnetworks, one corresponding to metal ions interconnected through bridging ligands, the other to π-electron donor or acceptor molecules, is appealing for the design of multifunctional materials. A series of organic/inorganic radical cation salts exhibiting electron delocalization, and even a metallic-like behavior, together with a magnetic response, have been obtained through the combination of inorganic molecular species, like polyoxometalates, with TTF-type organic donors. Recent reviews detail the richness of these compounds which appear well adapted for the design of polyfunctional materials, but in turn are characterized by weak molecular contacts and accordingly are potentially limited to achieve high temperature ordering [1].

The first example of extended 2D architecture involving organic species has probably been achieved in layered perovskites, but the most promising molecule-based compounds are the series of tris-oxalato transition metal complexes, whose structure and properties are described in Sect. 8.2. From a structural point of view, such materials may be described as composites with organic species ionically bound to the inorganic layers [2].

Gómez-Romero: Organic-Inorganic Materials. Pierre Rabu

ISBN: 3-527-30484-3

Section 8.3 is devoted to the MPS_3 intercalation compounds [3], whose interlayer spacing contains an exchangeable cation controlling the distance between the inorganic layers. Mostly, no real chemical bond exists in cation-exchangeable layered materials between the host and guest sub-networks. The cohesion is basically due to electrostatic or van der Waals interactions.

Finally, we will show in the last section that the presence of covalent interactions between the subunits, such as for instance in the transition metal phosphonates or the series of hydroxide-based compounds, favors a synergy between the properties of the organic and inorganic components. The design of such materials may be useful for applications in the field of multifunctional systems.

8.2 Molecule-based Materials with Extended Networks

The approach consisting in building solids from molecular bricks is powerful for the design of organic-inorganic compounds with potentially interesting properties, but, usually, the presence of weak interactions between magnetic units – Van der Waals or electrostatic-like – makes these compounds unsuitable for studying cooperative phenomena. In turn, the self-assembling of polyatomic ligands and magnetic centres into 2D extended architectures is appealing for the investigation of new properties. This was first reported for layered perovskite halides, and more recently in two-dimensional oxalate-based compounds [4, 5]. For the latter, a real crystal engineering has been developed on the basis of the self assembling of the oxalate bridging ligand, which is known to be a good mediator of both ferro- and antiferromagnetic interactions between metal ions.

8.2.1 Transition Metal layered Perovskites

The self-assembly of polyatomic ligands and magnetic metal ions into 2D architectures is observed in the layered perovskites $(RNH_3)_2MX_4$ and $(H_3N(CH_2)_nNH_3)_2MX_4$ where M is a divalent metal ion, R an organic molecule and X a halogen. Basically, the layered structure consists of corner-sharing MX_6 units, well separated by alkane-di-ammonium or monoammonium cations weakly bonded by van der Waals interactions [4]. The magnetic properties, whose 2D character is closely related to the length of the organic spacers, have attracted interest for a long time [6]. Briefly, in the Cu^{II} and Cr^{II} systems, the in-plane interaction is ferromagnetic and a very weak coupling occurs between the layers leading to various magnetic orderings. The Cu^{II} compounds are mainly 3D antiferromagnetic (AF) but canted antiferromagnetism (CAF) has also been reported with critical temperatures ranging between 6 K and 20 K, while tetrahalogenochromates order ferromagnetically below 42 K. On the other hand, the Mn^{II} and Fe^{II} derivatives are typical examples of quasi-2D antiferromagnets, with a small canting in the ordered state below ca. 40 K and 90 K respectively [7, 8].

Particular interest has been paid recently to the copper(II) series whose layer structure (Fig. 8.1) is governed by the cooperative Jahn-Teller effect: each octahedron is prolonged along the Jahn-Teller z axis, that lies in the CuX plane and has nearly orthogonal relations with its neighbors. The intralayer magnetic interaction is ferromagnetic, because of the orthogonal relation between the magnetic dx^2-y^2 orbitals. It has been shown that these structural features are sensitive to high pressure, the Jahn-Teller distortion being modified in a cooperative mode, thus modifying the magnetic behavior of the perovskite sheets [9, 10]. In particular, the study of the (p-cyanoanilinium)$_2$CuCl$_4$ compound [11], which shows one of the strongest ferromagnetic interactions in the series, has shown that such compounds might be candidates for switchable or controllable magnets governed by pressure. The pressure-induced enhancement of the ferromagnetic interaction in $[CuCl_4]^{2-}$ layers is analyzed as a decrease of the longest CuCl bond distance in the Cu-Cl-Cu superexchange pathway.

These compounds have also attracted recent interest as model systems related to the high T_C superconductors and organic-inorganic multifunctional materials [12].

8.2.2
Bimetallic Oxalate-bridge Magnets

The tris-oxalato transition metal complexes exhibit either 2D or 3D structures that are formally composed of $[M^{z+}(C_2O_4)_3]^{(6-z)-}$-building blocks, which in principle can polymerize in two ways. When building blocks of opposite chirality are alternately linked, the metal ions are forced to lie within layers, and form a 2D honeycomb network with stoichiometry $[A]^+[M^{II}M^{III}(C_2O_4)_3]^-$, where A is an organic cation, most

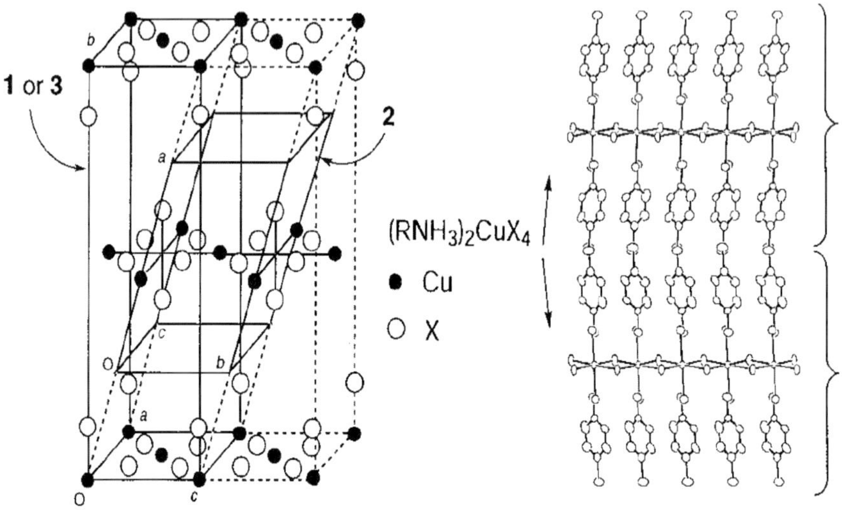

Fig. 8.1 The schematic comparison between the unit cells of **1** ($C_{12}H_{14}N_2C_{l2}CuBr_4$), **2** ($C_{12}H_{14}N_2CuCl_6$) or **3** ($C_{12}H_{14}N_4O_4CuCl_4$) ; the side views of the layered structures are shown for **1**

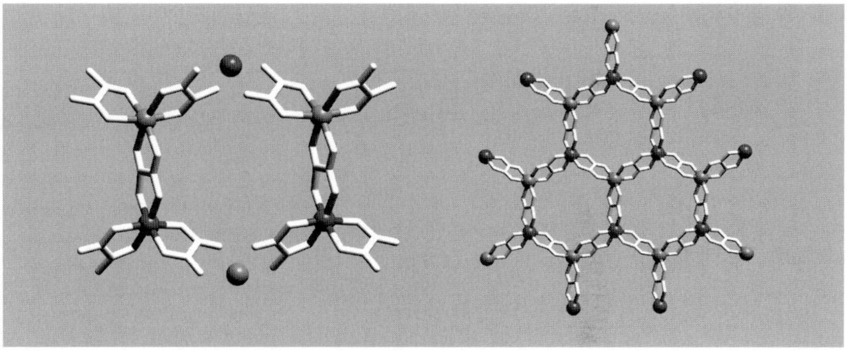

Fig. 8.2 (a) Two dimeric units of the alternating chirality type are necessary to form a closed hexagon ring; (b) the resulting planar network motif

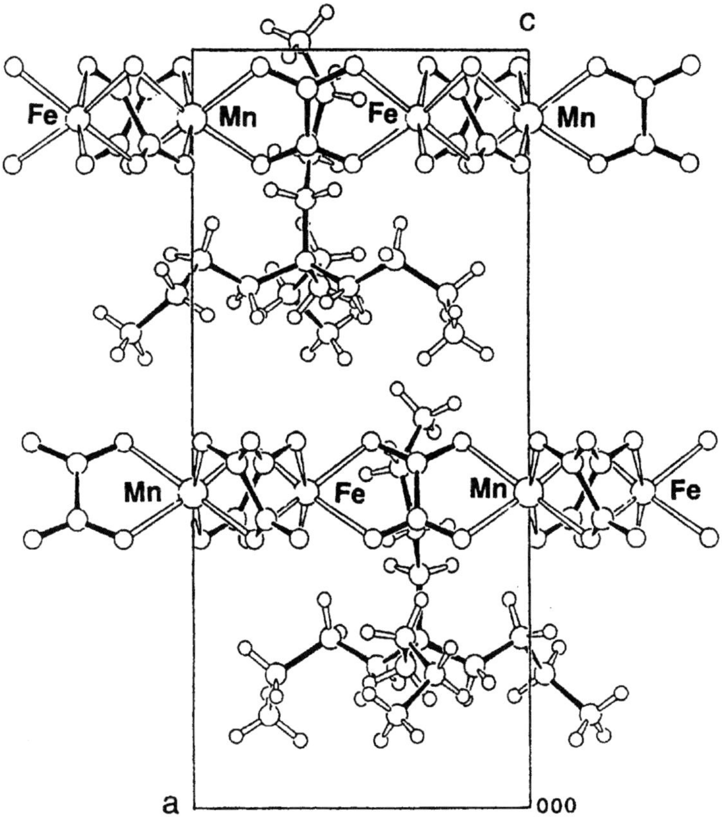

Fig. 8.3 Structure of the $[N(Bu)_4][Mn^{II}Fe^{III}(ox)_3]$ layered compound. (a) [001] projection; (b) [010] projection

usually a bulky ammonium or phosphonium $[XR_4]^+$ (Fig. 8.2) [13]. In turn, an assembly of building units with the same chirality induces the formation of an infinite 3D structure, such as $[Fe^{II}(bpy)_3][Mn^{II}_2(C_2O_4)_3]$ [14].

The crystal structure of a two-dimensional oxalate-bridge bimetallic compound, first reported for $[NBu_4][Mn^{II}Cr^{III}(C_2O_4)_3]$, reveals an anionic two-dimensional network of μ-oxalato bridged Mn^{II} and Cr^{III} ions, with NBu_4^+ cations located in between the layers [15]. The distance between adjacent layers is 8.95 Å. Further, Decurtins and colleagues obtained single crystals of $[PPh_4][Mn^{II}Cr^{III}(C_2O_4)_3]$ (space group R3c), whose structure comprises an alternating assembly of chiral Λ and Δ-type $[M(C_2O_4)_3]$ units, forming a two-dimensional honeycomb $[Mn^{II}Cr^{III}(C_2O_4)_3]_n^{n-}$ network with a basal spacing of 9.55 Å [5].

With a view to studying the influence of the organic cation [A] on the physical properties, a series of compounds of the type $[A][M^{II}M'^{III}(C_2O_4)_3]$ where A = NPr_4, NBu_4, $N(n\text{-}C_nH_{2n+1})_4$, PBu_4, PPh_4, $NBu_3(C_6H_5CH_2)$, Ph_3PNPPh_3, $AsPh_4$, M^{II} = Mn, Fe and M'^{III} = Cr, Fe, have been investigated [16,17,18,19]. Table 8.1 summarizes the structural features of some Mn^{II} compounds of this series.

The magnetic properties are, as expected, closely related to the spin configuration of the metal ions, the way in which they are connected, and the nature of the organic cation which may induce significant lattice distortions.

Thus, for the $[Mn^{II}Fe^{III}]$ compounds, the magnetic susceptibility mimics the behavior of classic 2D antiferromagnets, with a broad maximum at 55 K. This behavior arises because both metal ions have the same ground state, namely $3d^5$ ($S = 5/2$), and, as demonstrated by neutron diffraction experiments, the moments are mainly aligned along the *c*-axis [18].

Antiferromagnetic interactions between nearest neighboring metal ions are also observed for the $[Fe^{II}Fe^{III}]$ series. They behave as 3D ferrimagnets with T_C values ranging between 33 and 48 K, and furthermore some of them exhibit the typical features of a compensation temperature, namely a crossover from positive to negative magnetization, upon cooling down in very low field (Fig. 8.4). For the series with A = $(n\text{-}C_mH_{2m+1})_4$ and $m = 3 - 5$, T_C increases with interlayer separation, unlike the expected variation, and the low-temperature magnetization changes from positive ($m = 3$) to negative ($m = 4, 5$). It is shown elsewhere that this variation may be explained by the influence of dipolar interactions which are no longer negligible in such 2D systems.

Tab. 8.1 Crystallographic data for 2D transition metal oxalate based compounds

Compound	A (Å)	B (Å)	C (Å)	Space group	Z	Basal spacing
$[NBu_4][Mn^{II}Cr^{III}(C_2O_4)_3]$	9.414	9.414	53.662	R3c	6	8.95
$[PPh_4][Mn^{II}Cr^{III}(C_2O_4)_3]$	18.783	18.783	57.283	R3c	24	9.55
$[NPr_4][Mn^{II}Cr^{III}(C_2O_4)_3]$	9.363	9.363	49.207	R3c	6	8.20
$[NBu_4][Mn^{II}Fe^{III}(C_2O_4)_3]$	9.482	9.482	17.827	$P6_3$	2	8.91
$[N(n\text{-}C_5H_{11})_4][Mn^{II}Fe^{III}(C_2O_4)_3]$	9.707	16.140	19.883	$C222_1$	4	10.16
$[Fe(Cp^*)_2][Mn^{II}Fe^{III}(C_2O_4)_3]$	9.0645	17.143	9.215	C2/m	2	9.21

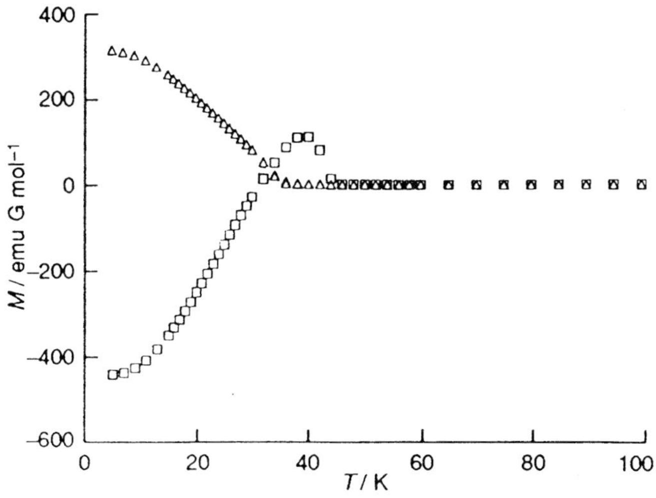

Fig. 8.4 Plot of magnetization against temperature for $AFe^{II}Fe^{III}(ox)_3$ [$A=NBu_4^+$ (square); NPr_4^+ (circle)] cooled in a field of 100 G

The origin of a negative magnetization at low temperature has been more specifically studied for $NBu_4Fe^{II}Fe^{III}(C_2O_4)_3$. According to Néel's model, the observed behavior was related to an initially steeper ordering for the Fe^{II} sublattice, due to the strong local anisotropy of the moments [18].

Unlike the above compounds, the $[Mn^{II}Cr^{III}]$ system shows ferromagnetic in-plane interactions and a long-range order at $T_C < 14$ K. The magnetic structure of $[P(C_6D_5)_4][Mn^{II}Cr^{III}(C_2O_4)_3]$ studied by Decurtins et al. [20] by elastic neutron scattering is described by a collinear alignment of the moments along the *c*-axis, in agreement with single-crystal magnetization results. In addition, specific heat and AC magnetic measurements of $[PPh_4][Mn^{II}Cr^{III}(C_2O_4)_3]$ have demonstrated the low dimensional magnetic character of these systems, and provided evidence for the onset of 3D magnetic ordering at low temperature [21]. In contrast, experimental results for the Fe^{II} derivative indicate a spinglass-like behavior, which may be assigned to the effect of structural disorder prevalent in this family of compounds.

In 1998, Kahn et al. [23] reported the magnetic properties of the ruthenium derivatives $[NBu_4][M^{II}Ru^{II}(C_2O_4)_3]$, ($M^{II}$ = Mn, Fe, Cu). The Mn^{II} compound shows ferromagnetic in-plane interactions, but no long-range order down to 2 K. The Fe^{II} derivative exhibits a magnetic ordering at $T_C = 13$ K, while the susceptibility and magnetization data for M = Cu^{II} was not well understood.

Furthermore, a series of compounds has also been obtained by combining the oxalate-bridge honeycomb network $[M^{II}M'^{III}(C_2O_4)_3]$ (M^{II} = Cr, Mn, Fe, Co, Cu; M'^{III} = Cr, Fe) with paramagnetic decamethylferrocenium [22]. The crystal structure, solved for the $[Mn^{II}Fe^{III}]$ derivative, shows that the $[Fe(Cp^*)_2]^+$ cation intercalates the oxalate bridge layers without entering into vacancies. Furthermore, unlike the $[XR_4]^+$ analogues, all layers are identical and the metal ions of adjacent layers are in registry.

For all these compounds, spontaneous magnetization was observed related to ferro- ferri- or canted antiferromagnetic order, with T_C ranging from 5 to 44 K.

From EPR and Mössbauer spectroscopy experiments, it was pointed out that no significant interaction occurs between the two subnetworks, and hence the properties are somewhat similar to those of the $[MX_4]^+$ derivatives. A simple spin polarization of the paramagnetic $[Fe(Cp^*)_2]$ species in the internal field originated by the inorganic layers is observed.

8.2.2.1 **Magnetism and Conductivity**

A strategy for achieving multiproperty materials is to insert functional cations, such as the organic π electron donor BEDT-TTF, into the bimetallic oxalato complex described above. The idea is to design infinite sheets of oxalate-based magnetic networks whose charge is counterbalanced by conducting BEDT-TTF cations, resulting in both ferromagnetic and metallic materials. Starting from the 2D oxalate based compounds $[XR_4]^+[M^{II}M^{III}(C_2O_4)_3]^-$, it is possible to substitute the electronically "inactive" organic cation by a functional one [24a]. This is achieved by electrocrystallization of a methanol/benzonitrile/dichloromethane solution containing the tris-oxalate Cr^{III} complex and the Mn^{II} ion, together with a suspension of BEDT-TTF, giving $[BEDT\text{-}TTF]_3\,[MnCr(C_2O4)_3]$ [24b].

The structure is triclinic and consists of layers of BEDT-TTF cation alternating with honeycomb layers of the bimetallic oxalato anion $[MnCr(C_2O_4)_3]^-$ (Fig. 8.5). The BEDT-TTF cations are tilted with respect to the inorganic layer by 45°.

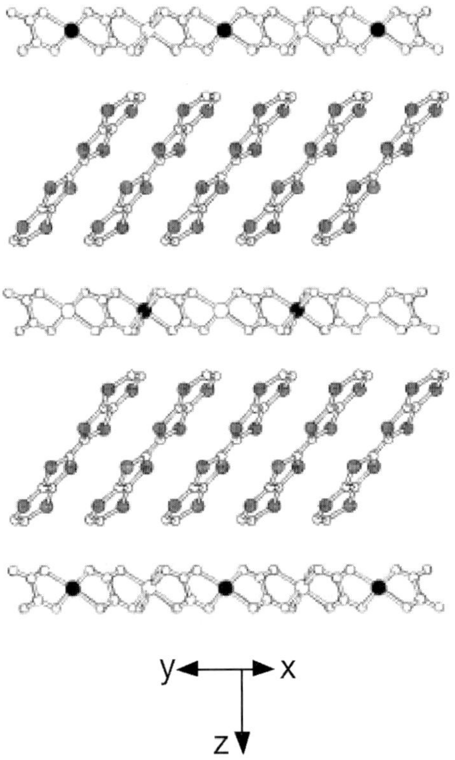

Fig. 8.5 Structure of $[BEDT\text{-}TTF]_3\,[MnCr(C_2O4)_3]$: packing of the BEDT-TTF molecules in between the oxalate layers

Actually, the inorganic oxalate-based layers are crystallographically disordered, because of the presence of stacking faults.

Both the AC magnetic susceptibility and magnetization measurements indicate that the compound is ferromagnetic-like with T_c =5 .5 K (Fig. 8.6). In fact, the saturation of magnetization at 2 K (7.1 μ_B instead of ~8 μ_B) points to the presence of a spin canted structure in the ferromagnetic state[25]. Small coercive fields are observed (Hc < 5–10 Oe) at low temperatures and accordingly the features are similar to those already reported for the other $Mn^{II}Cr^{III}$ oxalato layered compounds.

On the other hand, the BEDT-TTF compound exhibits a metallic behaviour down to 2 K. The resistivity decreases by a factor ~20 as temperature is decreased from 300 K to 2 K. No real interplay between the properties of the organic and inorganic subnetworks has been evidenced. Likewise, no superconductivity is observed down to 2 K, but it was not clear whether this is intrinsic to the organic sublattice or due

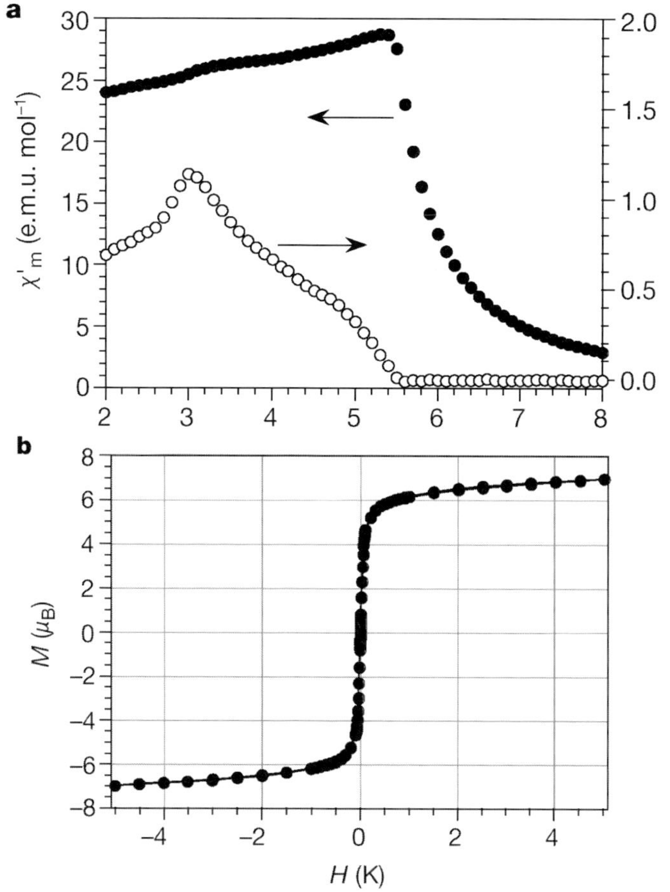

Fig. 8.6 (a) Magnetic AC susceptibilities of $[BEDT\text{-}TTF]_3[MnCr(C_2O4)_3]$ at 110 Hz; (b) plot of magnetization versus field at 2 K

to the presence of the ferromagnetic oxalato layer. Furthermore, application of a magnetic field (up to 5 T) perpendicular to the layers gives rise to a negative magnetoresistance below ca. 10 K. This was related to the influence of the internal field generated by the ferromagnetic layers at low temperature.

8.2.2.2 **Magnetism and Non-linear Optics**

The design of hybrid non-linear optic (NLO) magnetic materials has also been considered through a strategy aiming at replacing the A^+ cation by a highly polarizable species. A first work concerned the insertion of DAMS (4-[4-(dimethylamino)-α-styryl]-methylpyridinium) but no NLO properties were observed [26]. Later on, Bénard and co-workers reported on a new compound consisting of methoxyheptyl stilbazolium (MHS) cations embodied in $[Mn^{II}Cr^{III}(ox)_3]^-$ layers [27]. This derivative exhibits a second harmonic generation (SHG) efficiency 100 times that of urea (at 1.907 μm) and a net moment below 5.7 K. This compound was assumed to be isostructural to the NBu_4 [28] and PPh_4 [29] analogues. Actually, the noncentrosymmetric *R*3*c* space group is consistent with the observation of a bulk NLO response in the hybrid material. Recently, works on the $[A]^+[M^{II}Cr^{III}(C_2O_4)_3]^-$ series with M = Mn, Fe, Co, Ni, Cu, and A being stilbazolium-shaped organic dyes, have shown that these compounds are ferromagnetic below T_c values ranging between 6 and 13 K, and in addition exhibit SHG efficiencies [30]. Single crystals of two NLO inactive members of the family have confirmed the centrosymmetric space group $P2_1/c$ of these systems. $DAZOP[MnCr(C_2O_4)_3] \cdot 0.5\ CH_3CN$ (DAZOP is a DAMS analogue, where the central C=C core is replaced by an azo core), whose structure was solved by powder X-ray diffraction [31], exhibits NLO activity (ca. 100 times that of urea at 1.9 μm) and orders ferromagnetically below 6 K. This compound crystallizes in the non-centrosymmetric space group $P2_1$. The DAZOP chromophores are aligned in a polar manner, with their long axis parallel to the *b* direction and a tilt angle of ca. 48° with respect to the anionic layers (Fig. 8.7). Preliminary results suggest a similar packing in $DAZOP[FeCr(C_2O_4)_3]$ and $DAMS[MnCr(C_2O_4)_3]$ derivatives.

The main issue for the above mentioned hybrid compounds is the possible interplay between magnetism and NLO properties and particularly the incorporation of switchability into quadratic NLO systems [32, 33]. Up to now, very few reports have pointed out that NLO switches could be induced by magnetic phenomena [34], and the effect has been suggested but not observed yet. A recent paper reveals that the $S = 0 \leftrightarrow 2$ spin crossover in a Fe^{II} complex should result in an increase of the quadratic NLO response of about 25% [35]. In the present hybrid structures, the magnetic and NLO components are fairly well separated, and it can be assumed that the magnetization has no effect on the NLO tensor β. In turn, the propagation of the light is usually affected by magnetic moment, which should result in a modification of the NLO response, like Faraday rotation.

Experiments performed on a powdered sample of $Mn_{0.86}PS_3DAMS_{0.28}$ (see the following section) demonstrated that the internal magnetic field causes a decrease of the SHG intensity of about 10%, as the temperature is lowered below T_c [36]. This suggests that the above layered organic-inorganic magnets are interesting for the study of the magnetic-NLO property relationship.

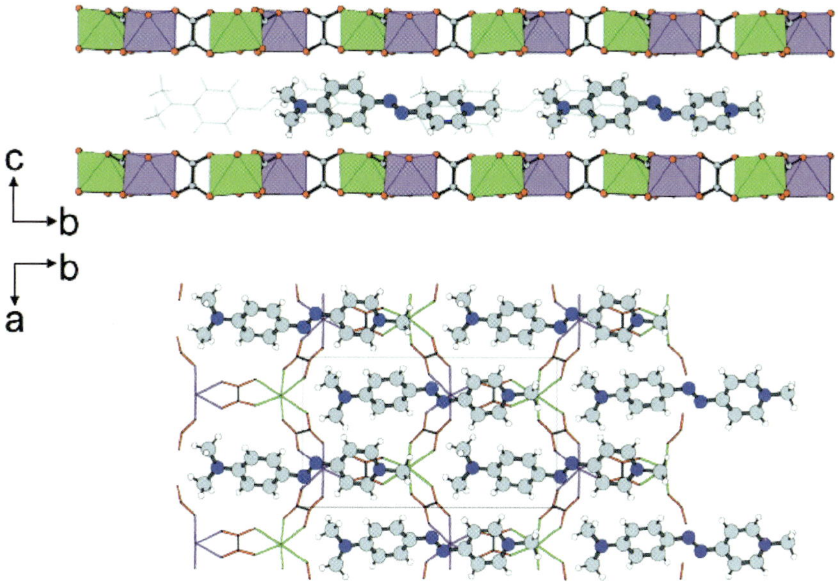

Fig. 8.7 Structure of the DAZOP: perspective view and top view of a layer showing the polar arrangement of the organic molecules

8.3
The Intercalation Compounds MPS_3

The past quarter of a century has witnessed substantial efforts in the development of intercalation chemistry, with the combination at the nanometer scale of molecular species and inorganic layered networks. The ability of the inorganic network to undergo redox or acido-basic reactions, and that of the intercalated organic molecule to undergo selective, controllable and reversible reactions, have originated attractive applications from a chemical viewpoint [37, 38]. This section is devoted to the MPS_3 series which received much attention, due to the unusual reactivity of these compounds and their exciting physical properties.

8.3.1
Ion-exchange Intercalation in MPS_3

The MPS_3 compounds have a $CdCl_2$ type structure well described by inorganic layers of M^{II} ions held together by $(P_2S_6)^{4-}$ bridging ligands (Fig. 8.8) [39]. They are suitable for intercalation chemistry that involves either redox or cation-exchange reactions. The basic observation of ion-exchange intercalation is the spontaneous reaction, at room temperature, of $MnPS_3$ with solutions of a number of ionic compounds such as KCl, tetramethylammonium or cobalticinium chloride. These reactions lead to compounds $Mn_{1-x}PS_3(A)_{2x}(H_2O)_y$ where the positive charge of the guest cations A^+ is counterbalanced by the loss of an equivalent charge of the inorganic layer [40, 41]. This process appears rather unusual, as it implies that manganese cations are able to leave their intralamellar sites under very mild conditions.

Bulky species can be inserted in two steps: insertion of hydrated alkali metal ions followed by exchange of the alkali ions with the bulky species. For example, potassium ions have been exchanged with $Ru(2,2'bipy)_3^{2+}$. Other intercalation processes are available, but the reaction often requires the assistance of a complexing agent coordinating the departing M^{II} cations [42]. Thus, $FePS_3$ inserts pyridinium or methylviologen cations under mild conditions, but the presence of EDTA is required to insert a tetraethylammonium species. The intercalation process afforded by the MPS_3 compounds is unique in the field of inorganic layered materials, and was described by Clément et al. as a local microdissolution and subsequent reconstruction of the host lattice, thanks to a heterogeneous equilibrium between solid MPS_3 and the constituting solvated species M^{2+}_{aq} and $P_2S^{4-}_{6aq}$ [43]. Note that ion exchange is not observed with $NiPS_3$ in which the M-S bonding is more covalent, but that the compound can intercalate electron donating species through a redox process.

8.3.2
Properties of the $MnPS_3$ Intercalates

X-ray powder diffraction studies of $MnPS_3$ intercalates show that cation exchange leads to a superstructure with tripling of the in-plane lattice parameter of $MnPS_3$ (Fig. 8.9) [41]. Additional features have been evidenced depending on the intercala-

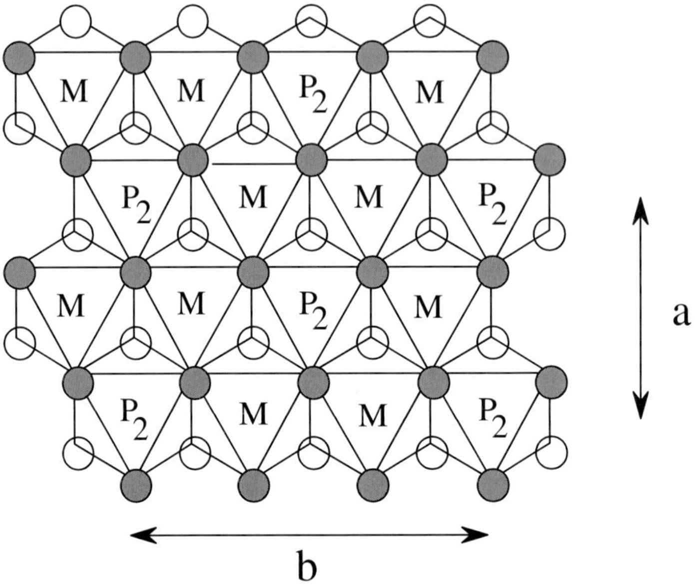

Fig. 8.8 Schematic top view of a single MPS_3 layer

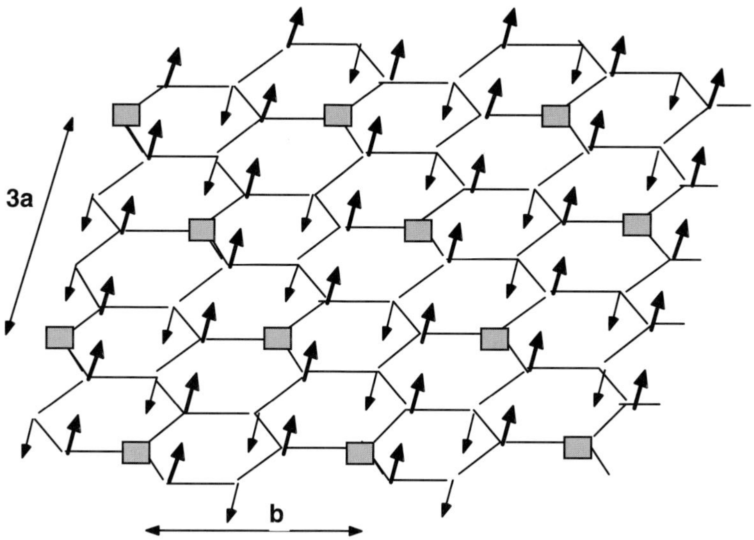

Fig. 8.9 Imbalanced spin arrangement in a ferrimagnetic host layer of the cobalticenium and tetramethylammonium $MnPS_3$ intercalates. Only the honeycomb lattice of spins is represented. The shaded squares represent the ordered manganese vacancies. The axes 3*a* and *b* of the superlattice cell are shown

ted cation [44]. Thus, in $Mn_{0.83}PS_3[Co(C_5H_5)_2]_{0.34}(H_2O)_y$, disorder in the layer stacking mode is observed, while in $Mn_{0.84}PS_3(Me_4N)_{0.32}(H_2O)_y$ the organic cation is much more ordered, resulting in an overall hexagonal symmetry with a unit cell parameter $a = 36.6$ Å.

The temperature-dependence of the magnetic susceptibility (measured with a Faraday magnetometer) of pure $MnPS_3$ and of several cation-exchanged intercalates is shown in Fig. 8.10 as $1/\chi = f(T)$ [40, 45]. Above 100 K, all the intercalates show a Curie-Weiss behavior with negative Weiss constant, which reveals that the interactions are antiferromagnetic (AF), and weaker than in MnPS3.

More striking results are observed at lower temperatures, where the susceptibility strongly increases below a critical temperature T_c ranging between 15 and 45 K. FC and ZFC magnetization measurements (Fig. 8.11) and narrow hysteretic effects (~200 Oe) are characteristic of the occurrence of ferrimagnetism or weak ferromagnetism in all the $MnPS_3$ intercalates.

Quite large remnant magnetization is observed, particularly for the cobalticenium intercalate $Mn_{0.83}PS_3[Co(C_5H_5)_2]_{0.34}(H_2O)_{\sim 0.3}$. The highest value of the saturation magnetization ($M_{sa} \sim 4000$ cm^3 Oe mol^{-1} of Mn) is reached for the tetramethylammonium and cobaltocenium intercalates, which value is however only a fraction of the theoretical value expected if all the spins were aligned (27900 cm^3 Oe mol^{-1} of Mn). Upon cycling the applied magnetic field, the intercalates show a narrow hysteresis loop characteristic of soft magnets.

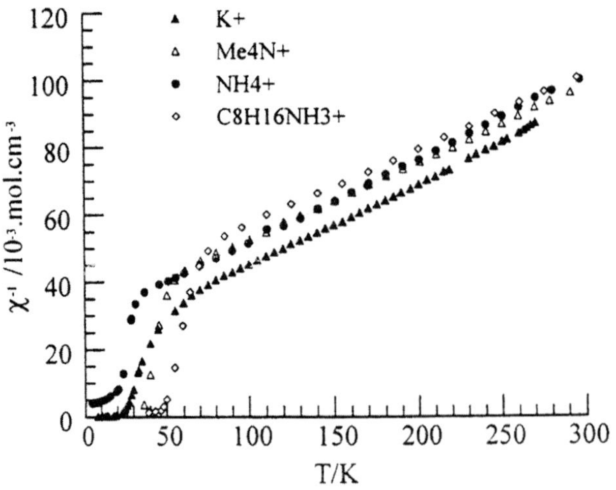

Fig. 8.10 Temperature-dependence of the reciprocal magnetic susceptibility χ^{-1} for different $MnPS_3$ intercalates

In addition, comparison between the variation of magnetization versus field parallel and perpendicular to the layers (Fig. 8.12) indicates a large anisotropy with an easy magnetization axis essentially perpendicular to the layers, as in pure $MnPS_3$.

The change from antiferro- to ferrimagnetic behavior upon intercalation into $MnPS_3$ is directly correlated to the structure of the cation exchanged derivatives showing a triple *a* axis compared to the parent $MnPS_3$ (see above). Similarly, powder neutron diffraction patterns recorded below T_c for $Mn_{0.83}PS_3(Co(C_5D_5)_{0.34}(D_2O)_y$ and $Mn_{0.84}PS_3[N(CD_3)_4]_{0.32}$ $(D_2O)_y$ [46] exhibit *hk0* magnetic reflections in the tripled *a* axis unit cell. The observation of only *hk0* reflections suggests that the magnetic order takes place mainly within each layer. Thus, the ferrimagnetic behavior of the exchanged products is strongly connected to the superstructure, and related to the ordering of the metal.

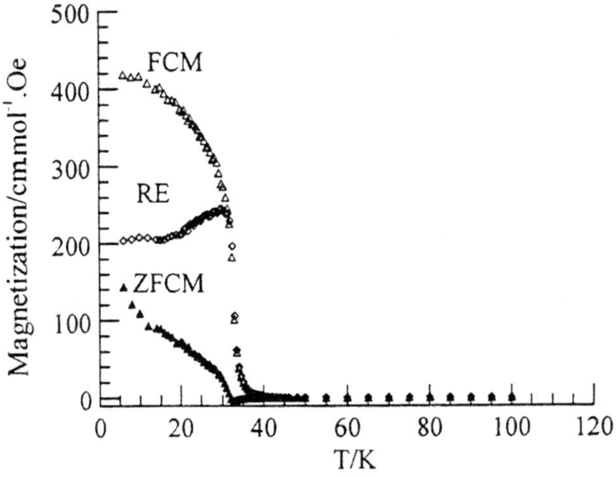

Fig. 8.11 Temperature dependence of the field-cooled magnetization (FCM), remanent magnetization (RE), and zero-field-cooled magnetization (ZFCM) of the $Mn_{0.8}PS_3(Me_4N)_{0.4}$ intercalate in a field of 30 Oe

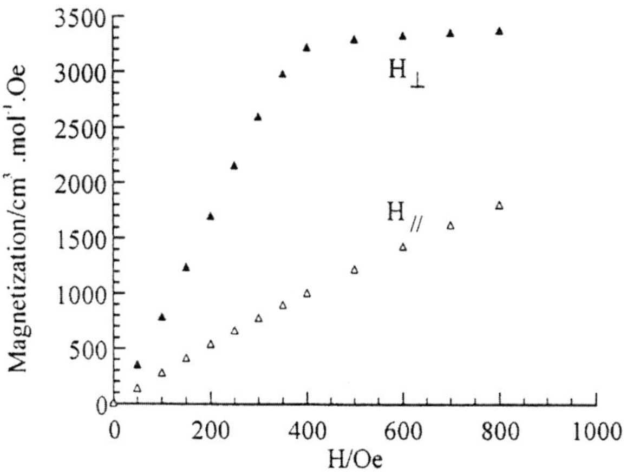

Fig. 8.12 Dependence of the magnetization at 10 K of a single platelet of the $Mn_{0.8}PS_3(Me_4N)_{0.4}$ intercalate on a magnetic field applied perpendicular or parallel to the plane of the platelets

Focusing now on the tetramethylammonium and cobaltocenium salts (Table 8.2), the exchange process leads to the loss of approximately one sixth of the metal ions, with an ordering of the metal vacancies. This arrangement results in a saturated magnetization of $1/6(Ng\ \mu_B\ S)$, i.e. 4000 cm^3 Oe mol^{-1} for the isotropic Mn^{II} system, which is very close to the experimental value.

Close examination of Table 8.2 shows that some intercalates exhibit lower value of M_{sat}. In fact, the saturated magnetization depends on the extent of metal loss and on the perfection of ordering of the created vacancies. In this model, indeed, no net magnetization is expected for randomly located vacancies. For instance, in octylammonium intercalate $Mn_{0.89}PS_3(n\text{-octyl-}NH_3)_{0.22}(H_2O)_{0.5}$ a much smaller metal loss is observed and no super-structure arises. Correlation with X-ray diffraction studies indicates that a small value of M_{sat} is characteristic of vacancy disorder.

The lowering of the ordering temperature after cation exchange is another feature of the intercalates. Actually, it is well known that the critical temperature in 2D magnetic systems is related to the number of interacting neighbors Z and the in-plane interaction J [47]. In the exchanged compounds, Z is lowered, and accordingly, the critical temperature is expected to be lower than 78 K as long as the interaction remains constant. It is worth noticing that higher T_c magnets were not reported in the $MnPS_3$ derivatives.

Tab. 8.2 Basal spacing, saturated magnetization at 10 K and critical temperature of $MnPS_3$ intercalates

Compounds	Spacing (Å)	M_S (Oe cm^3 mol^{-1})	T_c (K)
$Mn_{0.84}PS_3(Me_4N)_{0.32}(H_2O)$	11.45	4000	35
$Mn_{0.83}PS_3(CoCp_2)_{0.34}(H_2O)_{0.3}$	11.82	4000	35
$Mn_{0.81}PS_3(K)_{0.38}(H_2O)$	9.37	1000	20
$Mn_{0.89}PS_3(\text{octyl-}NH_3)_{0.22}(H_2O)_{0.5}$	10.38	1150	45
$Mn_{0.86}PS_3(pyH)_{0.28}(H_2O)_{0.7}$	9.65	80	35
$Mn_{0.80}PS_3(NH_4)_{0.4}(H_2O)$	9.38	3100	15

The methylviologen intercalate $Mn_{0.82}PS_3(MV)_{0.18}(H_2O)_y$ has a significantly different behavior, the magnetic susceptibility showing a sharp peak at 32 K, without remnant magnetization at low temperature. It was suggested that the increase of magnetization by cooling down to T_c is due to ferrimagnetism within the slabs, the decrease below 32 K being related to antiferromagnetic ordering between adjacent magnetic layers.

8.3.3 Properties of the $FePS_3$ Intercalates

When compared with $MnPS_3$, $FePS_3$ appears attractive because magnetic interactions are stronger ($T_N = 123$ K), and the electronic structure of the Fe^{II} is much more anisotropic and causes the magnetic behavior of $FePS_3$ to be Ising-like [48, 49]. However, ion-exchange intercalation is less easy because of a weaker ability of the lattice to release Fe^{II} ions, possibly due to a larger crystal field stabilization energy. Nevertheless, several iron deficient $Fe_{1-x}PS_3G_{2x}(solv)_y$ intercalates have been synthesized, with G = cobalticenium, tetraethylammonium, 4-picolinium, 3,5-lutidinium or N-methyl pyridinium, but most of them order antiferromagnetically. We focus here on the $FePS_3$ intercalates involving pyridinium or methylviologen guest cations which exhibit a net magnetization at low temperature.

For both compounds, the unit cell is similar to that of $FePS_3$, with an expansion of the lattice in the *c* direction of about 3.3 Å, because of the size of the guest species [50].

The temperature dependence of the field cooled magnetization ($H = 10$ Oe) of the pyridinium intercalate $Fe_{0.88}PS_3(pyH)_{0.24}(solv)_y$ [50] shows a rapid increase of the magnetization when T decreases below 100 K. From the results of the field dependent magnetization, it is deduced that a ferromagnetic-like state is stabilized with strong remnant magnetization and large hysteretic effect below the critical temperature $T_c = 90$ K.

A striking feature is observed in the $M(H)$ curves of $Fe_{0.88}PS_3(pyH)_{0.24}(solv)_y$ recorded between 40 K and T_c, as illustrated in Fig. 8.13. When the temperature is close to T_c, the magnetization increases rapidly for low values of the applied field, then reaches a saturation regime with a constant slope. However, as the temperature is lowered, a threshold applied field is needed before the magnetization increases. In other words, for a given magnetization to be reached, it is necessary to apply increasing values of the field as the temperature is cooled down, indicating a magneto-crystalline anisotropy field which depends strongly on the temperature.

Magnetic measurements have further been carried out close to T_c on single platelets oriented parallel or perpendicular to the applied field (Fig. 8.14) [50]. When the platelet is set up parallel to the applied field, the magnetization has a very weak variation, while a rapid increase occurs above ca. 800 Oe in the perpendicular orientation. This strongly suggests that the axis of easy magnetization is nearly perpendicular to the layers.

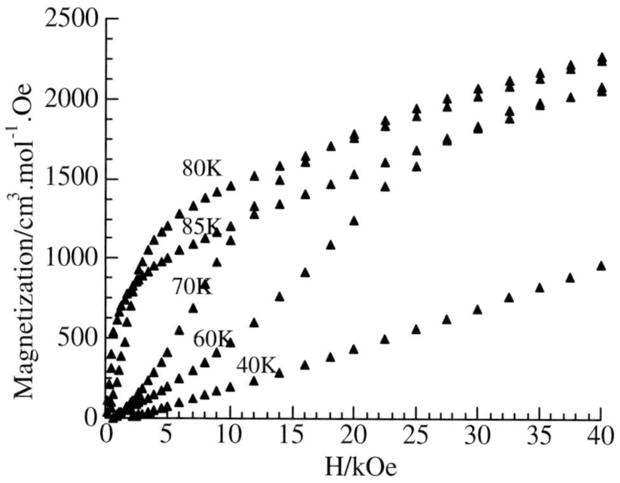

Fig. 8.13 Variation of the magnetization of $Fe_{0.88}PS_3(PyH)_{0.24}$ on applied magnetic field at different temperatures

Very similar behavior is evidenced for the methylviologen salt $Fe_{0.83}PS_3(MV)_{0.14}$ [50], with a lower critical temperature $T_c = 77$ K. Measurement of the field-dependent magnetization below T_c (Fig. 8.15) shows that it reaches more rapidly its saturation value (ca. 2000 cm^3 Oe mol^{-1}) which is about 10% of the expected value for aligned Fe^{II} spins. Moreover, the M(H) curve at 70 K exhibits a hysteresis loop with a width of ca. 200 Oe.

As already pointed out for the $MnPS_3$ derivatives, the lowering of T_c arises from a "spin dilution" effect due to the large amount of iron vacancies in $Fe_{0.83}P_{0.99}S_3(MV)_{0.14}$. Dilution of Fe^{II} ions is also thought to be responsible for the smaller anisotropic effect observed for the MV compound.

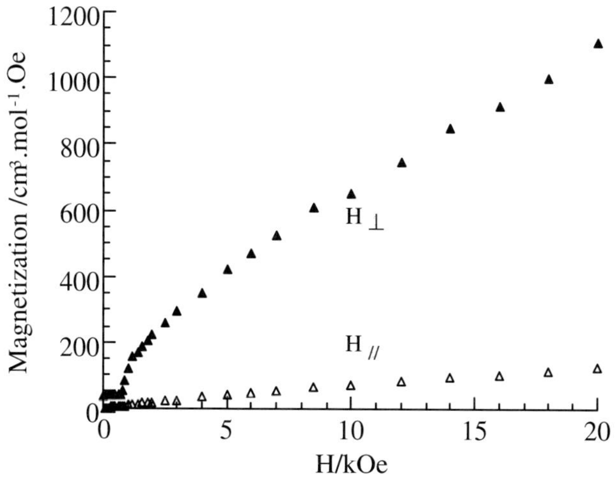

Fig. 8.14 Dependence of the magnetization of a single platelet of $Fe_{0.88}PS_3(PyH)_{0.24}$ on a magnetic field applied perpendicular or parallel to the magnetic planes, at 80 K

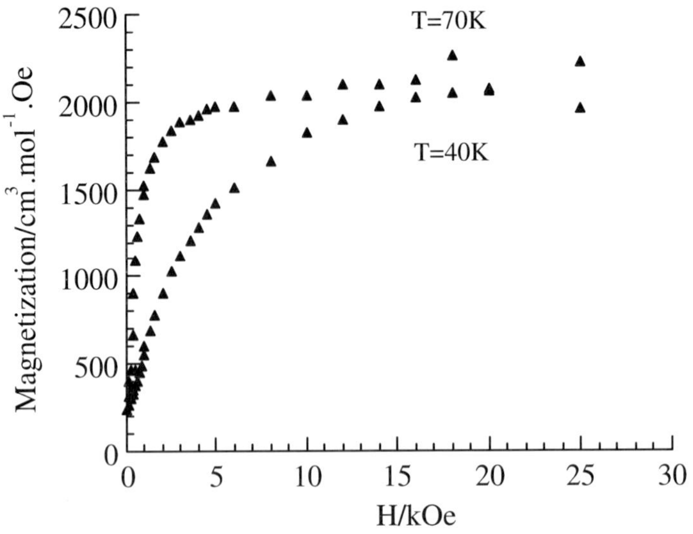

Fig. 8.15 Field dependence of the magnetization of $Fe_{0.83}PS_3(MV)_{0.14}$ at different temperatures

8.3.4
Magnetism and Non-linear Optics

The intercalation chemistry of optically transparent $MnPS_3$ compounds is appealing for the association of a host layered network with a net magnetization and a guest organic chromophore generating second-harmonic radiation. In this respect, the $Mn_{0.86}PS_3(DAMS)_{0.28}$ intercalate has been synthesized, where DAMS is the dimethylaminostilbazolium hyperpolarizable cation shown in Fig. 8.16 [51].

Although the $MnPS_3$ host lattice is centrosymmetric, this intercalate has a high efficiency for second harmonic generation, about 300 times that of urea ($\lambda = 1.34\ \mu m$), and becomes a magnet below 40 K. $Mn_{0.86}PS_3(DAMS)_{0.28}$ has been the first example, showing the coexistence of both NLO properties and spontaneous magnetization at quite high temperature ($T_c \approx 40$ K). Note that other stilbazolium chromophores reported more recently give similar effects. In such systems, the formation of very large *J*-type aggregates of the chromophores explains their non-centrosymmetric arrangement in the galleries [52]. Unfortunately, no real coupling between magnetism and NLO properties has been evidenced to date.

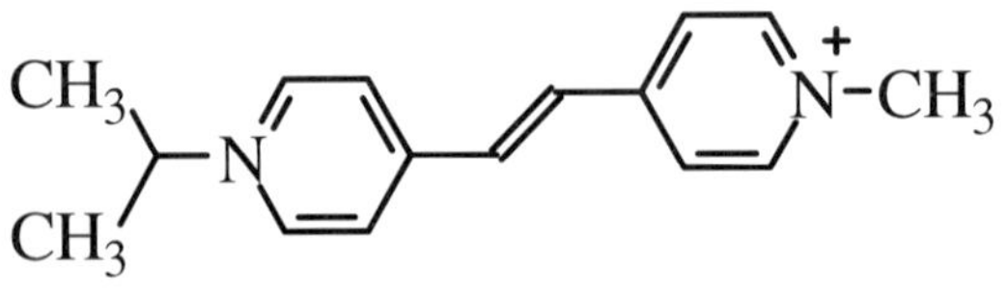

Fig. 8.16 Structure of the dimethylaminostilbazolinium (DAMS) hyperpolarizable cation

8.4 Covalently Bound Organic-inorganic Networks

Hybrid layered solids consisting of covalently bonded organic and inorganic layers are appealing from several points of view: (i) to get stable and well crystallized compounds for structural studies; (ii) to provide archetypical low-dimensional magnets; (iii) to promote, thanks to the hybridization of the organic and inorganic components, a synergy between the properties of both subnetworks. It is to be noted that the latter is a major target today in many laboratories. Examples are provided by the metal phosphonates, *i.e.* $M[C_nH_{2n+1}PO_3] \cdot H_2O$ and $M_2[O_3P{-}(CH_2)_n{-}PO_3]\ 2\ H_2O$, (M = Cu, Mn etc.), and the layered metal hydroxides $M_2(OH)_3X$, where X is an aliphatic carboxylate, in which the organic moieties are covalently bonded to the metal ion, thus favoring the thermal stability of the systems.

8.4.1 Divalent Metal Phosphonates

In the layered metal phosphonates $M[RPO_3] \cdot H_2O$ and $M_2[O_3P{-}(CH_2)_n{-}PO_3]\ 2\ H_2O$, ($R = C_nH_{2n+1}$, C_6H_5; M = Cu, Mn etc.), the metal ions are bridged by the oxygen atoms of the phosphonate ligands, and form sheets separated one from the other by the organic moieties [53]. The phosphonate ligands are particularly interesting, since they can be functionalised by various active groups, such as amine, carboxylic thio or phosphonic acid, $-NH_2$, $-CO_2H$, $-SH$, $-PO_3H_2$, to combine the physical properties of the individual subnetworks [54].

The crystal structure of $Mn[RPO_3] \cdot H_2O$ ($R = C_nH_{2n+1}$ and C_6H_5) was reported in 1988 by Cao et al. [55]. Two of the phosphonate oxygens chelate the metal atom, and further bridge across adjacent atoms in the same row. The third phosphonate oxygen bridges to an adjacent row, thus giving a layer structure (Fig. 8.17). The sixth position is occupied by the oxygen of the water molecule. The organic layer is made by the organic *R*-groups, which extend through the C–P bond perpendicular to the inorganic planes.

Similar structures have been observed for the Fe(II) derivatives, $Fe[C_2H_5PO_3] \cdot H_2O$ [56] and $Fe[C_6H_5PO_3] \cdot H_2O$ [57]. In turn, in $Co[(CH_3)_3C{-}PO_3] \cdot H_2O$ [58] the cobalt ions are located in tetrahedral and octahedral sites within layers. One cobalt is coordinated by three phosphonate oxygens and three water molecules, while tetrahedral coordination is observed for the other two cobalt positions, bonded only to the oxygens of the phosphonate ligand. The $(CH_3)_3C$- groups occupy the interlamellar space, roughly perpendicular to the corrugated layers.

The Co(II) phosphonate $Co_3[(O_2C{-}(CH_2)_2{-}PO_3]_2$ reported by Rabu et al. [59] has a pillared structure similar to that of $Zn_3[(O_2C{-}C_2H_4{-}PO_3]_2$ [60]. The organic groups extend perpendicular to the inorganic layers and are linked together via carboxylic moieties, thus leading to a 3D network.

The copper(II) phosphonates show a different coordination mode of the metal ion. The structure is still layered, but the copper ions are five coordinated and bridged by phosphonate groups, giving copper dimers within the plane [61, 62]. The layer structure of $Cu[C_6H_5PO_3] \cdot H_2O$ [63] is similar to that of the methyl-phosphonate (Fig. 8.18).

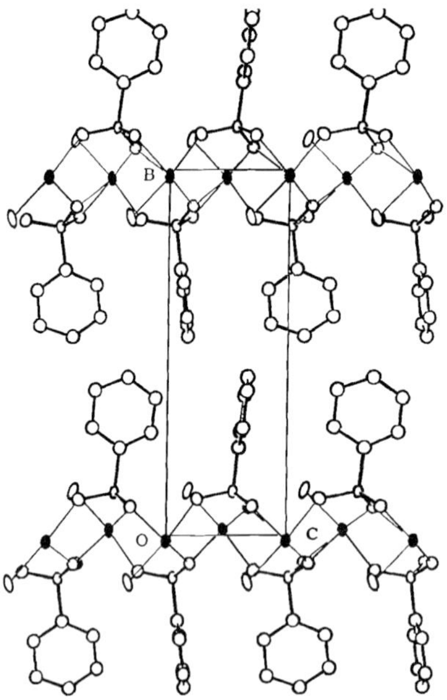

Fig. 8.17 The crystal structure of $M[C_6H_5PO_3]$ H_2O, M=Mg, Mn, viewed down the axis, showing the arrangement of the organic group

The layered $Mn[RPO_3] \cdot H_2O$ series has probably been the most extensively studied from a magnetic point of view [64, 65]. The high temperature regime indicates the presence of AF nearest-neighbour interactions within the Mn(II) layers, with an exchange constant in the range 2.5–3.0 K, while, below $T_N \approx 15$ K, a long-range AF 3D order takes place [64].

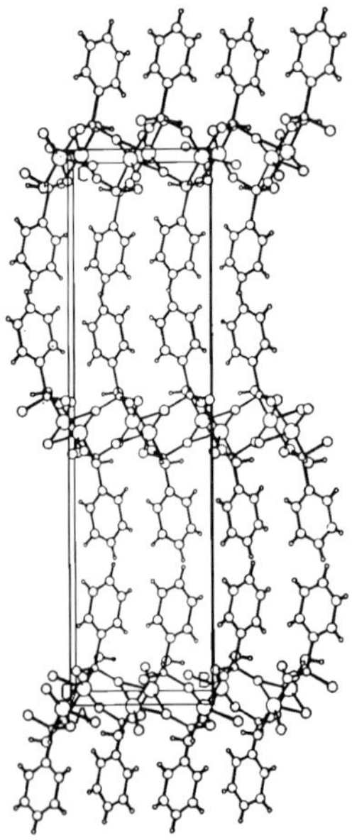

Fig. 8.18 The layer arrangement in $Cu[C_6H_5PO_3] \cdot H_2O$, as viewed down the *a*-axis

On close examination, it appears in fact that the magnetic susceptibility depends on the applied field, as observed for $Mn[C_4H_9PO_3] \cdot H_2O$ from field-cooled and zero-field cooled magnetization experiments (Fig. 8.19). This behavior, together with the field dependence of the magnetization, points to the stabilization of a non-colinear structure – namely a canted antiferromagnetic order – below T_N, due to the low crystalline symmetry of these compounds [64, 66]. This was confirmed by an antiferromagnetic resonance technique, permitting deduction of the canting angle from the frequency and field dependence of the AFMR [65a].

Furthermore, Langmuir-Blodgett films of Mn(II) octadecyl-phosphonate and Mn(II)4-(4′-tetradecyl-oxyphenyl)-butyl-phosphonate ordering at 13.5 K and 14.8 K, respectively, were isolated by Talham and coworkers [67]. Both show a canted AF order at low temperature.

The most recent studied phosphonate series refers to the Fe(II) ion [68–71]. Unfortunately, few crystal structures have been reported. All the compounds exhibit a long-range magnetic order at quite high temperature (21.5 to 26.0 K), and a T_N value which decreases slowly when the interlayer spacing becomes larger (see Table 8.3).

It is to be noted that the hysteretic effect observed in $Fe[C_6H_5PO_3] \cdot H_2O$ [71] (Fig. 8.20) is unusually large for this kind of molecule-based magnet. To our knowledge, only the cobalt(II) hydroxide-based magnets (see Sect. 8.4.2) exhibit higher coercive fields, which may be explained by the presence of very strong local distortions of the metal sites.

Furthermore, it can be emphasized that the critical temperature varies weakly with the interlayer distance, unlike what is expected in exchange-coupled systems.

Focusing now on the cobalt phosphonate $Co_3[O_3P–(CH_2)_2CO_2]_2$, AC and DC magnetic measurements as well as specific experiments show that the structure is again canted, with an AF long-range order at $T_N = 15.5$ K [59].

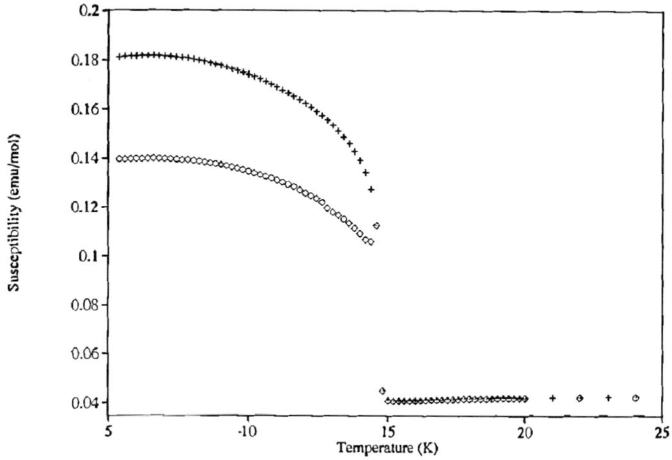

Fig. 8.19 ZFC (◊) and FC (+) χ vs T plots in a field of 100 G for $Mn[C_4H_9PO_3] \cdot H_2O$ in the temperature range 5–25 K

Tab. 8.3 Structural and magnetic parameters for Iron(II) phosphonates

Compound	T_N (K)	θ (K)	T_N/θ	Space group	Basal spacing (Å)
$(NH_4)FePO_4 . H_2O$	26.0	–65	0.40	$Pmn2_1$	8.8213(2)
$Fe[CH_3PO_3]\cdot H_2O$	25.0	–59	0.42	$Pna2_1$	8.77(1)
$Fe[C_2H_5PO_3]\cdot H_2O$	24.5	–43	0.57	P1n1	10.33(1)
$Fe[C_4H_9PO_3]\cdot H_2O$	25	–45	0.56		
$Fe[C_6H_5PO_3]\cdot H_2O$	21.5	–56	0.38	$Pmn2_1$	14.453(2)
$Fe_2[O_3P(CH_2)_2PO_3]\cdot 2H_2O$	25.0	–52	0.48		

The divalent metal diphosphonates, $M_2[O_3P–R–PO_3] \cdot 2H_2O$ (M = Co, Cu; R = CH_2, C_2H_4 etc.) show pillared structures. In both $Cu_2[O_3P–C_2H_4–PO_3(H_2O)_2]$ and $Cu_2[O_3P–C_3H_6–PO_3(H_2O)_2] \cdot H_2O$ the copper (II) ion is five-coordinated, four of the binding sites being oxygens from the phosphonates, the last one from water molecules [72]. The structure of $Co_2[O_3P–CH_2–PO_3] \cdot H_2O$ determined by single-crystal X-ray diffraction [73] contains cobalt atoms in tetrahedral and octahedral symmetry. The most striking feature is the presence of a microporous structure, with channels running along the *c*-axis; this compound is paramagnetic down to 5 K.

Among metal(II) *bis*-(phosphonates) derivatives, only the *bis*-hydrate form of the nickel(II) methylene-*bis*-(phosphonates) exhibits a ferromagnetic long-range order at T_c = 3.8 K [74].

8.4.2
Hydroxide-based Layered Compounds

Anion-exchangeable layered compounds appear well-adapted to favor covalent links between the organic and inorganic constituents, because the anions are bonded to the metal ions and form the framework of the crystal. This was achieved in the

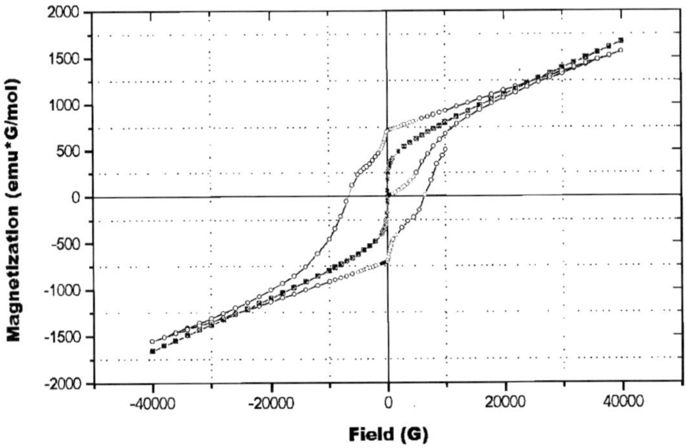

Fig. 8.20 Hysteresis loop for $Fe[C_6H_5PO_3]$ H_2O measured at T = 10 K

Botallackite-type layered compounds, $Cu_2(OH)_3X$ ($X = NO_3^-$, CH_3COO^-, etc.), whose structure, illustrated in Fig. 8.21 for $X = NO_3^-$, shows the copper(II) hydroxy layers and the X anions arranged in a zip-like fashion [75]. It is similar to the well-known structure of $Cd(OH)_2$, from which $Cu_2(OH)_3X$ is built by replacing periodically one-fourth of OH^- with X^- anions. Two crystallographically distinct Cu^{II} ions are evidenced lying in 4 + 2 and 4 + 1 + 1 elongated octahedra. From magnetic studies of $Cu_2(OH)_3(NO_3)$, there is clear evidence of AF in-plane interactions, and a long-range magnetic ordering below T_c =12 K [76]. Recent experimental and theoretical studies of the electronic (see Fig. 8.21) and spin densities on the nitrate groups confirm the 3D character of the long-range order [77].

The parent cobalt(II) compound, isostructural with the above compound, has a totally different magnetic behavior [78]. The temperature dependence of the χT product, and the field-dependent magnetization, are characteristic of ferromagnetic in-plane interactions. Clearly, the magnetic findings indicate two regimes: above the critical temperature T_c, ferromagnetic in-plane interactions dominate, while at lower temperature the interlayer interactions are no longer negligible, and a long-range AF order takes place.

Finally, if interlayer interactions are needed for the occurrence of long-range 3D ordering, the divergence of the in-plane correlation length ξ close to T_c appears to be the driving force, as given by the relationship $k\ T_c \sim \xi^2 j S^2$ [79].

8.4.2.1 **Anion-exchange Reactions**

Exchange of interlamellar anions by organic species (dicarboxylic acid anions) was first reported by Miyata and Kumura [80]. The interlayer arrangement of anionic surfactants has been extensively studied in layered double hydroxides (LDH's) $[M^{II}_{1-x}M^{III}_x(OH)_2]^{x+}A^-_x$, made of positively charged hydroxide layers separated by

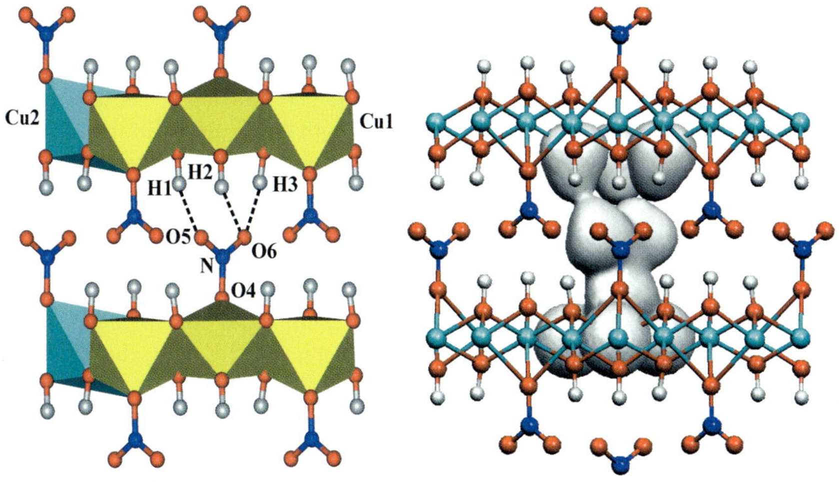

Fig. 8.21 View of the structure $Cu_2(OH)_3(NO_3)$ and electronic densities, showing the hydrogen bond interactions between adjacent layers

A^- anions counter-balancing the charge of the layers [81]. Likewise, Yamanaka et al. have investigated the exchange reactions in the layered compound $Cu_2(OH)_3(CH_3COO) \cdot H_2O$ in view of isolating highly active oxidation catalysts [82].

The acetate ion is totally exchanged after a few hours, giving X-ray diffraction patterns with intense (00l) reflections, in agreement with layered structures. The other reflections (*hkl* with h and $k \neq 0$) are much smaller, and exhibit the usual enlargement for disordered pillared compounds.

The use of *n*-alkyl sulfate or *n*-alkyl carboxylate anions appears to be convenient for tuning the basal spacing over a wide range of values [83,84], as in the $CuCl_4(n\text{-}C_mH_{2m+1}NH_3)_2$ series, whose properties are briefly discussed in Sect. 8.2.1 [6].

In such hybrid systems, the basal spacing is related to the carbon chain length m through the relationship $d(\text{Å}) = d_0 + \eta\,(1.27\,m \cos\theta)$, where $\eta = 1$ or 2, depending on the chain packing, and θ is the angle of the chains. The distance d_0 involves (i) the size of the bridging group, (ii) the van der Waals distance between either facing methyl groups or methyl groups and hydroxide layers, (iii) the thickness of the inorganic layer. For a homogeneous series, the molecular area of the chains being constant whatever the m value, a variation of d_0 will point to a change of the inorganic layer thickness. Likewise, a change of the tilt angle θ will usually be related to a slight structural deformation of the coordination polyhedra, and accordingly of the structure of the layers.

Basically, due to the nature of the chemical bond between the organic and inorganic sub-networks, such compounds are expected to differ from the LDHs. In particular, the connection between both subnetworks may have a strong influence on the bulk magnetic properties.

8.4.2.2 **Influence of Organic Spacers**

The $Cu_2(OH)_3X$ Series

Various intercalation compounds have been prepared by ion-exchange reaction with *n*-alkyl sulfate anions, corresponding to the general formulation $Cu_2(OH)_3(n\text{-}C_mH_{2m+1}SO_4) \cdot z\,H_2O$, with $z = 0$ or 1. The variation of the basal spacing increases linearly with the aliphatic chain length m, according to the relationship $d(\text{Å}) = 12.01 + 1.27m$. This implies unambiguously that the *n*-alkyl chains are organized in a zip-like fashion (monolayers) and most likely orientated normal to the inorganic layers (Fig. 8.22). In this configuration, the terminal methyl groups of the alkyl chains exhibit hydrophobic interaction with the hydrogen atoms of the hydroxide layers.

Likewise, the substitution of $n\text{-}C_mH_{2m+1}COO^-$ with $m \geq 4$ for CH_3COO^- gives two series of hybrid compounds (denoted α and β), depending on the synthesis conditions [83]. The α-type compounds are hydrated with the formulation $Cu_2(OH)_3(n\text{-}C_mH_{2m+1}COO) \cdot zH_2O$ (with $z = 0.3 - 0.5$), while the β-type compounds are anhydrous. The studied varieties show the same $Cu/OH/C_mH_{2m+1}COO$ ratios.

The variation of the basal spacing with the carbon chain length m for both carboxylate series, $d(\text{Å}) = d_0 + 2.54\,m \cos\theta$, agrees with a bilayer packing of the alkyl chains and a tilt angle $\theta = 22°$. The difference between d_0 values (about 5.4 Å) sug-

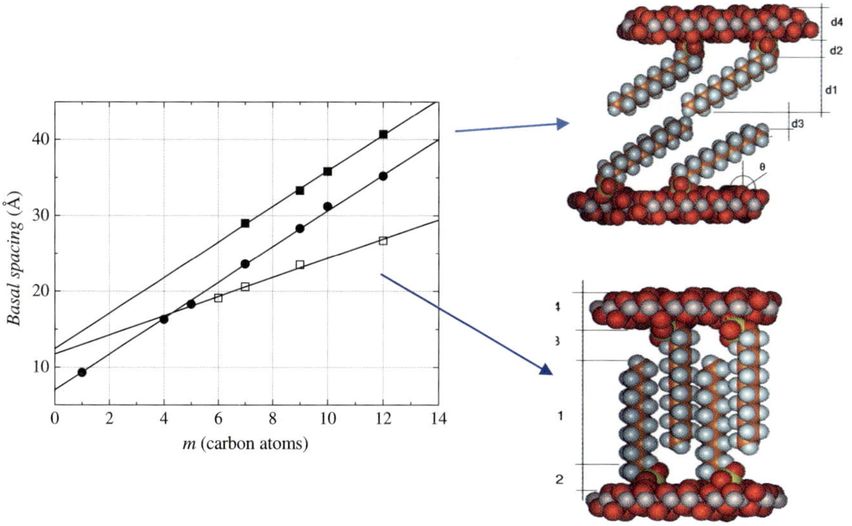

Fig. 8.22 Variation of the basal spacing with *m* (number of carbon atoms) in the copper(II) *n*-alkyl carboxylate (full squares and circles) and *n*-alkyl sulfate (open squares) series. The packing of alkyl chains is illustrated for both series

gests different structures for the inorganic layers. The d_0 value obtained for the α series agrees well with the layer thickness in copper(II) hydroxide nitrate [76], and accordingly a Brucite-like structure may be assumed. In turn, the β derivatives have a d_0 value very similar to that of *n*-alkyl sulfates, suggesting a large deviation from the Brucite layers. Clearly, the lack of crystal structure for these series is a limiting factor, but IR spectroscopy provides some insight, such as the coordination mode of COO^- groups with the metal ions and the neighboring OH^- ions. In fact, the character of the COO groups is shown to differ in the two series. Two carboxylate doublets are superimposed in the α derivatives, one with a unidentate character, the other with a bridging character. The former disappears completely in the β derivatives.

For more insight on the local structures, Cu K-edge EXAFS spectra of $Cu_2(OH)_3(NO_3)$ and $Cu_2(OH)_3(n\text{-}C_mH_{2m+1}COO)$ were achieved [85]. The comparison of the spectra indicates that the local structures of $Cu_2(OH)_3(NO_3)$ and the $m = 1$ short chain compound (α-type) are very similar, while significant differences are noticed for the β-type long chain materials. The change seems to result from the so-called "chemical pressure" produced by alkyl carboxylate chains, which causes the structural modification in the $[Cu_2(OH)_3]^+$ layers.

The magnetic susceptibility $\chi(T)$ of the β-type Cu(II) hydroxy carboxylate with $m = 10$, shown in Fig. 8.23, is representative of the long-chain compounds. M/H measured by cooling the sample in a DC field, and the AC susceptibility (dM/dH), recorded as 3.5 Oe, increase strongly on decreasing temperature and reach, respectively, a plateau or a sharp maximum, both limited by demagnetizing effects at a temperature close to 20.5 K where ferromagnetic order sets in. In the following, we limit the discussion to the high-temperature range, where the correction due to the demagnetizing field is small and can be neglected.

The existence of a magnetically ordered state is confirmed by the occurrence of an hysteretic effect with *Hc* as large as 1500 Oe at $T = 2$ K. There are several indications, including the magnitude of the spontaneous magnetization or the behavior of magnetization in high fields, which point towards a ferrimagnetic or a noncollinear spin configuration. A characteristic feature is the presence of a well-marked minimum in the temperature dependence of χT vs. T around 75 K, in the paramagnetic regime well above T_c (see inset in Fig. 8.23). This minimum is well fitted by the superposition of two exponentials: a high-temperature AF contribution which vanishes at absolute zero, and a low temperature ferromagnetic one, describing the strong variation in the temperature range 50–300 K. Note that the latter is fully justified for a 2D Heisenberg ferromagnet, whose low-temperature behavior is given by $\chi T = \exp(4\pi J S^2/kT)$ [86] where J is the in-plane exchange constant. Therefore, the data have been fitted with the expression [87]:

$$\chi T = C_1 \exp(\alpha J/kT) + C_2 \exp(\beta J/kT).$$

The best values of the parameters are $C_1 = 0.94$, $C_2 = 0.04$, $\alpha J = -58.1$ K and $\beta J = 116.2$ K, the sum $C_1 + C_2$ being nothing but the high temperature Curie constant. The driving interaction responsible for the initial high-temperature decay of

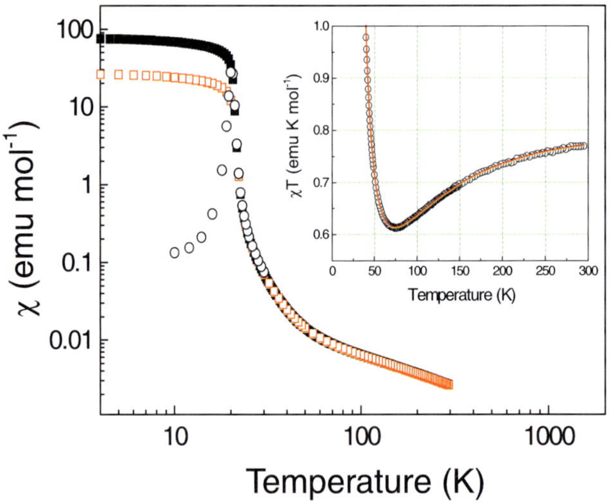

Fig. 8.23 Magnetic susceptibility *vs* temperature for $Cu_2(OH)_3(n\text{-}C_{10}H_{21}CO_2)$ measured in DC fields of 10 Oe (solid squares) and 50 Oe (open squares) and AC fields of 3.5 Oe (circles). The inset shows the $\chi T = f(T)$ plot in the paramagnetic regime, with the minimum characteristic of a ferrimagneticlike behavior

χT is antiferromagnetic (negative αJ). It concerns most of the moments and is attributed to the dominant in-plane interactions. In turn, the low-temperature behavior of χT is dominated by the second term that is ferromagnetic-like.

The $Co_2(OH)_3X$ Series

The parent cobalt(II) compounds exhibit very similar structure-property relationships. The compounds with X = n-$C_mH_{2m+1}COO)_x \cdot zH_2O$ (m = 1, 7, 9, 10, 12) or n-$C_mH_{2m+1}SO_4)_x \cdot zH_2O$ (m = 6, 9, 12) show intense (001) reflections, in agreement with layered structures. As might be anticipated, the basal spacing is closely related to the n-alkyl chain length, while the in-plane parameters remain nearly constant [83,88]. All the compounds show hexagonal symmetry, with space group P3m, similarly to $Co(OH)_2$.

Focusing on the variation of the basal spacing with the length of the coordinating anion, $d(Å) = 17.98 + 0.781\,m$, a monolayer structure of the organic chains is deduced, corresponding to a tilt angle of the chains, $\theta = 52°$, with respect to the c axis.

Owing to the chemical composition of the studied compounds, namely $Co_5(OH)_8(C_mH_{2m+1}COO)_2 \cdot z'H_2O$ (z' = 2.5 to 3.8), it can be stated that the structure derives from that predicted for the zinc analogue $Zn_5(OH)_8(NO_3)_2$ [89, 90] illustrated in

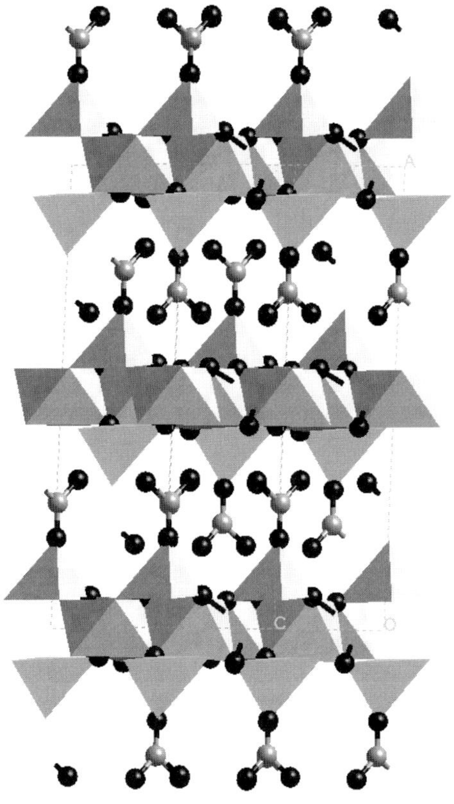

Fig. 8.24 Structural model for $Zn_5(OH)_8(NO_3)_2$ showing the packing of NO_3 groups in the interlayer spacing

Fig. 8.24. The presence of cobalt(II) in T_d and O_h sites is further confirmed from UV-visible spectra [88]. In such a model, the aliphatic chains are connected through the carboxylate moieties to the Co(Td) metal ions, and occupy the space in between the cobalt layers.

For the *n*-alkyl sulfate series the variation of the basal spacing with *m*, $d = 14.02 + 0.88m$, agrees with a monolayer stacking of the chains and a tilt angle $\theta = 46°$. The thickness of the inorganic layer is shown to be 8.6 Å, to be compared to ca. 4 Å for the Brucite-like layers [91], indicating that the structure likely involves, as above, cobalt(II) ions in O_h and T_d surroundings.

The cobalt(II) compounds with long chain anions, $C_mH_{2m+1}SO_4^-$ or $C_mH_{2m+1}COO^-$, exhibit very similar magnetic behaviors. The χT variations upon cooling down show a sharp increase, pointing to a ferro- or ferrimagnetic ground-state (Fig. 8.25). The maximum χT value rises from ca. 55 emu K mol^{-1} to 220 emu mol^{-1} as the basal spacing increases from 12.7 Å ($n = 1$) to 27.4 Å ($n = 12$). At very low temperature, the susceptibility depends closely on the applied field; the more the basal spacing increases, the more this effect is pronounced. Characteristic hysteretic effects are observed in the magnetization curves for the long chain compounds, with a large spontaneous magnetization and coercive field (inset in Fig. 8.25). Owing to the structural and magnetic findings, it can be thought that the ground-state is ferrimagnetic-like, due to the non compensation of spin sub-networks corresponding to cobalt(II) ions in O_h and T_d symmetries.

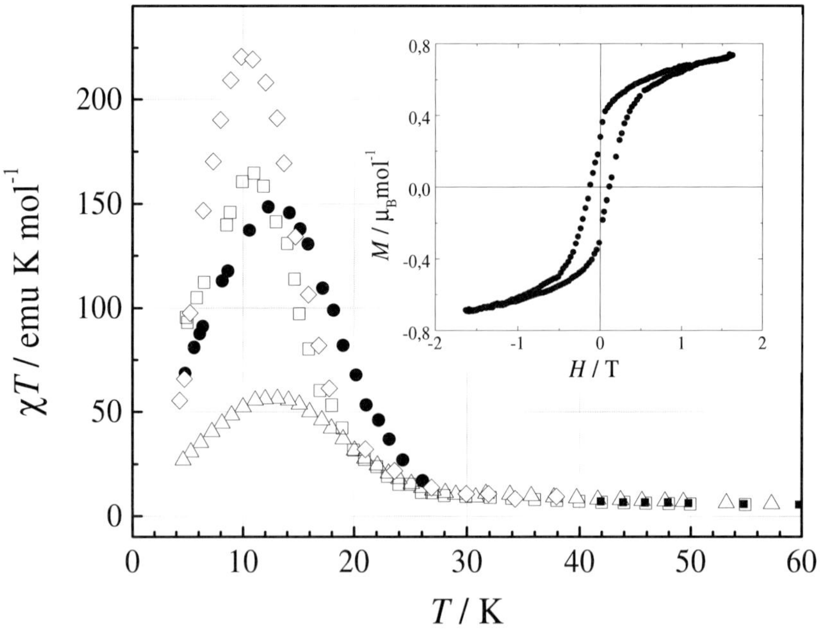

Fig. 8.25 Temperature dependence of χT for the layered cobalt(II) *n*-alkyl carboxylates with different *m* values (1 (triangles), 7 (circles), 9 (squares) and 12 (diamonds))

Finally, it is pointed out in the above compounds that for ferro- or ferrimagnetic correlations within the layers, the situation depends to a large extent on the interlayer spacing. For small spacing (less than ca. 10 Å), the through-bond interlayer interactions stabilize a 3D AF order. When the spacing is made larger, the direct exchange between unpaired electrons in adjacent layers becomes negligible. Nevertheless, the compounds exhibit a spontaneous magnetization and a characteristic hysteresis cycle. Clearly, the large T_c values and the weak dependence on the basal spacing can hardly be related to quantum interlayer interactions. So, the question which arises deals with the dimensionality of the magnetic network, namely, to what extent 2D long-range order may explain the observed behavior. The origin of the magnetic behavior is discussed hereafter.

8.4.2.3 **Origin of the Phase Transition**

The nature of the phase transition in layered magnets is currently a matter of debate. If the distance between magnetic layers and the pertinent exchange pathways can be determined, and even modulated by synthesis strategies, an essential feature is the dimensionality of the spin system. Depending on the spin symmetry (Ising, XY or Heisenberg), either 2D or 3D magnetic order may be stabilized.

For the systems under consideration, the intermediate interlayer distance and the nature of the bridging ligands claim in favor of a 3D long-range order. However, in the case of the cobalt(II) compound showing a strong anisotropy, a 2D magnetic order cannot be ruled out. For more insight, we discuss here the behavior of the copper(II) hydroxide-based compounds in the framework of the scaling theory of phase transitions [92].

This theory assumes that length of the spin correlations diverges like $\xi = \xi_0(1 - T_c/T)^{-\gamma}$, wherefrom it follows that the χT product may be written as [87]:

$$\chi T = C\,(1 - T_c/T)^{-\gamma} = C(1 - T_c/T)^{-\theta/T_c} \quad \text{with} \quad \theta = \gamma T_c \quad (1)$$

where C is the high temperature Curie constant. Near T_c, when ξ diverges, the exponent γ takes universal values which depend only on the spin and space dimensions. Notice that depending on the sign of θ, Eq. (1) describes either situations where χT increases upon cooling down, namely ferromagnets ($\theta > 0$), or situations where it decreases, namely antiferromagnets ($\theta < 0$).

In the copper(II) hydroxy carboxylates, the variation of χT is not a monotonic function of temperature, due to competing in-plane interactions. It may be well approximated by the sum of two exponentials, as seen above, with ferromagnetic $(\chi T)_F$ and antiferromagnetic $(\chi T)_{AF}$ contributions.

Further, the apparent susceptibility must be corrected for the demagnetizing field, which limits the effect of the external field. The actual susceptibility, χ_{act}, is then related to χ by the relation $M = \chi H = \chi_{act}\,(H - \alpha\chi H)$, whereby it follows that $1/\chi_{act} = 1/\chi - \alpha$ with $\alpha = 1/\chi_{max}$. The pertinent quantity is then:

$$\chi_{act} T = T/(\chi^{-1} - \chi_{max}^{-1}) \quad (2)$$

Rather than fitting χT data from the above equations, it is more convenient to differentiate and calculate:

$$d\ln T/d\ln(\chi T) = -(T - T_c)/\gamma T_c \tag{3}$$

which only contains two parameters. By plotting $d\ln T/d\ln(\chi T)$ vs. T, we obtain a straight line in the whole range where the model is valid. The line intersects the T axis at T_c, and the $T = 0$ axis at $1/\gamma$.

On this basis, $Cu_2(OH)_3(n\text{-}C_{10}H_{21}CO_2)$ appears as a system just sitting at a lower critical dimensionality. The exponential regime $C_2 \exp(\beta J/kT)$, characteristic of a 2D Heisenberg system (Fig. 8.26), is evidenced by a straight line which aims at the origin $T_c = 0$ and is shown to persist down to $T \sim 30$ K, well below the temperature window where C_2 and βJ were initially determined. The striking new feature, which shows up in this representation, is the spectacular "crossover" to a different scaling regime: below 30 K, a different straight line is apparent pointing to a finite $T_c = 21.05$ K and a critical exponent $\gamma \sim 1.36$, typical of three-dimensional ferromagnetic ordering ($\gamma \sim 1.24$ for Ising, 1.32 for XY, and 1.385 for Heisenberg spins). The results show clearly the crossover which occurs at about 30 K. The same crossover is also very apparent on the Arrhenius plot $\log \chi T$ vs. $1/T$ shown in Fig. 8.27 where it amounts to a sudden departure from linearity when one leaves the exponential 2D regime, to enter the 3D regime which obeys the power law $\chi T = 0.40(1 - 21.05/T)^{-1.36}$.

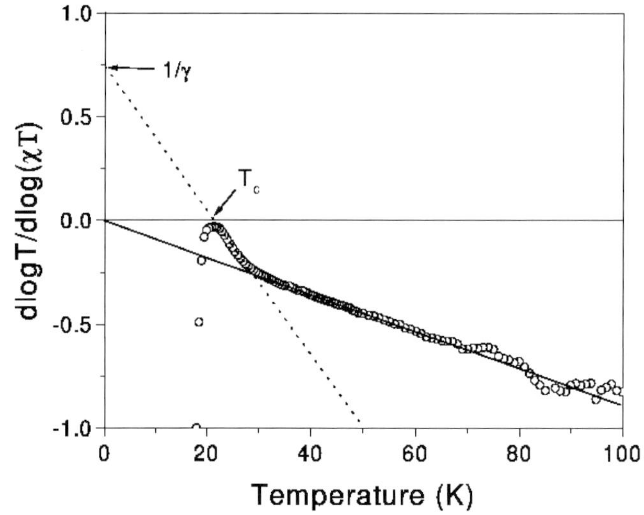

Fig. 8.26 Plot of the magnetic data as d ln T/d ln(χT) *vs* T for $m = 10$ showing the change of regime between the high-temperature 2D (exponential law) and low-temperature 3D (power law) behaviors. The crossover occurs at about 30 K. The intercepts with both axes give T_C and $1/\gamma$

8.4.2.4 **Interlayer Interaction Mechanism**

Owing to the nature of the organic anions and the distance between $[Cu_2(OH)_3]^+$ layers, the alkyl chains are very unlikely to participate in the interlayer coupling. Therefore we consider now the possibility that the interaction responsible for the 3D ordering would be only the dipolar interaction.

Whenever ferromagnetic in-plane interaction dominates, it is now well-documented that the situation depends to a large extent on the interlayer spacing. For small spacing (less than 10 Å), small through-bond interlayer interactions usually stabilize a 3D AF order, while for larger spacing dipolar through-space interaction dominates [93]. This one varies as r^{-3} between individual spins, but the effective coupling decreases more slowly for interacting ferromagnetic layers.

At nonzero temperature, the spins are only correlated on a finite distance ξ which, for isotropic spins, is related to the in-plane exchange constant J by the relationship [86]:

$$\zeta^2 = (JS/kT)\ \exp\ (4\pi JS^2/kT) \tag{4}$$

The basic idea is that, despite the fact that the dipole interaction between individual spins lying in different layers is very small, the dipole interaction between correlated spins of size ξ^2 becomes sizeable as the temperature decreases and leads to 3D order as soon as the in-plane correlation length ξ reaches a threshold value, giving giant effective moments $<\mu> = \xi^2 gS$ [93]. The overall energy is minimized when the order between layers is ferromagnetic for an axial anisotropy, antiferromagnetic for an in-plane anisotropy. Such a model has been worked out to explain the magnetic ordering in copper(II) compounds, characterized by large interlayer spacing.

The through-space dipolar field created by the magnetic layers has been computed by considering each layer at nonzero temperature as a checkerboard of spin blocks, each block containing ξ^2 parallel correlated spins. The net moment of each spin block is $\xi^2 gS = \xi^2\mu$ and is assumed oriented antiparallel to its neighbors, thus prefiguring a multidomain structure in the 3D ferromagnet below T_c. Accordingly, the expected transition is not towards a uniform but a multidomain ferromagnet. The average dipolar field acting on a spin block (SB), and arising from adjacent layers (ALs) is calculated from the expression:

$$H_{dip}=(1/\xi^2)\ <\mu>\ \Sigma_{SB}\Sigma_{AL}[3\ \cos^2\theta_{ij} - 1]/r_{ij} \tag{5}$$

which takes into account the spin fluctuations through the thermal average of the moment and the spatial extension of the spin block through the sum over its ξ^2 individual spins. The calculated value of H_{dip} converges "rapidly" when the sum over j is performed over distances of a few $\xi(T)$'s. It involves about 4×10^6 spins close to T_c. In the frame of the molecular field approximation, the thermal average of the moment depends on the internal field according to the classical Langevin function:

$$<\mu> = \mu\ [\coth(x)\ -1/x] \tag{6}$$

where here $x = \xi^2 \mu H_{dip}/kT$. This equation is linearized around $<\mu> = 0$, and we then deduce the implicit expression for T_c [87]:

$$T_c = \xi^2 \mu H_{dip}(T_c)/3k \tag{7}$$

It must be emphasized that this approach differs substantially from the standard mean-field approximation in that the thermal variation of the interaction energy is largely dominated by the exponential divergence of the effective interacting moments. Therefore the calculated T_c depends essentially on the rate of divergence of ξ, i.e. on the in-plane exchange interaction.

The computation has been performed for β-$Cu_2(OH)_3(n\text{-}C_{10}H_{21}CO_2)$ using the net moment per Cu^{II} given by the saturation magnetization at low T and the measured lattice parameters.

Using the above relation for the thermal variation of ξ and the preferred value of J deduced from $\chi T(T)$, we obtain the temperature dependence of the dipolar energy in the vicinity of T_c. By solving the implicit equations above, the critical temperature is found to range from 15.3 K to 18.2 K, within the uncertainties in the βJ value, which is to be compared to the experimental value $T_c = 20 \pm 0.5$ K. At T_c, the number of correlated spins is about 10^4. If we assume now a uniaxial anisotropy (Ising-like) for the moments, $<\mu>$ reduces to μ tanh(x), and the calculated ordering temperature increases by about 10% to 17.1–20.1 K for the same βJ values.

Although the model involves some crude approximations, it shows unambiguously that dipole-dipole interaction between correlated spins through the interlayer space can give rise to 3D ordering with a critical temperature close to the experimental observation.

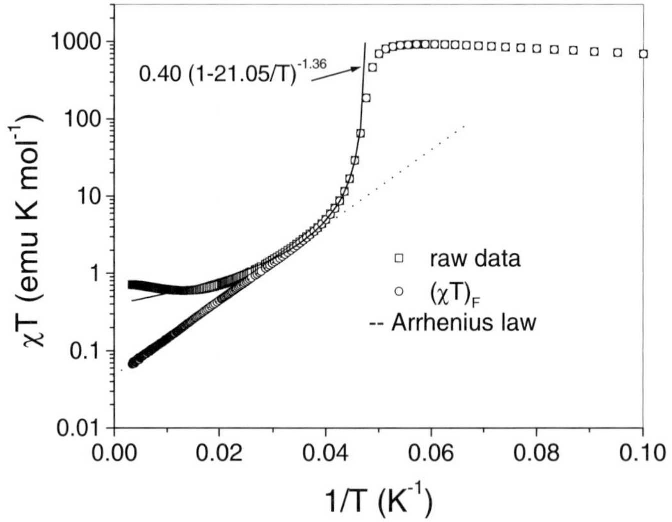

Fig. 8.27 Variation of (χT) *vs* $1/T$ for $m = 10$, showing the departure from the Arrhenius law (2D regime) below 30 K. The solid line corresponds to a power law

8.4.2.5 Difunctional Organic Anions

The compounds $Cu_2(OH)_{4-2x}A_{x'} \cdot z\,H_2O$ where A is a dicarboxylate anion have been prepared by ion exchange, starting from $Cu_2(OH)_3(OAc) \cdot H_2O$. The acetate is exchanged for alkanedioate anion, $CO_2XCO_2^{2-}$ with $X = (CH_2)_m$ and m ranging from 1 to 8, or alkenedioate anion ($X = C_2H_2$, C_4H_6 or C_4H_4), using the usual procedure [94].

The crystal structure of these compounds is not known at the moment, but the study of the series with different lengths of carbon chains (m) gives some insight into the structural packing [94]. All compounds exhibit a layered structure, characterized from XRD patterns by intense 001 reflections, up to at least the third order.

The intercalation of organic molecules is confirmed from IR spectra, from which it is deduced that the carboxylate moieties are bonded to copper(II) ions.

The basal spacing is related to the carbon chain length m, and varies from 9.20 Å for $m = 1$ to 15.6 Å for $m = 8$, with a step-like variation due to the alternating change of orientation of the carboxylate moieties in the bridging unit [94]. The tilt angle of the bridging units with respect to the c axis differs for odd and even m values (43° and 26° respectively). Further, the basal spacing for the extrapolated $m = 0$ value agrees with a monolayer packing of the metal ions, as evidenced previously for the monocarboxylate analogues from EXAFS experiments [85]. The same model holds for unsaturated bridging units, characterized by a weak variation of the basal spacing (0.2 Å step aside).

The terephthalate-based compound $Co_2(OH)_2(C_8H_4O_4)$ and the copper(II) analogue were synthesized by hydrothermal synthesis [95]. For the former, diaminopropane and piperazine were used as structural-directing agents.

The structures of $Co_2(OH)_2(tp)$ (with $tp = C_8H_4O_4$) and the copper(II) analogue have been solved by an *ab initio* XRPD method [95,96]. The former is monoclinic ($C2/m$) with unit cell parameters $a = 19.943$ Å, $b = 3.2895$ Å, $c = 36.2896$ Å and $\alpha = 95.746°$, while the latter is triclinic ($P-1$) with parameters $a = 10.1423$ Å, $b = 6.3388$ Å, $c = 3.4841$ Å, $\alpha = 99.17°$, $\beta = 96.56°$ and $\gamma = 98.76°$.

The structures of both compounds are very close, and essentially differ by the Jahn-Teller distortion always observed for the copper(II) derivative. The inorganic layers are 9.92 Å apart for the cobalt(II) compound, and 9.81 Å for the copper(II) derivative. The structure of $Co_2(OH)_2(tp)$ is illustrated in Fig. 8.28.

The terephthalates are pillared and coordinated to the metal layers, thus forming a three-dimensional network. The structure is characterized by octahedrally coordinated Co(II) ions bearing both triply bridging hydroxyls and $(\mu_3-\eta^1,\eta^2)$ carboxylate groups. Two crystallographically independent Co(II) ions are present: one (Co1) bound by four hydroxyls (O3) and two carboxylic oxygen atoms (O2), the other (Co2) bearing two hydroxyls (O3) and four tp oxygen atoms (O1).

Note that the interlayer distance of the terephthalate-based compounds is close to that of the alkenedioate derivatives with $m = 4$. Only the nature of the bridging unit differs significantly.

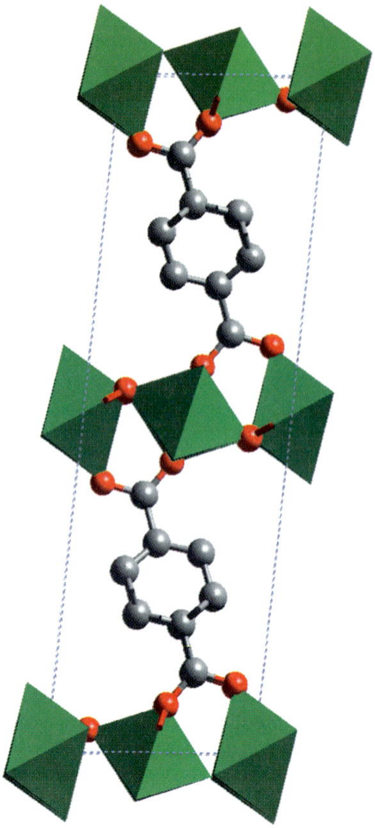

Fig. 8.28 Crystal structure of $Co_2(OH)_2(tp)$

Magnetic Properties

The magnetic properties of the alkane and alkenedioate-based compounds are very sensitive to the nature and spatial extension of the organic chains. The most striking results are observed for $m = 4$ with saturated ($X = C_4H_8$) and unsaturated ($X = C_4H_6$ and $X = C_4H_4$) aliphatic chains, which show a net moment within the layer. They are referred to as 4_0, 4_1 and 4_2, according to the number of carbon atoms and double bonds (0 to 2) within the anionic spacer.

The regular increase of χT upon cooling down (Fig. 8.29) is characteristic of in-plane ferromagnetic interactions, with a maximum at 15.7 K for 4_0, 13.0 K for 4_1 and 16.0 K for 4_2. According to the 2D Heisenberg model, χT is well described in the low temperature range by the expression $\chi T \sim \exp(4\pi JS^2/kT)$ [86], giving, as in-plane exchange constant, $J = 17$ K. The main difference occurs at low temperature, since 4_0 and 4_2 show an AF long range order with a metamagnetic transition for $H_c = 2250$ Oe and 2040 Oe respectively, while 4_1 exhibits a ferromagnetic order. This is evidenced from the strong divergence of χT upon cooling (factor 10 with respect to 4_0), and the out-of-phase signal (χ'') at the transition (inset in Fig. 8.29). A hysteretic effect of about 100 Oe, indicative of a weak anisotropy of the magnetic moments, is observed at 4 K.

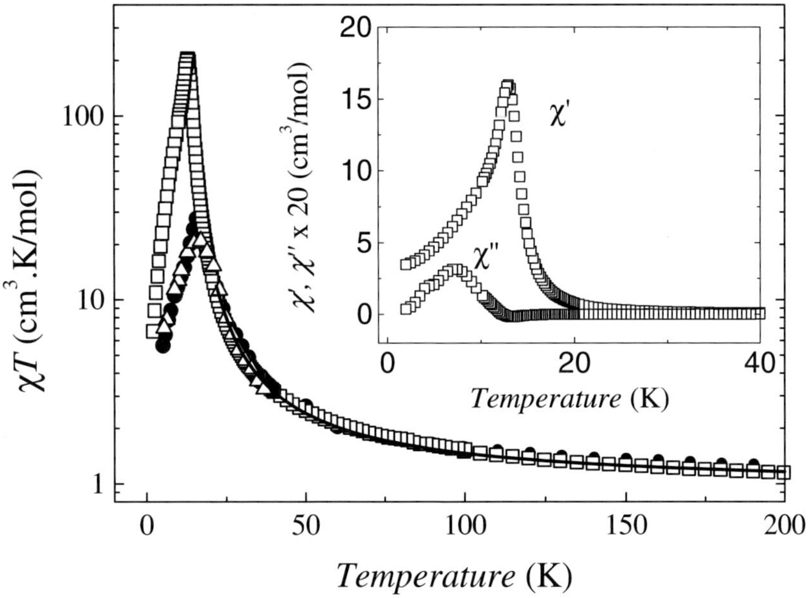

Fig. 8.29 Temperature dependence of χT for the copper(II) hydroxy dicarboxylates with saturated (4_0 (circles)), and unsaturated (4_1 (squares) and 4_2 (triangles)) bridging anions. The inset shows the metamagnetic transition of M(H) for 4_0

The temperature dependent magnetic susceptibility of $Co_2(OH)_2$(tp), measured at $H = 50$ Oe, agrees with the presence of a high-spin Co^{2+} ion. The χT product exhibits upon cooling a strong increase up to a sharp maximum at 48.0 K, then a decrease to zero (Fig. 8.30). The real part (χ') of the AC magnetic susceptibility ($H = \pm 3.5$ Oe) shows a maximum at $T = 47.8$ K, in agreement with the above findings. The absence of any out-of-phase signal (χ'') at this temperature points to a long-range antiferromagnetic order. However, $\chi'' = f(T)$ shows a peak at 44.0 K, which is the signature of a net moment associated with a partly canted antiferromagnetic structure.

The thermal variation of the magnetic anisotropy stabilizes three possible spin arrangements that are nearly degenerated: (i) a plain antiferromagnetic (AF) state between 48 and 45 K, corresponding to fully compensated ferromagnetic layers; (ii) a weakly canted antiferromagnetic (WCA) state between 45.0 and 30.0 K, due to nearly compensated ferromagnetic layers; and (iii) a strongly canted antiferromagnetic state (SCA) or weak ferromagnet with a large magnetic moment below 15 K (Fig. 8.31). Between 30 and 15 K, the last structure is responsible for a large remnant magnetization, but it is likely metastable in zero field. At very low temperature, a giant hysteretic effect is observed ($H_c = 5.9$ T at 4 K), pointing to a very large magnetic anisotropy, in relation with the spin-orbit coupling effect and the influence of dicarboxylate anions on the site symmetry.

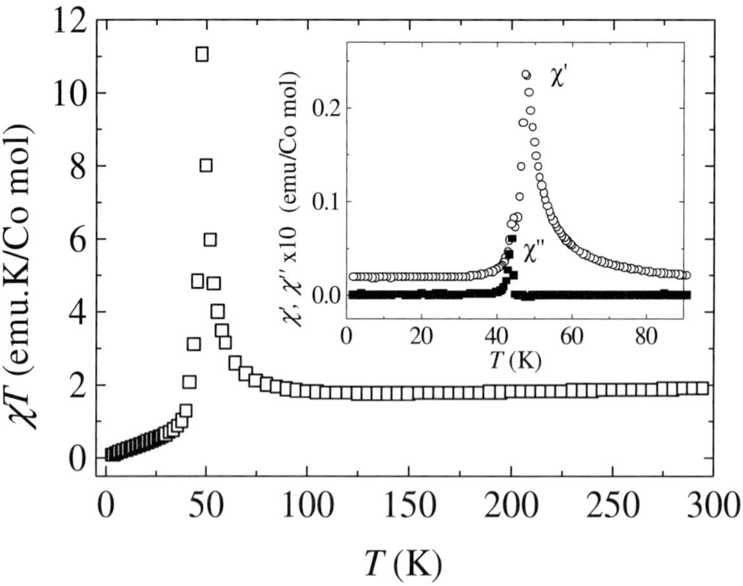

Fig. 8.30 Temperature dependence of the static χT product at 50 Oe and of the AC susceptibility under a 3.5 Oe alternating field of frequency 20 Hz for $Co_2(OH)_2$(tp)

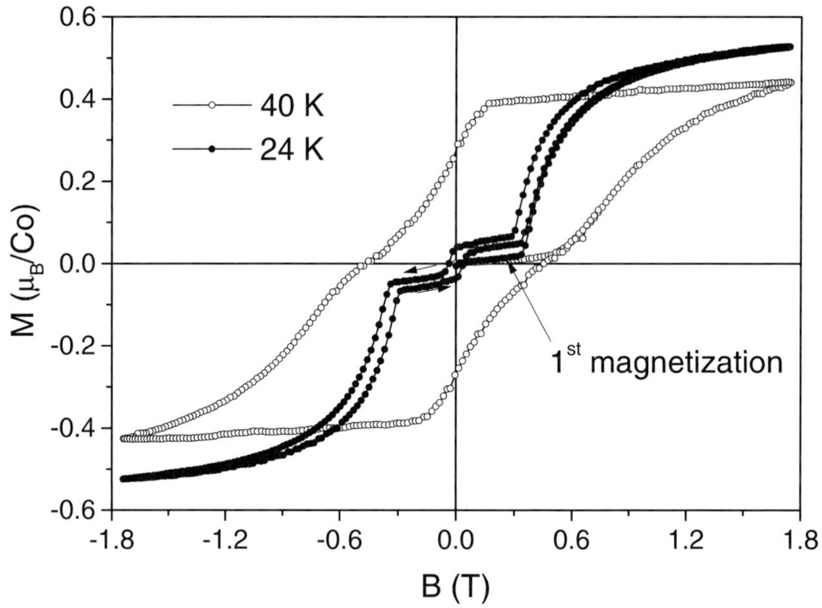

Fig. 8.31 Field-dependent magnetization of $Co_2(OH)_2$(tp), at 40 K and 24 K. The high field reversible part of the curve is not reached at 24 K

On close examination, the magnetic findings for the copper analogue $Cu_2(OH)_2(tp)$ exhibit significant changes with respect to those of the previous system. The χT variation upon cooling is, as for the cobalt(II) compound, characteristic of ferromagnetic in-plane interactions, with a value determined from the high temperature series expansions, $J = 5.5$ K. In turn, the nature of the magnetic transition differs. Indeed, a net moment occurs at the ordering temperature, $T_c = 4.5$ K, characterized by a maximum of χ' and a peak of the out-of-phase (χ'') signal (Fig. 8.32). The M(H) variation at lower temperature shows clearly a transition to a magnetized state, whose magnetic moment agrees with that of a ferromagnetic system (~1 μ_B/Cu(II)). Hence, the spin configuration below T_c is a collinear state. A very small hysteretic effect is observed, in relation with the isotropic character of the spin carriers.

The X-band EPR spectra of 4_0, 4_1 and 4_2 show a single line, with nearly Lorentzian shape at room temperature, which narrows upon cooling down to 150 K.

In W-band ($f = 95$ gHz) the spectra are typical of polycrystalline powders with axially symmetrical g-tensors whose axes are statistically distributed with respect to the direction of the external field (Fig. 8.33). An additional structure is observed in the perpendicular signal, witnessing the presence of a low symmetry component (orthorhombic) of the crystal field.

At 150 K, the values of the g-tensor in the axial symmetry approximation are $g_{||} = 2.058$ and $g_{\perp} = 2.187$. Upon cooling down, a small variation is observed for both components down to about 50 K, then upon approaching the phase transition, a striking discrepancy with a decrease of $g_{||}$ and an increase of $g_{\perp}$, witness to a co-

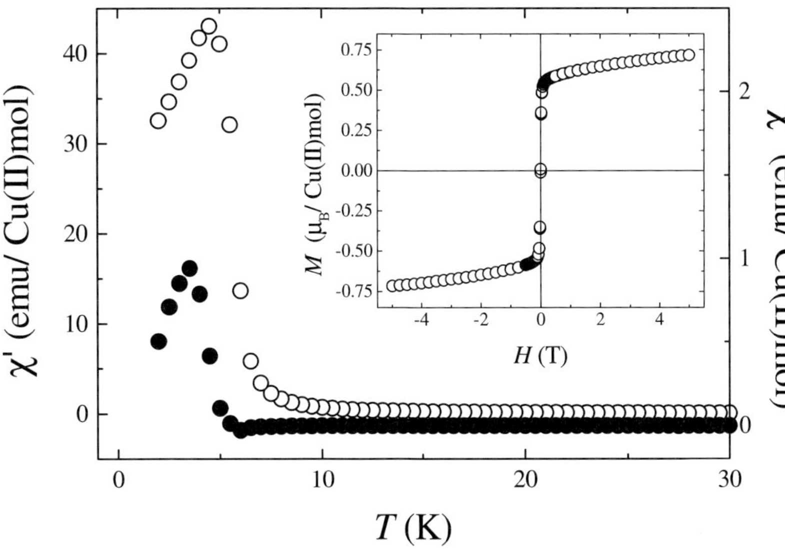

Fig. 8.32 Temperature dependence of the AC susceptibility under a 3.5 Oe alternating field of frequency 20 Hz for $Cu_2(OH)_2(tp)$. The field dependence of the magnetization is shown in the inset

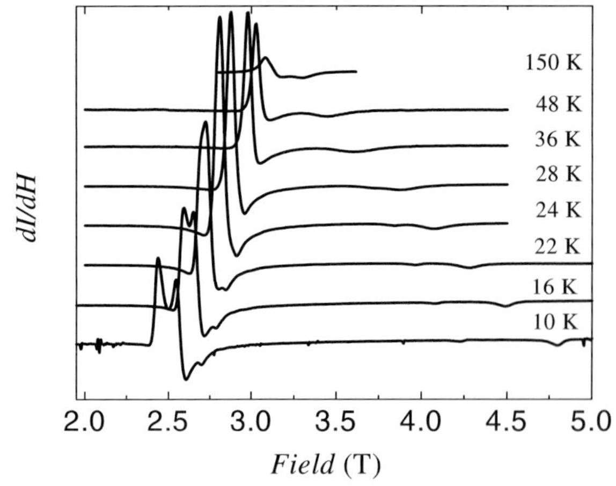

Fig. 8.33 Power spectra of 4_1 in the W-band at different temperatures

operative effect. The mean value of the g-factor is significantly higher at 10 K (2.266) than at 150 K (2.144).

It is worth noting that the single line in the X-band has the same features as the perpendicular component, $g_\perp$, in the W-band. The effective g-values characterizing the perpendicular component do not depend on the measuring field down to ca. 25 K, while at lower temperature the relative shifts are much larger in the X-band than in the W.

For terephthalate-based compounds, it can be noted that $Co_2(OH)_2$(tp) is EPR silent whatever frequency and temperature, while $Cu_2(OH)_2$(tp) exhibits a signal at low temperature only. Below 10 K, it displays a spectrum in the W-band which is similar to that of aliphatic chain compounds, $m = 4$, i.e. with axially symmetrical g-tensors. However, the effective g-value of the perpendicular part is much higher ($g_\perp = 6.57$ at 5 K) than those found for aliphatic chains (Fig. 8.34).

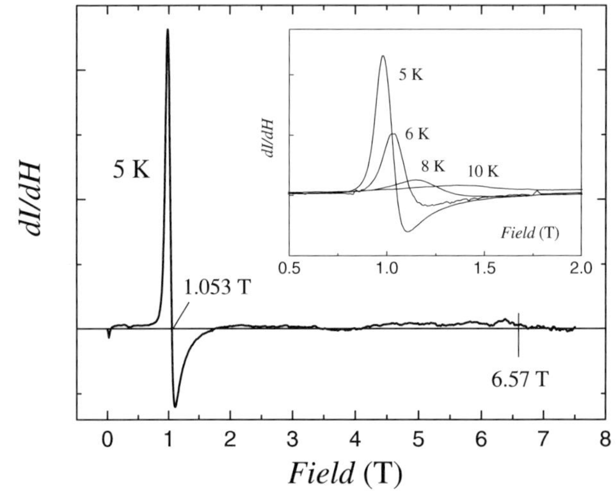

Fig.8.34 Axially symmetrical power spectrum of $Cu_2(OH)_2$(tp) at 5 K in the W-band. Inset: Temperature dependence of the perpendicular part of the spectrum between 5 K and 10 K

The other discrepancy is the disappearance of the spectra a few degrees above the ordering temperature, $T_c = 4.5$ K. This can be explained by a smaller ferromagnetic in-plane interaction ($J = 5.5$ K), compared to the aliphatic chain derivatives ($J = 17$ K). As a result, the 2 D spin correlations in the paramagnetic regime are too small to promote a strong uniform mode above 10 K.

Interlayer Interaction

As a result of the nature of the bridging units, and the distance between magnetic layers, different interlayer interactions, i.e. exchange, spin-polarisation, dipolar-like, must be considered in the investigated dicarboxylate-based compounds.

In the frame of the model developed above (see Sect. 8.4.2.4), the through-space dipolar field created by the copper(II) layers has been computed for the dicarboxylate series. Taking into account the temperature dependence of $<\mu>$, the ordering temperature has been deduced by solving the self-consistent set of equations:

$$<\mu_z> = \mu\,[\coth(x) - 1/x] \quad \text{with} \quad x = \xi^2 \mu H_m/kT \tag{8}$$

and

$$H_m = \lambda(\xi)\,<\mu_z>$$

where $\lambda(\xi)$ is the dipolar coupling coefficient, which depends on structural data.

Using the in-plane exchange constant, $J = 17$ K, deduced for the aliphatic chain compounds, an AF order is expected to occur at 7.4 K for an in-plane anisotropy. Clearly, the dipolar interaction only cannot explain the AF order observed for 4_0 and 4_2 ($T_N = 15.7$ K and 16 K respectively). Through-bond interaction must be assumed to describe these compounds. Further, the non-saturated derivative 4_1 exhibits a ferromagnetic order at 13 K, which is significantly higher than the calculated value ($T_c = 9.8$ K), when assuming an axial anisotropy. It is deduced that π electrons of the bridging unit have a worthwhile influence on the magnetic ordering. A spin polarization mechanism through the aliphatic chain is probably responsible for the stabilization of the magnetized state. This assumption is also supported by ^{13}C MAS-NMR results which show a significant shift of the signals for the unsaturated dicarboxylate bridges (4_1 and 4_2), indicating a noticeable spin density on the bridging unit, while there is no shift for the saturated dicarboxylate, 4_0 [97].

Finally, the fact that the non-saturated bridge compounds have different behaviors is explained from the alternation of spin density along the carbon chain for 4_1, while spin polarization, due to a weak delocalization of π electrons, is expected to dominate for 4_2.

The situation is similar for the cobalt terephthalate derivative. The dominant AF interlayer interaction is probably mediated through the terephthalate ligand, in agreement with the spin density alternation. The occurrence of a canted state close to the AF ground-state demonstrates that a competition between different mechanisms cannot be ruled out.

Finally, it can be emphasized that magnetic materials involving unsaturated dicarboxylate ligands differ markedly from the classical layered systems, mostly because they provide organic-inorganic multilayer magnets with possible tuning of the exchange between spin layers mediated by π electrons.

8.4.2.6 **Metal-radical Based Magnets**

The intercalation of 3-carboxy-2,2,5,5-tetramethyl-1-pyrrolidine-1-oxyl (proxyl) in between $[Cu_2(OH)_3]^+$ layers has been reported by Fujita [98]. $Cu_2(OH)_{3.5}$(proxyl)$_{0.5}$ · H_2O has been prepared by ion-exchange in copper hydroxide acetate with proxyl sodium salt. X-ray diffraction patterns point to the strong increase of the basal spacing, from 9.3 Å (hydroxide acetate) to 16.1 Å with radical intercalation. The temperature dependence of χT agrees with dominant AF in-plane interaction between neighboring spins, unlike the copper(II) hydroxide acetate.

Similarly, intercalation of para- and meta-imino nitroxide benzoate anions (*p*-INB and *m*-INB, respectively) within layered hydroxides has been carried out [99]. The most striking results have been obtained with the cobalt(II) layer compounds, which appear more suitable to give ferrimagnetic properties. Three compounds were obtained by ion-exchange reaction in $Co_2(OH)_3NO_3$ with *p*-INB, *m*-INB and the diamagnetic precursor bis-hydroxyimidazolidine (Fig. 8.35). From chemical analysis, the composition $Co_2(OH)_{3.5}(X)_{0.5}$, $2H_2O$ with X=*p*-INB (**1**), *m*-INB (**2**) is deduced. X-ray diffraction patterns show a layer structure with basal spacing ranging from 20.2 Å for **1** to 22.8 Å for **2**.

The magnetic properties are given in Fig. 8.36, and compared to those of the hydroxide nitrate. Upon cooling from room temperature, the observed χT product shows for the radical-based compounds a minimum around 60 K, and a strong increase up to a maximum at 10 K. Unlike the hydroxide nitrate [76], both compounds order ferromagnetically, as evidenced by the out-of-phase signal in an AC external field (Fig. 8.19). From the maximum of χ', the ordering temperatures are found to be 6.0 K and 7.2 K for **1** and **2** respectively. It is to be noted that the bis-

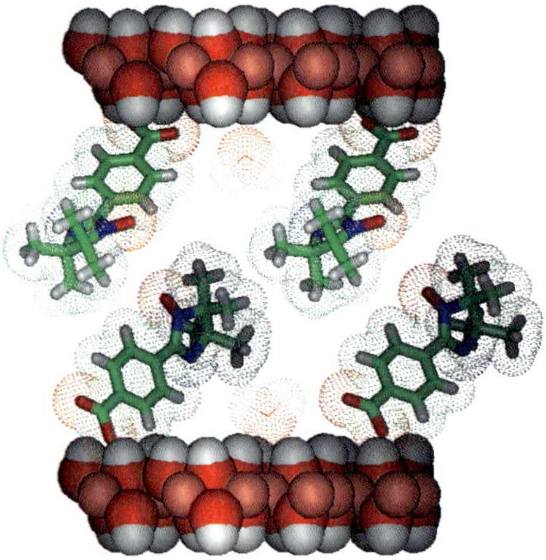

Fig. 8.35 Proposed structure of $Co_2(OH)_{3.5}(X)_{0.5}$, $2H_2O$ where X is the radical anion *p*-INB

hydroxyimidazoline diamagnetic anion compound has a much higher ordering temperature ($T_c = 15.3$ K). The M(H) curve for compound **2** exhibits a hysteresis loop characteristic of a ferromagnetic-like 3D order, with a coercive field of 510 Oe at $T = 1.8$ K (340 Oe for **1**). In turn, the high field magnetization is lower than expected for a complete spin alignment. The competition between in-plane exchange interactions and/or the spin-orbit coupling effect for cobalt(II) ions may induce such a behavior. The fact that the ordering temperature is strongly affected by the presence of a radical indicates that a significant metal-radical exchange interaction, rather than a simple polarization, takes place.

The magnetic transition is clearly displayed in the first derivative of the EPR absorption spectra given in Fig. 8.37 for compound **2**. As expected, the Lorentzian-shaped signal of the radical is located close to the free electron g-value at room temperature, and remains nearly unchanged down to T_c. Drastic changes occur close to T_c, as sketched in the low temperature spectra. A strong broadening of the EPR line and a large shift of the resonance field (see the inset) are simultaneously observed with the onset of bulk ferromagnetism. Therefore, the organic radical is actually probing the cooperative alignment of the neighboring cobalt(II) moments. Clearly, the radical spins do not simply line up within the internal field created upon ordering the cobalt(II) layers. The observed behavior is most likely due to the interlayer exchange interactions $J_{\pi\text{-Co}}$ and $J_{\pi\text{-}\pi}$ counteracting the ferromagnetic

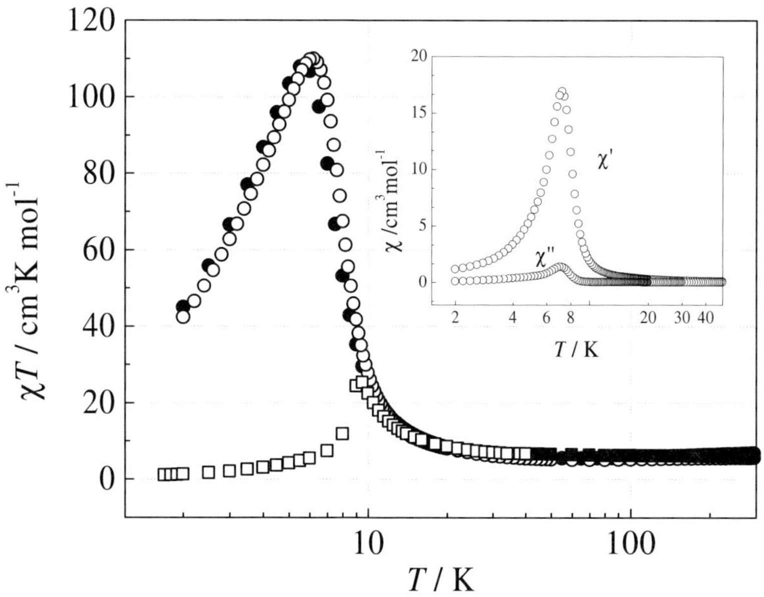

Fig. 8.36 Temperature dependence of χT for the radical-based compounds with X = p-INB (full circles), m-INB (open circles) and the starting hydroxide nitrate (squares). The susceptibility of the latter plotted in the inset shows the variation of the in-phase (χ') and out-of-phase (χ'') signals

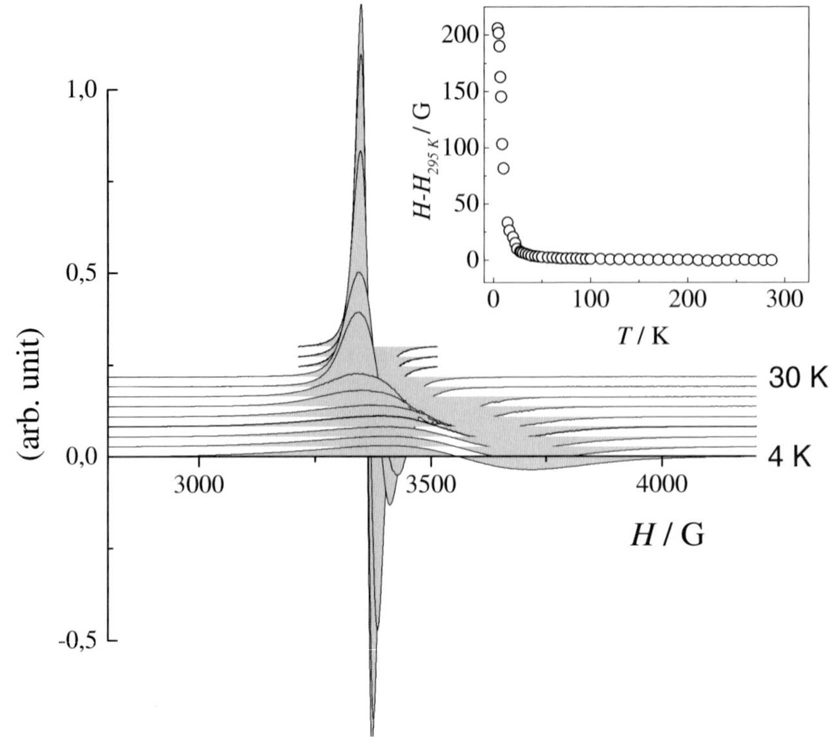

Fig. 8.37 EPR spectra of *m*-INB showing the evolution of the line shape between 4 K and 30 K. The temperature-dependent shift of the free radical line is given in the inset

through-space dipolar coupling. In this respect, this class of hybrid magnets differs from the classical intercalated layer compounds, mostly because both subnetworks (inorganic and organic) are in close interaction.

8.4.2.7 Solvent-mediated Magnetism

There is a considerable interest in the packing and dynamics of long-chain polymethylene compounds intercalated in layered compounds, because of their relevance to the functionality of biological membranes, liquid crystals and so on [100, 101]. In addition, some of these compounds exhibit structural transformations, such as a trans-gauche transition of the alkyl chains or a mono-bilayer phase transition, triggered by heating or soaking in solvents. If the magnetic properties of the inorganic layer are affected by a structural modification of the organic layer introduced by a stimulus, the hybrid compound can be regarded as a controllable and/or switchable magnetic system. In this section, we describe the magnetic properties of $Cu_2(OH)_3$(8-((*p*-(phenylazo)phenyl)oxy) octanoate) [102]. A reversible phase transition, which takes place in two organic solvents and results in a drastic change in the magnetic properties, is reported.

The intercalation compound was obtained from the acetate by ion exchange. The distance between layers, 20.7 Å, obtained by subtracting the calculated thickness of the inorganic layer from the basal spacing of 25.5 Å, well corresponds to the calculated anion height, i.e. 21.7 Å. This points to organic monolayers with chains aligned nearly perpendicular to the inorganic layers (Fig. 8.38). The compound was dispersed in acetonitrile, and the mixture was stirred for two days. The XRPD reflections of the original phase decrease remarkably in intensity and disappear after 48 hours. New (001) reflections appear which indicate that the basal spacing is 38.7 Å, which is almost twice as large as the molecular height of the anion. After soaking, the organic anion is considered to form a membrane-like bilayer, as shown in Fig. 8.38. In turn, it is found that the bilayer phase switches to the monolayer one in hot methanol.

Notice that the observed mono-bilayer transformation is similar to the soluto-induced interdigitation in phospholipid assemblies. Phospholipid molecules, which are principal lipid components of biomembranes, assemble spontaneously in aqueous solution to give a bilayer structure, while forming an interdigitated monolayer in presence of alcohol [103, 104]. The mechanism of interdigitation depends on alcohol concentration, lipid chain length and temperature.

The temperature dependence of χT for the original monolayer phase in shown in Fig. 8.39, and compared to that of the bilayer compound. For the former, χT gradually decreases upon cooling, pointing to the predominance of AF interactions.

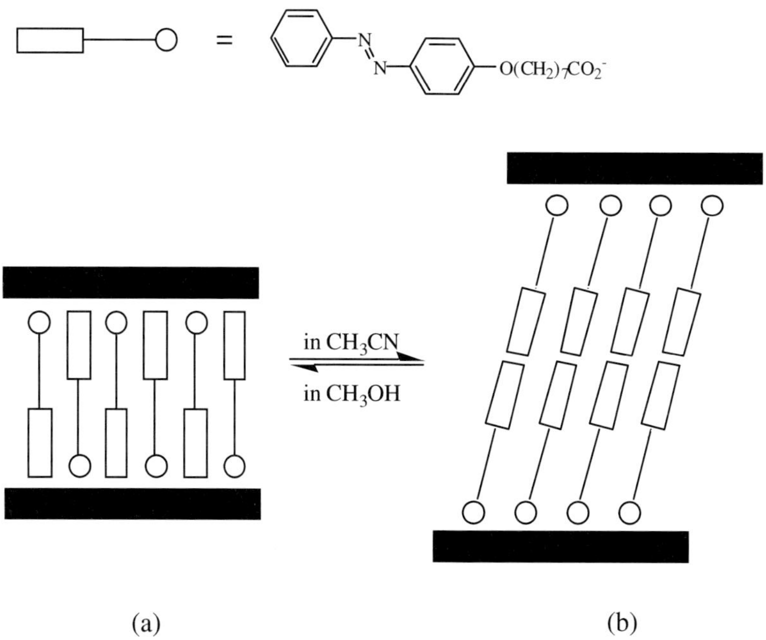

Fig. 8.38 Packing of the organic layer in the compound $Cu_2(OH)_3(ppo)$ in methanol (a) and acetonitrile (b)

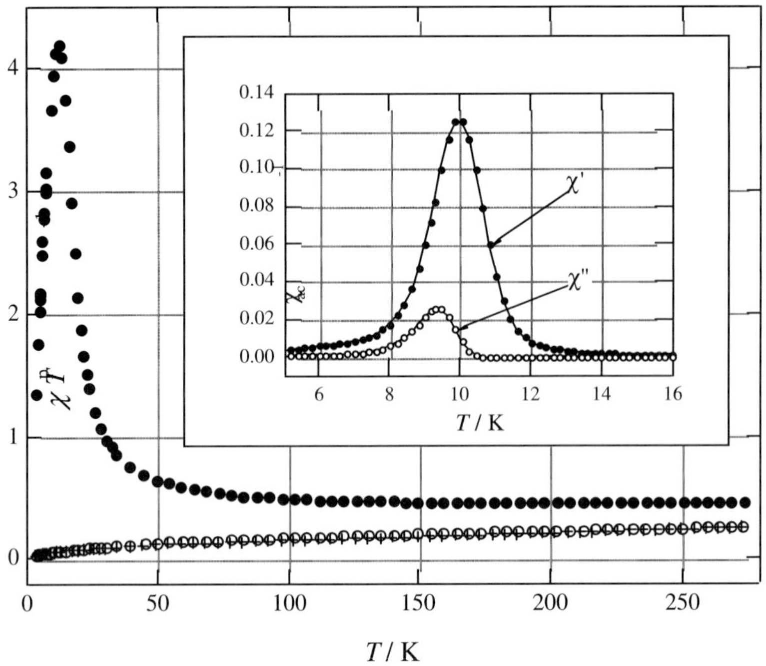

Fig. 8.39 Temperature dependence of the paramagnetic susceptibilities for the original (open circles) and revived (crosses) monolayer phases and for the bilayer phase (full circles). Inset: temperature dependence of the AC susceptibilities for the bilayer phase

In turn, the bilayer compound, obtained in acetonitrile, shows a striking increase of χT below 100 K, indicating a ferromagnetic 2D regime at high temperature, and a maximum at around 10 K, corresponding to a transition towards a higher dimensionality.

The temperature dependence of the AC susceptibilities (circles and squares stand for χ' and χ'', respectively) show an anomalous peak, characteristic of a ferromagnetic order at $T_c = 10.8$ K. From the magnetization curve at 4.5 K, it is deduced that this compound is a weak ferromagnet, with a canting angle of about 14° between moments.

Finally, such a reversible mono-bilayer transition, associated with a change from paramagnetic to ferromagnetic behavior, is quite unusual in organic/inorganic layered materials. Such a transition is driven by the modification of the organic layer, due to organic solvent, which affects the structure of the inorganic layer and accordingly the magnetic properties. This suggests for such systems the possibility of switchable and/or controllable physical properties induced by a chemical stimulus.

8.5 Concluding Remarks

The aim of this review was to demonstrate the versatility of hybrid organic-inorganic layered compounds for the design of new magnetic materials with tunable properties.

The most promising systems are probably those in which the organic and inorganic subnetworks are in close relation through covalent bond interactions. Another result of interest is that the distance between magnetic layers is not a critical feature for stabilising high T_c magnets, because of the existence of through-space dipolar interactions which have a long-range effect. So very large organic molecules, with specific physical properties, may be intercalated in between the inorganic layers without significant decrease of the magnetic features.

The future in this field lies in the exploitation of these host-guest solids, where each component is able to contribute its own set of physical characteristics. The search for synergistic properties within this class of multifunctional materials is currently an active and ongoing area of research and should provide many novel and interesting classes of compounds in the future.

References

1 E. Coronado, C. J. Gómez-Garcia, *Chem. Rev.* **1998**, *98*, 273; L. Ouahab, *Chem. Mat.* **1997**, 9,1909.

2 P. Day , *Phil. Trans. Roy. Soc. London* **1985**, *A314*, 145.

3 D. O'Hare in *Inorganic Materials*, D. W. Bruce and D. O'Hare, John Wiley & Sons, New York, 1992, 165.

4 R.Willett, H. Place, M. Middleton, *J. Am. Chem. Soc.* **1988**, *110*, 8639.

5 S. Decurtins, H. W. Schmalle, H. R. Oswald, A. Linden, J. Ensling, P. Gütlich, A. Hauser, *Inorg. Chim. Acta* **1994**, *216*, 65.

6 L. J. de Jongh, W. D. van Amstel and A. R. Miedema, *Physica* **1972**, *58*, 277.

7 R. D.Willett, in *Magneto-Structural Correlations in Exchange Coupled Systems*; eds R. D. Willett, D. Gatteschi and O. Kahn, Reidel Publishing Company, Dordrecht 1985.

8 R. Willett, H. Place, M. Middleton, *J. Am. Chem. Soc.* **1988**, *110*, 8639.

9 C. P. Landee, K. E. Halvorson, R. D. Willett, *J. Appl. Phys.* **1987**, *61*, 3295.

10 R. D. Willett, F. H. Jardine, I. Rouse, R. J. Wong, C. P. Landee, M. Numata, *Phys. Rev.* **1981**, *B 24*, 5372.

11 T. Sekine, T. Okuno, K. Awaga, *Mol. Cryst. Liq. Cryst.* **1996**, *279*, 65.

12 D. B. Mitzi, *Progr. Inorg. Chem.* **1999**, *48*, 1.

13 M. Pilkingon, S. Decurtins, in *Magnetism: Molecules to Materials* II, eds J. S. Miller, M. Drillon, Wiley VCH, Weinheim, 2001, 339.

14 S. Decurtins, H. W. Schmalle, P. Schneuwly, J. Ensling, P. Gütlich, *J. Am. Chem. Soc.* *116*, **1994**, 9521.

15 L. O. Atovmyan, G. V. Shilov, R. N. Lyubovskaya, N. S. Ovanesyan, Yu. G. Morozov, S. I. Pirumova, I. G. Gusakovskaya, *Pis'ma Zh. Eksp. Teor. Fiz.* **1993**, *58*, 818.

16 H. Tamaki, M. Mitsumi, K. Nakamura, N. Matsumoto, S. Kida, H. Okawa, S. Iijima, *Chem. Letts.* **1992**, 1975.

17 H. Tamaki, N. Matsumoto, S. Kida, M. Koikawa, N. Achiwa, Y. Hashimoto, H. Okawa, *J. Am. Chem. Soc.* **1992**, *114*, 6974.

18 C. Mathonière, S. G. Carling, D. Yusheng, P. Day, *J. Chem. Soc. Chem. Commun.* **1994**, 1551.

19 C. J. Nuttall, C. Bellito, P. Day, *J. Chem. Soc. Chem. Commun.* **1995**, 1513.

20 R. Pellaux, H. W. Schmalle, R. Huber, P. Fischer, T. Hauss, B. Ouladdiaf, S. Decurtins, *Inorg. Chem.* **1997**, *36*, 2301.
21 G. Antorrena, F. Palacio, M. Castro, R. Pellaux, S. Decurtins, *J. Magn. Magn. Mat.* **1999**, *196*, 581.
22 M. Clemente-Leon, E. Coronado, J. Galan-Mascaros, C. Gomez-Garcia, *J. Chem. Soc. Chem. Commun.* **1997**, 1729, ; E. Coronado, J. Galan-Mascaros, C. Gomez-Garcia, J. Ensling, P. Gütlich, *Chem. Eur. J.* **2000**, 6, 552.
23 J. Larionova, B. Mombelli, J. Sanchiz, O. Kahn, *Inorg. Chem.* **1998**, *37*, 679.
24 a) E. Coronado J.-R. Galán-Mascaròs, C. J. Gómez-García, *Mol. Cryst. Liq. Cryst.***1999**, *334*, 679.
24 b) E. Coronado, J.- R. Galán-Mascarós, C. J. Gómez-García, V. Laukhin, *Nature* **2000**, *408*, 447.
25 H. Tamaki, Z. J. Zhong, N. Matsumoto, S. Kida, M. Koikawa, N. Achiwa, Y. Hashimoto, H. Okawa, *J. Am. Chem. Soc.* **1992**, *114*, 6974.
26 Z. Gu, O. Sato, T. Iyoda, K. Hashimoto, A. Fujishima, *Mol. Cryst. Liq. Cryst.* **1996**, *286*, 147.
27 S. Bénard, P. Yu, T. Coradin, E. Rivière, K. Nakatani, R. Clément, *Adv. Mater.* **1997**, *9*, 981.
28 L. O. Atovmyan, G. V. Shilov, R. N. Lyubovskaya, E. I. Zhilyaeva, Y. G. Morozov, *JETP Lett.* **1993**, *10*, 766.
29 S. Decurtins, H. W. Schmalle, H. R. Oswald, A. Linden, J. Ensling, P. Gütlich, A. Hauser, *Inorg. Chim. Acta* **1994**, *216*, 65.
30 S. Bénard, P. Yu, J. P. Audière, E. Rivière, R. Clément, J. Guilhem, L. Tchertanov, K. Nakatani, *J. Am. Chem. Soc.* **2000**, *122*, 9444.
31 J. S. O. Evans, S. Bénard, P. Yu, R. Clément, *Chem. Mater.* **2001**, *13*, 3813.
32 B. J. Coe, *Chem. Eur. J.* **1999**, *5*, 2464.
33 J. A. Delaire, K. Nakatani, *Chem. Rev.* **2000**, *100*, 1817.
34 a) J.-F. Létard, S. Montant, P. Guionneau, P. Martin, A. Le Calvez, E. Freysz, D. Chasseau, R. Lapouyade, O. Kahn, *Chem. Commun.* **1997**, 745; b) J.-F. Létard, L. Capes, G. Chastanet, N. Moliner, S. Létard, J. A. Real, O. Kahn, *Chem. Phys. Lett.* **1999**, *313*, 115.
35 F. Averseng, C. Lepetit, P. G. Lacroix, J.-P. Tuchagues, *Chem. Mater.* **2000**, *12*, 2225.
36 P. G. Lacroix, I. Malfant, S. Bénard, P. Yu, E. Rivière, K. Nakatani, *Chem. Mater.* **2001**, *13*, 441.
37 D. O'Hare, *Inorganic Materials,* eds D. W. Bruce, D. O'Hare) John Wiley & Sons, Chichester, 1993, chapter 4.
38 R. Brec, D. Schleich, G. Ouvrard, A. Louisy, J. Rouxel, *J. Inorg. Chem.* **1979**, *18*, 1814.
39 a) W. Klingen, R. Ott, H. Hahn, *Z. Anorg. Allg. Chem.* **1973**, *396*, 271; b) W. Klingen, G. Eulenberger, H. Hahn, *Z. Anorg. Allg. Chem.* **1973**, *401*, 91.
40 R. Clément, *J. Amer. Chem. Soc.* **1981**, *703*, 6998.
41 R. Clément, A. Léaustic, in *Magnetism: Molecules to Materials* II, eds J.S. Miller, M. Drillon, Wiley VCH, Weinheim, 2001, p. 397.
42 R. Clément, O. Garnier, J. Jegoudez, *Inorg. Chem.* **1986**, *25*, 1404.
43 R. Clément, M. Doeuff, C. Gledel, *J. Chim. Phys.* **1988**, *85*, 1053.
44 J. S. O. Evans, D. O'Hare, R. Clément, *J. Am. Chem. Soc.* **1995**, *777*, 4595.
45 R. Clément, J. P. Audière, J. P. Renard, *Rev. Chim. Min.* **1982**, *79*, 560.
46 J. S. O. Evans, D. O'Hare, R. Clément, A. Léaustic, P. Thuery, *Adv. Mater.* **1995**, *7*, 735.
47 H. Stanley, T. A. Kaplan, *Phys. Rev. Lett.* **1996**, *77*, 913.
48 G. Le Flem, R. Brec, G. Ouvrard, A. Louisy, P. Segransan, *J. Phys. Chem. Solids* **1982**, *43*, 455.
49 P. A. Joy, S. Vasudevan, *Phys. Rev. B,* **1992**, *46*, 5425.
50 A. Léaustic, J. P. Audière, D. Cointereau, R. Clément, L. Lomas, F. Varret, H. Constant- Machado, *Chem. Mater.* **1996**, *8*, 1954.
51 P. G. Lacroix, R. Clément, K. Nakatani, J. Zyss, L. Ledoux, *Science* **1994**, *263*, 658.
52 T. Coradin, R. Clément, P. G. Lacroix, K. Nakatani, *Chem. Mater.* **1996**, *8*, 2153.
53 C. Bellito, in *Magnetism: Molecules to Materials* II, eds J.S. Miller, M. Drillon, WileyVCH, Weinheim, 2001, p.425.
54 M. Ishaque Khan, J. Zubieta, *Progr. Inorg. Chem.* **1995**, *143*, 1.

55 a) G. Cao, H. Lee, V.M. Lynch, T. E. Mallouck, *Inorg. Chem.* **1988**, *27*, 2781; b) K. Martin, P. J. Squattrito, A. Clearfield, *Inorg. Chim. Acta* **1989**, *155*, 7.

56 B. Bujoli, O. Pena, P. Palvadeau, J. LeBideau, C. Payen, J. Rouxel, *Chem. Mat.* **1993**, *5*, 583.

57 A. Altomare, C. Bellitto, F. Federici, S. A. Ibrahim, R. Rizzi, *Inorg. Chem.* **2000**, *39*, 1803.

58 J. LeBideau, A. Jouanneaux, C. Payen, B. Bujoli, *J. Mater. Chem.* **1994**, *4*, 1319.

59 P. Rabu, P. Janvier, B.Bujoli, *J. Mater. Chem.* **1999**, *9*, 1323.

60 S. Drumel, P. Janvier, P. Barboux, M. Bujoli-Doeuff, B. Bujoli, *Inorg. Chem.* **1995**, *34*, 148.

61 Y. Zhang, A. Clearfield, *Inorg. Chem.* **1992**, *31*, 2821.

62 J. LeBideau, B. Bujoli, A. Jouanneaux, C. Payen, P. Palvadeau, J. Rouxel, *Inorg. Chem.* **1993**, *32*, 4617.

63 Y. Zhang, A. Clearfield, *Inorg. Chem.* **1992**, *31*, 2821.

64 S. G. Carling, P. Day, D. Visser, R. K. Kremer, *J. Solid State Chem.*, **1993**, *106*, 111.

65 a) G. E. Fanucci, J. Krzystek, M. W. Meisel, L. C. Brunel, D. L. Talham, *J. Am. Chem. Soc.* **1998**, *120*, 5469; b) C. T. Seip, G. E. Granroth, M. W. Meisel, D. L. Talham, *J. Am. Chem. Soc.* **1997**, *119*, 7084; c) H. Bird, J. K. Pike, D. R. Talham, *Chem. Mater.* **1993**, *5*, 709.

66 a) S. G. Carling, P. Day, D. Visser, R. K. Kremer, *Inorg. Chem.* **1995**, *34*, 3917; b) S. G. Carling, P. Day, D. Visser, J. Deportes, J. Appl. Phys. **1991**, *69*, 6016.

67 a) G. E. Fanucci, M. A. Petruska, M. W. Meisel, D. L. Talham, *J. Solid State Chem.* **1999**, *145*, 443; b) H. Bird, J. K. Pike, D. R. Talham, *J. Amer. Chem. Soc.* **1994**, *116*, 7903.

68 a) C. Bellitto, F. Federici, S. H. Ibrahim, *J. Chem. Soc. Chem. Commun.* **1996**, 759; b) C. Bellitto, F. Federici, S. A. Ibrahim, *Chem. Mater.* **1998**, *10*, 1076.

69 B. Bujoli, O. Pena, P. Palvadeau, J. LeBideau, C. Payen, J. Rouxel, *Chem. Mater.* **1993**, *5*, 583.

70 J. LeBideau, C. Payen, B. Bujoli, P. Palvadeau, J. Rouxel, *J. Magn. & Magn. Mater.* **1995**, *140*, 1719.

71 C. Bellitto, A. Altomare, F. Federici, R. Rizzi, S. A. Ibrahim, *Inorg. Chem.* **2000**, *39*, 1803.

72 D. M. Poojary, B. Zhang, A. Clearfield, *J. Amer. Chem. Soc.* *119*, **1997**, 12550.

73 D. L. Lohse, S. C. Sevov, *Angew. Chem. Int. Ed. Engl.* **1997**, *36*, 1619.

74 Q. Gao, N. Guillou, M. Nogues, A. K. Cheetham, G. Ferey, *Chem. Mater.* **1999**, *11*, 2937.

75 H. Effenberger, *Z. Krist.* **1983**, *165*, 127.

76 M. Drillon, C. Hornick, V. Laget, P. Rabu, F. M. Romero, S. Rouba, G. Ulrich, R. Ziessel, *Mol. Cryst. Liq. Cryst.* *1995*, *273*, 125.

77 C. Massobrio, P. Rabu, M. Drillon, C. Rovira, *J. Phys. Chem.* **1999**, *B103*, 9387; S. Pillet, M. Souhassou, C. LeComte, P. Rabu, C. Masobrio, M. Drillon, to be published.

78 P. Rabu, S. Angelov, P. Legoll, M. Belaiche, M. Drillon, *Inorg. Chem.* **1993**, *32*, 2463.

79 L.J. de Jongh, *Magnetic Properties of Layered Transition Metal Compounds*, Kluwer Academic Publishers, Dordrecht, 1990.

80 S. Miyata, T. Kumura, *Chem Lett.* **1973**, 843.

81 M. Bujoli-Doeuff, L. Force, V. Gadet, M. Verdaguer, K. el Malki, A. de Roy, J. P. Besse, J. P. Renard, *Mat. Res. Bull.* **1991**, *26*, 577.

82 S. Yamanaka, T. Sako, M. Hattori, *Chem. Lett.* **1989**, 1869; *Solid State Ionics*, **1992**, *53*, 527.

83 P. Rabu, S. Rouba, V. Laget, C. Hornick, M. Drillon, *J. Chem. Soc., Chem. Comm.* **1996**, 1107; P. Rabu, M. Drillon, K. Awaga, W. Fujita, T. Sekine, in *Magnetism : Molecules to Materials* II,eds J. S. Miller, M. Drillon, Wiley VCH, Weinheim, 2001, p.357.

84 W. Fujita, K. Awaga, *Inorg. Chem.*, **1996**, *35*, 1915.

85 W. Fujita, K. Awaga, T. Yokoyama, *Inorg. Chem.* **1997**, *36*, 196.

86 F. Suzuki, N. Shibata, C. Ishii, *J. Phys. Soc. Jpn.* **1994**, *63*, 1539.

87 M. Drillon, P. Panissod, P. Rabu, J. Souletie, V. Ksenovontov, P. Gütlich, *Phys. Rev.* **2002**, *B65*, 104404.

88 V. Laget, C. Hornick, P. Rabu, M. Drillon, R. Ziessel, *Coord. Chem. Rev.* **1998**, *178*, 1533.

89 W. Stählin, H. R. Oswald, *J. Solid State Chem.* **1971**, *2*, 252.

90 M. Louër, D. Louër; D. Grandjean, *Acta Cryst.* **1973**, *B29*, 1696.

91 H. Effenberger, *Z. Krist.* **1983**, *165*, 127.

92 J. Souletie, *J. Phys. (Paris)* **1988**, *49*, 1211.

93 M. Drillon, P. Panissod, *J. Magn. Magn. Mat.* **1998**, *188*, 93 ; P. Panissod, M. Drillon, in *Magnetism: Molecules to Materials* IV, eds J. S. Miller, M. Drillon, Wiley VCH, Weinheim, 2002.

94 C. Hornick, P. Rabu, M. Drillon, *Polyhedron* **2000**, *19*, 259.

95 Z. L. Huang, M. Drillon, N. Masciocchi, A. Sironi, J. T. Zhao, P. Rabu, P. Panissod, *Chem. Mater.* **2000**, *12*, 2805.

96 S. Abdelouhab, M. François, P. Rabu, to be published.

97 H. Heise, F. H. Köhler, to be published.

98 W. Fujita, PhD thesis, University of Tokyo, 1996.

99 V. Laget, C. Hornick, P. Rabu, M. Drillon, P. Turek, R. Ziessel, *Adv. Mater.* **1998**, *10*, 1024.

100 G. Weissman, R. Claiborne, *Cell Membranes*, HP Publishing, New York, 1975.

101 T. Kajiyama, *J. Macromol. Sci. Chem. A25*, **1988**, 583.

102 W. Fujita, K. Awaga, *J. Am. Chem. Soc.* **1997**, *119*, 4563.

103 S. A. Simon, T. J. McIntosh, *Biochim. Biophys. Acta* **1984**, *773*, 169.

104 J. L. Slater, C.-H. Huang, *Prog. Lipid Res.* **1988**, *27*, 325.

9
Building Multifunctionality in Hybrid Materials

Eugenio Coronado, José R. Galán-Mascarós, and Francisco Romero

9.1
Introduction

The field of molecule-based materials has seen very rapid progress since the discovery of a variety of useful solid-state properties such as conductivity and superconductivity, non-linear optics, and ferromagnetism. After these discoveries, one of the most appealing aims is to create molecular materials that combine in the same crystal lattice two properties that are difficult or impossible to achieve in a conventional inorganic solid. This opens new possibilities for potential applications in molecular electronics. An attractive approach to this issue is based on the design of hybrid molecular materials formed by two molecular networks, such as anion/cation salts or host/guest solids, where each network furnishes distinct physical properties to the solid. This novel class of materials is interesting because it can give rise to the sought-for materials with coexistence of two properties, or to materials exhibiting improved properties with respect to those of the individual networks, or new unexpected properties, due to the mutual interactions between them.

We can imagine, for example, the combination of an extended inorganic magnetic layer furnishing the pathway for cooperative magnetism, with an organic radical. We may end up with a hybrid magnet that combines cooperative magnetism and paramagnetism. Other suitable combinations like electronic conductivity or superconductivity and magnetism, or non-linear optics and magnetism, can also be achieved from a wise choice of the constituent molecules. The interplay between the two properties can also provide interesting situations such as the control of one property by acting on the other with an external stimulus (light, pressure, magnetic field etc.). In the present contribution we report some relevant examples that illustrate the potentialities of this hybrid approach.

Gómez-Romero: Organic-Inorganic Materials. Eugenio Coronado

ISBN: 3-527-30484-3

9.2
Combination of Ferromagnetism with Paramagnetism

9.2.1
Magnetic multilayers

The use of paramagnetic molecules as templating agents for the formation of extended magnetic networks can provide the opportunity to obtain two network materials in which the cooperative magnetism of an extended inorganic network coexists with the paramagnetism of the inserted molecule. [1] Several examples of this possibility are furnished by the bimetallic complexes based upon the oxalato, $(C_2O_4)^{2-}$, ligand.

Attending only to geometric considerations, it is easy to envision the different types of networks that can be formed based on tris(oxalato)metallate complexes. If we build a polymer with oxalate metal complexes, in which each metal center is octahedrally coordinated by three bidentate oxalate ions with all oxalate ligands acting as bridging ligands, one possibility would be the formation of a 2D hexagonal network made out of six member rings (Fig. 1), with each metal linked to three nearest neighbors. In order to remain in the plane, an alternating chirality of the metal centers is needed. This means that the complexes in the structure will show alternatively and chirality. But if instead of a racemic mixture, only one of the stereoisomers is used to build an analogous network, it is clear that a 2D structure is impossible, since the oxalate from first neighbors of the same chirality will point up and down the plane to bind the second sphere of metals. Still, a chiral 3-connected 10-gon 3D net can be built this way (Fig. 9.2).

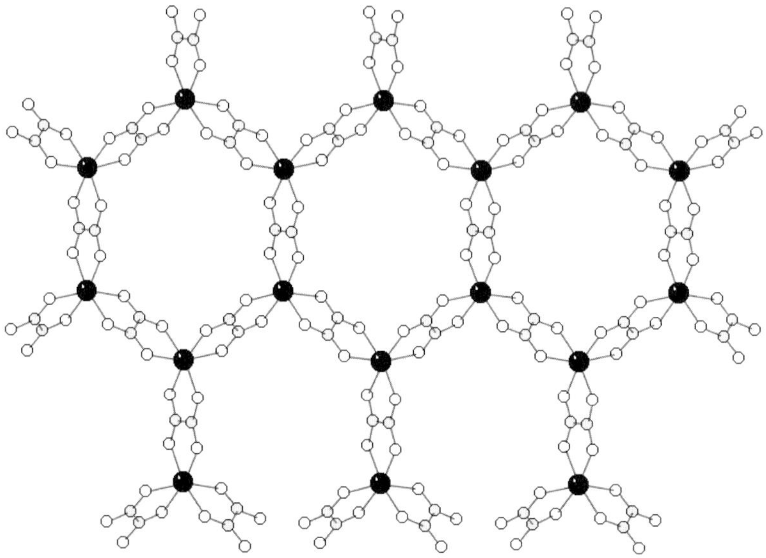

Fig. 9.1 Scheme of the 2D network of general formula $[M^{n+}M^{m+}(ox)_3]^{-(6-n-m)}$

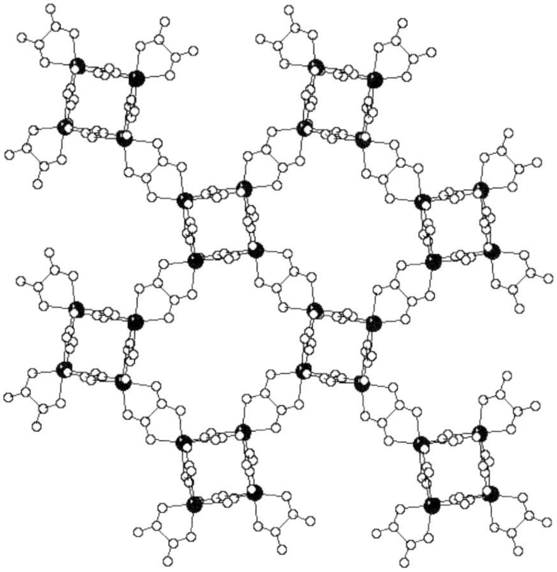

Fig. 9.2 Scheme of the 3D network of general formula $[M^{n+}M^{m+}(ox)_3]^{-(6-n-m)}$

Both types of networks exhibit identical magnetic connectivity, although different structural dimensionality, and analogous general formula: $[M^{+n}M^{+m}(ox)_3]^{-6+n+m}$. All the experimentally isolated compounds of these families are anionic, with many of the members of both families exhibiting cooperative magnetic phenomena due to the magnetic exchange interaction promoted by the oxalato bridges between paramagnetic metals. In each case the stabilization of one type of network or the other requires the use of the right templating cation.

Bulky monocations are able to template the formation of the 2D networks. These series were first prepared with "innocent" cations, such as tetraalkylammonium or tetraphenylphosphonium, [2] and characterized as molecular-based magnets of general formula $[cat][M^{II}M^{III}(ox)_3]$ (M^{II}= Mn, Fe, Co, Cu and Zn; M^{III} = Cr, Fe and Ru), as they order as ferro-, [3] ferri- [4] or canted antiferromagnets [5] with critical temperatures ranging from 5 K up to 44 K. These 2D networks have been used extensively for the preparation of magnetic multilayers, when alternating the anionic oxalate-based magnetic networks with paramagnetic cations, such as the organometallic monocation decamethylferrocenium $[FeCp^*_2]^+$. [6] The family of $[Z^{III}Cp^*_2][M^{II}M^{III}(ox)_3]$ (Z^{III} = Mn, Fe and Co; M^{II} = Mn, Fe, Co, Ni, Cu and Zn; M^{III} = Cr, Fe and Ru) [7] hybrids is a perfect example of how a chemical approach can also be used for the construction of relevant multifunctional systems, since it allows for the study of the effects that inserted paramagnetic ($[FeCp^*_2]^+$) or diamagnetic ($CoCp^*_2$) molecular layers have on ferro- or ferrimagnetic layered systems. The fact that these crystalline materials are perfectly ordered represents a big advantage over previous systems prepared by physical methods, where the defects and structural disorders are an important additional obstacle to reach final conclusions.

The first observation from these hybrid systems is that the magnets are mostly unaffected by the insertion of the paramagnetic layers, with all critical temperatures remaining unchanged within experimental error, and with only a small influence on the coercive fields observed, that decrease when a paramagnetic layer is present (Table 9.1). This result agrees with the lack of electronic interactions between the two networks, which are quasi-independent. Still, one can observe that below T_c the internal magnetic field created in the material acts on the spins of the paramagnetic molecules, which feel a magnetic field of 3000–4000 Oe.

This could be determined from EPR measurements, where the signal that corresponds to the $[FeCp_2^*]^+$ ions inserted in between the ferromagnetic bimetallic layers moves towards lower magnetic fields below the critical temperature of the magnets. For example, in the $[FeCp_2^*][FeCr(ox)_3]$ ferromagnet the $g_\perp$ component moves from 8000 G at 15 K to 3500 G at 4.3 K, when T_c is 13 K (Fig. 9.3). This g-shift indicates that the $S = 1/2$ of the $[FeCp_2^*]^+$ units "sees" the internal magnetic field that is added to the applied magnetic field so as to decrease the resonance field.

A step further into the understanding of complex layered magnetic systems can also be achieved by this series of materials. Since the sign of the magnetic interaction is controlled by the pair of metals used in the preparation, with Cr^{III} favoring ferromagnetic interactions, as observed in the $M^{II}Cr^{III}$ series of ferromagnets, and with Fe^{III} favoring antiferromagnetic interaction as observed in the $M^{II}Fe^{III}$ series of ferrimagnets, the preparation of solid solutions $M^{II}Cr_x^{III}Fe_{1-x}^{III}$ $(0 < x < 1)$ enables the study of the effect that competing interactions within the layers have on the magnetic properties. While the changes in critical temperatures are almost linear, with intermediate values between the pure Cr and pure Fe compounds, a huge increase in the coercive fields of these magnets accompanied by spin glass behavior [8] is obtained and it can be easily tuned by altering the chemical composition of the material. For example, for $[CoCp_2^*][Fe^{II}Fe_x^{III}Cr_{1-x}^{III}(ox)_3]$ the coercive field increases from values of 750 Oe $(x = 0)$ and 90 Oe $(x = 1)$, to a value as high as 16700 Oe (1.67 Teslas) [9] for $x = 0.5$ (Fig. 9.4), which is comparable to those observed in hard magnets. This strategy can give even better results when Cr^{III} is substituted by Ru^{III} in these solid solutions. This way one maintains a situation with competing ferro-

Tab. 9.1 Magnetic Parameters for the 2D Series $[ZCp_2^*][M^{II}M^{III}(ox)_3]$: critical temperature ($T_c$), saturation magnetization at 2 K and 5 T (M) and coercive field at 2 K (H_{coer})

M^{III}	M^{II}	T_c (K)	$[CoCp_2^*]$		$[FeCp_2^*]$		
			M(μ_B)	H_{coer} (mT)	T_c (K)	M(μ_B)	H_{coer} (mT)
Cr	Mn	5.1	7.0	4	5.3	8.1	2
	Fe	12.7	4.6	194	13.0	6.2	110
	Co	8.2	5.0	25	9.0	5.5	13
	Cu	6.7	3.6	20	7.0	4.9	4
Fe	Mn	25.4	0.4	15	28.4	1.3	12
	Fe	44.0	0.5	10	43.3	1.4	37

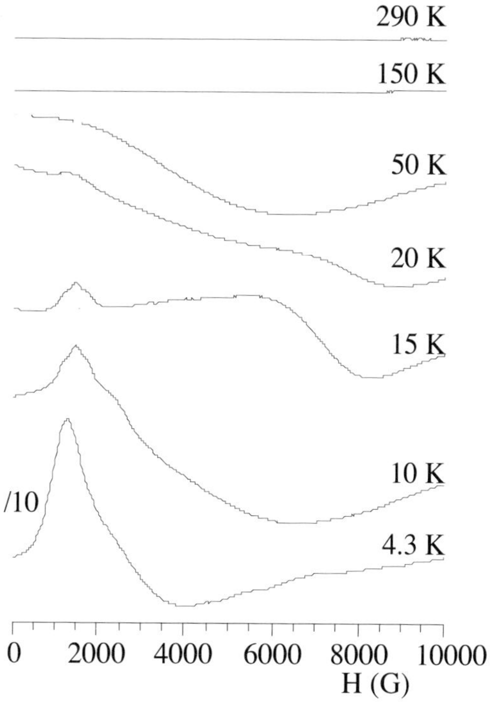

Fig. 9.3 Temperature dependence of the ESR spectra of the salt $[FeCp^{*}_{2}][FeCr(ox)_3]$

and antiferromagnetic interactions, promoted in this case by Ru^{III} and Fe^{III} respectively, but at the same time one increases the local ion anisotropy, with important effects on the hysteresis loops for these materials. The maximum in coercivity for the $[CoCp^{*}_{2}][Fe^{II}Fe^{III}_{x}Ru^{III}_{1-x}(ox)_3]$ was observed for $x = 0.47$, for a value of 22000 Oe (2.2 T; Fig. 9.5). [10]

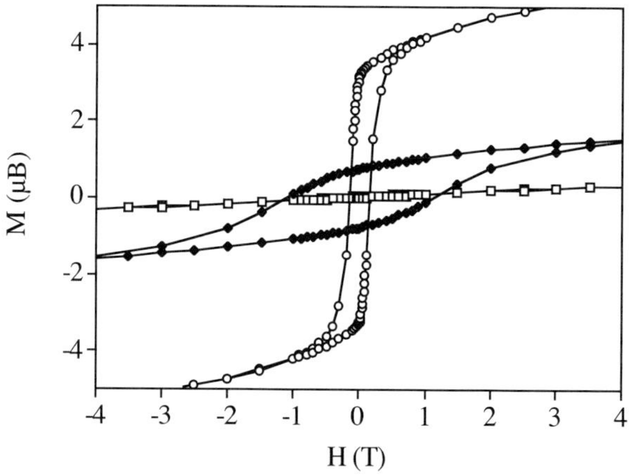

Fig. 9.4 Increasing the coercivity in the 2D $[NBu_4][Fe^{II}Fe^{III}_{x}Cr^{III}_{1-x}(ox)_3]$ layered magnets: $x = 1$ (empty squares), $x = 0$ (empty circles), and $x = 0.66$ (full squares)

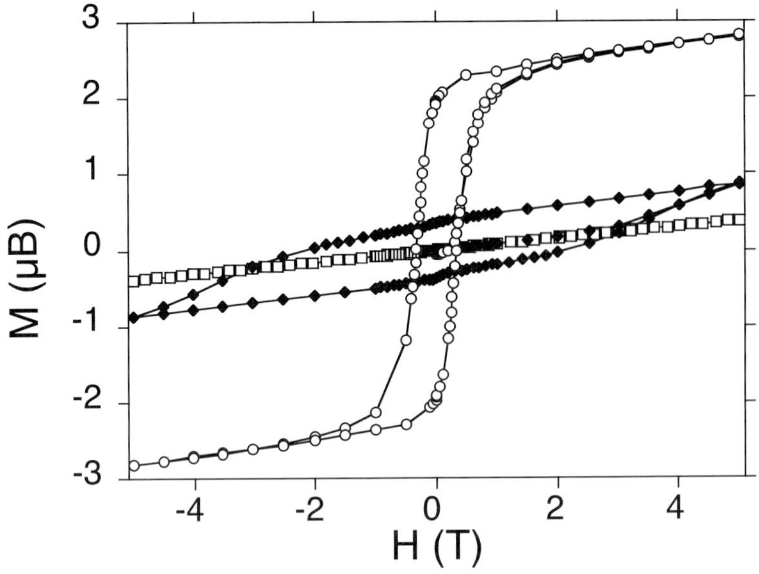

Fig. 9.5 Increasing the coercivity in the 2D $[NBu_4][Fe^{II}Fe^{III}_xCr^{III}_{1-x}(ox)_3]$ layered magnets: $x = 1$ (empty squares), $x = 0$ (empty circles), and $x = 0.50$ (full squares)

9.2.2
Host-guest 3D Structures

In order to obtain the analogous 3D series, chiral metallic complexes such as $[M(bpy)_3]^{n+}$ need to be used. It is noteworthy that these compounds self-assemble from solution, and that resolution of both isomers is not needed for the preparation. Racemic starting materials will form chiral crystals by cationic-anionic host-guest self-recognition. Of course, polycrystalline samples will contain crystallites of both chiralities, but each individual crystal will be chiral. The series of ferromagnets $[Z^{II}(bpy)_3][ClO_4][M^{II}Cr^{III}(ox)_3]$ (Z^{II} = Ru, Fe and Co; M^{II} = Mn, Fe, Co, Ni, Cu and Zn) has been well characterized [11] and even an enantiomerically pure synthesis has been described. [12] This series shows some differences with its 2D analogues. While the magnetic interactions remain ferromagnetic, as in the 2D Cr^{III} series, the interactions are heavily dependent on the nature of the counterions, with big changes observed in T_c (Table 9.2). For example, in the $[Z^{II}(bpy)_3][ClO_4][CoCr(ox)_3]$, the T_c changes from 2.8 K (Z^{II} = Ru) up to 6.6 K (Z^{II} = Fe). This effect is more related to the size of the cation that forces an expansion of the network, lengthening the metal-metal distances and therefore weakening the magnetic interactions, than to electronic interactions between the two sub-lattices, which seem to be quasi-independent. Still, one can take advantage of the possibility of modulating the size of the cavities to induce a variation in the properties of the cation. This peculiar characteristic has been exploited by Hauser and coworkers in order to induce a thermal spin transition in the compound $[Co(bpy)_3][LiCr(ox)_3]$. The ground state of $[Co(bpy)_3]^{2+}$ is usually the high-spin $^4T_1$ state. Inserting this complex inside the $[LiCr(ox)_3]^-$ network results in an increase of the chemical pressure and the ^2E low-spin state becomes the actual ground state. In fact, $[Co(bpy)_3][LiCr(ox)_3]$ shows a gradual spin transition with $T_{1/2} = 161$ K, as has been shown by magnetic measurements and single-crystal absorption spectroscopy. [13]

An alternative organic-inorganic hybrid salt approach towards the introduction of paramagnetic units in magnetically ordered materials is the use of cationic nitronyl-nitroxide (NN) radicals of the N-alkylpyridinium type [14] (rad^+, see Fig. 9.6) as templating agents in the formation of the extended oxalato-based networks. We have shown that upon combination of the salt $rad_3[Cr(ox)_3]$ with hexa aquo divalent metal precursors it is possible to obtain under kinetic control (fast precipitation) a family of compounds of formula $rad[MCr(ox)_3] \cdot 2\,H_2O$, showing the classical 2D oxalate network. All the compounds of this series order ferromagnetically with coercive fields ranging between 0.05 and 4.6 kG. The thermal variation of the in-phase component of the AC magnetic susceptibility for the MnCr

Tab. 9.2 Magnetic Parameters for the 3D Series $[Z^{II}(bpy)_3][ClO_4][M^{II}Cr^{III}(ox)_3]$: critical temperature ($T_c$), saturation magnetization at 2 K and 5 T (M) and coercive field at 2 K (H_{coer})

Z^{II}	M^{II}	T_c (K)	$M(\mu_B)$	H_{coer} (mT)
Ru	Mn	<1.9	6.7	0
	Fe	2.5	5.0	1.4
	Co	2.8	4.9	0.8
	Ni	6.4	4.0	2.2
	Cu	1.9	3.9	1.4
Fe	Mn	3.9	6.0	0
	Fe	4.7	4.5	8.0
	Co	6.6	5.0	5.5
Ni	Mn	2.3	7.7	1.3
	Fe	4.0	5.8	2.8
Co	Mn	2.2	7.5	1.3

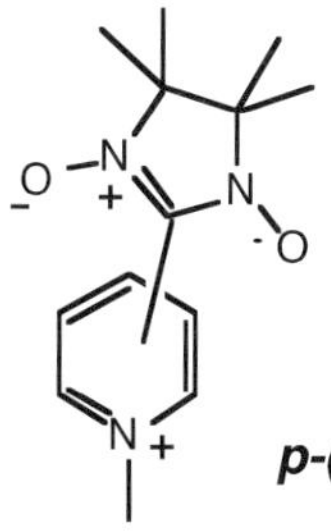

p-(m-)rad$^+$

Fig. 9.6 Cationic free radicals: rad^+

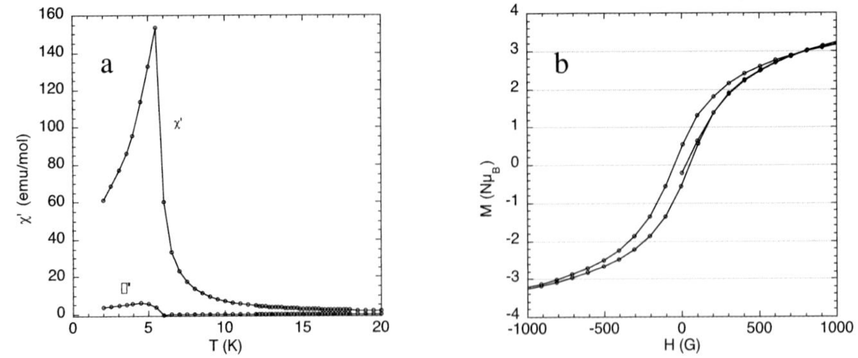

Fig. 9.7 (a) Thermal variation of the AC magnetic susceptibility at 997 Hz and (b) field dependence of the magnetization at 2 K for (rad)[$MnCr(ox)_3$] · 2 H_2O

derivative (Fig. 9.7) shows an abrupt peak at 6 K. Below this temperature, the onset of an out-of-phase signal is observed. The field dependence of the magnetization shows a relatively low coercive field ($H_C \sim 50$ G).

However, the combination of diluted solutions of $[Mn(H_2O)_6](ClO_4)_2$ and p-rad$_3$ $[Cr(ox)_3]$ (thermodynamic control) afforded the hybrid product p-rad$[Mn(H_2O)Cr(ox)_3]$ · 2 H_2O. The crystal structure [15] of this compound (Fig. 9.8) shows the presence of μ-oxalato bridges that organize alternating Mn-Cr linear arrays along the *c* axis. Within each array, the absolute configuration of the different building blocks follows the sequence ΔΔΛΛ ... The Mn-Cr chains are interconnected through another oxalate anion that binds adjacent Cr^{III} and Mn^{II} centers in a bidentate-monodentate mode. This results in the formation of an unprecedented achiral three-dimensional (10, 3) network.

The magnetic interaction between Mn^{2+} and Cr^{3+} cations connected through a μ-oxalato bridge is expected to be ferromagnetic, while the coupling through the

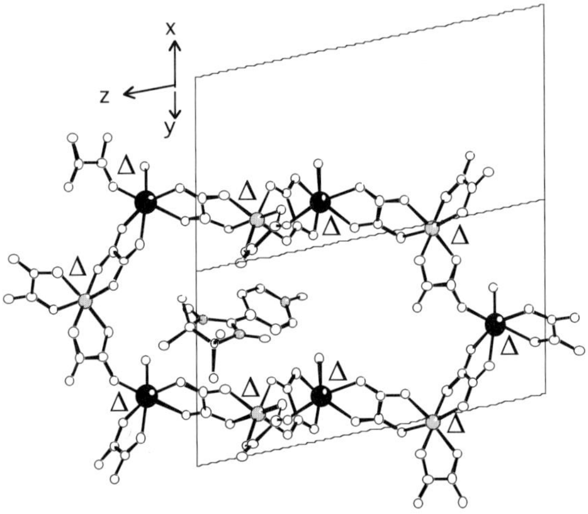

Fig. 9.8 View of the crystal structure of p-rad$[Mn(H_2O)Cr(ox)_3]$ · 2 H_2O showing the structural motif of the (10, 3) network

oxalate anion acting in a lower symmetry mode might be antiferromagnetic. Indeed, the analysis of the magnetic properties of a crystalline sample of this material is indicative of competing magnetic interactions, and the compound orders as an antiferromagnet at $T_N \sim 6$ K.

9.3 Hybrid Molecular Materials with Photophysical Properties

9.3.1 Photo-active Magnets

The coexistence of ferromagnetism with photophysical properties is one of the areas where molecular-based magnets acquire special relevance. Most of the classic magnets are also conductors, and therefore opaque. Light is absorbed by conducting electrons, with no possibilities for further studies. Molecular-based magnets being insulators, and most of them transparent, open a new door for the preparation of hybrid systems built from magnetic networks and optically active hosts, whose properties can now be carefully studied.

The very same oxalate-based bimetallic networks discussed above are flexible enough to be used in the preparation of this type of multifunctional system. Optically-active organic monocations have been used to prepare a series of multilayered materials of general formula $[D][M^{II}Cr^{III}(ox)_3] \cdot n$. Solvent where D is a chromophore. Bimetallic oxalate-based layers, described above and responsible for the magnetic ordering, alternate with layers of the optically active cations. These series are very useful for the study of interactions between magnetic ordering and the optical properties of the cationic layers. When D is a NLO active molecule such as hyperpolarizable stilbazolium-shaped A chromophores and others (Fig. 9.9), [16] it was reported that these materials exhibit unchanged magnetic properties and also Second Harmonic Generation (SHG) capability. Factors that allow for noncentrosymmetric packing arrangement of the chromophores in between the centrosymmetric magnetic layers are crucial for NLO activity, such as the matching between cationic size and interlayer separation. This approach yielded SHG active ferromagnets with SHG up to 100 times that of urea.

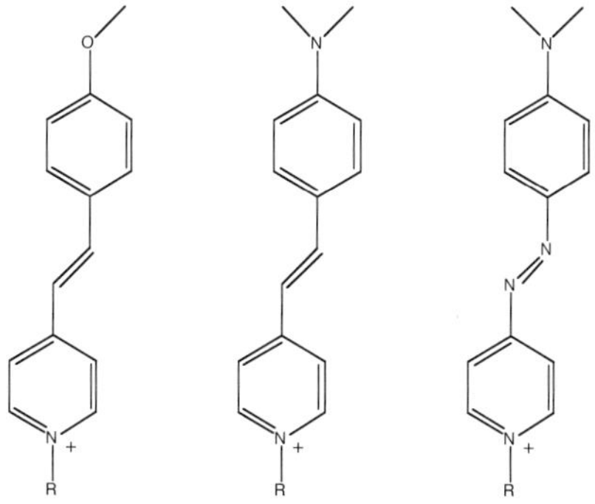

Fig. 9.9 Cationic chromophores (R = alkyl group)

When D is a bi-stable chromophore, such as a cationic spiropyran able to change its molecular structure by the action of UV radiation (Fig. 9.10), a more interesting effect was observed. [17] In this case, by irradiation with UV light the magnetic properties of the material change irreversibly. In particular the coercivity observed at 2 K changes from 40 G up to 290 G. These changes have been attributed to photoinduced crystal structure defects through the photoreaction of the chromophore. Still, some aspects are unclear with respect to this system, since the irreversibility of this effect could also be related to permanent damage by the UV radiation on the material, probably promoted by the chromophore, since in its absence no similar effects were observed.

Complexes able to exhibit spin crossover [18] are also convenient bi-stable building blocks that present a metastable long-lived state that can be accessed thermally, by irradiation with light or by pressure. [19] The most common complexes that show spin crossover are usually Fe complexes, either neutral or cationic. [20] Some attempts have been made to incorporate such cations into a magnetic anionic host to obtain two-network hybrid materials. In addition to the synthetic difficulties, the fact that the spin transition depends on the solid-state interactions makes it difficult to achieve such a goal. A complex able to exhibit spin transition, $[FeL_2]^+$, $L = [O(C_6H_4)N = CH(C_5NH_4)]^-$) was successfully inserted in bimetallic oxalate-based bidimensional magnets, [21] but the $[FeL_2][MnCr(ox)_3]$ did not exhibit spin crossover in addition to the ferromagnetic ordering observed below 6 K.

The oxalate-based bimetallic 3D systems are especially relevant in the search for photoactive magnets, from two different perspectives. The fact that these ferromagnets are chiral opens the possibility to study magneto-chiral effects in crystalline molecular-based materials. Magneto-chiral studies have been successfully performed on paramagnetic inorganic solids, [22] but no chiral magnets have been studied to date. Bulk ferromagnetism could amplify this effect, since the magnetic moment in a chiral magnet will be much larger than the individual spins of a paramagnetic sample. Therefore chiral molecule-based magnets are of great interest in this regard, with a few other examples found in cyanide chemistry [23] and organic chemistry. [24]

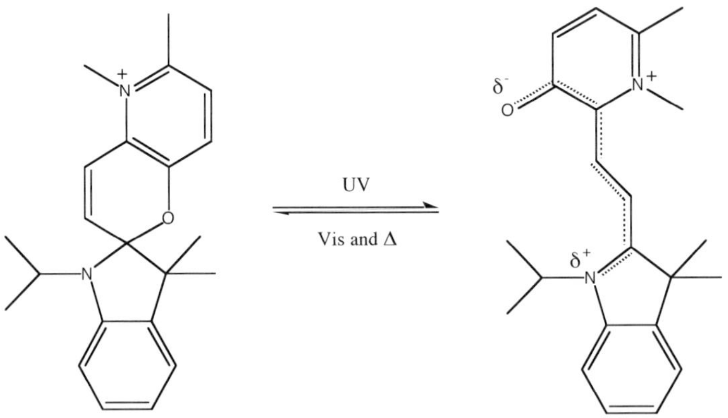

Fig. 9.10 Photoinduced equilibrium of the spiropyran cation

Another perspective involves the use of photoactive $[M(bpy)_3]^{n+}$ cations to study the synergetic effects between the two sublattices, the magnetic network and the luminescent network. Photophysical studies [25] have already been reported for the series $[Ru(bpy)_3][NaAl_{1-x}Cr_x(ox)_3]$ and $[Ru_{1-x}Os_x(bpy)_3][NaAl(ox)_3]$, although in this case these compounds lack cooperative magnetic ordering because of the presence of diamagnetic metals (Na or Al) as part of the network . Still, excitation transfer processes, with the photosensitizer $[Ru(bpy)_3]^{2+}$ as donor and $[Os(bpy)_3]^{2+}$ and $[Cr(ox)_3]^{3-}$ as acceptors, have been observed. The extension of these studies to the analogous magnets $[Ru(bpy)_3][ClO_4][M^{II}Cr(ox)_3]$ is under progress.

9.3.2
Photo-active Conductors

The interest of using such bi-stable molecules in the construction of two-network molecular-based materials resides in the possibility to access the excited state, which usually includes changes in the magnetic properties and more especially in the structural features of these molecules. Thus, these changes induced by external optical stimuli on the photochromic molecules would induce structural changes in the materials that could have a profound effect on the physical properties of the second network.

The possibility to obtain charge transfer salts of this kind has been explored for the nitroprusside anion, $[Fe(CN)_5NO]^{2-}$. This complex exhibits a photoinduced transition to an extremely long-lived metastable state upon irradiation with light in the wavelength range 350–580 nm at temperatures below ca. 160 K. [26] This electronic transition is accompanied by a possible change in the geometry of the complex. A BEDT-TTF salt of this anion was reported as exhibiting superconductivity below 7 K, [27] although detailed studies by different research groups indicated that this observation was probably an artifact. [28]

Other radical salts of this anion have been reported. θ-$(BEDT\text{-}TSF)_4[Fe(CN)_5NO]$ [29] exhibits metallic behavior down to 40 K where it undergoes a first-order metal-to-semiconductor phase transition. $(TTF)_7[Fe(CN)_5NO]$ is a semiconductor [30] and DSC measurements on irradiated samples have suggested the existence of the metastable state at temperatures as high as 240 K. [31] This may indicate the presence of strong interactions between the TTF molecules and the nitroprusside anions. Indeed, the material is formed by alternating layers of the anions and the cations, with the NO group "penetrating" partially into the organic layer with short S–O contacts. Since the metastable state is related to the isomerization of the NO group, this strong interaction in the solid state should be responsible for the stabilization of the excited state that usually is observed at much lower temperatures. There is still an open question on this system: how the population of the metastable state affects the electrical conductivity. In any case, the effect is very small because of the small fraction of nitrosyl complexes accessible to light excitation. The black color of this material indicates that most of the photons are absorbed by the conducting network, reducing the penetration depth of the irradiation.

9.4 Combination of Magnetism with Electric Conductivity

This striking possibility, in which magnetic properties are combined with conducting or superconducting properties in the same crystal lattice, constitutes a contemporary challenge in materials science, and molecule-based materials offer a unique approach. There are at least two main reasons that justify the intense effort currently devoted to prepare and physically characterize this type of material. The first one is to investigate if a metallic conductivity can stabilize indirect exchange interactions among the localized magnetic moments mediated by the conducting electrons. This exchange interaction resembles the so-called RKKY mechanism proposed in the solid state to explain the magnetic interactions in transition and rare-earth metals and alloys. [32] The studies with the molecular materials have led to a new insight, namely intermolecular d-π coupling between the two networks, in contrast with the intra-atomic coupling between the unpaired electrons of the metal (d or f) and the free (mainly s) electron carriers. Furthermore, in molecular conductors the electrons are strongly correlated, while the RKKY model assumes free electrons. A second impetus is that of bringing together in the same crystal lattice unusual combinations of physical properties, as for example ferromagnetism and superconductivity, which have long been considered as mutually exclusive. [33]

The organic radicals of the TTF family (Fig. 9.11), well-known for being the largest family of organic conductors and superconductors, [34] present many advantages to be used as building blocks for hybrid molecular conductors. This type of organic molecules tends to stack in the solid state to form segregated 2D cationic layers, alternating with anionic layers that act as counterions. The goal consists of electrochemically preparing solids formed by stacks of the partially oxidized BEDT-TTF π-electron donor, that support electronic conductivity or even superconductivity, with magnetic complexes as charge-compensating anions (Fig. 9.12).

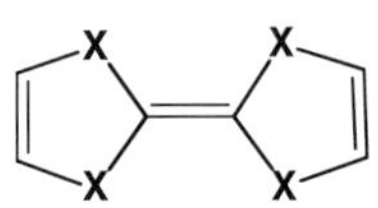

(X = S) Tetrathiofulvaleno (TTF)
(X = Se) Tetraselenofulvaleno (TSF)

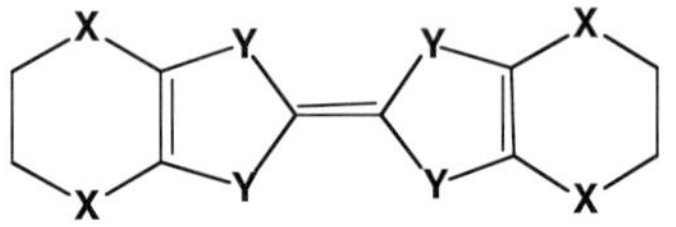

(X = Y = S) BEDT-TTF (ET)
(X = S; Y = Se) BEDT-TSF (BETS)
(X = Se; Y = S) BEDS-TTF (BEST)
(X = O; Y = S) BEDO-TTF (BEDO)

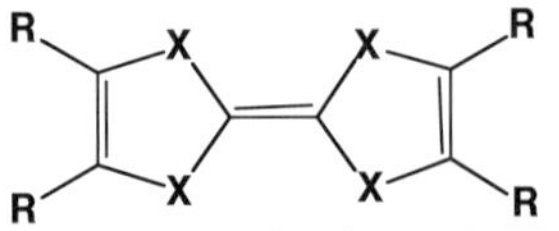

(X = S; R = -CH_3) TM-TTF
(X = Se; R = -CH_3) TM-TSF

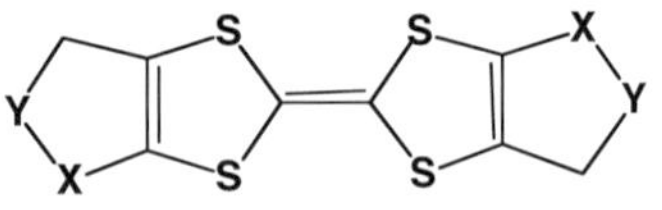

(X = S; Y = -CH_2) BET-TSF (BET)
(X = -CH_2; Y = S) BMDT-TTF

Fig. 9.11 Organic donors of the TTF family

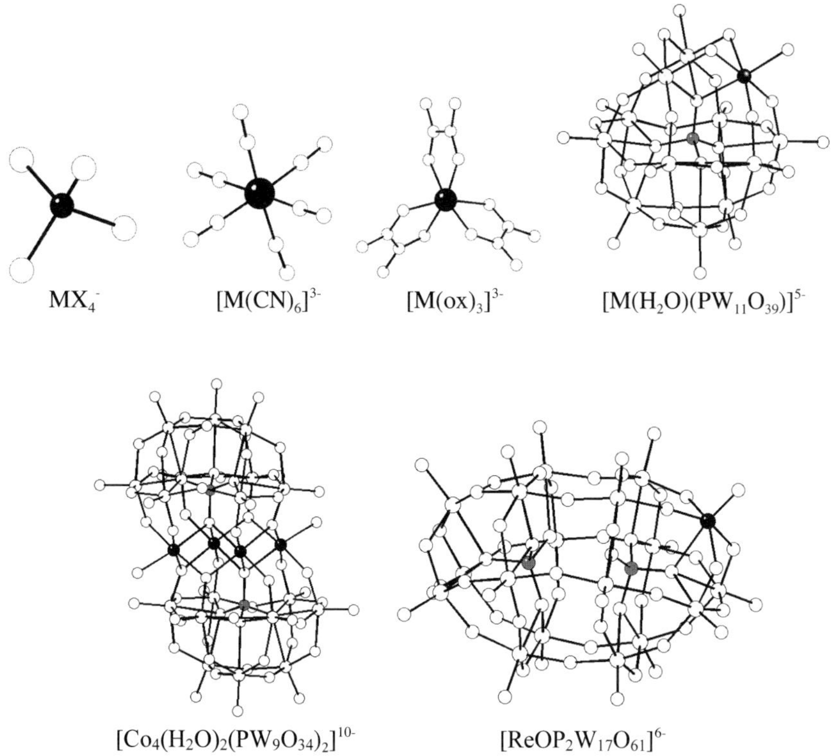

Fig. 9.12 Several paramagnetic anions used as counterions in hybrid conductors with organic donors

9.4.1
Paramagnetic Conductors from Small Inorganic Anions

A wide choice of small paramagnetic anions of different geometries and sizes has been used in the preparation of radical salts of the TTF family: tetrahedral complexes $[MX_4]^{n}$- (M=Fe^{III}, Cu^{II}; X^- = Cl, Br, SCN); octahedral complexes $[M(X)_6]^{3-}$ (X^- = Cl, CN, SCN; M^{III} = Fe, Cr, Co) and $[M(ox)_3]^{3-}$ (ox = oxalate; M^{III} = Cr, Fe, Al, Ga, Rh); and also mixed ligand complexes and combinations of all these. This simple approach has been very successful, with two paramagnetic series of superconductors reported to date.

The first one was obtained by P. Day and co-workers in the series (BEDT-TTF)$_4$[$A^IM^{III}(ox)_3$].solvent (A^+ = H_3O, K, NH_4; M^{III} = Cr, Fe, Co, Al; solvent = C_6H_5CN, $C_6H_5NO_2$, C_5H_5N). [35] Although with the same formula, two different structural types are found for this family. Most of these materials crystallize in an orthorhombic space group, while others crystallize in a monoclinic space group. The cationic layers in the orthorhombic systems present charge localization, with dimers of completely oxidized (+1) BEDT-TTF molecules surrounded by neutral ones, which explains the semiconducting behavior observed experimentally for these phases. By contrast, in the monoclinic materials, the organic radicals stack in a delocalized real 2D structure with short S . . . S distances, reminiscent of the well-known β″ phase, responsible for the appearance of superconductivity at low temperatures. $(BEDT\text{-}TTF)_4[(H_3O)Fe(ox)_3] \cdot C_6H_5CN$ (Fig. 9.13) was the first example of a molecular superconductor containing a lattice of magnetic ions. [36] It shows a conductivity of ~102 S cm^{-1} at 200 K, with the resistance decreasing monotonically by a factor of about 8 down to temperatures between 7–8 K where it becomes superconducting (Fig. 9.14). Another member of this family exhibits a lower critical temperature at 3 K ($A^+ = H_3O$; $M^{III} = Cr$; solvent = $C_6H_5NO_2$). [37]

The second paramagnetic superconductor was prepared on the basis of a remarkable observation: the contrasting behavior in the two isomorphous salts λ-$(BETS)_2[MCl_4]$ (M^{III} = Fe, Ga). The structural features being identical, with layers of BETS-TTF molecules alternating with layers of the $[MCl_4]^-$ anions, the Ga compound exhibits a metal-superconductor transition at 8.5 K, while the Fe compound, on the other hand, undergoes antiferromagnetic order accompanied by a sharp metal-insulator transition. [38] Thus, both salts being isostructural, it was possible to dope the Ga compound with Fe without destroying the superconducting transition, and preparing in this way a paramagnetic superconductor. [39]

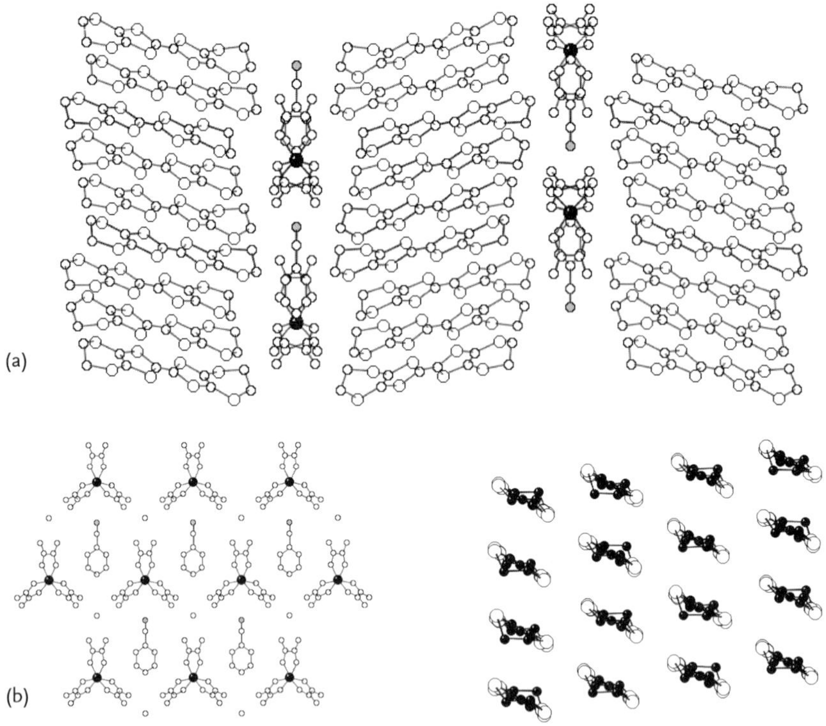

Fig. 9.13 Representation of the crystal structure of the $(BEDT\text{-}TTF)_4[(H_3O)Fe(ox)_3] \cdot C_6H_5CN$ salt: (a) multilayer structure, (b) top view of the inorganic and organic layers

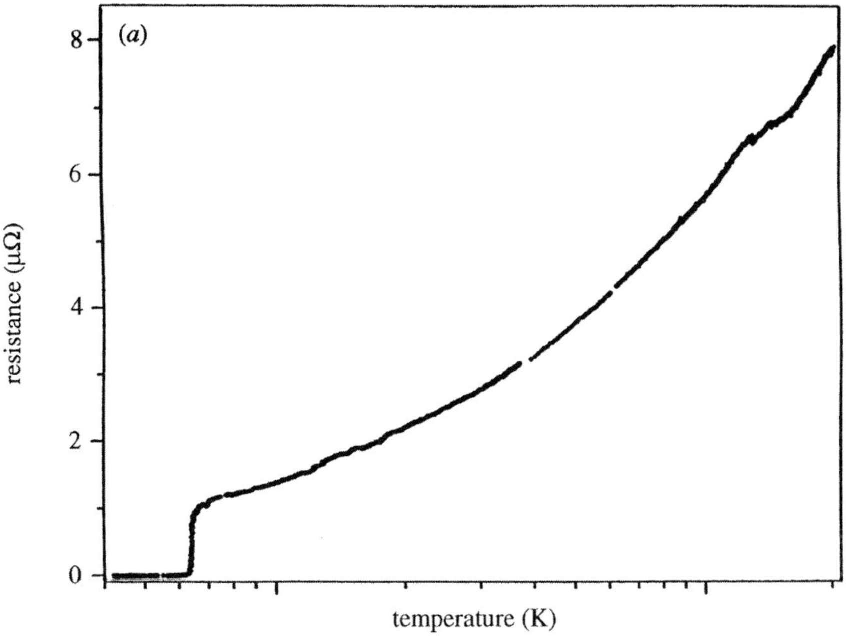

Fig. 9.14 Temperature dependence of the resistance of $(BEDT\text{-}TTF)_4[(H_3O)Fe(ox)_3] \cdot C_6H_5CN$

Further studies on the $\lambda(BETS)_2[FeCl_4]$ salt have discovered many interesting features, such as magnetic-field-induced superconductivity under very high magnetic fields ($H > 18$ T). [40], [41] This striking result demonstrates how by applying a magnetic field the superconducting transition can be triggered, while the opposite effect is usually observed in most superconductors. This very same effect has also been observed in the related κ $(BETS\text{-}TTF)_2FeBr_4$. [42]

Some examples of paramagnetic metals have also been reported, such as $\alpha(BEDT\text{-}TTF)_2Cs[Co(NCS)_4]$, [43] which undergoes a metal-insulator transition at 20 K, and the compounds $(BET\text{-}TTF)_2[MCl_4]$ (M^{III} = Ga, Fe) [44] which undergo also a metal-insulator transition at 30 K and 20 K respectively. All these compounds present a layered structure, with layers of the organic radicals alternating with layers of the inorganic anions, and with little contact between the two networks.

Nevertheless, most of the radical salts obtained with small paramagnetic complexes show semiconducting behavior, with relatively high room-temperature conductivity. [45] This is the case for the salts α-$(BEDT\text{-}TTF)_4[ReCl_6] \cdot C_6H_5CN$ and $(BEDT\text{-}TTF)_2[IrCl_6]$, that magnetically they show paramagnetic behavior dominated by the anions. Similar overall behavior is observed in the β phases $(BEDT\text{-}TTF)_5[Fe(CN)_6] \cdot 10\,H_2O$ [46] and $(BEST\text{-}TTF)_4[Fe(CN)_6]$; [47] and the κ phases $(BEDT\text{-}TTF)_4(NEt_4)[Fe(CN)_6] \cdot 3\,H_2O$ [48] and $(BET\text{-}TTF)_4(NEt_4)_2[Fe(CN)_6] \cdot 10\,H_2O$ [49]. Magnetically, appreciable Pauli paramagnetism was observed in only a few of these salts, added to the paramagnetism of the anions.

A common trend for all these materials, paramagnetic metals and semiconductors is the absence of correlations between itinerant (conducting) electrons and localized electrons (paramagnetic centers). Even when thiocyanide complexes were used with the aim of promoting interactions with the sulfur-containing organic molecules, no close magnetic interactions between the two sub-lattices were observed.

Some attempts have been made to increase the solid-state magnetic interactions between the two networks. One strategy takes advantage of modified TTF-type donors by substitution of component atoms or functional groups. By introduction of the desired functional groups, one can favor the interactions with the metal complexes in the solid state. One example is that of the donor C_1TET-TTF (4,5-ethylene-dithio-4′,5′-bis(methylthio)tetrathiafulvalene (Fig. 9.15). While the ethylenedithio groups from the more common radical BEDT-TTF prefer interactions along the short side of the molecule, [50] and usually are involved in short contacts inside the organic network, the methylthio groups lead to interactions along the long side, [51] and can be involved in short contacts with halogen atoms. This favors strong magnetic coupling between the organic radicals and halogen complexes of paramagnetic transition metals. The salts (C_1TET-TTF)FeX_4 (M = Cl, [52] Br [53]) have been reported as insulators, probably due to the fact that the organic donors appear completely oxidized. Although their structures are almost identical, the

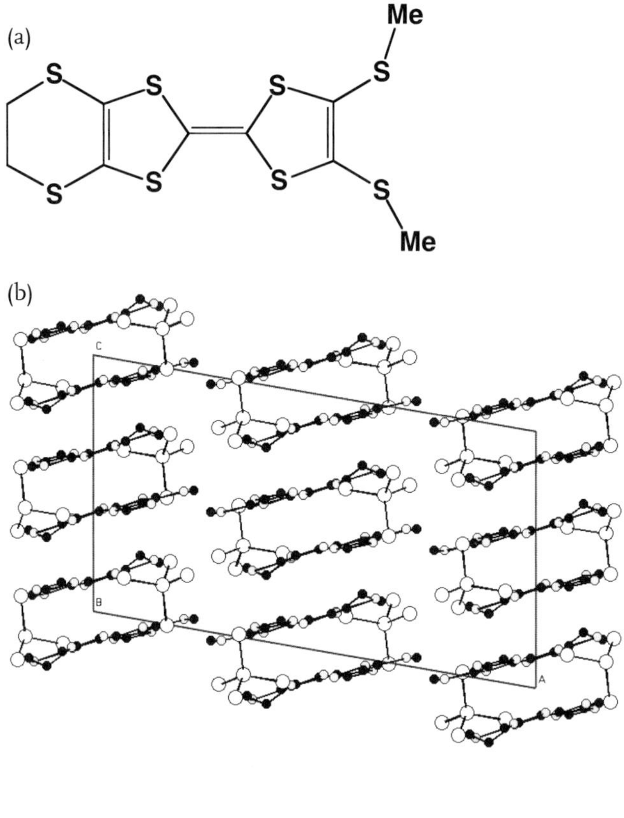

Fig. 9.15 (a) Scheme of the donor C_1TET-TTF and (b) view of the structure of the salt (C_1TET-TTF)$FeBr_4$

magnetic behavior is quite different. A triangle-based spin ladder lattice in both cases gives an antiferromagnet in the case of $FeCl_4^-$, and a weak ferromagnet in the case of $FeBr_4^-$. The origin of this difference in magnetic behavior was related to the competition between dipolar anisotropy and single-ion anisotropy, as observed in detailed studies on the mixed salts $(C_1TET\text{-}TTF)(FeX_4)_{1-x}(FeCl_4)_x$. [54]

Following another possibility of increasing the π-d interactions, materials where the organic radicals are covalently linked to metals are being investigated. Organic donors have been functionalized with groups able to coordinate metal centers to form stable complexes. Functionalized TTF, with pyridine-based organic ligands added to the TTF core, have been used to coordinate metal complexes, as in $Cu(hfac)_2(TTF\text{-}py)_2$ (hfac = hexafluoroacetylacetonate; TTF-py = 4-(2-tetrathiafulvalenyl-ethenyl)pyridine) (Fig. 9.16). [55] In this complex the radicals are neutral and therefore an insulating behavior is found. Although the ligands can be oxidized, giving stable species, chemical or electrochemical partial oxidation of this type of complex has not been successful, with the target being salts where the paramagnetic centers would be covalentely linked to the conducting network. In addition to synthetic problems, these complexes are very demanding in terms of packing, since the metal restrains the options for the organic donors to stack. Metal complexes of TTF derivatives functionalized with phosphines are also known, [56] and present similar characteristics.

Striking results were obtained when the interactions between the organic donors and anions are strong enough to destroy the formation of segregated low-dimensional structures. Instead, both networks appear interpenetrated. This is the case for salts of the anion $[M(NCS)_4(C_9H_7N)_2]^-$ (M = Cr, Fe), a trans-complex where two of the thiocyanide ligands have been substituted by isoquinoline to form the salts $(BEDT\text{-}TTF)[M(NCS)_4(C_9H_7N)_2]$ and $(TTF)[M(NCS)_4(C_9H_7N)_2]$. [57] All these salts show bulk ferrimagnetic ordering due to antiparallel alignment of the spins in the organic and the inorganic sublattices. The critical temperatures for these systems are as high as 8.9 K in the $(TTF)[Cr(NCS)_4(C_9H_7N)_2]$ derivative. When the isoquinoline ligand is substituted by 1,10-phenantroline, and then the anion adopts a cis-configuration, the analogous salt obtained, $(TTF)[M(NCS)_4(phen)]$, still shows ferrimagnetic behavior with $T_c = 9$ K. [58]

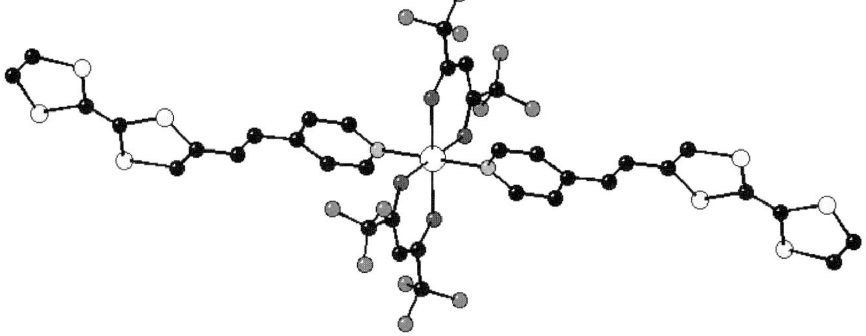

Fig. 9.16 Molecular structure of *trans*-$Cu(hfac)_2(TTF\text{-}py)_2$

It is worthwhile to mention that these compounds are remarkable because the magnetic interactions between and d electrons are usually so weak as very rarely to yield bulk magnetic properties. In this case, the strong interactions between spin carriers are promoted by close p-p contacts between the organic donors and the organic ligands from the metal complexes (isoquinoline and phenantronile). Similar interpenetrated structures where the metal complexes do not possess this kind of conjugated ligands, such as $(TTF)_{11}[Fe(CN)_6]_3 \cdot 5\,H_2O$ [59] and $(BEST\text{-}TTF)_3[Fe(CN)_6]_2 \cdot H_2O$,[47] exhibit negligible magnetic interactions of the same order.

Since the organic radicals do not present close interactions to each other forming low dimensional stacks able to promote electrical conductivity, these materials are insulators. We need to stress that the magnetic interactions are present between localized unpaired electrons in the organic donors and in the inorganic complexes, but not between localized and itinerant electrons, as the scope of this section was proposing.

9.4.2
Paramagnetic Conductors from Polyoxometalates

The more complex and bulky paramagnetic anions, such as polyoxometalates of molybdenum or tungsten encapsulating paramagnetic transition metals, have also been used as counterions of radical salts of the BEDT-TTF type [60] and more recently of other organic donors such as perylene. [61] The interest in the use of polyoxometallates (POMs) resides in their large size and charge that can induce new packings in the organic sublattice, and in many cases charge localization.

The first hybrid salts of organic donors with large inorganic clusters were reported in the 80's, [62] with polyoxometallates being one of the most successful examples. [63] Simple polyoxometallates were used in the preparation of hybrid radical salts, such as $[M_6O_{19}]^{2-}$ (Lindqvist), $\beta\text{-}[Mo_8O_{26}]^{4-}$.

All the salts prepared with the TTF organic donor presented semiconducting behavior, such as for example those prepared with different polyoxometallates with Lindqvist $[TTF]_3[M_6O_{19}]$; [64] or the Keggin structures of general formula $[TTF]_6[XM_{12}O_{40}][NEt_4]$ (M = W, Mo; X = P, Si), [65] even with the polyoxometallates in their reduced form (polyblues).

Better conducting properties were observed for radical salts of other donors, such as BEDT-TTF, because of the increasing dimensionality of the organic network. The same Lindqvist anion forms various types of BEDT-TTF radical salts with different stoichiometries, [66] the most interesting being the 5:1 phases $[BEDT\text{-}TTF]_5[VW_5O_{19}] \cdot 5\,H_2O$ and $[BEDT\text{-}TTF]_5[V_2W_4O_{19}] \cdot 5\,H_2O$ [67] that were reported as exhibiting metallic behavior at room temperature, with a transition into a semiconducting state below 250 K. Now a further step consisted of the use of magnetic polyoxometallates, with the aim of preparing magnetic conductors.

Many POMs containing paramagnetic centers have been used in these regard, many of them with the Keggin structural type (Fig. 9.12). The first example was the family of salts $\alpha\text{-}(BEDT\text{-}TTF)_8[XM_{12}O_{40}]$ ($X = Co^{II}$, Cu^{II} and Fe^{III}; $M = Mo^{VI}$ and W^{VI}).

[68] In this case the paramagnetic centers are located in the interior tetrahedral cavity of the Keggin anions, and therefore quite isolated from the conducting electrons. These salts are semiconductors, with the magnetic properties being the sum of the contributions of the paramagnetic ions and the organic sublattice, that in this case behaves as antiferromagnetically coupled $S = 1/2$ chains, although the structure is two-dimensional.

Maintaining the same structure for the POMs, another series of analogous salts was reported with the paramagnetic centers occupying the external octahedra, and therefore closer to the organic sublattice. The series α-(BEDT-TTF)$_8$[XZ(H_2O)$M_{11}O_{39}$] ($X = P^{V}$ and Si^{IV}; $Z = Fe^{III}$, Cr^{III}, Co^{II}, Ni^{II} and Cu^{II}; $M = Mo^{VI}$ and W^{VI}; Fig. 9.17) [69] showed identical behavior to the previous salts based in the Keggin anions.

A special case is α-(BEDT-TTF)$_8$[PMn(H_2O)$W_{11}O_{39}$], where the POMs polymerize to form one-dimensional chains, while the organic layer maintains the same structure (Fig. 9.18). [70] In this case weak magnetic coupling was observed between the MnII ions, but was determined to occur through the POM's chain, and not related to interactions with the delocalized electrons in the organic layer.

Among the few examples of radical salts based on POMs that show metal-like conductivity are those prepared using the Dawson anion. [71] The electrical properties are maintained also when the anion contains a paramagnetic center as in the [BEDT-TTF]$_{11}$[ReO$P_2W_{17}O_{61}$] · 3 H_2O salt. It contains as inorganic component

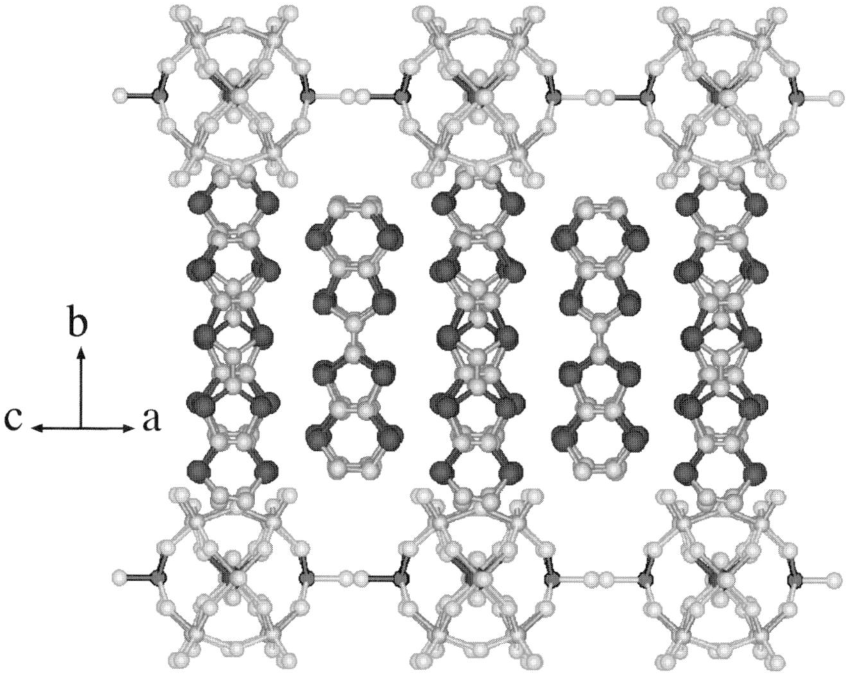

Fig. 9.17 View of the multilayer structure of the series α-(BEDT-TTF)$_8$[XZ(H_2O)$M_{11}O_{39}$]

(a)

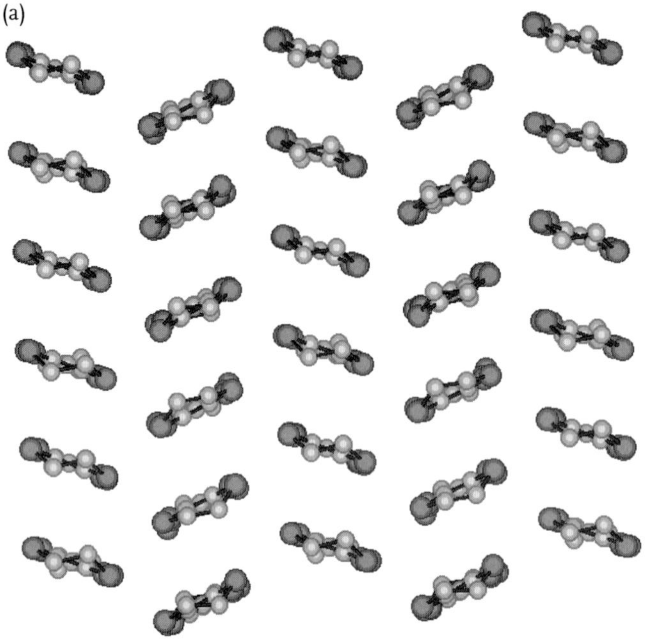

(b)

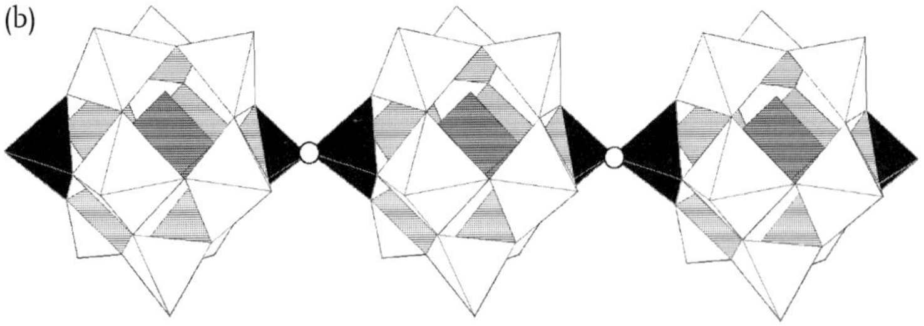

Fig. 9.18 View of (a) the organic layer for the series α-$(BEDT\text{-}TTF)_8[XZ(H_2O)M_{11}O_{39}]$; (b) 0 the inorganic one-dimensional chain of Keggin anions found in α-$(BEDT\text{-}TTF)_8[PMn(H_2O)M_{11}O_{39}]$

the polyoxoanion $[ReOP_2W_{17}O_{61}]^{6-}$, with a paramagnetic Re^{VI} ion (S = 1/2). Its structure consists of layers of polyoxoanions alternating with layers of -stacked BEDT-TTF radical cations (Fig. 9.19). [72] This salt shows a metallic-like behavior in the region 250–300 K with an increase in the conductivity from ca. 11.3 S cm^{-1} at room temperature to 11.8 S cm^{-1} at 250 K. Below 250 K the salt becomes a semiconductor, reaching a value of 1.9 S cm^{-1} at 80 K, with very low activation energy. Even in its metallic-like state, this material is paramagnetic, in agreement with the presence of a spin S = 1/2 on each Re^{VI}. Thus, this hybrid salt provides a good example of co-existence of localized magnetic moments with highly delocalized electrons.

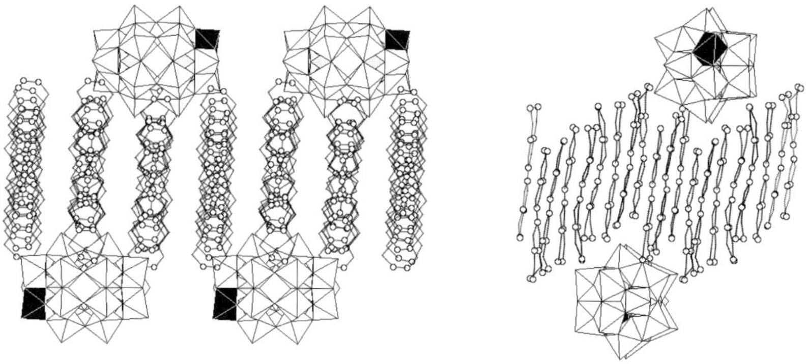

Fig. 9.19 Two views of the crystal packing in the $[BEDT\text{-}TTF]_{11}[ReOP_2W_{17}O_{61}] \cdot 3\ H_2O$ salt

Magnetic clusters can also be encapsulated by POMs and used in the preparation of charge transfer salts. The anions $[M_4(H_2O)_2(PW_9O_{34})_2]^{10-}$ (Fig. 9.12), which contain a tetranuclear cluster M_4O_{16} with ferromagnetic (M^{II} = Fe, Co and Ni) or antiferromagnetic coupling (M^{II} = Mn and Cu), have been combined with the BEDT-TTF in radical salts with a 6 : 1 stoichiometry. [73] ESR and magnetic measurements on these samples confirmed the presence of the magnetic clusters and of conducting organic sub-lattices as independent networks.

POMs can be easily reduced by one or more electrons to generate the so-called "polyblues". In these species the electrons are confined in the POMs, but at the same time delocalized over all the metal centers. Hybrid salts with two different types of itinerant electrons, conducting electrons in the organic sub-lattice and delocalized electrons in the POMs, can be prepared with the Keggin anion $[PMo_{12}O_{40}]^{4-}$and TTF, [74] TM-TSF, [75] BEDT-TTF [76] and BEDS-TTF. [77] In all cases the reduced anions show the presence of delocalized electrons that become localized at low temperatures, as confirmed by ESR and magnetic susceptibility measurements. No interactions between the two sub-lattices were observed.

POMs have also been used as counterions for other organic radicals. One interesting study is that of the perylene salts. Three isostructural anions of different charges, the Lindquist POMs $[M_6O_{19}]^{2-}$ (M = Mo, W) and $[VW_5O_{19}]$,$^{3-}$ were used to prepare the series of salts $(per)_5[M_6O_{19}]$ [78] that present dimerized chains of perylene separated by orthogonal neutral perylene molecules and the anions (Fig. 9.20). The same structure incorporates the POMs with different charges. The electrical conductivity is different depending on the charge of the anion, that imposes a different electronic population of the conducting band. While the dianion salts behave as semiconductors with low r.t. conductivities (≈ 0.2 S cm^{-1}), high activation energies (≈ 200 meV) and positive Seebeck coefficients (200–600 μV K^{-1}), the trianion salt shows an electrical conductivity higher by an order of magnitude at room temperature (3 S cm^{-1}), a lower activation energy (≈ 120 meV) and a negative Seebeck coefficient (–350 μV K^{-1}). This shows the versatility of polyoxometallates, and their possibilities in this field towards a better understanding of structural correlations.

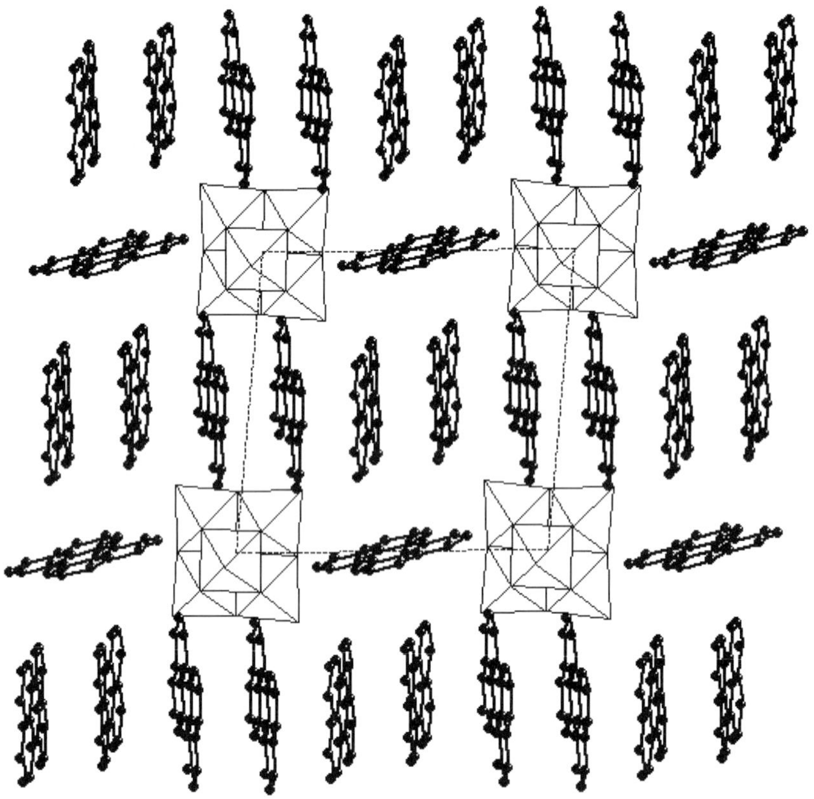

Fig. 9.20 Crystal structure of the salt $(per)_5[M_6O_{19}]$

9.4.3
Coexistence of Electrical Conductivity and Magnetic Ordering

In most of the reported cases no interactions between the localized electrons mediated by the conducting electrons are observed, so the two sub-lattices, paramagnetic and conducting, are quasi-independent. This opens up the question about the possibility for superconductivity and magnetic ordering to coexist in one material if both sub-lattices are independent. To achieve cooperative magnetic behavior, extended lattices instead of discrete molecules are a better choice. And in this context, the 2D oxalate-based bimetallic phases are perfect hosts for this purpose. These anionic networks could alternate in the solid with cationic layers of organic conductors.

The first approach to this type of material is represented by the series (BEDT-TTF)$_4$[$A^IM^{III}(ox)_3$].solvent ($A^+ = H_3O$, K, NH_4; M^{III} = Cr, Fe, Co, Al; solvent = C_6H_5CN, $C_6H_5NO_2$, C_5H_5N), already described in Section 9.4.1. Still, in these series of semi- or superconductors the anionic networks are formed by a hexagonal arrangement of the tris(oxalate)metallates alternating with small and "innocent" monocations. So, all these compounds remain paramagnetic, since the presence of the A^+ cation instead of a paramagnetic metal while building the 2D honeycomb net destroys the possibility for long-range magnetic ordering. Down to very low temperatures, the compounds present a Curie-Weiss paramagnetic behavior, with an additional and strong paramagnetic contribution in the superconducting regime. As in previous cases, both networks are quasi-independent, as shown by the presence of two distinct signals, one from each network, in the EPR spectra.

In an attempt to increase the nuclearity of the oxalate complexes, charge transfer salts of dimeric anions with an oxalate bridge between the metal centers were prepared. The salts $(TTF)_5[Fe_2(ox)_3] \cdot 2\ PhMe \cdot 2\ H_2O$, [79] $(TM\text{-}TTF)_4[Fe_2(ox)_3]$ [79] and $(BEDT\text{-}TTF)[Fe_2(ox)_3]$ [80] are insulators and showed a magnetic behavior dominated by the intramolecular antiferromagnetic exchange promoted by the oxalate anion. Structurally, the organic layers appear to be forming dimers with close contacts that behave as spin-paired monopositive cations and do not allow effective electron delocalization. A more interesting case is that of the partially substituted oxalate-bridge dimer $[M_2(ox)(NCS)_8]^{4-}$ ($M = Fe^{III}$, Cr^{III}). The phases $(BEDT\text{-}TTF)_5\ [M_2(ox)(NCS)_8]$ [81] are semiconductors. In this case the organic layers have a two-dimensional character, with dimers of BEDT-TTF interleaved by monomers, in an array reminiscent of the -phases. The magnetism is dominated by the inorganic anions, and, in the case of the $(BEDT\text{-}TTF)_5[FeCr(ox)(NCS)_8]$ salt, by ferromagnetic interactions leading to a $S = 4$ ground state. The same building blocks form a different phase in the case of the Cr anion: $(BEDT\text{-}TTF)_8[Cr_2(ox)(NCS)_8]$, [82] a metallic conductor with a 0.1 S cm^{-1} conductivity at room temperature. The anion shows antiferromagnetic coupling between the Cr centers, and no contribution from the organic lattice.

The $(TTF)_4\{M^{II}(H_2O)_2[M^{III}(ox)_3]_2\}.14H_2O$ (M^{II} = Mn, Fe, Co, Ni, Cu, Zn; M^{III} = Cr, Fe) series [83] was the first step towards molecular conductors containing bimetallic magnetic layers. These compounds contain alternating layers of the TTF radical with trinuclear anions formed by a central M^{II} with trans-oxalate-bridges to two M^{III} (Fig. 9.21). Most of these trimers have high spin ground states stabilized by the magnetic interactions promoted by the oxalato bridge: ferromagnetic for the Cr^{III} family and antiferromagnetic for the Fe^{III} family. The trimers are connected via hydrogen bonding to build a pseudo-hexagonal network, that can be seen as a precursor of the bimetallic phases. The organic layer exhibits an unprecedented packing in the TTF layer similar to a κ-phase, although the interdimer distances are different in the two directions. These materials behave as semiconductors, with low electrical conductivity due to the localization of the charges in the organic layer.

Coexistence of ferromagnetism with metallic conductivity has recently been achieved for the first time in the hybrid compound β-$(BEDT\text{-}TTF)_3[MnCr(ox)_3]$. [84] This structure shows alternating layers of the BEDT-TTF radicals and of the polymeric 2D $[MnCr(ox)_3]^-$ anionic network. The organic layer is formed by a rough hexagonal arrangement of BEDT-TTF molecules with close S–S contacts (Fig. 9.22). The inorganic layer has the typical bimetallic 2D honeycomb structure, and therefore this compound behaves as a ferromagnet with T_c of 5.5 K (Fig. 9.23). This compound also exhibits metallic conductivity down to 0.3 K, without becoming

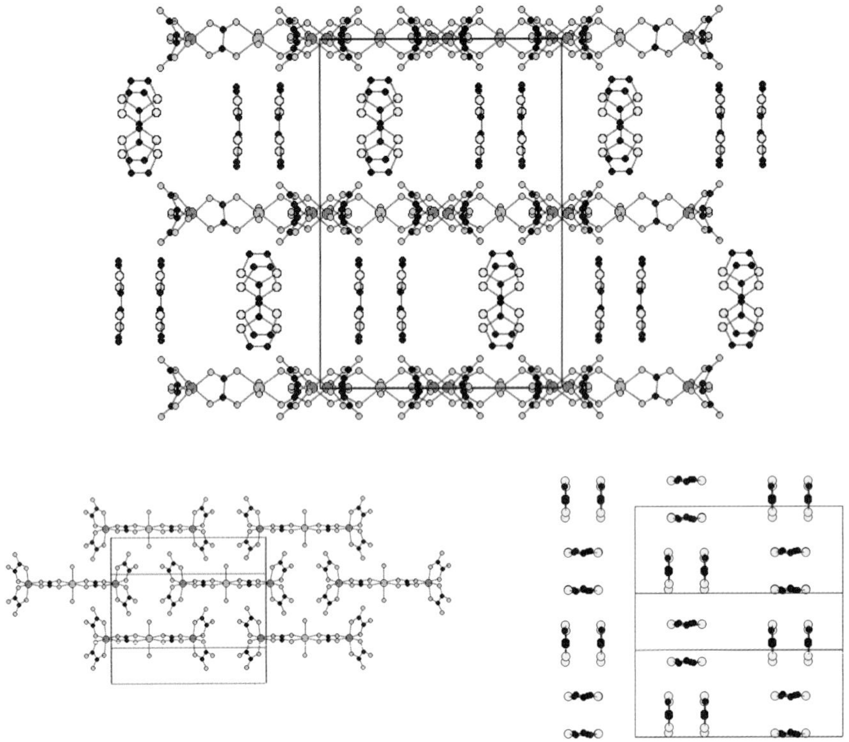

Fig. 9.21 Representation of the crystal structure of the $(TTF)_4\{M^{II}(H_2O)_2[M^{III}(ox)_3]_2\} \cdot 14\ H_2O$ salts: (a) multilayered structure, (b) top view of the inorganic and organic layers

superconducting. If the absence of a transition to a superconducting state is intrinsic to the organic layer, or if it is effected by the presence of the internal field created by the bimetallic network, is not clear, and more studies need to be done in this direction. Interestingly, this compound exhibits an interplay between conductivity and magnetism. Thus, a negative magneto-resistance is observed at low temperatures along the organic/inorganic layers when a small external magnetic field is applied perpendicular to the layers. Such an effect is highly anisotropic, in agreement with the structure. In fact, when the field is applied parallel to the layers the magneto-resistance becomes positive, and the resistance shows a very small field dependence. The discovery of such a ferromagnetic metal opens the way to explore the synthesis of other layered materials with other combinations of metals in the inorganic network and other BEDT-TTF type donors. This is another of the advantages of molecular materials. Several new conducting magnets have

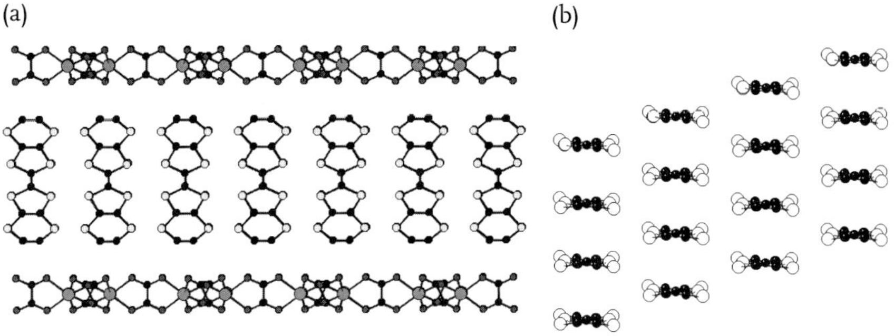

Fig. 9.22 Representation of the crystal structure of $(BEDT\text{-}TTF)_3[MnCr(ox)_3]$: (a) multilayered structure, (b) packing of the organic layer

been prepared using this strategy with either $[MnCr(ox)_3]^-$ or $[CoCr(ox)_3]^-$ layers and the donors BEDT-TTF, BEST, BETS and BET-TTF (Fig. 9.12). In the MnCr derivatives the critical temperature is 5.5–6.0 K while in the CoCr ones it is 9–10 K. [85] Regarding the conducting properties, not all the different donor types give metal-like conductivity. Different packings of the organic layer have been found. For instance, the BETS-TTF derivative shows semiconducting behavior in all temperature ranges, but it forms an -type packing instead of β. Detailed studies and correlations between structure and physical properties for this fascinating series of materials will allow us to answer important questions for the field of multifunctional materials.

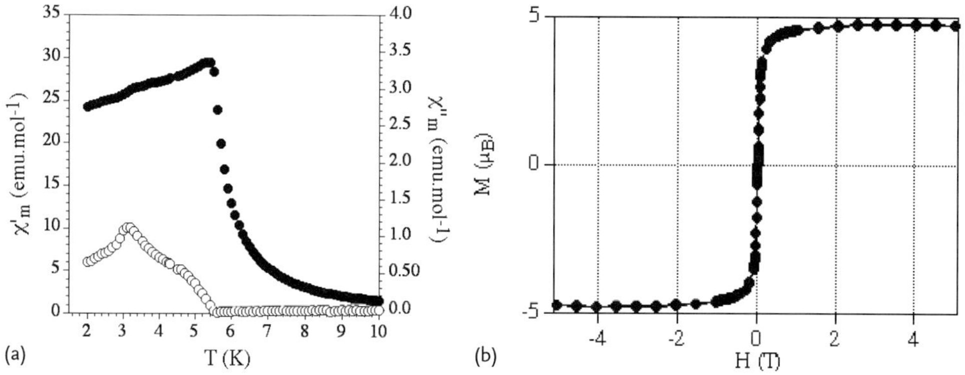

Fig. 9.23 Magnetic properties of $(BEDT\text{-}TTF)_3[MnCr(ox)_3]$: (a) in-phase and out-of-phase AC magnetic susceptibility, (b) hysteresis loop at 2 K

9.5 Conclusions

In this contribution we have illustrated with some examples the wide possibilities offered by molecular chemistry for preparation of novel classes of molecular materials formed by the combination of an organic component with an inorganic one. So far, this hybrid approach has been quite useful to increase the complexity of the system by building up two-network materials in which both networks play an electronic role. Some examples of this concept are the coexistence of an extended network of interacting spins, which undergoes a magnetic ordering below T_c, with a molecular network of insulated spins, or the coexistence of localized spins and delocalized electrons. Although these hybrid materials cannot be considered as bi-functional, they represent a critical step towards bi-functionality. To achieve bi-functionality is a much more challenging goal. The crystal engineering required to design a solid exhibiting two cooperative properties such as ferromagnetism and metallic conductivity, or ferromagnetism and non-linear optics, is still at the beginning. In fact, only very recently the first ferromagnetic molecular metal has been obtained. More examples of bi-functional materials are expected to be created in the future using this hybrid approach. In this sense, it should be noted that this approach to multifunctionality is not restricted to crystalline materials. Hybrid films can also be constructed using molecular chemistry that can exhibit multifunctionality. Some promising results have recently been obtained using the Langmuir-Blodgett technique. In so doing, organized films formed by monolayers of inorganic POM clusters alternating with monolayers of organic molecules of the TTF type have been obtained. More precisely, we have used, as TTF derivative, SF-EDT and the polyoxometalate $[Co_4(H_2O)_2(P_2W_{15}O_{56})_2]^{16-}$ that contains the ferromagnetic cluster Co_4O_{16} (Fig. 9.24). This LB film has been prepared by dipping the substrate

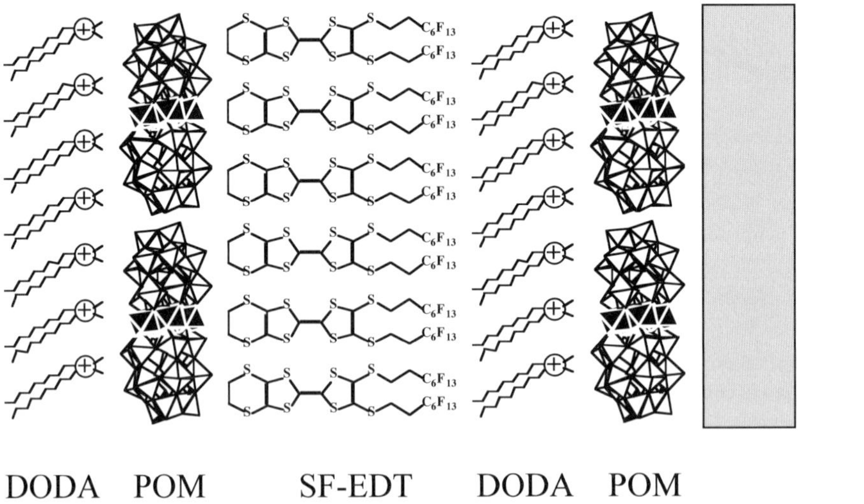

Fig. 9.24 Structure of the POM/DODA/SF-EDT LB film

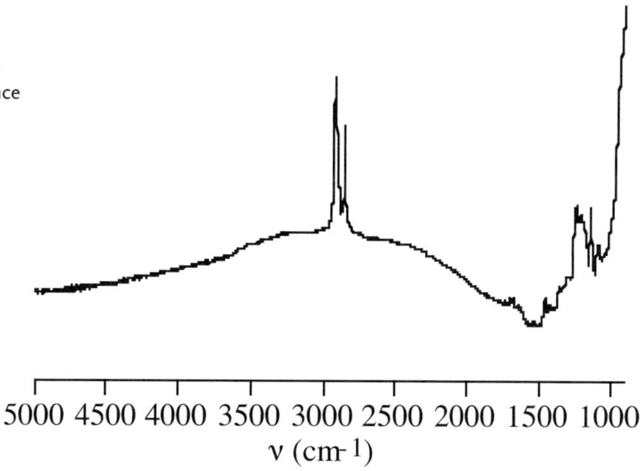

Fig. 9.25 IR spectrum of the oxidized POM/DODA/Sf-EDT LB film showing the presence of the oxidized cation radical

alternately in two different types of Langmuir films, one containing the positively-charged DODA cations in which the polyoxometalate is adsorbed, and the other one containing the neutral SF-EDT molecules. By doing so, a LB film formed by successive monolayers of POM/DODA/SF-EDT is obtained which can be easily oxidized by iodine to afford a mixed-valence delocalized state in the SF-EDT layer (Fig. 9.25). [86] This represents the first attempt to obtain a LB film with coexistence of delocalized electrons and localized magnetic moments. This new direction in molecular chemistry constitutes a promising synthetic approach toward the design of multifuctional molecular materials.

Acknowledgments
We wish to thank our colleagues and co-workers for their important contributions to the work reported herein. Their names appear in the references. Financial support from the Spanish MCYT and FEDER (Grant MAT2001–3507) is acknowledged.

References

1 K. Awaga, E. Coronado, M. Drillon, *MRS Bull.* **2000**, *11*, 52.
2 S. Decurtins, H. W. Schmalle, H. R. Oswald, A. Linden, J. Ensling, P. Gütlich, A. Hauser, *Inorg. Chim. Acta* **1994**, *216*, 65.
3 H. Tamaki, Z. J. Zhong, N. Matsumoto, S. Kida, M. Koikawa, N. Achiwa, Y. Hashimoto, H. Okawa, *J. Am. Chem. Soc.* **1992**, *114*, 6974.
4 a) H. Tamaki, M. Mitsumi, K. Nakamura, N. Matsumoto, S. Kida, H. Okawa, S. Iijima, *Chem. Lett.* **1992**, 1975; b) J. Larionova, B. Bombelli, J. Sanchiz, O. Kahn, *Inorg. Chem.* **1998**, *37*, 679.
5 C. Mathonière, J. Nuttall, S.G. Carling, P. Day, *Inorg. Chem.* **1996**, *35*, 1201.
6 M. Clemente León, E. Coronado, J. R. Galán-Mascarós, C. J. Gómez-García, *Chem. Commun.* **1997**, 1727.
7 a) E. Coronado, J. R. Galán-Mascarós, C. J. Gómez-García, J. Ensling, P. Gütlich, *Chem. Eur. J.* **2000**, 6, 552; b) E. Coronado, J. R. Galán-Mascarós, C. J. Gómez-

García, J. M. Martínez-Agudo, E. Martínez-Ferrero, J. C. Waerenborgh, M. Almeida, *J. Solid State Chem.* **2001**, *159*, 391.
8 A. Battacharjee, S. Iijima, *Phys. Stat. Sol. (a)* **1997**, *159*, 503.
9 E. Coronado, J. R. Galán-Mascarós, C. J. Gómez-García, J. M. Martínez-Agudo, *Adv. Mater.* **1999**, *11*, 558.
10 E. Coronado, J. R. Galán-Mascarós, C. J. Gómez-García, J. M. Martínez-Agudo, *Synth. Met.* **2001**, *122*, 501.
11 E. Coronado, J. R. Galán-Mascarós, C. J. Gómez-García, J. M. Martínez-Agudo, *Inorg. Chem.* **2001**, *40*, 113.
12 R. Andres, M. Brissard, M. Gruselle, C. Train, J. Vaisserman, B. Malezieux, J. P. Jamet, M. Verdaguer, *Inorg. Chem.* **2001**, *40*, 4633.
13 R. Sieber, S. Decurtins, H. Stoeckli-Evans, C. Wilson, D. Yufit, J. A. K. Howard, S. C. Capelli, A. Hauser, *Chem. Eur. J.* **2000**, *6*, 361.
14 A. Yamaguchi, T. Okuno, K. Awaga, *Bull. Chem. Soc. Jpn.* **1996**, *69*, 875.
15 G. Ballester, E. Coronado, C. Giménez-Saiz, F.M. Romero, *Angew. Chem. Int. Ed.* **2001**, *40*, 792.
16 S. Bénard, P. Yu, J. P. Audière, E. Rivière, R. Clément, J. Guilhem, L. Tchertanov, K. Nakatani, *J. Am. Chem. Soc.* **2000**, *122*, 9444.
17 S. Bénard, E. Rivière, P. Yu, K. Nakatani, J. F. Delouis, *Chem. Mater.* **2001**, *13*, 159.
18 P. Gütlich, A. Hauser, H. Spiering, *Angew. Chem. Int. Ed. Engl.* **1994**, *33*, 2024.
19 P. Gütlich, *Struct. Bonding* **1981**, *44*, 83.
20 P. Gütlich, J. Jung, H. A. Goodwin, in *Molecular Magnetism: From Molecular Assemblies to the Devices*, eds. E. Coronado, P. Delhaès, D. Gatteschi, J. S. Miller. NATO ASI Series, Kluwer Academic Publishers, Dordrecht, **1996**, *E321*, 327.
21 E. Coronado, J. R. Galán-Mascarós, C. J. Gómez-García, *Mol. Cryst. Liq. Cryst.* **1999**, *334*, 679.
22 G. L. J. A. Rikken, E. Raupach, *Nature* **1997**, *390*, 493.
23 a) F. Bellouard, M. Clemente-León, E. Coronado, J. R. Galán-Mascarós, C. J. Gómez-García, F. Romero, K. R. Dunbar, *Eur. J. Inorg. Chem.* **2002**, 1603;
b) K. Inoue, H. Imai, P. S. Ghalsasi, K. Kikuchi, M. Ohba, H. Okawa, J. V. Yakhmi, *Angew. Chem. Int. Ed.* **2001**, *40*, 4242.
24 a) M. Minguet, D. B. Amabilino, J. Cirujeda, K. Wurst, I. Mata, E. Molins, J. J. Novoa, J. Veciana, *Chem. A Eur. J.* **2000**, *6*, 2350; b) M. Minguet, D. B. Amabilino, K. Wurst, J. Veciana, *J. Solid State Chem.***2001**, *159*, 440.
25 a) S. Decurtins, H. W. Schmalle, R. Pellaux, P. Schneuwly, A. Hause, *Inorg. Chem.* **1996**, *35*, 1451; b) A. Hauser, H. Riesen, R. Pellaux, S. Decurtins, *Phys. Lett.* **1996**, *261*, 313.
26 a) T. Woike, W. Krasser, P. S. Bechthold, S. Haussühl, *Phys. Rev. Lett.* **1984**, *53*, 1767; b) M. Carducci, D. V. Fomitchev, P. Coppens, *J. Am. Chem. Soc.* **1997**, *119*, 2669.
27 H. Yu, D. Zhu, *Physica C* **1997**, *282*, 1893.
28 a) L. Kushch, L. Buratov, V. Tkacheva, E. Yagubskii, L. Zorina, S. Khasanov, R. P. Shibaeva, *Synth. Met.* **1999**, *102*, 1646; b) M. Gener, E. Canadell, S. S. Kashanov, L. V. Zorina, R. P. Shibaeva, L. A. Kushch, E. B. Yagubskii, *Solid State Commun.* **1999**, *111*, 329.
29 M.-E. Sanchez, M.-L. Doublet, C. Faulmann, I. Malfant, P. Cassoux, L. A. Kushch, E. B. Yagubskii, *Eur. J. Inorg. Chem.* **2001**, 2797.
30 M. Clemente-León, E. Coronado, J. R. Galán-Mascarós, C. J. Gómez-García, E. Canadell, *Inorg. Chem.* **2000**, *39*, 5394.
31 H. Zöllner, W. Krasser, T. Woike, S. Haussühl, *Chem. Phys. Lett.* **1989**, *497*, 161.
32 J. R. Elliott in *Magnetism, vol. IIA*, eds Rado and Suhl, Academic Press, New York, **1965**, p. 385.
33 T. Matsubara, A. Kotani, *Superconductivity in Magnetic and Exotic Materials, Springer Series in Solid State Science*, Springer, Berlin, **1984**, Vol. 52.
34 J. M. Williams, J. R. Ferraro, R. J. Thorn, K. D. Carlson, U. Geiser, H. H. Wang, A. M. Kini, M. H. Whangbo, in *Organic Superconductors (Including Fullerenes). Synthesis, Structure, Properties and Theory*, ed R. N. Grimes, Prentice Hall, Englewood Cliffs, **1992**.
35 a) M. Kurmoo, A. W. Graham, P. Day, S. J. Coles, M. B. Hursthouse, J. M. Caulfield, J. Singleton, L. Ducasse, P. Guionneau, *J. Am. Chem. Soc.*, **1995**, *117*, 12209;

b) L. L. Martin, S. S. Turner, P. Day, P. Guionneau, J. A. K. Howard, D. E. Hibbs, M. E. Light, M. B. Hursthouse, M. Uruichi, K. Yakushi, *Inorg. Chem.*, **2001**, *40*, 1363; c) S. Rashid, S. S. Turner, D. Le Pevelen, P. Day, M. E. Light, M. B. Hursthouse, S. Firth, R. J. H. Clark, *Inorg. Chem.*, **2001**, *40*, 5304.

36 A. W. Graham, M. Kurmoo, P. Day, *J. Chem. Soc., Chem. Commun.* **1995**, 2061.

37 S. Rashid, S. S. Turner, P. Day, J. A. K. Howard, P. Guinneau, E. J. L. McInnes, F. E. Mabbs, R. J. H. Clark, S. Firth, T. Biggs, *J. Mater. Chem.* **2001**, *11*, 2095.

38 H. Kobayashi, H. Tomita, T. Udagawa, T. Naito, A. Kobayashi, *Synth. Met.* **1995**, *70*, 867.

39 H. Kobayashi, H. Tomita, T. Naito, A. Kobayashi, F. Sakai, T. Watanabe, P. Cassoux, *J. Am. Chem. Soc.* **1996**, *118*, 368.

40 S. Uji, H. Shinagawa, T. Terashima, T. Yakabe, Y. Terai, M. Tokumoto, A. Kobayashi, H. Tanaka, H. Kobayashi, *Nature* **2001**, *410*, 908.

41 V. Jaccarino, M. Peter, *Phys. Rev. Lett.* **1962**, *9*, 290.

42 H. Fujiwara, H. Kobayashi, E. Fujiwara, A. Kobayashi, *J. Am. Chem. Soc.* **2002**, in the press.

43 H. Mori, I. Hirabayashi, S. Tanaka, T. Mori, Y. Manyama, *Synth. Met.* **1995**, *70*, 789.

44 E. Coronado, L. R. Falvello, J. R. Galán-Mascarós, C. Giménez-Saiz, C. J. Gómez-García, V. N. Lauhkin, A. Pérez-Benitez, C. Rovira, J. Veciana, *Adv. Mater.* **1997**, *9*, 984.

45 T. Mallah, C. Hollis, S. Bott, M. Kurmoo, P. Day, *J. Chem. Soc., Dalton Trans.* **1990**, 859.

46 P. LeMagueres, L. Ouahab, P. Briard, J. Even, M. Bertault, L. Toupet, J. Ramos, C. J. Gómez-García, P. Delhaes, *Mol. Cryst. Liq. Cryst.* **1997**, *305*, 479.

47 M. Clemente-León, E. Coronado, J. R: Galán-Mascarós, C. Giménez-Saiz, C. J. Gómez-García, J. M. Fabre, *Synth. Met.* **1999**, *103*, 2279.

48 M. Rahal, D. Chasseau, J. Gaultier, L. Ducasse, M. Kurmoo, P. Day, *Acta Cryst.* **1997**, B53, 159.

49 M. Clemente-León, E. Coronado, J. R. Galán-Mascarós, C. Giménez-Saiz, C. J. Gómez-García, E. Ribera, J. Vidal-Gancedo, C. Rovira, E. Canadell, V. Lauhkin, *Inorg. Chem.* **2001**, *40*, 3526.

50 T. Mori, A. Kobayashi, Y. Sasaki, H. Kobayashi, G. Saito, H. Inokichi, *Bull.*4, *57*, 627.

51 P. Wu, T. Mori, T. Enoki, K. Imaeda, G. Saito, H. Inokuchi, *Bull. Chem. Soc. Jpn.* **1986**, *59*, 127.

52 M. Enomoto, A. Miyazaki, T. Enoki, *Synth. Met.* **2001**, *120*, 977.

53 M. Enomoto, A. Miyazaki, T. Enoki, *Mol. Cryst. Liq. Cryst.* **1999**, *335*, 293.

54 M. Enomoto, A. Miyazaki, T. Enoki, *Bull. Chem. Soc. Jpn.* **2001**, *74*, 459.

55 F. Iwahori, S. Golhen, L. Ouahab, R. Carlier, J.-P. Sutter, *Inorg. Chem.* **2001**, *40*, 6541.

56 B. W. Smucker, K. R. Dunbar, *J. Chem. Soc., Dalton Trans.* **2000**, *8*, 1309.

57 S. S. Turner, C. Michaut, D. Durot, P. Day, T. Gelbrich, M. B. Hursthouse, *J. Chem. Soc. Dalton Trans.* **2000**, 905.

58 S. S. Turner, D. Le Pevelen, P. Day, K. Prout, *J. Chem. Soc. Dalton Trans.* **2000**, 2739.

59 S. Bouherour, L. Ouahab, O. Pena, J. Padiou, D. Grandjean, *Acta Cryst.* **1989**, *C45*, 371.

60 E. Coronado, C. J. Gomez-Garcia, *Chem. Rev.* **1998**, *98*, 273.

61 M. Clemente-León, E. Coronado, C. Giménez-Saiz, C. J. Gómez-García, E. Martínez-Ferrero, M. Almeida, E. B. Lopes, *J. Mater. Chem.* **2001**, *11*, 2176.

62 a) L. Ouahab, P. Batail, A. Perrin, C. Garrigou-Lagrange, *Mater. Res. Bull.* **1986**, *21*, 1223; b) P. Batail, L. Ouahab, A. Penicaud, C. Lenoir, A. Perrin, *C. R. Acad. Sc. Paris. Ser. II* **1987**, *304*, 1111.

63 L. Ouahab, M. Bencharif, D. Grandjean, *C. R. Acad. Sci. Paris Ser II* **1988**, *307*, 749.

64 S. Triki, L. Ouahab, J. F. Halet, O. Peña, J. Padiou, D. Grandjean, C. Garrigou-Lagrange, P. Delahes, *J. Chem. Soc., Dalton Trans.* **1992**, 1217.

65 a) L. Ouahab, M. Bencharif, A. Mhanni, D. Pelloquin, J. Halet, J. Amiell, P. Delhaes, *Chem. Mater.* **1992**, *4*, 666. b) D. Attanasio, C. Bellito, M. Bonamico,

V. Fares, S. Patrizio, *Synth. Met.* **1991**, *41–43*, 2289.
66 S. Triki, L. Ouahab, D. Grandjean, *Acta Cryst.* **1991**, *C47*, 645.
67 S. Triki, L. Ouahab, D. Grandjean, R. Canet, C. Garrigou-Lagrange, P. Delhaes, *Synth. Met.* **1993**, *55–57*, 2028.
68 a) C. J. Gómez-García, L. Ouahab, C. Giménez-Saiz, S. Triki, E. Coronado, P. Delhaes, *Angew. Chem., Int. Ed. Engl.* **1994**, *33*, 223; b) C. J. Gómez-García, C. Giménez-Saiz, S. Triki, E. Coronado, P. LeMagueres, L. Ouahab, L. Ducasse, C. Sourisseau, P. Delhaes, *Inorg. Chem.* **1995**, *34*, 4139.
69 E. Coronado, J. R. Galán-Mascarós, C. Giménez-Saiz, C. J: Gómez-García, S. Triki, *J. Am. Chem. Soc.* **1998**, *120*, 4671.
70 J. R. Galán-Mascarós, C. Giménez-Saiz, S. Triki, C. J: Gómez-García, E. Coronado, L. Ouahab, *Angew. Chem., Int. Ed. Engl.* **1995**, *34*, 1460.
71 E. Coronado, J. R: Galán-Mascarós, C. Giménez-Saiz, C. J. Gómez-García, *Adv. Mater.* **1996**, *8*, 801.
72 E. Coronado, M. Clemente-León, J. R. Galán-Mascarós, C. Giménez-Saiz, C. J. Gómez-García, E. Martínez-Ferrero, *J. Chem. Soc., Dalton Trans.* **2000**, 3955.
73 M. Clemente-León, E. Coronado, J. R. Galán-Mascarós, C. Giménez-Saiz, C. J. Gómez-García, T. Fernández-Otero, *J. Mater. Chem.* **1998**, *8*, 309.
74 a) L. Ouahab, M. Bencharif, A. Mhanni, D. Pelloquin, J. F. Halet, O. Peña, D. Grandjean, C. Garrigou-Lagrange, J. Amiell, P. Delahes, *Chem. Mater.* **1992**, *4*, 666; b) C. Bellito, G. Staulo, R. Bozio, C. Pecile, *Mol. Cryst. Liq. Cryst.* **1993**, *234*, 2670.
75 L. Ouahab, D. Grandjean, M. Bencharif, *Acta Cryst.* **1991**, *C47*, 2670.
76 C. Bellito, M. Bonamico, V. Fares, F. Federici, G. Righini, M. Kurmoo, P. Day, *Chem. Mater.* **1995**, *7*, 1475.
77 E. Coronado, J. R. Galán-Mascarós, C. Giménez-Saiz, C. J. Gómez-García, L. R: Falvello, P. Delhaes, *Inorg. Chem.* **1998**, *37*, 2183.
78 E. Ojima, H. Fujiwara, K. Kato, H. Kobayashi, H. Tanaka, A. Kobayashi, M. Tokumoto, P. Cassoux, *J. Am. Chem. Soc.* **1999**, *121*, 5581.
79 E. Coronado, J. R. Galán-Mascarós, C. J. Gómez-García, *J. Chem. Soc., Dalton Trans.* **2000**, 205.
80 S. Rashid, S. S. Turner, P. Day, M. E. Light, M. B. Hursthouse, *Inorg. Chem.* **2000**, *39*, 2426.
81 S. Triki, F. Berezovsky, J. S. Pala, C. J. Gómez-García, E. Coronado, K. Costuas, J. F. Halet, *Inorg. Chem.* **2001**, *40*, 5127.
82 S. Triki, F. Berezovsky, J. S. Pala, A. Riou, P. Molini, *Synth. Met.* **1999**, *103*, 1974.
83 a) E. Coronado, J. R. Galán-Mascarós, C. Ruiz-Pérez, S. Triki, *Adv. Mater.* **1996**, *8*, 737; b) E. Coronado, J. R. Galán-Mascarós, C. Giménez-Saiz, C. J. Gómez-García, *Synth. Met.* **1997**, *85*, 1677.
84 E. Coronado, J. R. Galán-Mascarós, C. J. Gómez-García, V. Laukhin, *Nature* **2000**, *408*, 447.
85 A. Alberola, E. Coronado, J. R. Galán-Mascarós, C. Giménez-Saiz, C. J. Gómez-García, F. M. Romero, *Synth. Metals* **2002**, in press.
86 M. Clemente-León, E. Coronado, P. Delhaes, C. J. Gómez-García, C. Mingotaud, *Adv. Mater.* **2001**, *13*, 574.

10
Hybrid Organic-Inorganic Electronics

David B. Mitzi

10.1
Introduction

The potential for new technological opportunities and reduced product cost has fueled recent efforts to replace crystalline silicon as the active component in electronic devices. The advent of hydrogenated amorphous silicon (a-Si:H), for example, as a useful semiconducting material, has enabled the relatively convenient low-cost fabrication (particularly with respect to large-area applications) of solar cells, image sensor arrays, and active-matrix flat-panel displays [1]. Recently, the effort to introduce semiconducting organic materials within practical electronic devices has opened the possibility of depositing semiconducting components at near room temperature and therefore on a range of substrates, including those that are large area, organic-based and flexible [2]. The organic-based materials, when fully developed, are expected to enable such technologies as electronic newspapers, smart cards and fabric, inexpensive ID tags and similar flexible, low-cost products.

Among the numerous important criteria to be considered when designing new semiconductors for electronic devices, the carrier (electron/hole) mobility, μ, is defined as the proportionality constant between the applied electric field and the corresponding average carrier drift velocity and it is therefore, in a more practical sense, related to the maximum switching speed achievable within a given electronic device geometry. Fig. 10.1 depicts the improvement in reported field-effect mobility as a function of year for thin-film organic semiconductors, as well as the types of applications enabled at each level of mobility [3]. The advances in mobility have been enabled by the development of new organic semiconductors, as well as by new techniques to better order the molecules on surfaces. After approximately 25 years of development, the room-temperature mobility values in organic thin film devices have reached ~1 $cm^2 V^{-1} s^{-1}$ (approximately the value of mobility for the best a-Si material, but still several orders of magnitude lower than crystalline silicon). While great progress has been made in improving the mobility of organic materials, the room-temperature mobility appears to be limited by the weak intermolecular interactions (i.e., generally van der Waals interaction) in molecular mate-

Gómez-Romero: Organic-Inorganic Materials. David B. Mitzi

ISBN: 3-527-30484-3

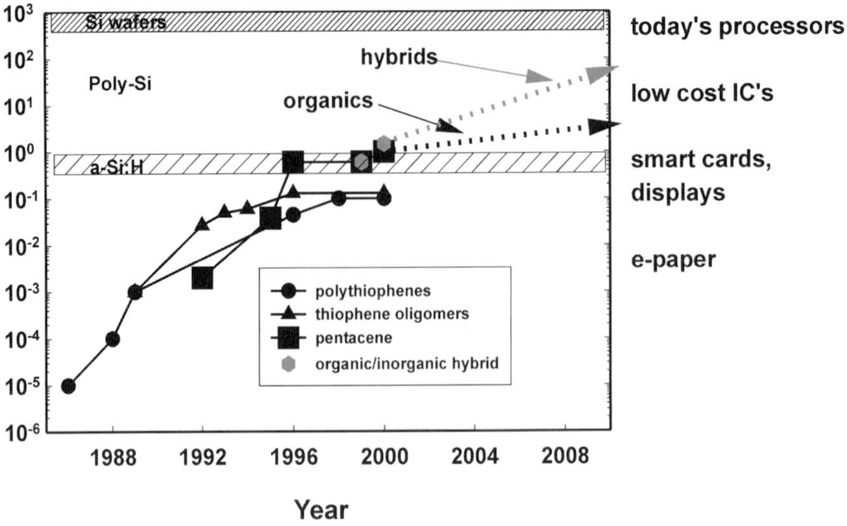

Fig. 10. 1 Yearly performance improvements in the field-effect mobility of thin film transistors with semiconducting channels consisting of organic small molecules, short-chain oligomers, long-chain polymers, and organic-inorganic hybrid materials. The dotted-line arrows for organics and hybrids represent an expected projection beyond year 2000. The right-side captions detail the required field-effect mobilities for targeted electronic applications. (reproduced with permission from Ref. 3)

rials and the ability to organize and order the chains in polymer-based materials [2].

The band gap is also an important semiconductor parameter since it directly relates to the intrinsic carrier density at room temperature, as well as to the wavelength of any possible emission that might be induced from the material (either photon or electrically-induced). The bandgap, E_g, of most useful semiconductors is less than 2.5 eV and is sometimes as small as a few tenths of an eV. The intrinsic carrier density varies as exp $(-E_g/2k_BT)$, a factor which has a value (at room temperature) of order 10^{-1} for $E_g = 0.1$ eV and of order 10^{-26} for $E_g = 3$ eV. Consequently, as channel materials for thin film field-effect transistors (FETs), semiconductors with too small a band gap will render the device difficult to shut off, while too large a band gap will make it impractical to inject sufficient charge in the channel to switch the device on. Note that injection of charge into the semiconductor depends on the relative positioning of the conduction and valence bands with respect to the Fermi energy of the metal contacts, as well as on the absolute value of the band gap. Other potentially important material parameters include the dielectric constant and the nature of trap states within the semiconductor (which often has more to do with surface properties and crystalline imperfections than with the intrinsic character of the semiconductor).

In addition to electronic properties, the ability to conveniently and inexpensively process the semiconductors as films is critical for potential applications. Amorphous silicon, for example, is typically grown using a plasma chemical vapor deposition (CVD) technique, in which silane gas (SiH_4) is decomposed in a plasma excited by radio-frequency power [1]. Despite the convenience of this process for large area applications, for truly low-cost processing, techniques that are performed at near-ambient temperatures and without vacuum requirements are desirable. Solution-based techniques are particularly attractive for this reason (e.g. spin coating, inkjet printing), perhaps enabling new applications for semiconducting materials (e.g. deposition on flexible plastic substrates, incorporation within fabrics). Melt processing is another useful capability, providing compatibility with reel-to-reel, lamination, capillary filling, and extrusion techniques.

Given the above considerations, organic-inorganic hybrids offer attractive opportunities for many electronic applications, because they can combine useful attributes of both organic and inorganic materials within a single molecular scale composite. Organic components provide, for example, lightweight and flexible mechanical properties, ease of deposition, and the ability to tailor efficient luminescence. Inorganic materials, on the other hand, may contribute higher electrical mobility (as a result of the strong covalent bonding within these systems), band gap tunability, and thermal/mechanical stability. As potential thin film transistor materials, the combination of simple processing and higher carrier mobilities can be envisioned, leading to new technological opportunities (Fig. 10.1). Numerous other useful combinations can also be devised, providing opportunities for electronic and optical devices.

Bonding between the organic and inorganic components of the hybrids may take on several different forms, ranging from weak van der Waals type interactions to hydrogen bonded, ionic, and finally covalently bonded compounds. Compounds within the most weakly bonded category of hybrids are generally formed by reversibly intercalating organic species (guest molecules) into the interlayer space of a preexisting inorganic layered structure (host), and have largely been observed and studied in layered silicates, transition metal oxide, chalcogenide, and phosphate systems [4, 5]. The structure of the inorganic framework is little affected by the guest molecule, highlighting the small degree of interaction (e.g. generally van der Waals) between the two components. In ionic compounds, which have well-defined ratios between organic and inorganic components, the organic ion is an intimate part of the overall structure and the structure actually depends on the organic component for overall charge neutrality. Transition metal, main group, and rare-earth metal halides with perovskite and related structures [6, 7] form one important family of ionic hybrids. Although there are numerous reported compounds with covalent bonding between organic and inorganic units, most feature isolated molecules or clusters. A few examples, however, including $ZnTe(NH_2C_nH_{2n}NH_2)_{1/2}$ ($n = 2, 3$) [8], $Zr(HPO_3)_{1.2}(O_3P{-}R{-}PO_3)_{0.4}$ ($R{=}(CH_3)_2H_2C_6C_6H_2(CH_3)_2$) [9], CH_3ReO_3 [10], $(CH_3)_2SnF_2$ [11], and $C_nH_{2n+1}BiI_2$ [12], contain extended inorganic chains or layers.

While the magnetic, optical, structural and electrical properties of organic-inorganic hybrids have been of interest for many years, the device potential of the hybrids for electronic and optical components has only recently begun to develop [13]. A high gain active planar waveguide based on a laser-dye-doped glass film and prepared using a sol-gel technique, for example, has recently been reported [14]. The high gain of the waveguide renders it suitable for solid-state lasers [15]. An electro-optic modulator based on nonlinear organic-silica hybrid films has also been described [16]. Another interesting optical material is created by the successive vacuum deposition of amorphous copper phthalocyanine and titanium oxide layers, yielding a photoconductive composite with a response significantly larger than that observed in single layers of copper phthalocyanine, presumably because of the greater probability of charge separation at the organic-inorganic interface [17]. A similar concept has recently been used to create relatively efficient, hybrid CdSe nanorod-polymer solar cells [18]. Electrostatically layered hybrid multilayer structures consisting of alternating layers of positively charged (protonated) poly(allylamine) or PPV and negatively charged CdSe nanocrystals have yielded a broad (nearly white) electroluminescence (EL) spectrum [19], which can be further refined and tailored by selecting nanocrystals of different size [20]. An electrically rectifying device (Zener diode) has similarly been prepared using a two-layer configuration consisting of a p+-doped semiconducting polymer [polypyrrole or poly(3-methylthiophene)] layer and an n-type multilayer structure of CdSe (capped with trioctylphosphine oxide) and 1,6-hexanedithiol. By controlling the level of doping in the p-doped polymer, an asymmetrically doped junction can be created, leading to rectifying behavior under forward bias and Zener breakdown in reverse bias [21].

In each of these examples, the functional organic-inorganic hybrid consists of at least one amorphous component. While the family of potential organic-inorganic hybrids is vast [22], this review will primarily focus on the metal halide-based perovskite family, since this is currently one of the most thoroughly studied crystalline hybrid systems with respect to potential electronic device applications. Crystalline hybrids are particularly interesting because of the ability to conveniently correlate structure-property relationships, thereby providing a mechanism for understanding and tailoring properties. Additionally, the good solubility and relatively low-melting and boiling points for the metal halide component of the perovskites provide for a range of low-cost film deposition options.

10.2 Organic-Inorganic Perovskites

10.2.1 Structures

The basic AMX_3 hybrid perovskite structure consists of a three-dimensional network of corner-sharing metal (M) halide (X) octahedra, with the A^+ organic cations occupying the 12-fold coordinated sites between the octahedra (Fig. 10.2). For a perfectly packed cubic structure, the ionic radii of the A, M, and X ions satisfy $(R_A + R_X) = t\sqrt{2}\,(R_M + R_X)$, where R_A, R_M, and R_X are the corresponding ionic radii and the tolerance factor $t \sim 1$. Generally, it is found that $0.8 \leq t \leq 0.9$ for cubic perovskites, although the exact end-points of this range depend on the criteria chosen for determining the ionic radii and there is an extended range for distorted perovskite structures [6, 23]. The three-dimensional (3D) confinement for the A^+ cation site limits the types of organic cations that can be incorporated within the 3D perovskite structure. Taking, for example, a system with one of the largest possible values for $R_M + R_X$, with M=Pb and X=I [$R_{Pb} = 1.19$ Å and $R_I = 2.20$ Å] [24] and using $t = 1$, R_A is limited to approximately 2.6 Å (i.e., organic cations with at most two to three C–C or C–N bonds) [6]. Several examples of cubic or distorted organic-inorganic perovskites include $(CH_3NH_3)PbX_3$ (X = Cl, Br, I) [25, 26], $(CH_3NH_3)SnI_3$ [27, 28], $(CH_3NH_3)SnBr_3$ [29], and $(NH_2CH{=}NH_2)SnI_3$ [30].

Layered perovskite structures can also be formed and are structurally derived from the 3D AMX_3 structure by taking n-layer thick cuts from along a particular crystallographic direction of the structure and stacking these slabs in alternation with organic cation layers (Fig. 10.3). Several possibilities for the perovskite slab orientation include <100>, <110>, and <111> cuts [7]. The single layer ($n = 1$) <100>-oriented layered perovskites are the simplest and most widely studied

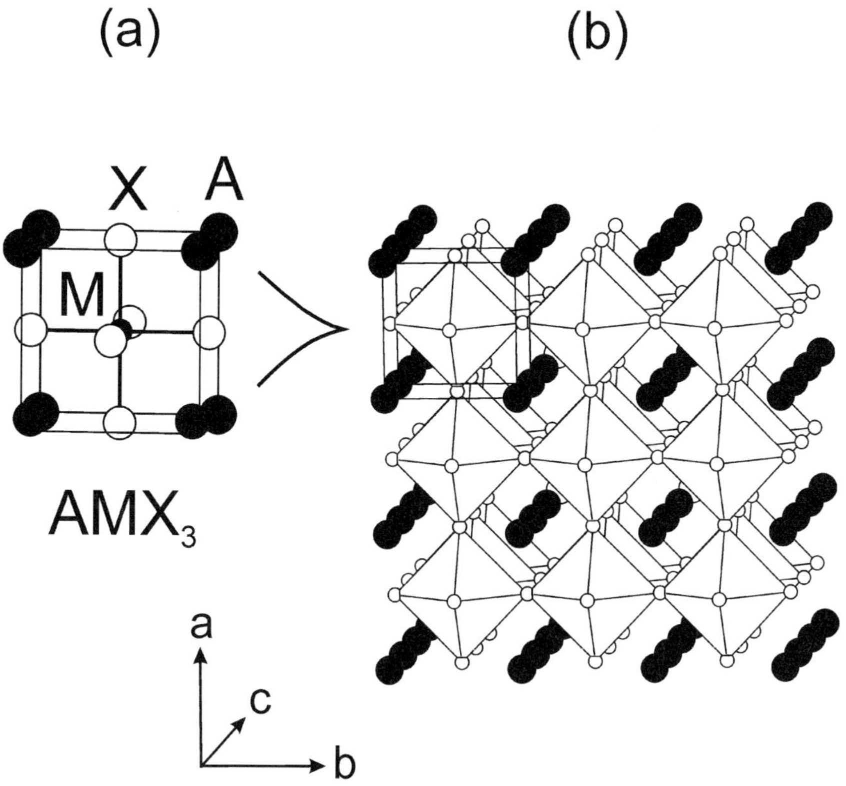

Fig. 10.2 (a) Ball and stick model of the basic AMX_3 perovskite unit cell and (b) polyhedral representation of how the structure extends in three dimensions (from [7], reproduced by permission of the Royal Society of Chemistry)

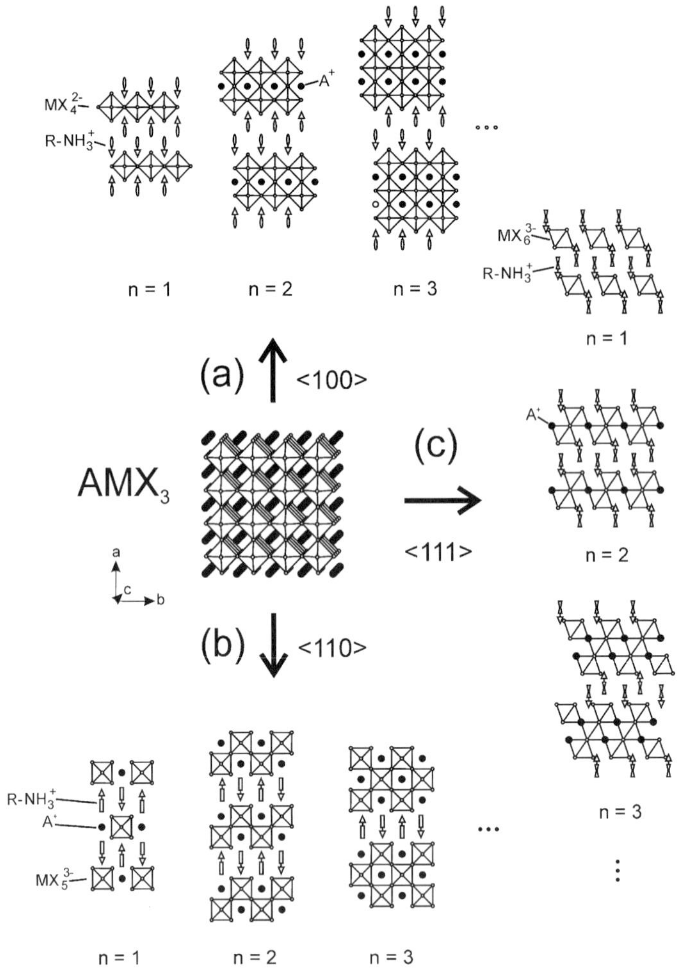

Fig. 10.3 Three families of layered perovskites derived from the basic 3D AMX_3 structure by taking cuts from along three different crystallographic directions. The different families can be written (a) $(R\text{-}NH_3)_2A_{n-1}M_nX_{3n+1}$ (<100>-oriented), (b) $(R\text{-}NH_3)_2A_nM_nX_{3n+2}$ (<110>-oriented), and (c) $(R\text{-}NH_3)_2A_{n-1}M_nX_{3n+3}$ (<111>-oriented)

perovskites and contain single MX_4^{2-} layers, where M is a divalent metal (e.g. Cu^{2+}, Ni^{2+}, Co^{2+}, Fe^{2+}, Mn^{2+}, Cr^{2+}, Pd^{2+}, Cd^{2+}, Ge^{2+}, Sn^{2+}, Pb^{2+}, Eu^{2+}, or Yb^{2+}) and X is a halide (e.g. Cl^-, Br^-, I^-). Recently, the single-layer perovskite family has been extended to include higher-valent metal halide sheets (e.g. $Bi_{2/3}I_4^{2-}$, $Sb_{2/3}I_4^{2-}$), with appropriately chosen organic cation layers templating the metal-deficient inorganic layers [31]. The large repertoire of inorganic frameworks enables tailoring of the electronic (e.g. insulating, semiconducting, metallic), optical (e.g. luminescent, nonlinear optical) and magnetic (e.g. ferromagnetic, antiferromagnetic) properties [6].

The organic cation layer alternates with the inorganic sheets and is generally comprised of molecules containing either one (R-NH_3^+) or two (^+H_3N-R-NH_3^+) ammonium groups that tether to the halides in the inorganic sheets and an organic tail (R) that extends within the space between layers. For the monoammonium cation, the ammonium group interacts with a single inorganic layer through hydrogen and ionic bonding, with the organic tail extending into the space between the layers and a van der Waals gap separating the bilayer of organic tails from adjacent layers. For diammonium cations, each organic cation spans the entire distance between layers, interacting with two adjacent inorganic layers through the ammonium groups attached at each end. While the length of the organic cation can in principle be arbitrarily long, the width or cross-sectional area of the organic moiety is limited by potential steric interaction with nearest neighbor organic cation sites. Given the distances between nearest-neighbor organic cation sites within the sheets of corner-sharing metal halide octahedra, the cross sectional area is limited to ~40 Å^2 for inorganic frameworks based on some of the larger metal cations (e.g. Pb^{2+}) and halides (e.g. I^-). The organic R-group most commonly consists of an alkyl chain or a single-ring aromatic group. These simple organic layers help to define the degree of interaction between, and the properties arising in, the inorganic layers. In addition to the simple organic cations, the structural constraints on the organic component provide for significant flexibility to incorporate more complex, functional organic cations within the hybrid structure. Recent examples include 5,5‴-bis(2-aminoethyl)-2,2′:5′,2″:5″,2‴-quaterthiophene [32], 1-pyrenemethylamine [33], 2-naphthylmethylamine and 2-anthrylmethylamine [34]. These and related molecules provide important opportunities for luminescence, electrical conduction, non-linear optical properties, and structural templating. While most examples of organic cations involve protonated primary amine groups, tin(II) and lead(II) iodide perovskites containing the non-primary ammonium cation, biimidazolium, have also recently been reported [35].

Another important opportunity for tailoring the properties of hybrid perovskite systems using the organic component of the structure involves the incorporation of photoreactive (e.g., diene or diyne) groups within the organic layer of the structure, which can undergo solid-state polymerization within the perovskite framework [36–38]. In one example, the hydrochloride salt of 6-amino-2,4-*trans,trans*-hexadienoic acid, within a cadmium(II) chloride framework, polymerizes under ultraviolet (UV) or γ-irradiation [36]. The polymerization occurs through a 1,4 addition mechanism, leading to a well-ordered polymer layer. The same photoreactive organic cation in a copper(II) chloride framework does not undergo polymerization. The structural differences in the inorganic framework, induced by the Jahn-Teller distortion of the copper(II) chloride octahedra, lead to an unfavorable (with respect to the polymerization process) configuration of the monomer within the organic layer. Consequently, the inorganic framework can be used to control whether or not a monomer-containing hybrid system is susceptible to polymerization.

The van der Waals gap between organic cation layers in monoammonium-based <100>-oriented layered perovskite structures also appears ideally suited for intercalation reactions. In one of the few studies examining this possibility, Dolzhenko et al. [39] examined the interaction of 1-chloronaphthalene, *o*-dichlorobenzene and hexane with cadmium(II) chloride- and lead(II) iodide-based perovskites containing long aliphatic cations. While powder X-ray diffraction data provide evidence of intercalation in the form of a new set of diffraction peaks with enhanced d-spacing, the diffraction patterns quickly revert back to that observed for the parent (unintercalated) compound after the source of the intercalating species is removed. More recently, a relatively stable hybrid structure was developed [40], which employs the fluoroaryl-aryl interaction to bind the intercalating species more firmly within the structure. Two of these intercalated systems, $(C_6H_5C_2H_4NH_3)_2SnI_4 \cdot C_6F_6$ and $(C_6F_5C_2H_4NH_3)_2SnI_4 \cdot C_6H_6$, were characterized by single crystal structure refinement (Fig. 10.4), thermal analysis and optical measurements, unambiguously establishing intercalation in the hybrids. The observed offset face-to-face interaction between the organic cation and the intercalated molecule, the high dissociation temperature for the intercalating species relative to the boiling point for this molecule, and the fact that attempts to form the related hybrids, $(C_6F_5C_2H_4NH_3)_2SnI_4 \cdot C_6F_6$ and $(C_6H_5C_2H_4NH_3)_2SnI_4 \cdot C_6H_6$ (i.e. with either both

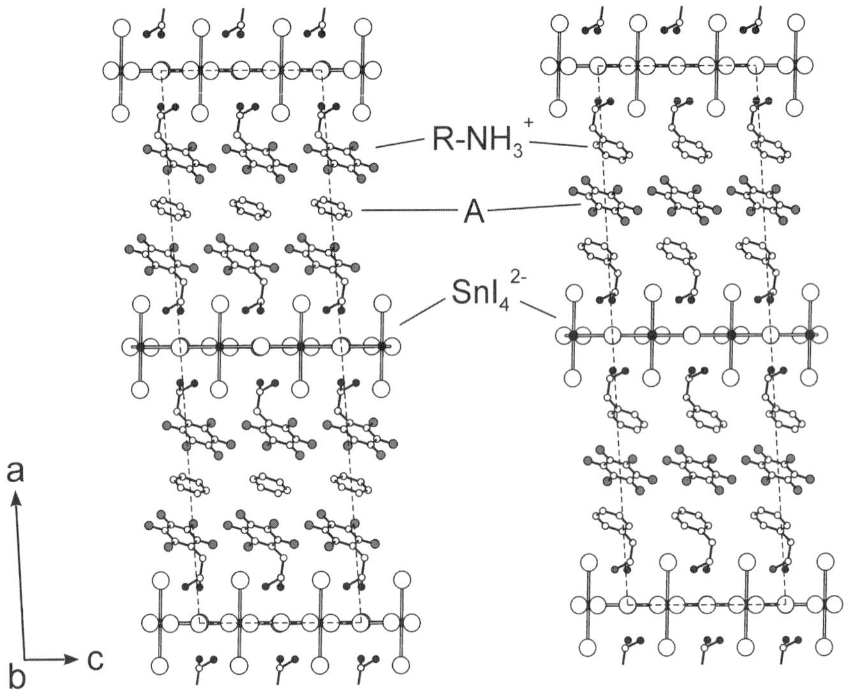

Fig. 10.4 Crystal Structures for the intercalated $(R\text{-}NH_3)_2SnI_4 \cdot A$ hybrid perovskites (a) $(C_6F_5C_2H_4NH_3)_2SnI_4 \cdot (C_6H_6)$ and (b) $(C_6H_5C_2H_4NH_3)_2SnI_4 \cdot (C_6F_6)$, each stabilized by fluoroaryl-aryl interaction. Dashed lines depict the unit cell outlines. (reproduced with permission from Ref. 40. Copyright 2002 American Chemical Society)

fluorinated or both unfluorinated organic cation and intercalating molecule), were not successful, all point to the importance of the fluoroaryl-aryl interaction in stabilizing the intercalated hybrid structure and demonstrate an important new example of chemistry within the van der Waals gap of the hybrid perovskites. It is expected that interactions between the organic cation and intercalated species (e.g., charge transfer) or among intercalated species themselves may provide important opportunities for tailoring the properties of the hybrid perovskite systems for both device applications, as well as for evoking interesting physical properties.

10.2.2
Properties

The flexible structural and chemical characteristics of the organic-inorganic perovskites provide for a wide range of interesting and potentially useful physical properties. Interesting magnetic properties, for example, arise in many of the layered hybrid perovskites with first-row transition metals in the inorganic layers, as a result of the low (and tunable) dimensionality of the structures [6, 41]. While magnetic and structural properties are quite notable for the hybrids, this review will focus on optical and electrical properties since these are currently most relevant for electronic device applications.

Many of the layered hybrid perovskites, especially those containing germanium(II), tin(II), or lead(II) halide sheets, resemble multilayer quantum well structures, with semiconducting inorganic sheets alternating with wider band gap (i.e. HOMO-LUMO gap) organic layers (Fig. 10.5) [13]. Making substitutions on either the metal or halogen site modifies the band gap of the inorganic layers (well depth), while the width of the barrier and well layers can easily be adjusted by changing the length of the organic cations and the number of perovskite sheets between each organic layer, respectively. In the hybrid structures with semiconducting inorganic sheets and simple organic cations (i.e. aliphatic or single ring aromatic moieties), the conduction band of the inorganic layers is generally below that of the organic layers and the valence band of the inorganic layers is similarly above that of the organic layers (Fig. 10.5 a). Therefore, the inorganic sheets act as quantum wells for both electrons and holes. Alternatively, if larger band gap metal halide sheets are integrated with more complex, conjugated (i.e. smaller HOMO-LUMO gap) organic cations, the well and barrier layer roles can be reversed (Fig. 10.5 b) [13]. Given appropriate modifications of the chemistry of the organic and inorganic layers, the band gaps for the organic and inorganic layers can also be offset (Fig. 10.5 c), leading to a type II heterostructure, in which the wells for the electrons and holes are in different layers. The self-assembling structures share many of the interesting electronic characteristics of quantum well structures prepared by "artificial" techniques, such as molecular beam epitaxy (MBE), and are therefore expected to yield similarly interesting optical and electrical characteristics [42]. Examples of the different types of hybrid quantum well structures, along with details of optical and electrical characteristics, are given below.

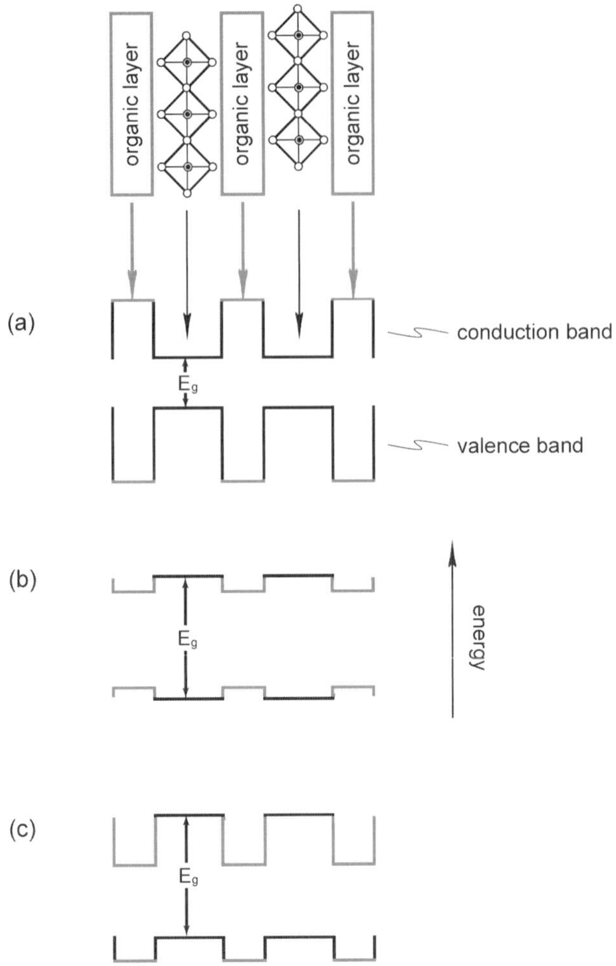

Fig. 10.5 Schematic organic-inorganic perovskite structure and several possible energy level schemes that can arise within these structures. The most common arrangement is shown in (a), where semiconducting inorganic sheets alternate with much wider band gap (HOMO-LUMO separation) organic layers, resulting in a type I quantum well structure. In (b), wider band gap inorganic layers and organic cations with a smaller HOMO-LUMO gap result in the well/barrier roles of the organic and inorganic layers being switched. In (c), by shifting the electron affinity of the organic relative to the inorganic layers, a staggering of the energy levels leads to a type II quantum well structure. (reproduced with permission from D. B. Mitzi, K. Chondroudis, C. R. Kagan, *IBM J. Res. & Dev.* **2001**, *45*, 29–45)

10.2.2.1 **Optical Properties**

The band gaps associated with lead(II) and tin(II) halide perovskite layers (derived from optical absorption measurements) range from 1.6 eV to >3.5 eV [43], which is a useful range for use in semiconducting devices. Perovskite quantum well structures with semiconducting inorganic sheets and large HOMO-LUMO gap organic

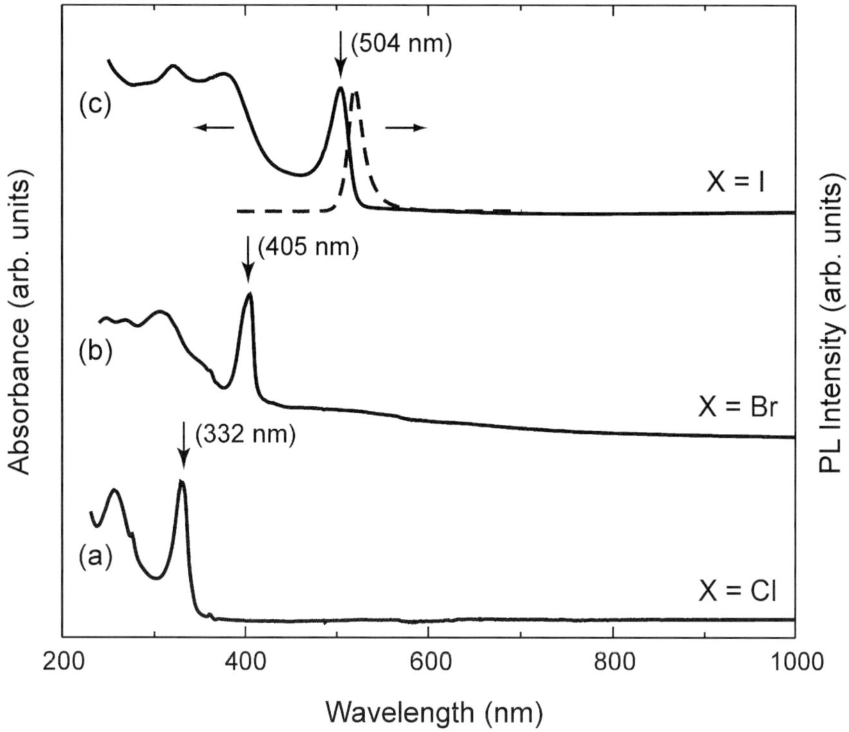

Fig. 10.6 Room temperature UV-vis absorption spectra for spin-coated thin films (on quartz disks) of $(C_4H_9NH_3)_2PbX_4$ with (a) X = Cl, (b) X = Br, and (c) X = I. In each spectrum, the arrow indicates the position of the exciton absorption peak (with the wavelength in parentheses). In (c), the corresponding photoluminescence (PL) spectrum ($\lambda_{ex} = 370$ nm) is indicated by the dashed curve (reproduced with permission from D. B. Mitzi, K. Chondroudis, C. R. Kagan, *IBM J. Res. & Dev.* **2001**, *45*, 29–45)

cations exhibit an exciton resonance (Fig. 10.6) below the band edge in room temperature optical absorption spectra. The exciton band is associated with the band gap of the inorganic framework and therefore the spectral positions of the transitions can be tailored across the visible spectrum by substitution on the metal or halide site or by changing the thickness of the inorganic layers [43, 44]. The strong binding energy of the excitons, which enables the optical features to be observed at room temperature, arises because of the two-dimensionality of the structure, coupled with the dielectric modulation among the organic and inorganic layers [45]. In addition to the sharp transition in the absorption spectra, the large exciton binding energy and oscillator strength lead to strong photoluminescence, non-linear optical effects [46], tunable polariton absorption [47] and biexciton lasing [48]. In addition to tailoring the optical features through substitutions on the metal or halogen sites, control over the specific structure of the metal halide sheets, through the appropriate choice of the organic cation, also provides a means of engineering the optical and electronic properties. This has been demonstrated [49] in the family of substituted phenethylammonium tin(II) iodide compounds, $(n\text{-XPEA})_2SnI_4$, where n-XPEA = n-halophenethylammonium and $n = 2$, 3, or 4 (Fig. 10.7). For X = F, as n is shifted from 2 to 3 and 4, the exciton peak shifts progressively from 588 nm to 599 nm and 609 nm, respectively, with the $n = 4$ structure yielding a virtually identical spectrum to that for the unsubstituted phenethylammonium system, $(PEA)_2SnI_4$. A corresponding shift in the band edge is also noted and a similar range of shifts occur in other related (X = Cl and Br) substituted phenethylammonium tin(II) iodide systems.

The lowest exciton state arises from excitations between the valence band, which consists of a hybridization of Sn(5 s) and I(5 p) states in the layered tin(II) iodide based perovskites, and the conduction band, which derives primarily from Sn(5 p) states. The shifts in optical spectra are therefore correlated with changes in the detailed structure of the tin(II) iodide sheets (such as Sn-I-Sn bond angle and average Sn-I bond length) as a function of which organic cation is substituted into the structure. For example, the average Sn-I-Sn bond angle varies from 156.5° for $(PEA)_2SnI_4$ to 148.7° for $(2\text{-BrPEA})_2SnI_4$, the two extremes in Fig. 10.7 with respect to the positions of the optical features. The control over the inorganic framework structure is provided by the organic cation through hydrogen bonding interactions, which occurs between the two components of the structure, as well as through steric constraints imposed by the organic cation on the inorganic sheets.

Organic cations with a longer conjugation length (smaller HOMO-LUMO gap) can also have a direct influence on the optical properties in the visible spectral range [32–34]. The series $(R\text{-NH}_3)_2PbCl_4$ (R = 2-phenylethyl, 2-naphthylmethyl, and

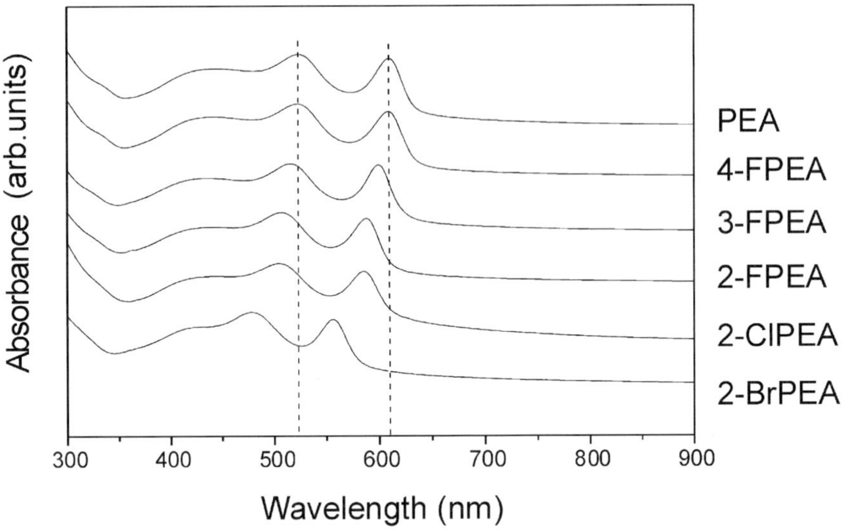

Fig. 10.7 Room-temperature UV-vis absorption spectra for spin coated thin films (on quartz disks) of various layered hybrid perovskites of the form $(R\text{-NH}_3)_2SnI_4$, where $R\text{-NH}_3^+$ is a halogen-substituted derivative of phenethylammonium (PEA). The dashed lines are guides for the eye to highlight the shift in spectral feature as a function of the organic cation

2-anthrylmethyl) is one interesting series because of the ability to tune the relative positions of the energy levels for the organic chromophore and the inorganic framework using the number of aromatic rings contained in the cation [34]. For R = 2-phenylethyl, the singlet and triplet states of the organic cation are higher in energy than the exciton state of the inorganic sheets, leading to emission from the inorganic exciton state in the photoluminescence spectrum. For R = 2-naphthylmethyl, the inorganic exciton state falls between the organic singlet and triplet states, and phosphorescence from the organic molecule dominates the emission spectrum. Finally, for R = 2-anthrylmethyl, the inorganic exciton state is relatively higher in energy than both the organic singlet and triplet states, and chromophore singlet emission is dominant.

The oligothiophene derivative, 5,5‴-bis(aminoethyl)2,2′:5′,2″:5″,2‴-quaterthiophene (AEQT) has also been incorporated within a lead(II) halide framework, $(H_2AEQT)PbX_4$ [32]. The modification of the halide from X = Cl to X = Br and I enables the energy levels of the inorganic framework to be conveniently tailored relative to those for the dye molecule, thereby facilitating a study of the interaction between the two components of the structure. For each X = Cl, Br, and I hybrid, absorption spectra demonstrate both the characteristic features of the inorganic framework (i.e. the exciton peak) and the π-π* transition for the organic cation at approximately 390 nm (Fig. 10.8). For X = Cl, with a larger lead(II) chloride band gap relative to the HOMO-LUMO separation for the dye molecule, only strong

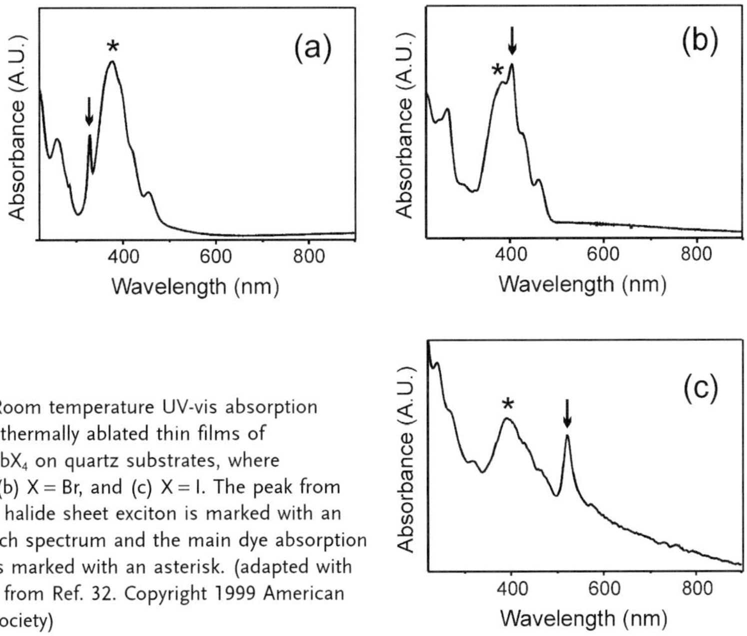

Fig. 10.8 Room temperature UV-vis absorption spectra for thermally ablated thin films of $(H_2AEQT)PbX_4$ on quartz substrates, where (a) X = Cl, (b) X = Br, and (c) X = I. The peak from the lead(II) halide sheet exciton is marked with an arrow in each spectrum and the main dye absorption transition is marked with an asterisk. (adapted with permission from Ref. 32. Copyright 1999 American Chemical Society)

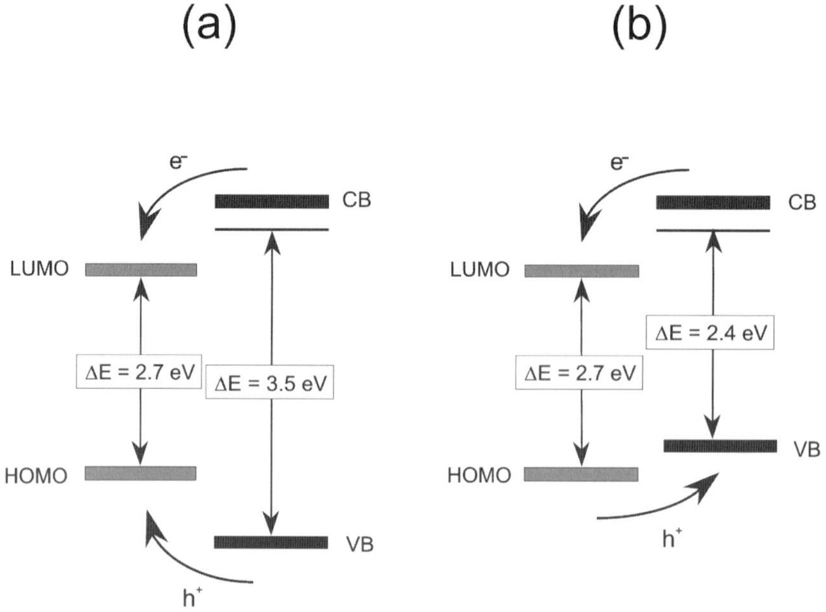

Fig. 10.9 Proposed schematic energy level diagrams for two $(H_2AEQT)PbX_4$ hybrid systems, highlighting the fact that the organic and inorganic components of the chromophore-containing hybrids are spatially separated, as well as the possibility of different energy level alignments. The type I heterojunction (a) is drawn for the X=Cl system and the type II heterojunction (b) is for the X = I system. (adapted with permission from Ref. 32. Copyright 1999 American Chemical Society)

singlet emission from AEQT is observed in the photoluminescence emission spectrum. By comparison, AEQT luminescence is progressively quenched across the series X=Cl → Br → I, as the band gap of the inorganic framework decreases in energy compared to the HOMO-LUMO splitting of the organic cation [32]. Although the details of the quenching mechanism are still not fully established, one likely explanation is the transition of the system from a type I to a type II multilayer structure (Fig. 10.9). In the case of the type II heterostructure, the physical separation of electron and hole at the organic-inorganic interface reduces the probability of radiative recombination and therefore quenches the luminescence observed in the material.

Another interesting means for generating structures with increased conjugation length in the organic component of semiconducting hybrids has also recently been described [38], involving the topochemical polymerization of diacetylene-containing cations within a lead(II) halide-based perovskite. The polymerization is actuated using a ^{60}Co γ-source at room temperature, yielding a significant shift in the absorption spectrum for the hybrid as the monomeric cations are converted to polymer.

10.2.2.2 **Electrical Transport Properties**

While most metal halides are good insulators, an unusual semiconductor-metal transition has been noted in the families $(R\text{-}NH_3)_2(CH_3NH_3)_{n-1}Sn_nI_{3n+1}$ and $[NH_2C(I){=}NH_2]_2(CH_3NH_3)_nSn_nI_{3n+2}$, as a function of increasing perovskite layer thickness (i.e. increasing n) (Fig. 10.10) [50, 51]. The tin(II) halide based perovskites exhibit high carrier mobilities and, in some cases, even metallic conduction. The three-dimensional perovskite, $CH_3NH_3SnI_3$, for example (i.e. the $n \to \infty$ member of both above-mentioned families), is a low carrier density p-type metal with a Hall hole density of $1/R_he \approx 2 \times 10^{19}$ cm^{-3} and a Hall mobility of $\mu \approx 50$ cm^2 V^{-1} s^{-1} at room temperature [28]. The high degree of conductivity derives from the large dispersion of the Sn(5s) band [hybridized with I(5p)] along the <111> direction of the cubic Brillouin zone, leading to a marginal crossing of the Sn(5s) and Sn(5p) bands near the R point $(2\pi[^1/_2, {}^1/_2, {}^1/_2]/a)$, with the Fermi energy falling roughly between the two bands. The relatively high mobility, combined with the ability to conveniently process the materials, renders the tin(II) iodide based perovskites an interesting family to consider for use as TFT channel materials and as the active semiconducting component in other electronic devices.

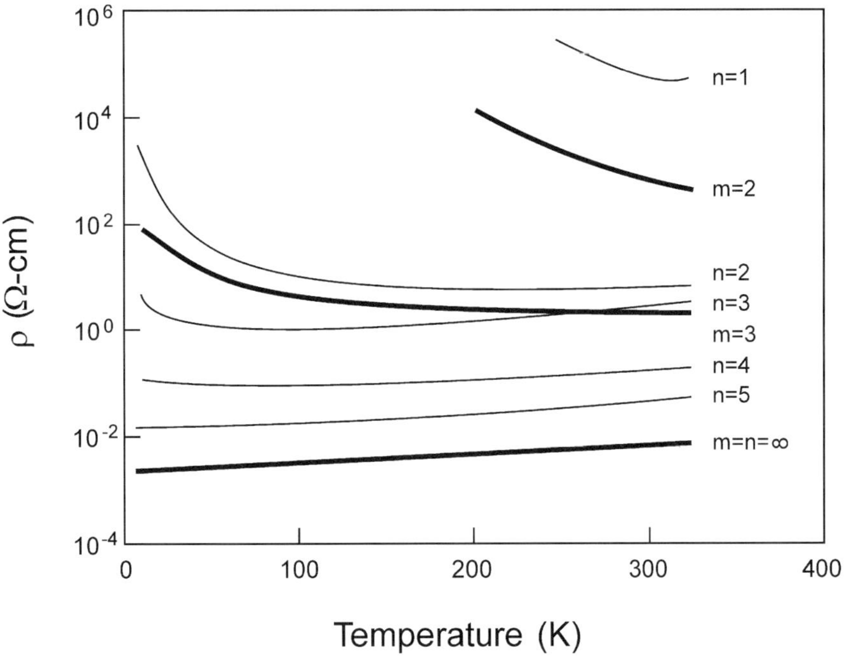

Fig. 10.10 Resistivity, ϱ, as a function of temperature for pressed-pellet $n = 1-5$ $(C_4H_9NH_3)_2(CH_3NH_3)_{n-1}Sn_nI_{3n+1}$ (<100>-oriented) and $m = 2, 3$ $[NH_2C(I){=}NH_2]_2(CH_3NH_3)_mSn_mI_{3m+2}$ (<110>-oriented) layered perovskites. The resistivity of $CH_3NH_3SnI_3$ (n or $m \to \infty$) is also shown for comparison. (adapted with permission from Ref. 51. Copyright 1995 American Association for the Advancement of Science)

In the systems mentioned above, the conducting properties are arising from the semiconducting nature of the inorganic framework since the organic cations are simple moieties with a large HOMO-LUMO separation (Fig. 10.5 a). As for optical properties, the electrical characteristics of the hybrid perovskites are likely to be substantially influenced by the incorporation of more complex organic cations. For organic cations with relatively small HOMO-LUMO gaps, transport in the organic cation layer may be achievable. In these systems, the inorganic framework should provide a means of templating the conformation and orientation of the oligomeric or dye-containing cation [7]. Given the importance of molecular ordering to the mobility and transport characteristics of organic semiconductors in TFTs and related devices, this templating capability should prove quite significant. Presently, however, the transport properties within the organic component of hybrid perovskites have not been rigorously explored.

10.2.3 Film Deposition

Films are typically the most desirable medium for use in semiconducting devices. The hybrid perovskites are often soluble in common polar solvents and both components generally volatilize at relatively low temperatures. In addition, the range of interactions both within and between the organic and inorganic components of the structure (e.g., covalent/ionic, hydrogen bonding, van der Waals) typically strongly favors the formation of the hybrid perovskite from the components. This enables the assembly of hybrid perovskite crystals or thin films using a number of simple processes [52], including vacuum thermal evaporation, solution-based techniques such as spin-coating and stamping, and even melt processing. Each of these options provides advantages for selected applications, thereby enabling convenient deposition on a range of substrates, including those envisioned for flexible plastic displays and low-cost electronic devices.

10.2.3.1 Thermal Evaporation

Gradual heating of an organic-inorganic perovskite in a vacuum or under an inert atmosphere would generally be expected to lead to the organic component decomposing or dissociating from the system at a significantly lower temperature or more rapidly than the metal halide component. This apparent temporal or thermal incompatibility with evaporation-based deposition techniques can be overcome using a two-source thermal evaporation process, as demonstrated by the successful deposition of the lead(II)-iodide-based perovskite films of $(C_6H_5C_2H_4NH_3)_2PbI_4$ and $(CH_3NH_3)PbI_3$ [53]. For the phenethylammonium compound, PbI_2 and $C_6H_5C_2H_4NH_2 \cdot HI$ were simultaneously deposited onto a fused quartz substrate under a base pressure of $\sim 10^{-6}$ Torr. During the deposition, the substrates were allowed to remain at room temperature. The resulting films demonstrated the characteristic absorption spectrum and X-ray pattern for the hybrid perovskite. Film growth is proposed to occur through the intercalation of organic ammonium iodide into the simultaneously deposited PbI_2 film. As a test of this mechanism, hybrid films have also been formed by the sequential deposition of the metal halide film followed by exposure of the film to a vapor of the organic salt [54]. Lead(II) bromide films, for example, were exposed to alkylammonium bromide, vaporized by heating at a base pressure of 10^{-5} Torr. As for two-source thermal evaporation, single phase films of the desired hybrid formed during the process.

While the metal halide component of the hybrid evaporates in a well-defined fashion in the two-source process, the organic salt deposition is often difficult

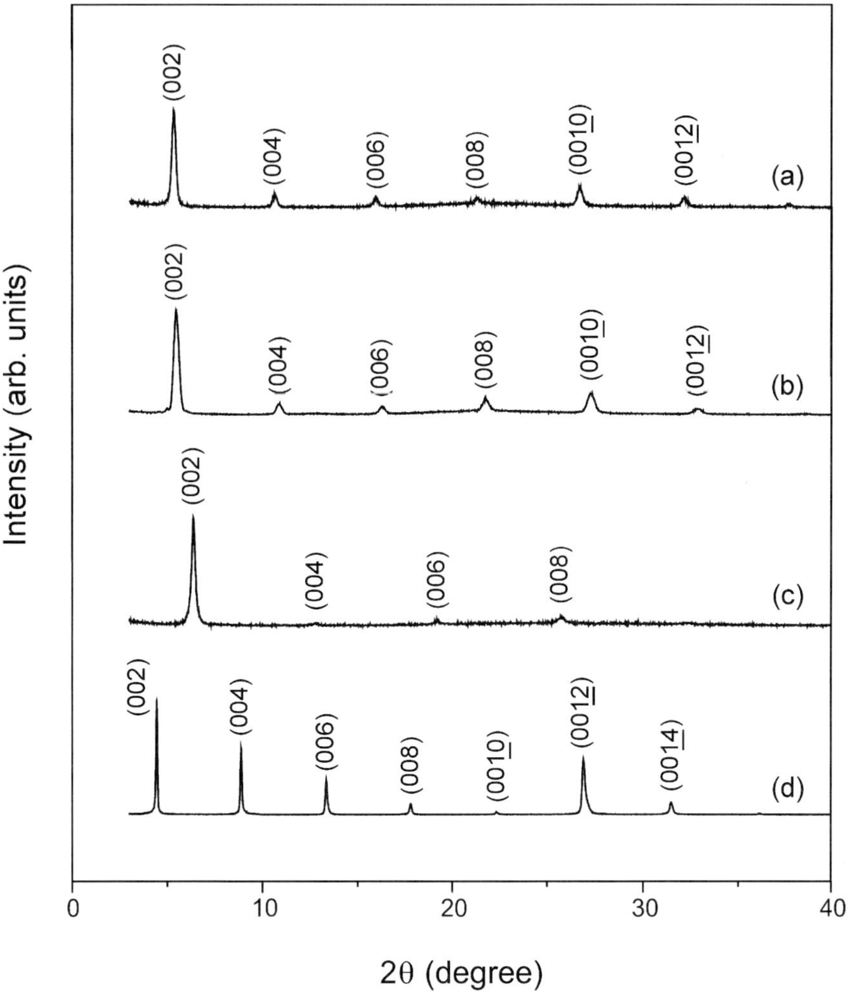

Figure 10.11 Room temperature X-ray diffraction data for unannealed SSTA-deposited films of the organic-inorganic perovskites (a) $(C_6H_5C_2H_4NH_3)_2PbBr_4$, (b) $(C_6H_5C_2H_4NH_3)_2PbI_4$, (c) $(C_4H_9NH_3)_2SnI_4$, and (d) $(C_4H_9NH_3)_2(CH_3NH_3)Sn_2I_7$. (reproduced with permission from Ref. 52. Copyright 2001 American Chemical Society)

to control. In addition, given the relatively long times employed for two-source evaporation, high vacuum conditions are generally required during these depositions. Recently, a vacuum-based single source thermal ablation (SSTA) technique has also been reported [55], which employs a single evaporation source and very rapid heating. A starting charge is deposited on the thin tantalum sheet heater in the form of crystals, powder, or a concentrated solution (which is allowed to dry before ablating). After establishing a suitable vacuum, a large current is passed through the heater, leading to essentially instantaneous ablation of the hybrid, generally before the organic cation has a chance to decompose. The organic and inorganic components reassemble on the substrates (positioned above the tantalum sheet) after ablation to produce optically clear films of the desired product. A number of organic-inorganic hybrids have been deposited using this technique, including $(C_6H_5C_2H_4NH_3)_2PbBr_4$, $(C_6H_5C_2H_4NH_3)_2PbI_4$ and $(C_4H_9NH_3)_2SnI_4$ [55]. The successful deposition of $(C_4H_9NH_3)_2(CH_3NH_3)Sn_2I_7$ films demonstrates that mixed-organic-cation systems can be deposited. X-ray diffraction patterns (Fig. 10.11) demonstrate that the as-deposited films are single phase, crystalline, and highly oriented (as for spin-coated films of the same material). An atomic force microscope (AFM) study of an as-prepared (i.e. unannealed) $(C_6H_5C_2H_4NH_3)_2PbI_4$ film indicates a mean roughness of approximately 1.6 nm, similar to the values observed for spin-coated films of the same material and suitably smooth for device applications. The ablated films of the above systems (without annealing) generally exhibit a relatively small (<75 nm) grain size.

In many cases (especially with relatively simple organic cations), the as-deposited SSTA films are single phase and crystalline, indicating that the organic-inorganic hybrids can reassemble on the substrate at room temperature. Deposition of organic-inorganic perovskite films with more complex organic cations, however, especially those containing diammonium-based cations, often requires a short (~15 min) low-temperature (<250 °C) post-deposition anneal to achieve crystalline films. Films of $(H_2AETH)PbX_4$ {X = Br, I; AETH = 1,6-bis[5′-(2″-aminoethyl)-2′-thienyl]hexane}, for example, have been deposited using the SSTA technique [56]. The as-deposited films do not exhibit a diffraction pattern, indicating the very fine-grained or amorphous nature of the films. Low temperature annealing (<120 °C) leads to progressive grain growth (as observed in X-ray diffraction patterns and AFM studies), and a bathochromic shift of the characteristic perovskite layer exciton peak in the absorption data. Even for hybrids that are crystalline upon deposition, low-temperature annealing can be used to increase grain size.

10.2.3.2 **Solution Processing**

The range of common polar solvents (e.g. water, aqueous acid, alcohols, acetone, acetonitrile, N,N-dimethylformamide) available for the metal halide based hybrids enables a variety of solution-based film techniques, including spin coating, stamping and printing. Spin coating can be used to deposit thin films on a variety of substrates, including glass, quartz, sapphire, silicon and plastic [52]. The process involves finding a suitable solvent for the hybrid, preparing a solution with the desired concentration, applying a quantity of the solution to a substrate, and spinning the substrate. As the solution spreads on the substrate, it dries and leaves a film of the hybrid. Generally, the perovskite films are highly crystalline and oriented, with the plane of the perovskite sheets parallel to the substrate surface. The relevant parameters for the deposition include the choice of substrate, the solvent, the concentration of the hybrid in the solvent, the substrate temperature, and the spin speed. Pre-treating the substrate surface with an appropriate adhesion promoter may improve the wetting properties of the solution. Post-deposition, low-temperature annealing (T <200 °C) of the films is also sometimes employed to improve crystallinity and phase purity. Spin-processed hybrid films are generally very smooth (mean roughness ~1–2 nm), suggesting their suitability for use in device structures.

Grain structure is crucial for device properties and can be controlled by the choice of solvent, as well as through the choice of organic cation in the hybrid

Fig. 10.12 Atomic Force Microscope (AFM) topology images for spin-cast films of (a) $(3\text{-FPEA})_2SnI_4$, (b) $(PEA)_2SnI_4$, and (c) $(2\text{-FPEA})_2SnI_4$ on silicon substrates with 5000 Å of thermally grown oxide. In each case, the spinning solution consisted of 20 mg of the recrystallized hybrid in 1.6 mL of freshly dried and distilled methanol. The spin cycle consisted of a 1 s ramp to 3000 rpm and 30 s dwell at 3000 rpm. Each film was imaged within the channel region of a similar TFT device. (reproduced with permission from Ref. 49. Copyright 2001 American Chemical Society)

structure. In the $(R\text{-}NH_3)_2SnI_4$ perovskite family, for example, where $R\text{-}NH_3^+$ is a substituted phenethylammonium cation, the position of substitution has a pronounced effect on the grain structure of the films [49]. Fluorine substitution on the 3 and 4 positions of the phenethylammonium cation yields films with a larger grain structure (~150–200 nm) (Fig. 10.12a) relative to that for the unsubstituted phenethylammonium tin(II) iodide films (Fig. 10.12b) given similar spin conditions. In contrast, fluorine substitution at the 2 position leads to a significantly smaller grain structure (<100 nm) (Fig. 10.12c) than that for the unsubstituted system. The changes in film morphology in the current samples might either arise from the different solubility of the organic cations as a function of the position of the fluorine substitution (2-FPEA is more soluble than 4-FPEA in many polar solvents) or from variations in the cation-substrate interactions (in this case, the fluorine atom should interact with the surface in the 3-FPEA and 4-FPEA systems, but not in the PEA or 2-FPEA systems). Changing the solvent can also have a dramatic effect on the grain structure and size. Methanol solutions were chosen in the studies described in this chapter because the resulting films consisted of reasonably sized and densely packed crystals, and yielded the best results for the field-effect transport measurements.

Organic-inorganic perovskite films have also recently been formed using the Langmuir-Blodgett (LB) technique which, in principal, provides the possibility of thickness control at the monolayer level [57]. This technique involves compressing a close-packed monolayer of amphiphiles at a liquid (subphase)-gas surface using a moveable barrier and then mechanically transferring the monolayer assembly to a solid support that is passed through the surface. In one example [57], a docosylammonium bromide ($C_{22}H_{45}NH_3Br$) monolayer is spread on a subphase containing lead(II) bromide and methylammonium bromide. The docosylammonium cations are presumably arranged at the surface of the Langmuir trough, with the organic tails pointing out of the solution and the ammonium groups extending into the solution in association with the lead(II) bromide anions. After compressing the monolayer to 30 mN m^{-1}, vertical dipping is used to transfer the monolayer to a fused quartz substrate, which has been made hydrophobic by treatment with hexamethyldisilazane. Only the single layer perovskite structure is formed in this process (i.e. one in which a single lead bromide sheet alternates with a bilayer of organic cations), indicating that methylammonium cations are, for the most part, not incorporated into the structure.

For electronic circuit fabrication, the ability to pattern the semiconducting organic-inorganic hybrid thin films is important to define the active area of the device. A low-cost, low-temperature, parallel process (Fig. 10.13) has recently been demonstrated, using microcontact printed templates to pattern subsequently spin-coated hybrid materials [58]. The additive nature of this process (as opposed to traditional lithography, which generally requires patterning and removal of material) has the attractive feature that the organic-inorganic semiconducting materials are not exposed to potentially harmful post-deposition processing. The technique consists of preparing a polydimethylsiloxane (PDMS) stamp with desired topographies and "inking" the stamp with a molecular species comprised of a head group that binds to the substrate surface and a hydrophobic tail group that directs the deposition of the solution-deposited hybrid material. Alkyl- and fluorinated alkyltrichlorosilane "inks" are used on SiO_2. Alkylphosphonic acids are used on SiO_2 or ZrO_2 and hydroxamic acids are used on ZrO_2 [58]. The inked PDMS stamp is briefly (30–60 s) brought into conformal contact with the substrate surface and removed, transferring the chemical ink to the oxide surface in a pattern defined by the stamp topography. The hybrid semiconductor can then be deposited on the patterned surface using a spin coating (or other solution-based) process as described above. During spinning, regions of the substrate that are covered with the hydrophobic molecular ink de-wet while the hydrophilic bare oxide surface is effectively covered by the evaporating solution. In this way, effective patterning during solution processing can be achieved, with a current resolution of approximately 3 μm [58].

In contrast to the pure solution-processed systems described above, a suitable solvent sometimes cannot be found for the self-assembling organic-inorganic hybrid. This may arise when the organic and inorganic components of the structure have incompatible solubility requirements or when a good solvent for the

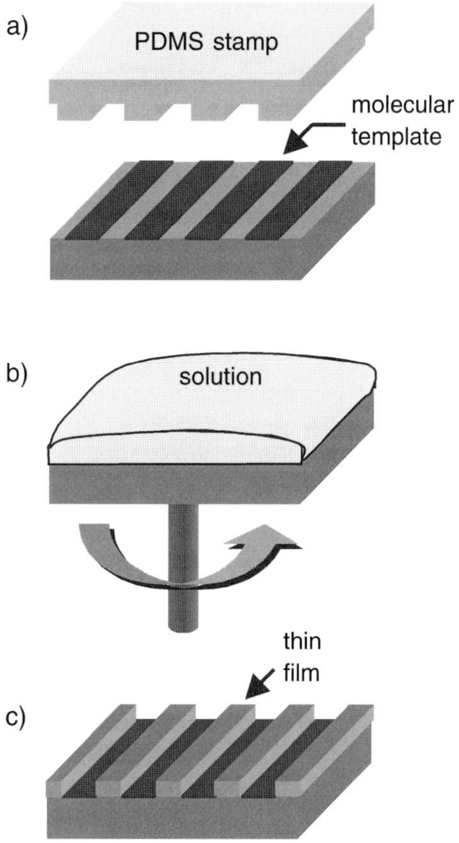

Fig. 10.13 Schematic representation of the steps used to form patterned solution-deposited films using microcontact printed molecular templates. (a) The stamp is "inked" by immersing it in a solution of chemical "ink" and brought into conformal contact with the substrate surface, depositing the molecular template on the surface in a geometry defined by the stamp topography. (b) The substrate patterned with the molecular template is flooded with a solution of the semiconducting material, covering both the bare substrate and the molecular layer. The substrate is spun, confining the self-assembly of hybrid films to the regions not on the hydrophobic molecular template (c). (Reprinted with permission from Ref. 58. Copyright 2001 American Institute of Physics)

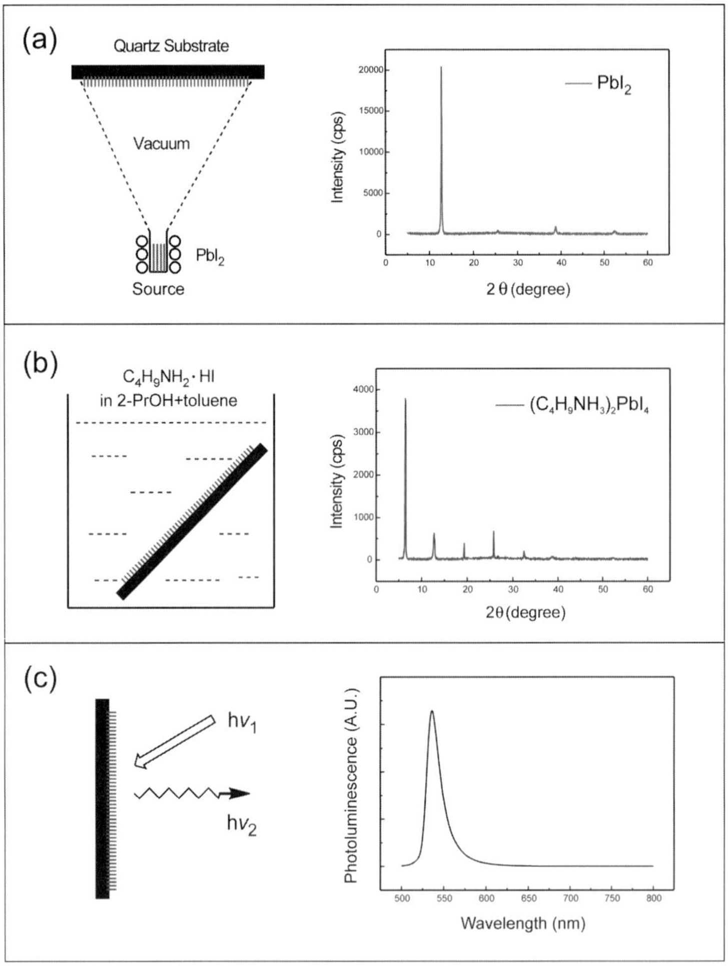

Fig. 10.14 Schematic representation of the two-step dipping technique. In (a), a film of the metal halide (in this case, PbI_2) is deposited onto a substrate using vacuum evaporation, yielding an ordered film with the characteristic X-ray pattern. The metal halide film is then (b) dipped into a solution containing the organic cation (in this case, $C_4H_9NH_3^+$). The resulting film after dipping has the characteristic X-ray pattern for the hybrid perovskite, as well as (c) the characteristic room-temperature photoluminescence spectrum. (reproduced with permission from Ref. 52. Copyright 2001 American Chemical Society)

hybrid does not wet the substrate surface. For such situations, a dip-processing technique [59] has been reported (Fig. 10.14) in which a pre-deposited (vacuum evaporation or solution deposited) metal halide film is dipped into a solution containing the organic cation. The solvent for the dipping solution is selected so that the organic salt is soluble, but the starting metal halide and the final organic-inorganic hybrid are not soluble. In the case of the organic-inorganic perovskites and related hybrids, for which there is generally a large driving force toward the formation of the perovskite structure relative to the organic and inorganic components, the organic cations in solution intercalate into and rapidly react with the metal halide on the substrate and form a crystalline film of the desired hybrid. Films of the perovskite family, $(R\text{-}NH_3)_2(CH_3NH_3)_{n-1}M_nI_{3n+1}$ (R = butyl, phenethyl; M=Pb, Sn; $n = 1, 2$) have been grown using this technique with a toluene/2-propanol solvent mixture for the organic salt and relatively short dipping times (several seconds to several minutes, depending on the system) [59]. The dipping technique also provides a promising pathway for patterning since, prior to the dipping step, the surface of the film can be coated with a resist so that only selected areas will be exposed to the organic cation in solution. A potential limitation of the technique is that during the reaction of the organic cation with the metal halide film, there is typically substantial grain growth, potentially leading to morphologically rough films.

10.2.3.3 **Melt Processing**

While the organic component of semiconducting hybrid perovskites generally decomposes or dissociates from the system at a lower temperature than that required for bulk melting, recently it has been reported [60] that the organic cation can be used to reduce the melting temperature to values below both the hybrid decomposition temperature, as well as below the glass transition temperature of selected flexible plastic substrates. The ability to tailor the thermal properties thereby enables the melt processing of hybrid films on flexible plastic (as well as rigid inorganic) substrates.

The initial melt-processing study [60] focused on the thermal properties of an isostructural tin(II)-iodide-based hybrid series, $(R\text{-}NH_3)_2SnI_4$, where several related phenethylammonium-based $R\text{-}NH_3^+$ cations were considered, including phenethylammonium (PEA), 2-fluorophenethylammonium (2-FPEA), 3-fluorophenethylammonium (3-FPEA), and 2,3,4,5,6-pentafluorophenethylammonium (5FPEA). Fig. 10.15 shows the thermogravimetric analysis (TGA) and differential scanning calorimetry (DSC) scans for each of the hybrids. In each system the peak rate of weight loss, corresponding to the dissociation or decomposition of the organic component, occurs at approximately the same temperature (255–261 °C). Within the DSC scan, there are three major transitions marked (1)–(3). Visual examination of the hybrid crystals as they are gradually heated confirms that transition (2) corresponds to melting, while the hybrid perovskites undergo bulk decomposition at (3). Temperature-dependent X-ray diffraction demonstrates that transition (1) corresponds to a structural transition [60]. Despite the relative invariance of the temperature at which most of the organic component dissociates from the samples and the similar tin(II) iodide framework, the temperatures of the structural transition (1) and the melting transition (2) progressively shift to lower temperature across the series $(R\text{-}NH_3)_2SnI_4$ (R=5FPEA, 3-FPEA, PEA, and 2-FPEA). While the $(5FPEA)_2SnI_4$ sample decomposes as it melts, as is typical for many of the hybrid perovskites, the melt becomes significantly more stable as the temperature of

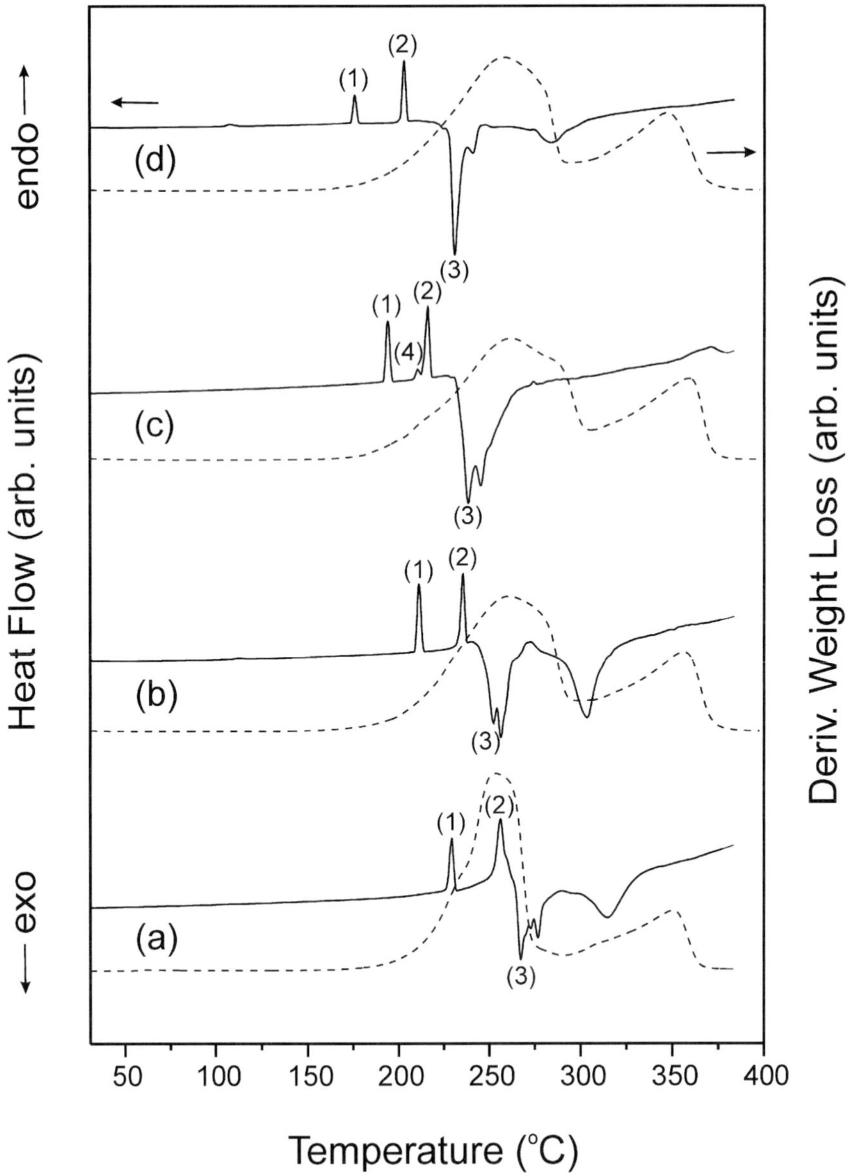

Fig. 10.15 Derivative of the weight loss (dashed line) and heat flow (solid line) as a function of temperature for (a) $(5FPEA)_2SnI_4$, (b) $(3\text{-}FPEA)_2SnI_4$, (c) $(PEA)_2SnI_4$, and (d) $(2\text{-}FPEA)_2SnI_4$. The scans were performed in a flowing nitrogen atmosphere using ground crystals of each compound and a 2 °C min^{-1} ramp rate. The significant transitions are marked with (1), (2) and (3). The origins of these transitions are discussed in the text. A minor structural transition (4), which is only fully resolvable at slower ramp rates (<5 °C min^{-1}), is also noted in the $(PEA)_2SnI_4$ data before melting. (reproduced with permission from Ref. 60. Copyright 2002 American Chemical Society)

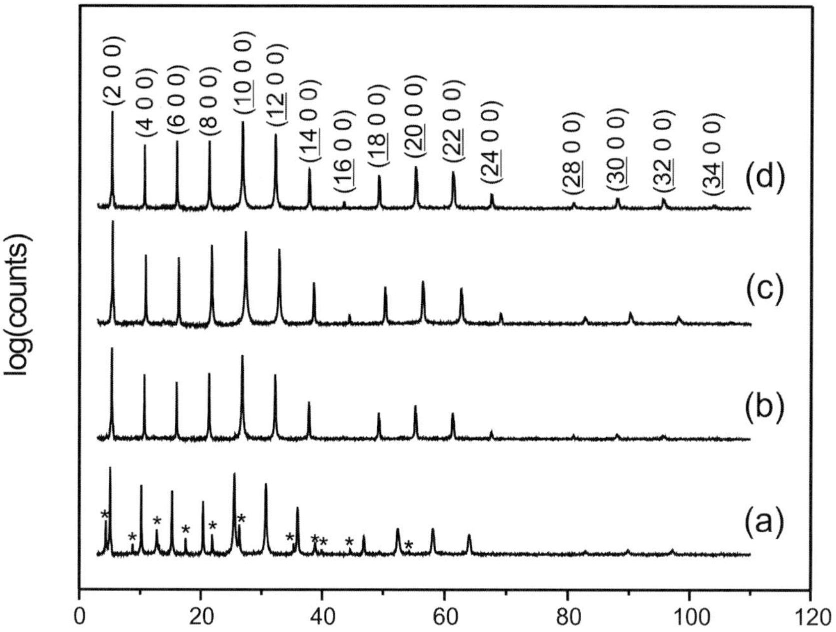

Fig. 10.16 Room temperature x-ray diffraction patterns (Cu Kα radiation; $\lambda = 1.5418$ Å) for the Kapton-laminated melt-processed films of (a) $(5FPEA)_2SnI_4$ [17.47(1) Å], (b) $(3\text{-}FPEA)_2SnI_4$ [16.63(1) Å], (c) $(PEA)_2SnI_4$ [16.31(1) Å], and (d) $(2\text{-}FPEA)_2SnI_4$ [16.64(1) Å], with the observed interlayer spacing between inorganic sheets noted in parentheses. The reflection indices are marked in (d) and the impurity peaks in pattern (a) (resulting from partial decomposition during melt processing) are noted using an *. The background signal from the Kapton was independently measured and subtracted from each pattern. (Reproduced with permission from Ref. 60. Copyright 2002 American Chemical Society)

melting transition (2), T_b, decreases across the substituted phenethylammonium cation series.

The ability to form a stable melt at low temperatures (~200 °C) provides an ideal opportunity to melt process films of the semiconducting hybrids on rigid or flexible substrates. In the reported examples [60], the films were formed in a nitrogen-filled glovebox by heating a Kapton® sheet approximately 10 °C above the expected melting temperature of the hybrid, placing hybrid crystals or powder on the sheet, and immediately depositing another Kapton sheet on top of the melting materials. The hybrid melt effectively wets the Kapton and capillary action spreads the melt uniformly between the sheets, leading to the formation of a film, the thickness of which depends on the initial quantity of hybrid (generally in the range 0.5 – 10 μm if no pressure is applied to the top sheet during melt processing). After allowing the film to spread, it is cooled below the melting temperature, resulting in a polycrystalline film laminated between the two Kapton sheets.

The x-ray diffraction pattern for the $(5FPEA)_2SnI_4$ sample (Fig. 10.16a) indicates substantial decomposition during the melting process, while the $(3\text{-}FPEA)_2SnI_4$

sample (Fig. 10.16b) has a slightly shifted spacing between the inorganic sheets relative to the bulk solution-grown crystals (16.63 Å vs. 16.79 Å), presumably due to the proximity of the melting and decomposition transitions. The diffraction patterns for the two lower melting point films (Figs. 10.16c and 10.16d), however, are identical with the initial crystals and the large number of higher order (*h* 0 0) reflections attest to the preferred orientation and high degree of crystallinity of the films. The resulting Kapton/hybrid/Kapton laminate is flexible and the Kapton sheets serve to partially protect the hybrid materials from the environment. The melt-processing technique is also attractive as a result of the potential for use of roll-to-roll techniques for fabricating laminated films and circuits. Finally, optimization of the film grain structure through controlled cooling of the melted films should also be possible, thereby providing a means of improving material or device performance.

10.3
Hybrid Perovskite Devices

The desirable electrical and optical properties exhibited by the hybrid perovskites, along with the availability of simple, low-cost processing techniques, provide unique opportunities for building devices. While this review will focus on devices based on the perovskite family, other hybrid systems are likely to also provide interesting technological opportunities.

10.3.1
Optical Devices

As previously mentioned (Sect. 10.2.2.1), excitonic transitions, associated with the inorganic sheets of the $(R\text{-}NH_3)_2MX_4$ (M = Ge, Sn, Pb; X = Cl, Br, I) perovskites, give rise to strong photoluminescence that can be tuned by incorporating different metal or halogen atoms in the structure. These useful optical characteristics make the perovskites attractive as potential emissive materials in electroluminescent (EL) devices. In fact, light-emitting devices (LEDs) have been prepared, with structures analogous to that of traditional organic LEDs (OLEDs), but with hybrid $(R\text{-}NH_3)_2PbX_4$ (X = Br or I) perovskite emitting layers replacing the organic emitting layer [61, 62]. In early examples, the perovskite organic layers contain optically inert (in the visible spectrum) phenethylammonium ($C_6H_5C_2H_4NH_3^+$) or cyclohexenylethylammonium ($C_6H_9C_2H_4NH_3^+$) cations. The heterostructure device (Fig. 10.17) consists of an indium tin oxide (ITO) anode, a spin coated $(R\text{-}NH_3)_2PbX_4$ emissive layer (that also acts as the hole transport layer), an evaporated 1,3,4-oxadiazole,2,2′-(1,3-phenylene)bis[5-[4-(1,1-dimethylethyl)phenyl]] (OXD7) electron transport layer, and a MgAg cathode. When the OILED (Organic-Inorganic Light-Emitting Diode) is driven with an applied voltage of 24 V at liquid nitrogen temperature, it exhibits intense and efficient EL, with a similar spectrum to that observed for the photoluminescence of the hybrid perovskite (Fig. 10.18). More recent devices [63] consist

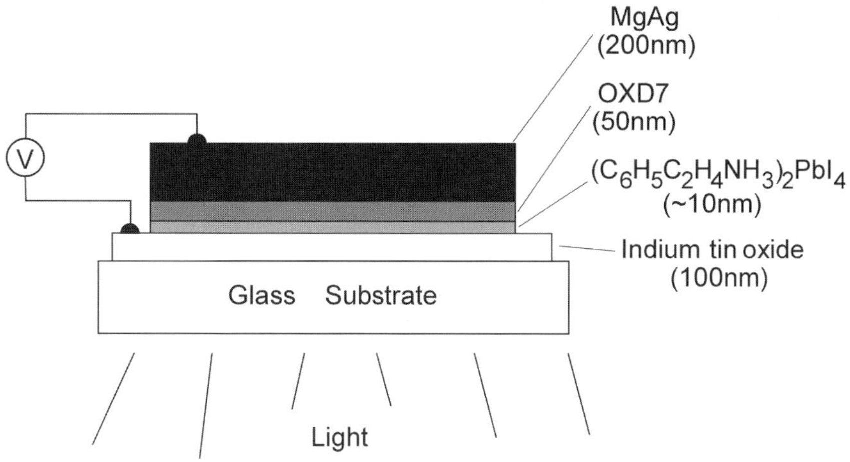

Fig. 10.17 Schematic structure of an electroluminescent device [61] based on the organic-inorganic perovskite $(C_6H_5C_2H_4NH_3)_2PbI_4$

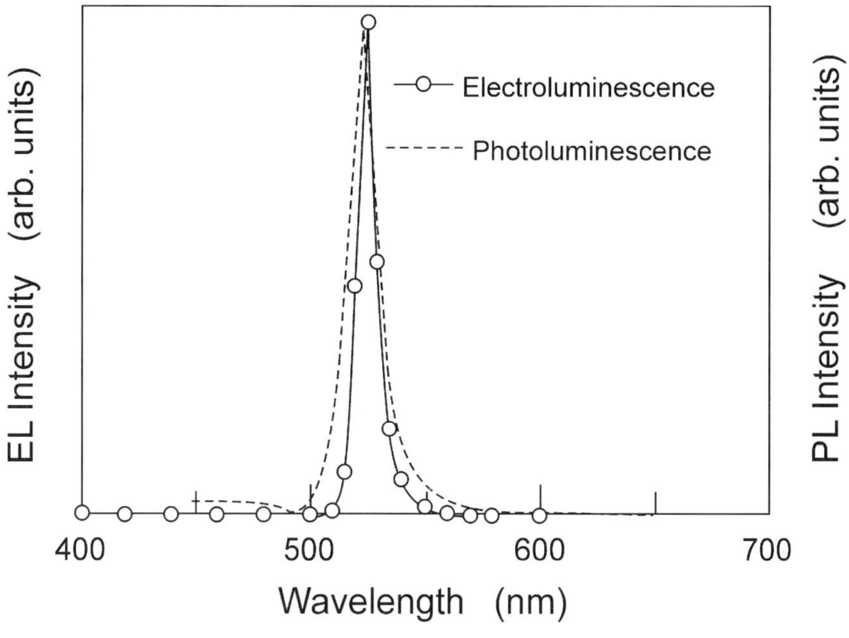

Fig. 10.18 Electroluminescence spectrum of the device in Fig. 10.17 at liquid nitrogen temperature. The emission layer of the device consists of a spin-coated $(C_6H_5C_2H_4NH_3)_2PbI_4$ film. The dotted line shows the photoluminescence spectrum of a similarly-prepared $(C_6H_5C_2H_4NH_3)_2PbI_4$ film. (Reprinted with permission from Ref. 61. Copyright © 1994 American Institute of Physics)

of either ITO anode/$(C_6H_9C_2H_4NH_3)_2PbBr_4$/OXD7/AlLi cathode or ITO anode/copper phthalocyanine (CuPc)/$(C_6H_9C_2H_4NH_3)_2PbBr_4$/OXD7/AlLi. The extra CuPc hole transport layer in the second set of devices substantially lowers the drive voltage for current injection and enhances the low-temperature EL efficiency of the device. Unfortunately, for each of the devices discussed above, the emission intensity drops off rapidly with increasing temperature, rendering such devices impractical for display applications. The dramatic reduction in EL efficiency near room temperature results, most probably, from thermal quenching of the excitons (i.e. the temperature dependence is very similar to that for the photoluminescence quantum yield).

In the above described hybrid devices, light emission arises from excitons within the inorganic component of the structure. To obtain room temperature EL from the hybrids, more complex dye cations, such as H_2AEQT^{2+}, were designed and synthesized to replace the optically inert alkyl or simple aromatic component typically used in layered perovskite structures (see Sect. 10.2.2.1) [6, 32]. The goal of this substitution was to incorporate additional functionality (i.e. high fluorescence efficiency) within the organic layer of the perovskite, thereby increasing the likelihood of efficient room-temperature emission. The device structure employed [64] can be

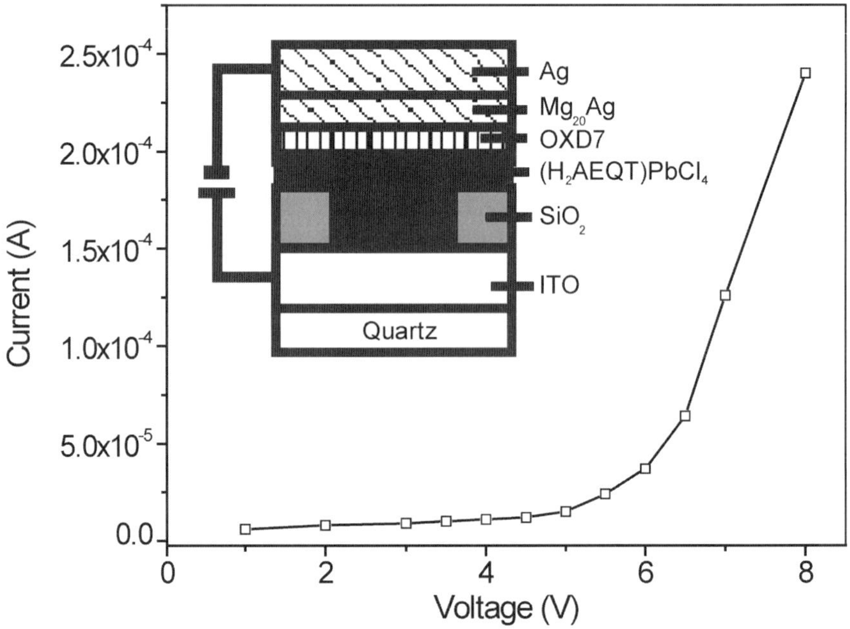

Fig. 10.19 Current-Voltage (*I-V*) curve for an OILED structure based on an $(H_2AEQT)PbCl_4$ emission layer. The inset shows a cross-sectional view of the device used in the study. (adapted with permission from D. B. Mitzi, K. Chondroudis, *Mol. Cryst. Liq. Cryst.* **2001**, *356*, 549–558)

seen in Fig. 10.19 (inset). It consists of an optically polished quartz substrate on which ITO has been e-beam deposited (1500 Å) as an anode. To avoid shorting between the anode and cathode, 1300 Å of SiO_2 is e-beam deposited on top of the ITO through contact masks, thereby defining four rectangular areas (3 × 1 mm) of exposed ITO. After cleaning the substrates using solvent-based and oxygen plasma processes, a patterned (i.e. to cover the exposed ITO areas) 3000 Å $(H_2AEQT)PbCl_4$ film is deposited using SSTA at 10^{-7} Torr, followed by a short (5 min), low-temperature (115 °C), inert-atmosphere annealing cycle. Subsequently, 200 Å of OXD7 is vacuum deposited as an electron-transport material. To complete the devices, an alloy of $Mg_{20}Ag$ is coevaporated (600 Å) through a contact mask to form the cathode, with an additional 1200 Å of Ag deposited on top to inhibit oxidation. The devices are maintained and prepared under either vacuum or inert atmosphere conditions and are crudely encapsulated using a cover glass and epoxy before evaluation in air.

Fig. 10.19 shows the current-voltage characteristics for the OILED structure, demonstrating a low turn-on voltage of about 5.5 V [64]. The *I-V* and EL-*V* curves are essentially superimposable, indicating a relatively balanced electron and hole injection. Bright green-yellow light ($\lambda_{max} \approx 530$ nm) is observed in a well-lit room when the devices are forward biased under ambient conditions (Fig. 10.20). The electroluminescence spectrum corresponds well to the photoluminescence spectrum of $(H_2AEQT)PbCl_4$ (Fig. 10.21), as well as to that of the dye salt AEQT · 2HCl, indicating that the EL arises from the quaterthiophene moiety. The maximum efficiency observed is 0.1 lm W^{-1} at 8 V and 0.24 mA (power conversion efficiency

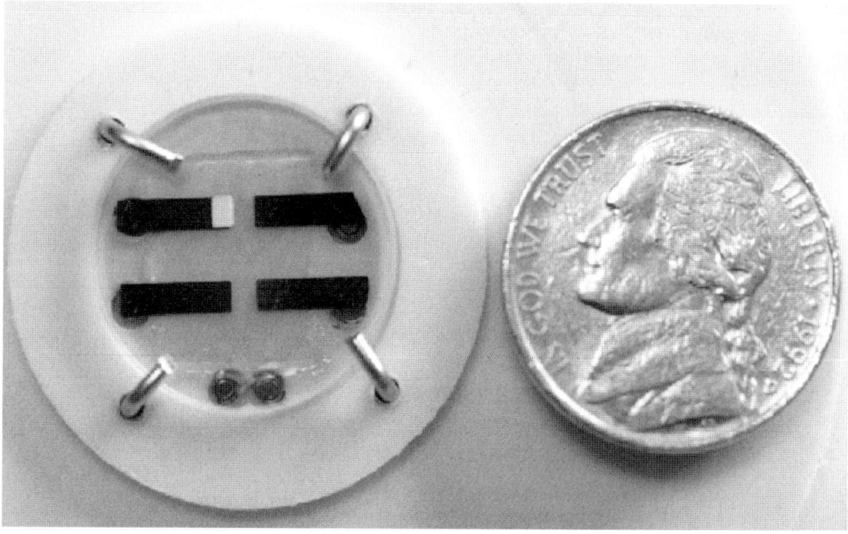

Fig. 10.20 Photograph of an operational OILED device based on $(H_2AEQT)PbCl_4$ as the emitting material. The nickel on the right provides a scale for size comparison. (reproduced with permission from Ref. 13)

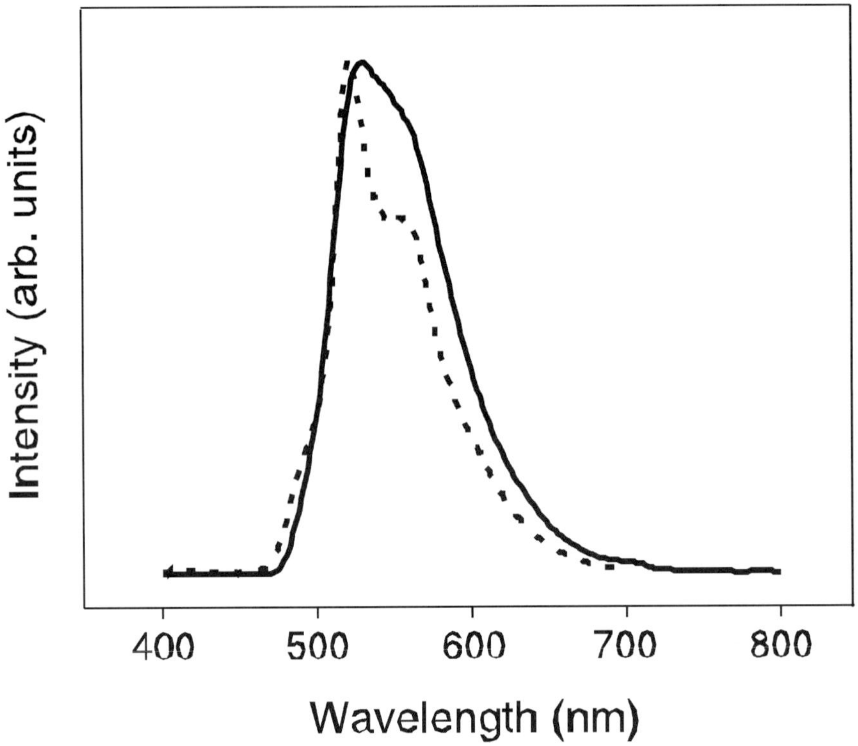

Fig. 10.21 Room temperature electroluminescence spectrum (solid line) for the OILED shown in Fig. 10.19 (inset), driven at 8 V. For comparison, the room temperature photoluminescence spectrum (excited at 360 nm) of $(H_2AEQT)PbCl_4$ is also shown (dashed line). (Reproduced with permission from Ref. 64. Copyright 1999 American Chemical Society)

0.11%). The fact that thicker emission layers can be used, compared to conventional OLEDs (typically <600 Å), should in principle enable improved device reliability and lifetime since it renders the devices less prone to pinholes and shorts.

The molecular-level sequencing of organic and inorganic layers may be advantageous for several reasons. First, the artificial layering (i.e. by successive evaporation) of organic dye molecules (e.g. Alq_3) and wide bandgap inorganic compounds (e.g. LiF) has been shown [65] to enhance the EL efficiency and device lifetime compared to systems with a single layer of the emitting dye. The improved device performance can apparently be attributed to better carrier injection, transport and electron-hole recombination. The self-assembling hybrid perovskites provide a much simpler way (i.e. a single evaporation step) of achieving this type of alternating structure. Interestingly, hybrid perovskite devices designed with lead bromide (smaller bandgap) inorganic sheets exhibit 20 to 30 times lower efficiencies than the ones with lead chloride sheets, whereas preliminary results for analogous

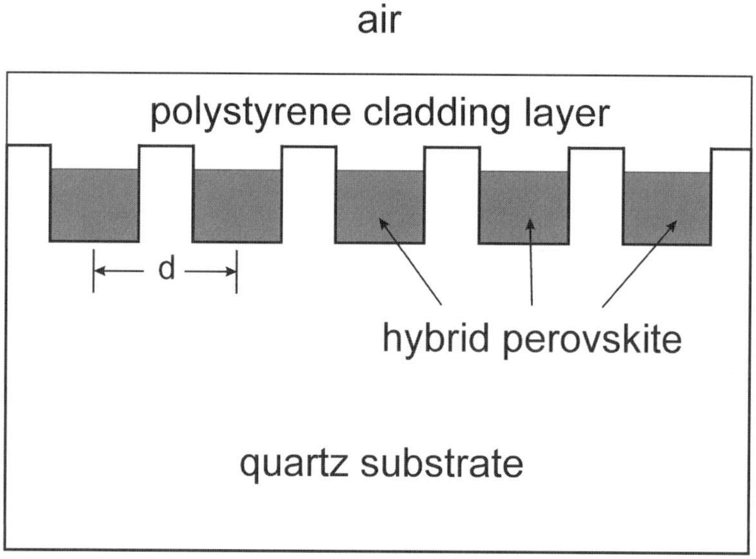

Fig. 10.22 Schematic structure of photonic crystal slab with $(C_6H_5C_2H_4NH_3)_2PbI_4$ active material in the square holes. The period of the array, *d*, may be varied to tailor the optical response of the slab. (adapted with permission from Ref. 70)

devices with cadmium chloride layers (higher bandgap) yield higher efficiencies of ~0.16 lm W^{-1} [13]. This further suggests that, as for the OILEDs made with alternating Alq_3/LiF layers, the self-assembling organic-inorganic structures can have a large impact on the luminous efficiencies exhibited by the emitting species within the structure.

The choice of oligothiophene dyes for this work was based on the ability to conveniently tune the electronic properties of the organic cation by tailoring the length of the oligothiophene moiety. However, ultimately the EL efficiencies are limited by the generally low fluorescence quantum yield of oligothiophenes as a result of effective intersystem crossing [66, 67]. The efficiency of the early hybrid perovskite devices (~ 0.2 lm W^{-1}) is therefore quite promising and higher values are expected with further fine-tuning of the device structure, and especially through the incorporation of chromophores with fundamentally better fluorescence quantum yields. Appropriate modification of a wide range of structurally compatible dye molecules should provide a repertoire of candidate cations and also enable the realization of different emission colors (e.g. blue, red) – a requirement for full color display applications [13].

Given the potentially high carrier mobilities and the tunable electronic structure achievable within the hybrid perovskite framework, these materials (even with optically inert organic cations) may also be employed as carrier transport layers within LED device structures. For example, $[C_6H_5(CH_3)CHNH_3]_2PbX_4$ (X = Cl, Br) perovskites have recently been demonstrated as hole transport layers in an OILED with a conventional organic emitter layer of poly(N-vinyl carbazole) (PVK) doped with the laser dye Coumarin 6 [68]. It was found that using lead(II) chloride inorganic perovskite sheets leads to almost a ten-fold improvement in efficiency compared to the case with lead(II) bromide layers. This improvement is attributed to the different band structures of the two perovskites and/or the better electron-blocking abilities of the lead chloride (higher bandgap) perovskite.

In addition to LED structures, the exciton features from the inorganic component of the framework and flexibility to incorporate functional organic cations in the organic layers provide a range of other opportunities for optical device architectures. Potentially useful attributes of the exciton resonance include non-linear optical effects [46] and biexciton lasing [48]. Recently, a photonic crystal slab was described [69, 70] employing the hybrid semiconductor contained within a periodic array of etched square holes on a quartz substrate (Fig. 10.22). The hybrid crystal slab may exhibit directional photoluminescence and optical non-linearity with possible application in ultrafast optical switches [70].

10.3.2
Electronic Devices

In addition to their potential applications in optical devices, semiconducting hybrid perovskites present exciting opportunities as channel materials for thin-film field-effect transistors (FETs) [13, 49, 71, 72]. The hybrid materials are interesting for this application because they may combine higher carrier mobilities of covalent/ionic inorganic semiconductors with the simple, low-temperature thin-film techniques of organic semiconductors. Fig. 10.23 depicts the typical bottom-gate organic-inorganic TFT test structure, with the first demonstrated hybrid devices employing $(C_6H_5C_2H_4NH_3)_2SnI_4$ as the semiconducting channel material [71, 72]. Thin films of the hybrid perovskite are spin coated from a methanol solution, in an inert atmosphere, onto thermally oxidized, degenerately n-doped silicon wafers (which also act as the gate electrode). The SiO_2 layer acts as the gate insulator and is typically 5000 Å thick. High work function metals, such as Pd, Au, or Pt (Pd has yielded the best results for the spin-coated devices tested so far), are deposited by e-beam evaporation through silicon membrane masks and serve both as the source and drain electrodes and to define the channel dimensions. While the organic-inorganic perovskite may be deposited either before or after metallization, depositing the source and drain electrodes before spin coating eliminates exposing the hybrid perovskite films to potentially harmful metal deposition conditions (e.g. high temperatures). Spin-coating tin(II) iodide based perovskites with simple monoammonium organic cations results in polycrystalline, highly-oriented (Fig. 10.24) thin films having < 300 nm in-plane grain size (see Sect. 10.2.3.2).

Each tin(II) iodide-based perovskite film forms a p-channel transistor, consistent with previous Hall measurements [28,50]. Fig. 10.25 shows representative device characteristics for a TFT with the organic-inorganic perovskite $(C_6H_5C_2H_4NH_3)_2SnI_4$ as the channel layer. Application of a negative bias to the gate electrode ($V_G < 0$) increases the number of majority holes in the semiconducting channel contribut-

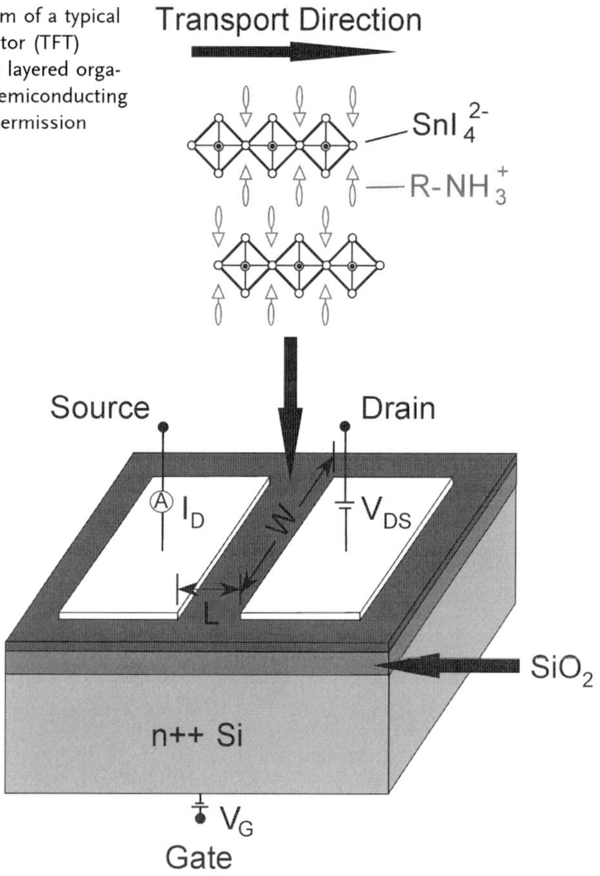

Fig. 10.23 Schematic diagram of a typical bottom-gate thin film transistor (TFT) device structure employing a layered organic-inorganic perovskite as semiconducting channel. (Reproduced with permission from Ref. 13)

ing to the drain current (I_D). These devices show typical transistor-like behavior, as I_D increases linearly at low source-drain voltage (V_{DS}) and then saturates as V_{DS} increases and the holes in the channel are pinched off near the drain electrode. Application of a positive gate bias ($V_G > 0$) depletes the holes in the channel, turning the device off.

The device characteristics of the organic-inorganic field-effect transistors (OIFETs) are modeled by the standard equations used for both inorganic and organic semiconducting channel materials. Fig. 10.25 b shows the dependence of I_D and $I_D^{1/2}$ versus V_G, at $V_{DS} = -100$ V. As determined from these curves, the saturation regime field-effect mobility for the device is $\mu = 0.61$ cm^2 V^{-1} s^{-1} and the I_{ON}/I_{OFF} ratio is $\sim 10^6$. These device characteristics are representative of $(C_6H_5C_2H_4NH_3)_2SnI_4$ and are comparable to the performance of amorphous silicon and the best organic semiconductors deposited in high vacuum. The reported I_{ON}/I_{OFF} ratio is achieved by crudely patterning the perovskite semiconductor using a solvent-soaked cotton swab to remove material from around each device. By isolating devices, leakage current between source/drain electrodes and the back gate

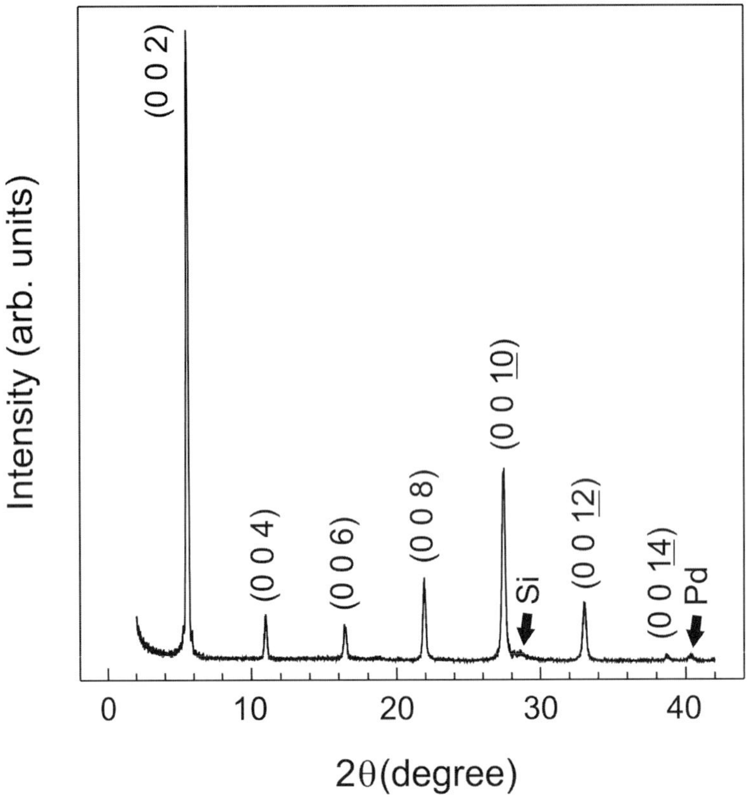

Fig. 10.24 X-ray diffraction pattern for a completed TFT with $(C_6H_5C_2H_4NH_3)_2SnI_4$ as the semiconducting channel and Pd source and drain electrodes. (Reprinted with permission from Ref. 72. Copyright 1999 American Association for the Advancement of Science)

is effectively avoided. Patterning can also be accomplished using a microcontact printed template to pattern subsequently spin-coated hybrid materials [58]. Unpatterned device structures show the same field-effect mobilities, but leakage current between gate and drain electrodes reduces the magnitude of current modulation. Similar device characteristics (i.e. μ and I_{ON}/I_{OFF}) are achieved at substantially lower applied voltages for devices fabricated using a thinner or high dielectric constant gate insulator [73].

It is important to note that the mobility discussed above represents the mobility in the saturation regime. In fact, the mobility is gate voltage dependent [72]. Generally, the mobility starts off small at low gate bias and increases until saturating at higher bias. This type of voltage dependence of the mobility has also been observed in strictly organic and amorphous silicon semiconductors and has been attributed to trap states in the band gap of the semiconductor [72, 74]. Namely, at

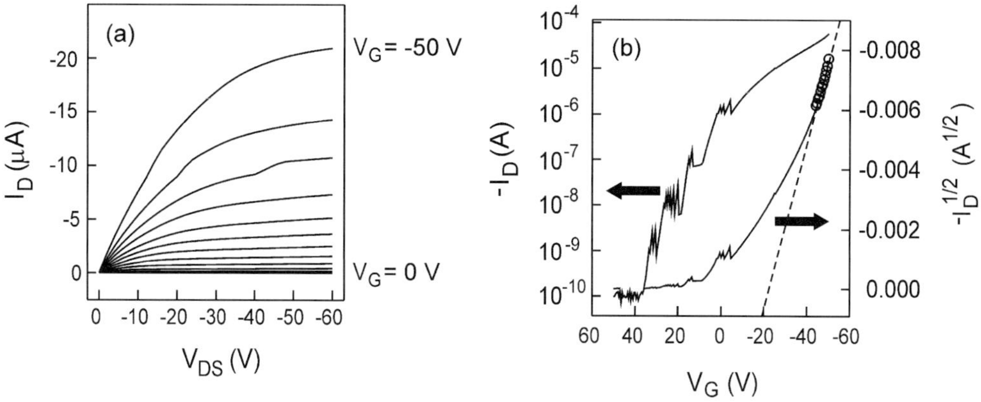

Fig. 10.25 (a) Drain current I_D versus source-drain voltage V_{DS} as a function of gate voltage V_G for a TFT with a spin-coated channel of $(C_6H_5C_2H_4NH_3)_2SnI_4$, a channel length L = 28 μm and a channel width W = 1000 μm. The gate dielectric is 5000 Å of SiO_2. (b) Plots of I_D and $I_D^{1/2}$ versus V_G at constant $V_{DS} = -100$ V, as used to calculate current modulation (I_{on}/I_{off}) and field-effect mobility (μ). (Reproduced with permission from Ref. 13)

low gate bias, there are more trap states than carriers and transport is via hopping between localized trap states. As the gate bias is made progressively more negative (for a p-channel device), the trap states are filled until ultimately all traps are filled and additional carriers can transport with the intrinsic mobility of the semiconductor. A second point of interest involves the discontinuities observed in the I_D versus V_G curve (Fig. 10.25 b). These discontinuities are reproducible over many measurements and are attributed (see below) to grain boundary effects.

Modification of the electronic structure of the $n = 1$ tin(II) iodide based perovskites can be achieved by making substitutions on the metal or halogen sites [6]. Recently, the effects of organic cation substitution on the electronic characteristics of TFTs based on the hybrid perovskites have also been explored [49]. TFT devices were constructed using fluorine-substituted phenethylammonium-based perovskites as the channel layer. Fig. 10.26 shows the TFT characteristics as a function of the position of fluorine substitution. Mobilities for the various fluorophenethylammonium tin(II) iodide channel layers are similar to those achieved in comparable phenethylammonium tin(II) iodide layers (typically 0.2–0.6 $cm^2\ V^{-1}\ s^{-1}$). The peak current in a device with a channel that is 15.4 μm long and 1500 μm wide decreases across the series $(4\text{-FPEA})_2SnI_4$, $(3\text{-FPEA})_2SnI_4$, and $(2\text{-FPEA})_2SnI_4$, from 160 mA to 20 mA, as the band gap of the hybrid increases. Perhaps equally interesting, the magnitude and frequency of the discontinuities in the I_D versus V_G curves decreases in the 4-FPEA and 3-FPEA samples relative to the previously discussed PEA_2SnI_4 samples. Furthermore, the 2-FPEA sample has a somewhat worse behavior in this respect. An atomic force microscope (AFM) examination of the morphology of the films used for these measurements (e.g. Fig. 10.12) indicates that despite the similar spin-coating conditions for the thin films deposited for this

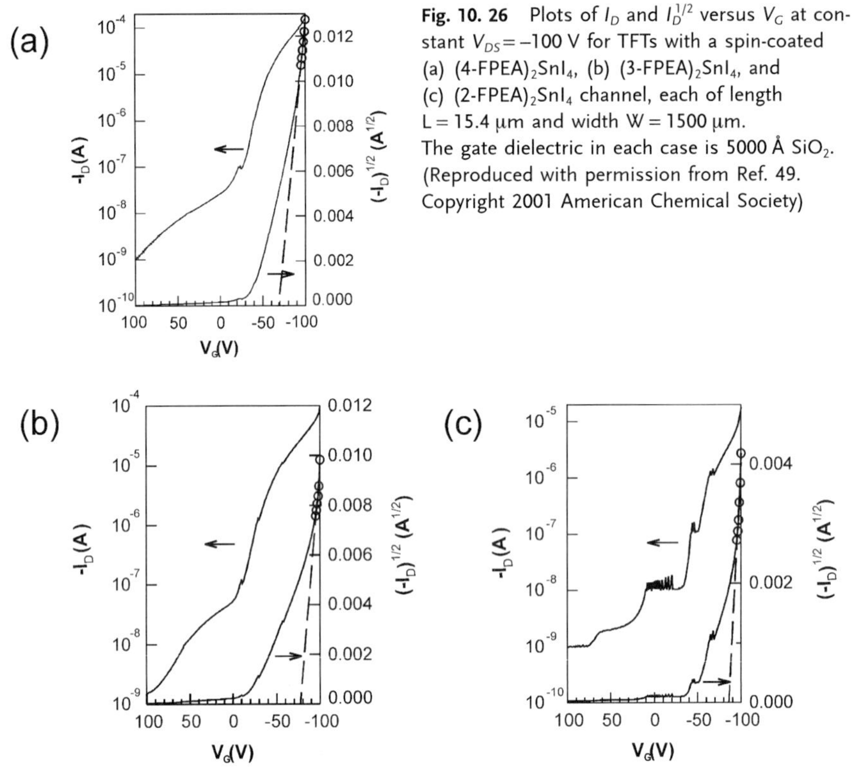

Fig. 10. 26 Plots of I_D and $I_D^{1/2}$ versus V_G at constant $V_{DS}=-100$ V for TFTs with a spin-coated (a) (4-FPEA)$_2$SnI$_4$, (b) (3-FPEA)$_2$SnI$_4$, and (c) (2-FPEA)$_2$SnI$_4$ channel, each of length L = 15.4 μm and width W = 1500 μm. The gate dielectric in each case is 5000 Å SiO_2. (Reproduced with permission from Ref. 49. Copyright 2001 American Chemical Society)

study, the grain structure is quite different for the related organic cations. The 4-FPEA and 3-FPEA based samples have the largest grains (~150–50 nm), whereas the 2-FPEA sample has substantially smaller grains (<100 nm). The PEA-based sample has an intermediate grain size when deposited using the same conditions. The variations in grain morphology correlate fairly well with the observation of discontinuities in the I_D versus V_G curves and highlight the importance of being able to control grain structure in the semiconducting hybrid films. The previously described (Sect. 10.2.3.2) melt-processing technique may provide an important opportunity to control grain size/texturing and recent measurements using melt-processed channel layers have yielded promising results [75].

The exploration of organic cation flexibility for tailoring the properties of the hybrid perovskites is in its early stages. In addition to the devices based on organic cations with a single ammonium-based tethering group, diammonium cations can be employed to more intimately link adjacent metal halide layers. Besides the simple aliphatic and single ring aromatic organic cations, more complex and potentially functional organic cations can also be employed. Organic cations with extended conjugated systems offer the possibility of charge transport in the organic layer of the structure instead of or in addition to the charge transport in the inorganic sheets.

10.4 Conclusions

While biologically important examples of organic-inorganic hybrids with superior mechanical properties (e.g., teeth, seashell, pearls and bone) have long been realized [76], use of hybrids as active materials in semiconducting devices has only recently begun to be extensively explored [13]. Organic-inorganic hybrids provide substantial opportunities with respect to both combining useful attributes of organic and inorganic components on a molecular scale, as well as potentially evoking new properties as a result of the interface between the components. An exceptionally diverse family of crystalline perovskite hybrids forms the basis of this review. The structures consist of 1-, 2-, and 3-D networks of corner-sharing metal halide octahedra alternating with organic cation layers. The inorganic layers can provide interesting and potentially useful electrical, optical, magnetic and dielectric properties. The organic cation layer enables control over the dimensionality and orientation of the inorganic framework, as well as the possibility of templating the formation of metal-deficient inorganic layers and intercalated molecular layers sandwiched between the organic cation bilayers. In addition, more complex organic cations can provide direct functionality to the material in the form of interesting optical or electrical properties. The absolute positioning of energy levels in the organic and inorganic components of the structure may further enable charge separation at the interface between two layers. The range of optical and electrical properties of the hybrid enables numerous opportunities for electronic devices.

As seen in this review, recent examples of hybrid perovskite devices include both LEDs and TFTs. For the LEDs, hybrids with both optically inert organic cations, which emit primarily at low temperature, as well as dye-containing systems, which operate at room temperature, have been demonstrated. For the dye-containing systems, the influence of the inorganic framework on the emission from the dye has been demonstrated. However, early devices operate at relatively low efficiency, presumably as a result of the low quantum yields for the dye molecules incorporated. It is expected that as more efficient dye molecules are incorporated, the efficiencies of the hybrids could rival those of their organic counterparts, while at the same time offering the potential benefits of bandgap tunability and desirable thermal and mechanical properties of an inorganic framework. The tin(II) iodide-based hybrids also offer an alternative spin-coatable or melt-processible TFT channel material with saturation regime mobilities comparable to that of amorphous silicon. Higher mobilities are envisioned for hybrid materials, as compared with strictly organic systems, as a result of the covalent/ionic character of the inorganic framework. In addition to increasing mobility, other important issues for hybrid TFT research include the need to improve the air/moisture stability and the search for n-type analogs to the tin(II) iodide systems (thereby enabling complementary logic). Both of these goals will likely require the extension of the semiconducting hybrids to include alternative inorganic frameworks (besides tin iodide).

The hybrid perovskites are a convenient model system for demonstrating the possibilities offered by organic-inorganic systems with respect to device opportunities. Their crystalline nature enables convenient correlation between structure and physical properties and the corresponding rich chemistry of the hybrid perovskites provides an extensive array of possibilities for tailoring the two components of the structure to the specific requirements of the device. However, as briefly outlined in the Introduction (and beyond the scope of this review), numerous other crystalline and amorphous hybrid systems provide equally intriguing opportunities for photovoltaic, LED, TFT, diode, non-linear optical, waveguide, laser, memory and other device configurations. As with organic-based devices, which have shown dramatic progress over the last 20 years, it is expected that hybrid device technology will continue to mature and provide opportunities beyond those possible for their strictly organic counterparts.

References

1 For a recent review, see: *Technology and Applications of Amorphous Silicon* (Ed.: R. A. Street), Springer-Verlag, Berlin, **2000**.

2 For a recent review, see: C. D. Dimitrakopoulos, D. J. Mascaro, *IBM J. Res. & Dev.* **2001**, *45*, 11–27.

3 J. M. Shaw, P. F. Seidler, *IBM J. Res. & Dev.* **2001**, *45*, 3–9.

4 M. Ogawa, K. Kuroda, *Chem. Rev.* **1995**, *95*, 399–438.

5 *Intercalation Chemistry* (Ed.: M. S. Whittingham, A. J. Jacobson), Academic, New York, **1982**.

6 For a recent review, see: D. B. Mitzi, *Prog. Inorg. Chem.* **1999**, *48*, 1–121.

7 D. B. Mitzi, *J. Chem. Soc., Dalton Trans.* **2001**, 1–12.

8 X. Huang, J. Li, H. Fu, *J. Am. Chem. Soc.* **2000**, *122*, 8789–8790.

9 G. Alberti, U. Costantino, F. Marmottini, R. Vivani, P. Zappelli, *Angew. Chem. Int. Ed. Engl.* **1993**, *32*, 1357–1359.

10 a) W. A. Herrmann, R. W. Fischer, *J. Am. Chem. Soc.* **1995**, *117*, 3223–3230; b) W. A. Herrmann, W. Scherer, R. W. Fischer, J. Blümel, M. Kleine, W. Mertin, R. Gruehn, J. Mink, H. Boysen, C. C. Wilson, R. M. Ibberson, L. Bachmann, M. Mattner, *J. Am. Chem. Soc.* **1995**, *117*, 3231–3243.

11 E. O. Schlemper, W. C. Hamilton, *Inorg. Chem.* **1966**, *5*, 995–998.

12 a) D. B. Mitzi, *Inorg. Chem.* **1996**, *35*, 7614–7619; b) S. Wang, D. B. Mitzi, G. A. Landrum, H. Genin, R. Hoffmann, *J. Am. Chem. Soc.* **1997**, *119*, 724–732.

13 D. B. Mitzi, K. Chondroudis, C. R. Kagan, *IBM J. Res. & Dev.* **2001**, *45*, 29–45.

14 Y. Sorek, R. Reisfeld, I. Finkelstein, S. Ruschin, *Appl. Phys. Lett.* **1995**, *66*, 1169–1171.

15 I. Finkelstein, S. Ruschin, Y. Sorek, R. Reisfeld, *Opt. Mater.* **1997**, *7*, 9–13.

16 Y. H. Min, J. Mun, C. S. Yoon, H.-K. Kim, K.-S. Lee, *Electron. Lett.* **1999**, *35*, 1770–1771.

17 J. Takada, H. Awaji, M. Koshioka, A. Nakajima, W. A. Nevin, *Appl. Phys. Lett.* **1992**, *61*, 2184–2186.

18 W. U. Huynh, J. J. Dittmer, A. P. Alivisatos, *Science* **2002**, *295*, 2425–2427.

19 M. Gao, B. Richter, S. Kirstein, H. Möhwald, *J. Phys. Chem. B* **1998**, *102*, 4096–4103.

20 V. L. Colvin, M. C. Schlamp, A. P. Alivisatos, *Nature* **1994**, *370*, 354–357.

21 T. Cassagneau, T. E. Mallouk, J. H. Fendler, *J. Am. Chem. Soc.* **1998**, *120*, 7848–7859.

22 see, for example, "Organic-Inorganic Nanocomposite Materials," special issue of *Chem. Mater.* **2001**, *13*, 3059–3809.

23 F. S. Galasso, *Structure, Properties and Preparation of Perovskite-Type Compounds*, Pergamon, New York, **1969**.

24 R. D. Shannon, *Acta Crystallogr., Sect. A* **1976**, *32*, 751–767.

25 D. Weber, *Z. Naturforsch.* **1978**, *33b*, 1443–1445.

26 A. Poglitsch, D. Weber, *J. Chem. Phys.* **1987**, *87*, 6373–6378.
27 K. Yamada, T. Matsui, T. Tsuritani, T. Okuda, S. Ichiba, *Z. Naturforsch.* **1990**, *45a*, 307–312.
28 D. B. Mitzi, C. A. Feild, Z. Schlesinger, R. B. Laibowitz, *J. Solid State Chem.* **1995**, *114*, 159–163.
29 K. Yamada, S. Nose, T. Umehara, T. Okuda, S. Ichiba, *Bull. Chem. Soc. Jpn.* **1988**, *61*, 4265–4268.
30 D. B. Mitzi, K. Liang, *J. Solid State Chem.* **1997**, *134*, 376–381.
31 D. B. Mitzi, *Inorg. Chem.* **2000**, *39*, 6107–6113.
32 D. B. Mitzi, K. Chondroudis, C. R. Kagan, *Inorg. Chem.* **1999**, *38*, 6246–6256.
33 M. Braun, W. Tuffentsammer, H. Wachtel, H. C. Wolf, *Chem. Phys. Lett.* **1999**, *307*, 373–378.
34 M. Braun, W. Tuffentsammer, H. Wachtel, H. C. Wolf, *Chem. Phys. Lett.* **1999**, *303*, 157–164.
35 Z. Tang, J. Guan, A. M. Guloy, *J. Mater. Chem.* **2001**, *11*, 479–482.
36 B. Tieke, G. Chapuis, *Mol. Cryst. Liq. Cryst.* **1986**, *137*, 101–116.
37 P. Day, R. D. Ledsham, *Mol. Cryst. Liq. Cryst.* **1982**, *86*, 163–174.
38 Y. Takeoka, K. Asai, M. Rikukawa, K. Sanui, *Chem. Commun.* **2001**, 2592–2593.
39 Y. I. Dolzhenko, T. Inabe, Y. Maruyama, *Bull. Chem. Soc. Jpn.* **1986**, *59*, 563–567.
40 D. B. Mitzi, D. R. Medeiros, P. R. L. Malenfant, *Inorg. Chem.* **2002**, *41*, 2134–2145.
41 see, for example, L. J. de Jongh, *Magnetic Properties of Layered Transition Metal Compounds*, Kluwer, Dordrecht, **1990**.
42 see, for example, *Physics and Applications of Quantum Wells and Superlattices* (Eds.: E. E. Mendez and K. von Klitzing), NATO ASI Series B: Physics Vol. 170, Plenum Press, New York, **1987**.
43 a) G. C. Papavassiliou, *Prog. Solid State Chem.* **1997**, *25*, 125–270; b) G. C. Papavassiliou, I. B. Koutselas, *Synth. Met.* **1995**, *71*, 1713–1714.
44 D. B. Mitzi, *Chem. Mater.* **1996**, *8*, 791–800.
45 X. Hong, T. Ishihara, A. V. Nurmikko, *Phys. Rev. B* **1992**, *45*, 6961–6964.
46 J. Calabrese, N. L. Jones, R. L. Harlow, N. Herron, D. L. Thorn, Y. Wang, *J. Am. Chem. Soc.* **1991**, *113*, 2328–2330.
47 T. Fujita, Y. Sato, T. Kuitani, T. Ishihara, *Phys. Rev. B* **1998**, *57*, 12428–12434.
48 T. Kondo, T. Azuma, T. Yuasa, R. Ito, *Solid State Commun.* **1998**, *105*, 253–255.
49 D. B. Mitzi, C. D. Dimitrakopoulos, L. L. Kosbar, *Chem. Mater.* **2001**, *13*, 3728–3740.
50 D. B. Mitzi, C. A. Feild, W. T. A. Harrison, A. M. Guloy, *Nature* **1994**, *369*, 467–469.
51 D. B. Mitzi, S. Wang, C. A. Feild, C. A. Chess, A. M. Guloy, *Science* **1995**, *267*, 1473–1476.
52 D. B. Mitzi, *Chem. Mater.* **2001**, *13*, 3283–3298.
53 M. Era, T. Hattori, T. Taira, T. Tsutsui, *Chem. Mater.* **1997**, *9*, 8–10.
54 M. Era, K. Kakiyama, T. Ano, M. Nagano, *Trans. Mater. Res. Soc. Jpn.* **1999**, *24*, 509–511.
55 D. B. Mitzi, M. T. Prikas, K. Chondroudis, *Chem. Mater.* **1999**, *11*, 542–544.
56 K. Chondroudis, D. B. Mitzi, P. Brock, *Chem. Mater.* **2000**, *12*, 169–175.
57 M. Era, S. Oka, *Thin Solid Films* **2000**, *376*, 232–235.
58 C. R. Kagan, T. L. Breen, L. L. Kosbar, *Appl. Phys. Lett.* **2001**, *79*, 3536–3538.
59 K. Liang, D. B. Mitzi, M. T. Prikas, *Chem. Mater.* **1998**, *10*, 403–411.
60 D. B. Mitzi, D. R. Medeiros, P. W. DeHaven, *Chem. Mater.* **2002**, *14*, 2839–2841.
61 M. Era, S. Morimoto, T. Tsutsui, S. Saito, *Appl. Phys. Lett.* **1994**, *65*, 676–678.
62 T. Hattori, T. Taira, M. Era, T. Tsutsui, S. Saito, *Chem. Phys. Lett.* **1996**, *254*, 103–108.
63 M. Era, T. Ano, M. Noto, in *Novel Methods to Study Interfacial Layers* (Eds.: D. Möbius, R. Miller), Elsevier, Amsterdam, **2001**, 166–173.
64 K. Chondroudis, D. B. Mitzi, *Chem. Mater.* **1999**, *11*, 3028–3030.
65 W. Riess, H. Riel, P. F. Seidler, H. Vestweber, *Synth. Met.* **1999**, *99*, 213–218.
66 S. Rentsch, J. P. Yang, W. Paa, E. Birckner, J. Schiedt, R. Weinkauf, *Phys. Chem. Chem. Phys.* **1999**, *1*, 1707–1714.

67 D. Grebner, M. Helbig, S. Rentsch, *J. Phys. Chem.* **1995**, *99*, 16991–16998.

68 T. Gebauer, G. Schmid, *Z. Anorg. Allg. Chem.* **1999**, *625*, 1124–1128.

69 T. Ishihara, T. Fujita, A. Seki, H. Nakashima, *Mol. Cryst. Liq. Cryst.* **2001**, *371*, 167–170.

70 T. Ishihara, *RIKEN Rev.* **2001**, *37*, 38–42.

71 a) K. Chondroudis, C. D. Dimitrakopoulos, D. B. Mitzi, unpublished work (1998); (b) K. Chondroudis, C. D. Dimitrakopoulos, C. R. Kagan, I. Kymissis, D. B. Mitzi, United States Patent US6180956, January 30, 2001.

72 C. R. Kagan, D. B. Mitzi, C. D. Dimitrakopoulos, *Science* **1999**, *286*, 945–947.

73 C. D. Dimitrakopoulos, C. R. Kagan, D. B. Mitzi, United States Patent US6344662, February 5, 2002.

74 C. D. Dimitrakopoulos, S. Purushothaman, J. Kymissis, A. Callegari, J. M. Shaw, *Science* **1999**, *283*, 822–824.

75 D. B. Mitzi, C. D. Dimitrakopoulos, J. Rosner, D. R. Medeiros, Z. Xu, C. Noyan, *Adv. Mater.* **2002**, *14*, 1772–1776.

76 see, for example, J. D. Currey, *J. Mater. Educ.* **1987**, *9*, 119–196.

11
Bioactive Sol-Gel Hybrids

Jacques Livage, Thibaud Coradin and Cécile Roux

11.1
Introduction

The development of biomaterials has long been focused on the realization of implants for medical applications. Such materials have to exhibit good mechanical properties and to be biocompatible in order to be implanted in the human body. Bioactive implants bind to living tissues and favor the regeneration of damaged tissues or bones. Such materials have been recently reviewed and will not be described here [1, 2]. We shall focus on new bioactive materials in which biospecies such as enzymes or whole cells are immobilized. These bioactive hybrids are not used as implants. They find applications in the field of biotechnology for the realization of biosensors and bioreactors. They take advantage of the high activity of enzymes, antibodies or micro-organisms to perform specific reactions that would not be possible with the usual chemical routes. Active biospecies are immobilized on or in a solid substrate in order to be reusable and protected from denaturation. Natural and synthetic polymers (polysaccharides, polyacrylamides, alginates, etc.) are then currently used for bio-immobilization via covalent binding or entrapment. However, inorganic materials such as glasses and ceramics would offer significant advantages over these polymers. They exhibit better mechanical strength together with improved chemical and thermal stability. Moreover they do not swell in most solvents, preventing the leaching of entrapped biomolecules.

Combining fragile biomolecules with tough materials then becomes an innovative research field for materials scientists, but glasses and ceramics are made at high temperature, often above 1000 °C. Such harsh conditions are not compatible with fragile biomolecules. Enzymes or cells can only be grafted onto the surface of porous solids and the inorganic carrier must be activated before immobilization via silanization. It is then treated with a coupling reagent, usually an organosilane such as γ-aminopropyltriethoxysilane, $H_2N{-}(CH_2)_3{-}Si{-}(OEt)_3$, that bears an organic functional group at one end and an alkoxy-silyl group at the other end (Fig. 11.1) [3]. This method has been successfully scaled up from laboratory to industry and several processes have now been commercialized for more than twenty years [4].

Gómez-Romero: Organic-Inorganic Materials. Jacques Livage

ISBN: 3-527-30484-3

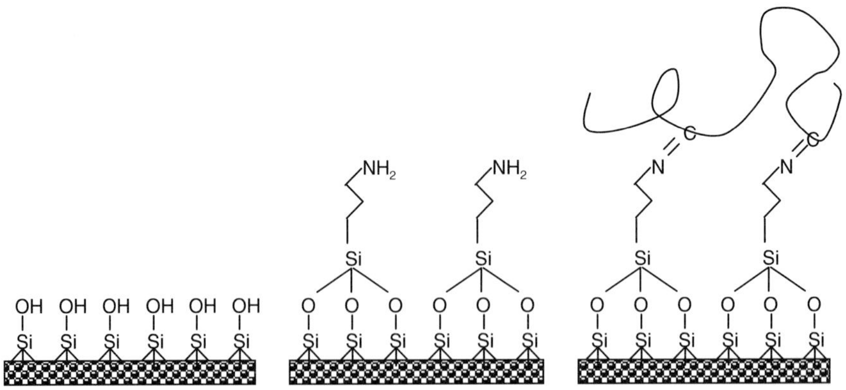

Fig. 11.1 Enzyme grafting on silica substrates. Glass surfaces (a) are first grafted with aminopropyltriethoxysilane (b), which can then bind the protein (c)

Surface attachment is currently performed onto porous glasses and ceramics. Such carriers have to be highly porous and the recent development of mesoporous silica opens new possibilities. These materials are synthesized via the condensation of molecular precursors such as $Si(OR)_4$ in the presence of surfactants. They exhibit high surface areas, above 1000 m^2/g, and mesopores ranging in size from 1.5 nm to several tens of nanometers [5, 6]. Enzymes can be immobilized within these pores via impregnation (Fig. 11.2). However, their interaction with the silica matrix is rather weak and the problem of enzyme desorption and leakage has still to be overcome [7–13]. Some authors take advantage of the slow release of adsorbed proteins for the realization of drug delivery systems [14–16].

Binding enzymes and cells onto the surface of inorganic substrates requires chemical modifications that may affect their bioactivity. Therefore physical entrapment would be a favored method for immobilization, particularly for enzymes having small substrates. However, this is not possible with ordinary glasses and ceramics

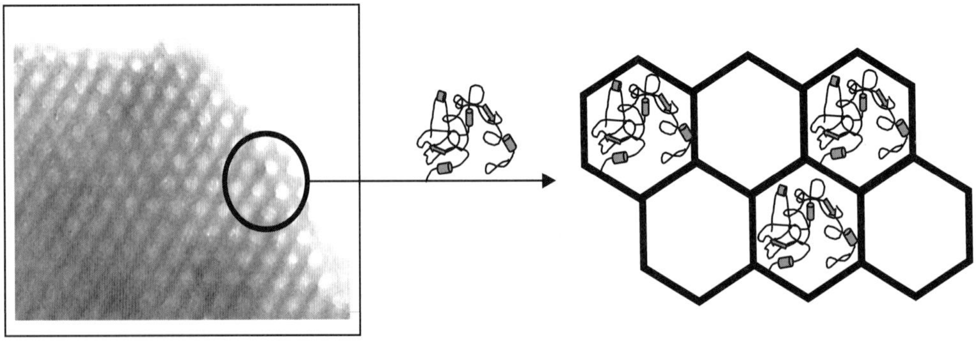

Fig.11.2 Enzyme immobilization within mesoporous hosts

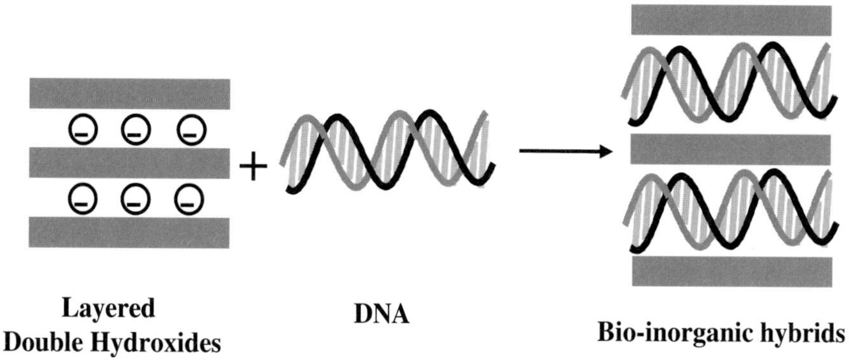

Fig.11.3 Intercalation of DNA in layered double hydroxydes (adapted from [21])

that have to be processed at high temperatures. It is not yet possible with mesoporous silica, since the synthetic routes to such materials still involve extreme pH, temperature treatment and surfactants that are known to induce cell membrane lysis.

Some authors report on the intercalation of biomolecules in layered double hydroxides (LDH). LDH can easily be synthesized by coprecipitation. They exhibit a brucite-like structure made of $M^{II}(OH)_2$ layers that can be positively charged via the partial substitution of the framework divalent cations with trivalent metal ions. Negatively charged amino acids can then be directly intercalated between the positively charged layers. They thus gain extra stabilization energy and LDH can play the role of a reservoir to protect intercalated amino acids [17,18]. Intercalated active compounds can be slowly released for drug delivery [19]. Very interesting experiments have been reported in which DNA was intercalated within LDH and used as carriers to favor its transfer into mammalian cells (Fig. 11.3) [20, 21].

Most significant advances in the field of bioactive inorganic hybrids have been made during the past decade using sol-gel glasses. The sol-gel route allows the low temperature synthesis of glasses and ceramics. It opens new possibilities in the field of bio-encapsulation. Enzymes and cells can now be mixed with molecular precursors in solution and trapped within the solid materials that grow around them.

11.2 Sol-gel Encapsulation

11.2.1 The Alkoxide Route

The traditional sol-gel process is based on the hydrolysis and condensation of alkoxide precursors $Si(OR)_4$ where R is an alkyl group such as methyl CH_3 (TMOS) or ethyl C_2H_5 (TEOS).

$$Si(OR)_4 + 2\,H_2O \Rightarrow Si(OH)_4 + 4\,ROH \tag{1}$$

$$Si(OH)_4 \Rightarrow SiO_2 + 2\,H_2O \tag{2}$$

Alkoxides are not miscible with water so that a common solvent, usually the parent alcohol, has to be added in order to have a clear precursor solution. Organic molecules can easily be trapped within sol-gel silica glasses, giving a wide range of hybrid organic-inorganic nanocomposites [22]. However, sol-gel chemistry is not mild enough for fragile biomolecules such as enzymes. Proteins are denatured by alcohol and have to be kept at a pH close to 7. The sol-gel process has then to be modified to fit with the requirements of biomolecules and encapsulation is currently performed in two steps [23].

1. Acid hydrolysis: In order to avoid alcohol, water is added directly to the alkoxide giving an emulsion that has to be vigorously shaken (often via sonication) for hydrolysis to take place. Some acid (HCl) is usually added to the water in order to increase hydrolysis rates. Alcohol released during hydrolysis behaves as a co-solvent, leading to a clear homogeneous solution after a few minutes.
2. Basic condensation around $pH \approx 7$: Proteins are kept in a buffered medium around $pH \approx 7$ and mixed with the acid aqueous solution of prehydrolyzed precursors $Si(OH)_4$. Basic catalysis around $pH \approx 7$ favors condensation and gelation occurs within a few minutes. A porous silica network is formed and biomolecules remain trapped within the growing oxide network. The pore size strongly depends on the sol-gel procedure (hydrolysis ratio, pH, aging, sonication etc.). It currently ranges between one and ten nanometers.

However, even in such conditions, alcohol is released during the hydrolysis step. Because of its high dielectric constant, methanol is less harmful than ethanol, therefore TMOS is usually taken as a precursor rather than TEOS. However, concentrations up to 8 M could be reached via the full hydrolysis of tetramethoxysilane $Si(OCH_3)_4$. Such an amount of alcohol can be harmful for enzymes and cells. It has been shown, for instance, that concentrations of methanol larger than 0.5 M lead to the lysis of *E. coli* bacteria [24].

Evaporation or distillation of the alcohol produced during hydrolysis appears as an efficient solution, as demonstrated by the gas-phase Biosil process [25]. However, silicon alkoxide precursors are usually not fully hydrolyzed during the first acid-catalyzed step, and alcohol can still be released during condensation and aging in the presence of the immobilized species.

The modification of the silicon alkoxide precursor may also take place *in situ* through esterification of its pre-hydrolyzed form [26, 27]. This strategy is all the more interesting as the alcohol group of alkoxide precursors, usually methanol or ethanol which are known to be cytotoxic, can be substituted by biocompatible polyols such as glycerol [26].

A more straightforward approach lies in the addition of the organic component to the reaction media. Synthetic (poly(vinyl alcohol) PVA, poly(ethyleneglycol) PEG, nafion, etc.) as well as natural (chitosan, alginate, gelatine, etc.) polymers have been used to form hybrid encapsulation matrices [28–33]. Apart from providing biocompatible media, such additives strongly modify the gel structure and its mechanical properties. For instance, adding PEG to a TEOS-based silica gel leads to hybrid materials with improved optical transmittance, better resistance to cracking and reduced pore surface polarity that enhances the enzymatic activity of entrapped lipases [30].

11.2.2
The Aqueous Route

The aqueous route has not yet often been used for bio-encapsulation despite the fact that it does not produce harmful by-products such as alcohol. It is based on the condensation of solute precursors via pH modification [34]. Ionic dissociation and hydrolysis occurs when a silicate salt is dissolved in water leading to more or less protonated silicate species $[H_nSiO_4]^{(4-n)-}$. In the case of sodium silicate Na_2SiO_3, a convenient commercial precursor for silica, a basic aqueous solution (pH ≈ 12) is obtained that contains Na^+ and $[SiO(OH)_3]^-$ species at pH ≈ 12.

As for the alkoxide route, a two-step procedure is also followed.

1. $[Si(OH)_4]$ neutral precursors are first formed by acidification of the sodium silicate solution. Protonation can be performed by adding an acid or, even better, via a proton exchange resin in order to remove sodium ions [35].
2. Condensation then occurs by mixing this solution with a suspension of enzymes in a buffer solution.

Colloidal silica particles can be added to increase silica concentration and get stronger gels [36]. Such sols then do not contain too many sodium counter-ions so

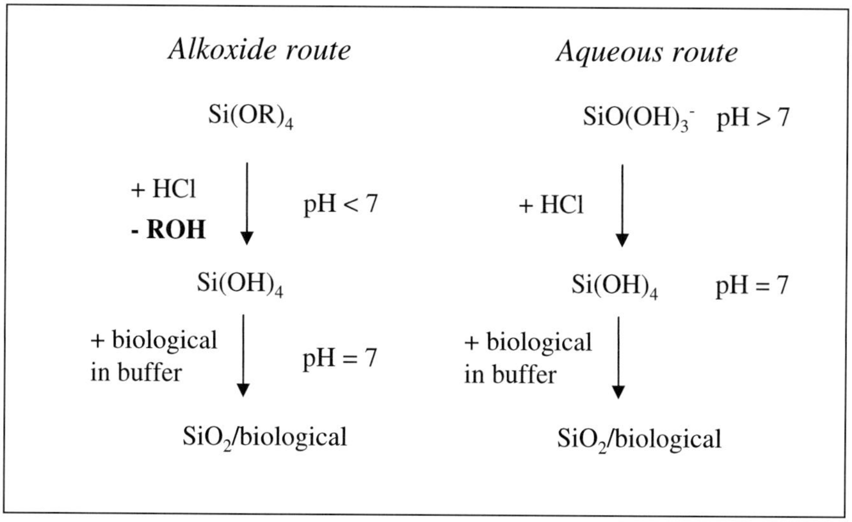

Fig. 11.4 Comparison of alkoxide and aqueous sol-gel routes for bioencapsulation

that pH decrease can be more easily obtained by adding HCl. A recent approach combines sodium silicate and silica sols as precursors [24]. This strategy affords a better control of reactant concentration and allows the design of an encapsulation process very similar to the traditional alkoxide pathway, thus providing a basis for comparison of both routes (Fig. 11.4).

11.3 Enzymes

Enzymes are biological catalysts that are responsible for the chemical reactions of living organisms. Their high specificity and huge catalytic power is due to the fact that the geometry of the active site can fit exactly that of the substrate according to the so-called "lock-and-key" model. Therefore even small changes in the enzyme conformation upon immobilization can drastically reduce their catalytic activity. Such catalysts could be very useful for industrial applications and the physical entrapment in silica gels would offer new possibilities for making biosensors and bioreactors. However, encapsulation could lead to the denaturation of the protein, and the pores of the sol-gel matrix have to be large enough to allow easy diffusion of the substrate and small enough to avoid the leaching of enzymes. Many enzymes have been trapped within sol-gel glasses and used for making biosensors or bioreactors [37–43].

In most cases, sol-gel encapsulation does not lead to denaturation during the formation of the silica network. Even after drying at room temperature, silica xerogels still contain enough water to provide a mainly aqueous environment so that the catalytic activity of entrapped enzymes may be close to that observed in aqueous suspensions. The kinetics of the reaction, limited by the diffusion of the substrate, are usually slower than with free enzymes. This is not a real drawback and the porous matrix may protect encapsulated proteins against denaturation. Environmental effects that otherwise destroy the catalytic activity may be greatly reduced by the limited diffusion of external reagents [44–46]. Moreover, the sol-gel matrix protects trapped enzymes against denaturation and leaching. For instance, the half-life of entrapped glucose oxidase or acid phosphatase was shown to be two orders of magnitude longer in gels than in solution [47]. Upon encapsulation, enzymes are trapped in a cage tailored to their size. The silica matrix constrains the motions of encapsulated proteins and may prevent structural deformations that would lead to irreversible denaturation [48, 49].

11.3.1 Glucose Biosensors

A large number of enzymes have already been trapped within sol-gel matrices but glucose oxidase (GOD) is by far the most studied enzyme. It is known to catalyze the oxidation of glucose by molecular oxygen and finds many applications in clinical analysis for the diagnosis of diabetes and in the food industry.

GOD has been extensively used in biosensors for the titration of glucose [50]. It provides a good example of the numerous possibilities offered by the sol-gel process. The oxidation of glucose by molecular oxygen gives gluconolactone and then gluconic acid as follows:

$$C_6H_{12}O_6 + O_2 = (GOD) \Rightarrow C_6H_{10}O_6 + H_2O_2$$

Glucose detection can be made via the consumption of O_2, the redox reaction at the active site of GOD or the formation of H_2O_2 (Fig. 11.5).

Sol-gel glasses are optically transparent so that the formation of hydrogen peroxide can be followed by optical measurements using the enzymatic oxidation of an organic dye catalyzed by a horseradish peroxidase (HRP). H_2O_2 molecules produced at the active site of GOD diffuse through the porous gel and react with the organic dye at the active site of HRP. The whole glass becomes colored when dipped into a glucose solution, showing that the enzymatic activity is distributed througout the gel and that the color change is not due to surface reactions only. Optical spectra are almost the same in the gel and in the solution, and a fair correlation is observed between optical density and the amount of glucose in the solution [51–53].

Optical detection is not very convenient for making reversible sensors. Therefore amperometric biosensors have also been made based on the electrochemical detection of redox reactions at the active site of GOD. However, due to steric hindrance, electrons cannot tunnel directly from the active site to the electrode and molecular mediators such as ferrocene have to be added to allow electron transfer. The reduced mediator is regenerated at the anode and the Faradaic current is proportional to the amount of converted glucose [54, 55] (Fig. 11.5b).

Such reactions can be performed with enzymes in silica gels, but electrochemical detection requires electronically conductive matrices [56]. Therefore carbon-ceramic composite electrodes (CCE), in which an enzyme-loaded carbon powder is mixed with the sol-gel solution, have been developed [57, 58]. These electrodes can be prepared in virtually any desired shape: thick films for disposable electrodes, micro-tips ($\approx 10\ \mu m$) [59] and disks or rods [60]. Hydrophobic electrodes have been made using hybrid silica precursors. Water penetration is limited to a very thin layer at the surface of the electrode that can be renewed by a simple polishing step [61, 62]. Printing inks have been developed in order to deposit CCE by screen-printing. They use ferrocene as a mediator and polyvinyl pyrrolidone (PVP) as a binder [63] (Fig. 11.5c).

A new kind of biosensor has recently been described which detects glucose in fruit juice, Coca-Cola™ and human blood serum. It is based on changes in enthalpy during the enzymatic oxidation of glucose recorded as a thermometric peak by a sensitive thermistor [64].

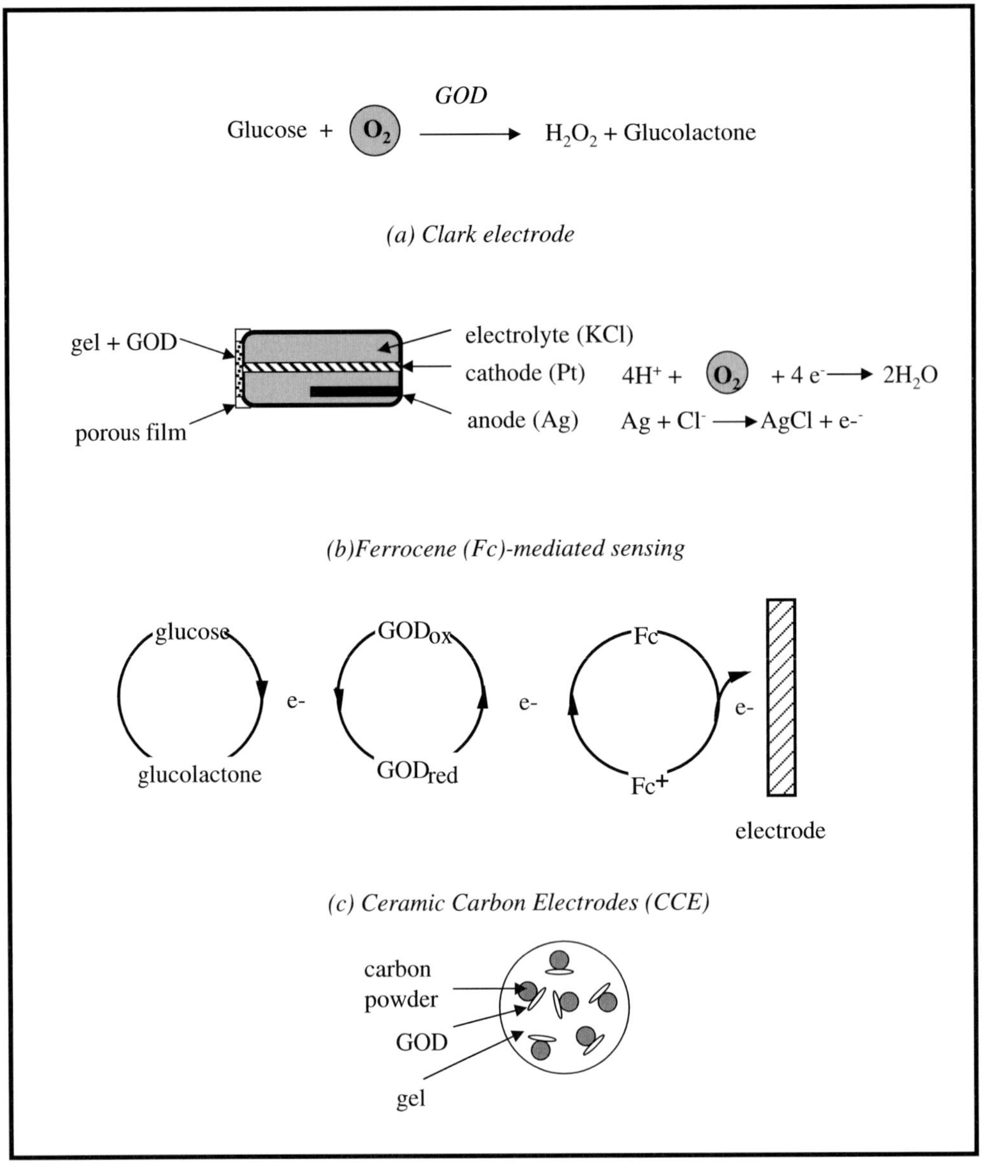

Fig.11.5 Glucose oxidase (GOD)-based electrochemical biosensors.
(a) Clark electrode measuring O_2 consumption during glucose oxidation.
(b) Glucose oxidation mediation using ferrocene.
(c) Carbon-ceramic composite electrodes (adapted from [42])

11.3.2
Bioreactors, Lipases

Most enzymes work in aqueous media. They require at least one water layer to retain their active configuration. A milestone for the use of enzymes in organic chemistry was the discovery that some of them retain their catalytic activity in non-aqueous media. This is the case for lipases that are known to catalyze hydrolysis-esterification reactions:

$$R\text{–}COOR' + H_2O \Leftrightarrow R\text{–}COOH + R'OH$$

In aqueous media they hydrolyze fats and oils into fatty acids and glycerol whereas in organic media esterification or transesterification reactions occur. Lipases find a lot of applications in synthetic organic chemistry, food industries and oil processing. However, lipases are not soluble in organic solvents; freeze-dried powders are currently used and their catalytic activity in organic solvents is often rather low. At least one hydration layer is required for enzymes to retain their bioactivity. Immobilization methods have then to be found that could increase the activity of lipases in organic solvents.

Actually it has been shown that confinement within silica gels could provide a chemical surrounding that favours the enzymatic activity of lipases [65]. Lipases are interfacially activated enzymes. In an aqueous solution, an amphiphilic peptidic loop covers the active site just like a lid. At a lipid/water interface, this lid undergoes a conformational rearrangement that renders the active site accessible to the substrate [66] (Fig. 11.6). The hydrophilic-hydrophobic nature of silica pores surrounding trapped enzymes has then to be carefully controlled in order to keep the active site accessible. The activity of lipases trapped within a hydrophilic silica

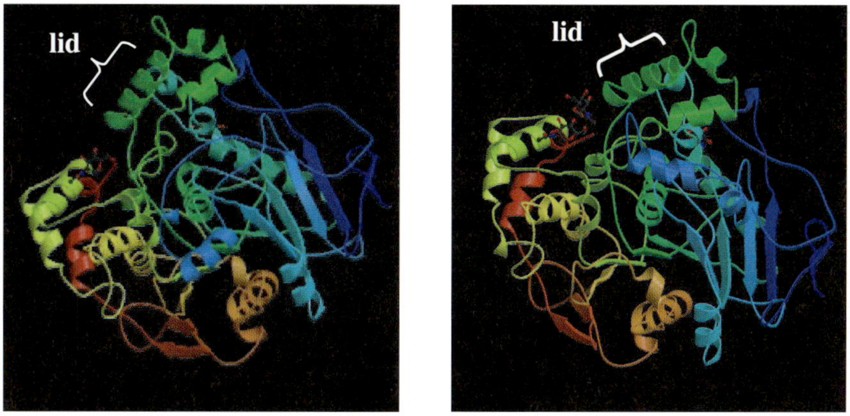

Fig. 11.6 Interfacial activation of lipases. Comparison of the closed (left) and open (right) conformations of lipase. Note the tipping of the lid releasing the enzyme active site

matrix is very poor, only 5% of the activity of free lipases [37]. Adding PVA increases the activity to 30%, but lipases can be almost 100 times more active when trapped within a hybrid silica matrix. The co-hydrolysis of $Si(OMe)_4$ and $RSi(OMe)_3$ precursors provides alkyl groups that offer a lipophilic environment that could interact with the active site of lipases and increase their catalytic activity. It was observed that for a TMOS: $R–Si(OCH_3)_3$ ratio of 1:1, the activity increases with R chain length. This was attributed to the possible conformational changes of lipase in hydrophobic media. Such changes allow uncovering of the enzyme active site, thus favoring its catalytic efficiency. Therefore, the increase of lipase activity is correlated with the enhancement of the hydrophobicity of the pores' surface by longer R chains. Such entrapped lipases are now commercially available.

11.4 Antibody-based Affinity Biosensors

The recognition of antigens by antibodies is highly specific and sensitive. Antibodies are large macromolecules, several hundreds of kDa in molecular weight, belonging to the immunoglobulin family. They are produced by the immune system to protect the body against external aggression by bacteria or viruses. They are widely used as immunosensors for medical applications, but they could also be used for the detection of chemicals. Small molecules (haptens) do not exhibit antigenic properties unless they are bound to a macromolecular carrier such as bovine serum albumin (BSA). When injected into a mammal (e.g. mouse, rabbit, goat) they induce an immune response and stimulate the production of antibodies by plasma cells (Fig. 11.7a). These antibodies are then able to recognize specifically the target analyte (hapten) and can be used as immunosensors for the detection of traces of chemical species.

Sol-gel encapsulated antibodies have been used for the detection of various chemicals such as fluorescein [67, 68], dinitrobenzene (DNB) [69], dinitrophenyl (DNP) [70]. The high binding constants of antibodies to antigens make them ideal for chromatographic separation. Immunoaffinity purification is a form of chromatography in which antibodies are used to bind a specific molecule from a mixture of compounds. The bound molecule is then released with an eluting solution (Fig. 11.7b).

Sol-gel immunosensors have been developed for the analysis of pesticides [71]. The extensive use of pesticides may cause environmental problems, mainly for drinking-water resources. The maximum level allowed for a single pesticide by European rules is 0.1 mg L^{-1} (0.1 ppb). Highly sensitive detection methods are then required and immunochemical reactions are good candidates for pesticide analysis. Sol-gel detection was proposed for atrazine, a widely spread herbicide. Monoclonal anti-atrazine antibodies produced by a mouse were trapped within a sol-gel silica chromatographic column. Nanograms of an atrazine solution were then poured through the column and up to 90% of the whole atrazine was fixed by trapped antibodies. The reaction is reversible and atrazine molecules can be eluted from the gel

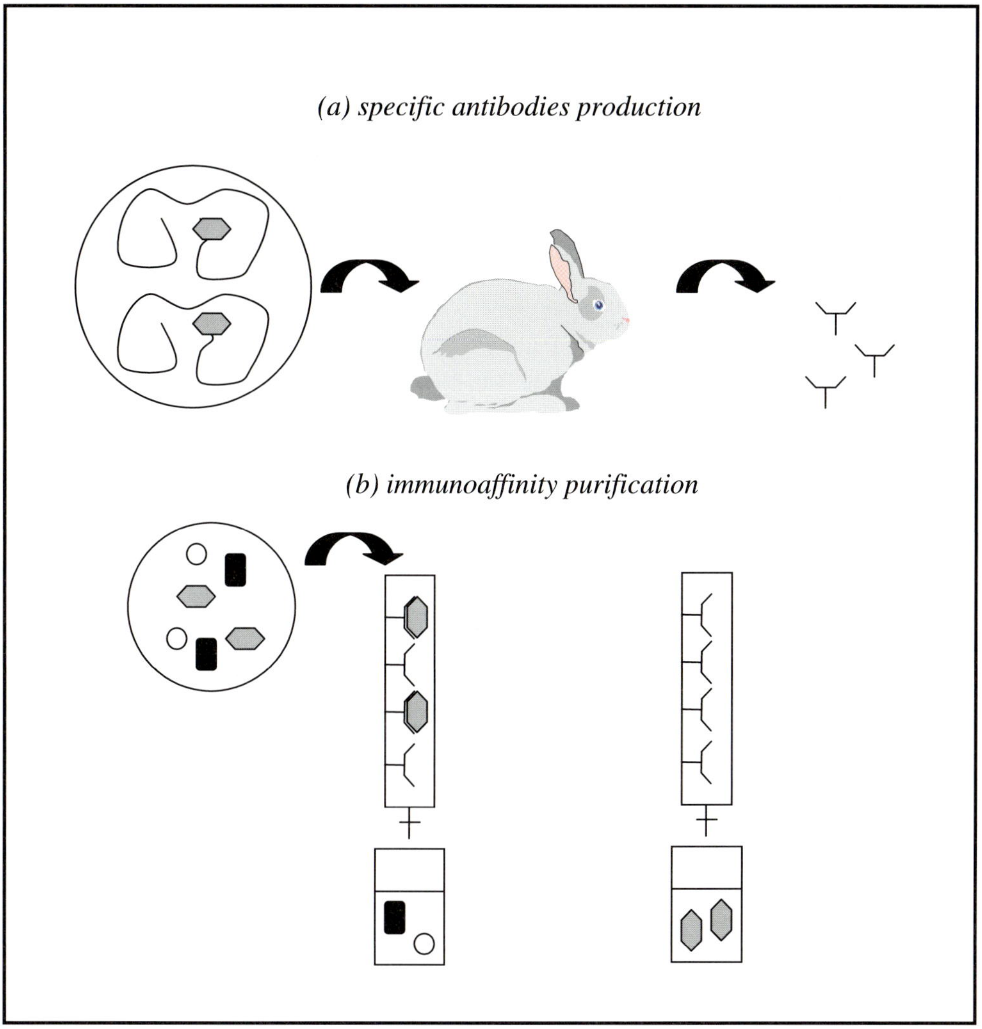

Fig. 11.7 Antibodies-based biosensors.
(a) Haptens bound to large proteins are injected in a rabbit producing corresponding antibodies.
(b) Entrapped antibodies are packed in a chromatography column and specifically retain corresponding haptens that can be recovered through elution

and titrated by the classical Enzyme Linked ImmunoSorbent Assays (ELISA) [72, 73].

More recently, optical detectors for trinitrotoluene (TNT), a commonly used explosive, have been described. Anti-TNT antibodies, trapped within optically transparent sol-gel glasses, retain their ability to bind TNT. Immunoassays, performed with a fluorescent probe, were able to detect traces of the order of ppm and even to differentiate between TNT and TNB (trinitrobenzene) [74].

Antibodies have also been used for electrochemical immunosensors. Antigen-containing thick-film electrodes were made by screen-printing with an alkaline phosphatase-labelled antibody. Encapsulated antigens are readily accessible to the labelled antibodies and the dispersed graphite powder offers facile detection of the product via cyclic voltammetry [75].

11.5 Whole Cells

11.5.1 Yeast and Plant Cells

The immobilization of cells is another challenge for biotechnology. Their metabolic activity can be used in a large variety of processes for the food industry, waste treatment, production of chemicals or drugs and even cell transplantation. As for enzymes or antibodies, whole cells can be attached to the surface of a carrier or trapped within a porous matrix [76].

Silica gels have been shown to be biocompatible. They can be used as a support for the culture of bacteria [77] or fungi [78]. The high porosity of silica gels favours water retention and nutrient diffusion.

The first sol-gel entrapment of whole cells was published more than ten years ago by G. Carturan *et al.*, with *Saccharomyces cerevisiae* cells [79]. These yeasts are involved in the conversion of sugar and carbohydrates into ethyl alcohol and CO_2 and are currently employed in the fermentation of beers and the raising of bread. They have been shown to exhibit almost the same activity in silica and in a solution. Silica gels can be reprocessed by solvent exchange in water to remove fermentation by-products and then impregnated with fresh food. No change was observed, suggesting that bioactivity can be sustained for repeated cycles during several months [80, 81]. Encapsulated yeast cells have also been used for environmental protection [82]. They fix heavy metals (Hg^{II}, Cd^{II}. etc.) from aqueous solutions [83, 84]. Entrapped plant cells have been used for the continuous production of enzymes (invertase). The bioactivity was monitored for two months, giving a stable productivity per biomass unit. Highly pure enzyme is produced that can be directly separated from the circulating solution by standard procedures [85].

Plant cells have also been immobilized by the sol-gel process. They have been shown to retain their ability to produce metabolites [86, 87].

11.5.2 Bacteria

Whole cell bacteria have recently been entrapped within sol-gel silica matrices in order to perform bio-catalytic experiments. *Escherichia coli* have been chosen as a model for sol-gel encapsulation. They divide every 20 minutes and large quantities of biomass can easily be produced. *E. coli* were induced in order to express the

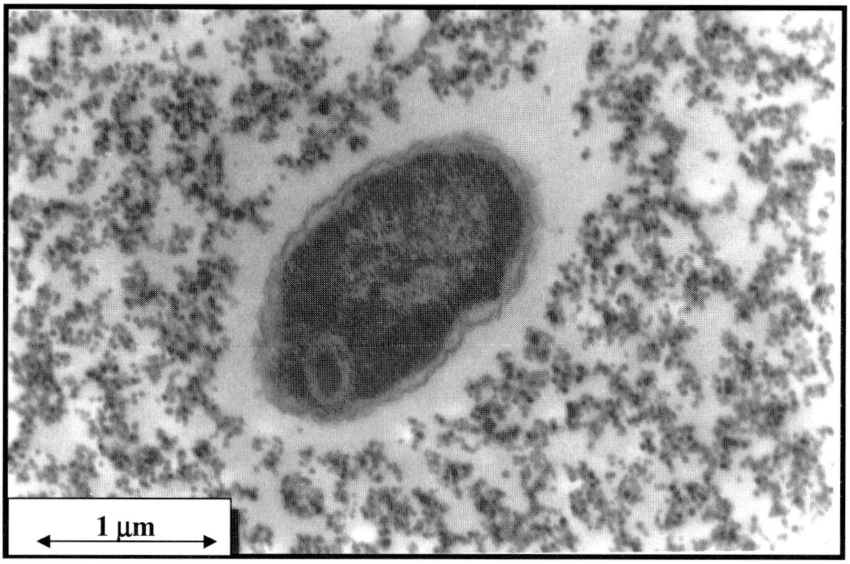

Fig. 11.8 *E. coli* bacterium immobilized in a silica matrix

β-galactosidase enzyme. Encapsulation is performed with bacteria extracted from the culture medium, suspended in a phosphate buffer solution and then mixed with a prehydrolyzed $Si(OH)_4$ solution at pH ≈ 7. A gel is formed around the cells within a few minutes. Transmission electron microscopy shows that bacteria are randomly dispersed within the silica matrix and that their cellular organization is preserved (Fig. 11.8) [88].

The β-galactosidase activity of entrapped *E. coli* follows a typical Michaelis behavior. The kinetics of the reaction increase rapidly with the substrate concentration and reach a maximum value $V_{max} = 1.1$ mmol/mn. The corresponding Michaelis constant is $K_M = 0.25$ mM. It is slightly better than for free bacteria ($V_{max} \approx 0.9$ mmol/mn, $K_M \approx 0.45$ mM). This increase in catalytic activity has been ascribed to some lysis of the membrane of bacteria cells allowing a better diffusion of the reactants [89]. Experiments have been performed in order to compare the behavior of bacteria trapped in alkoxide or aqueous silica gels with bacteria suspended in a buffer solution. In all cases the -galactosidase activity of bacteria increases with aging at room temperature. However, this effect is smaller for TMOS gels than for non-encapsulated bacteria. The effect increases in the order free bacteria > bacteria in alkoxide gel > bacteria in aqueous gel, suggesting that sol-gel entrapment may partially prevent the lysis of bacteria [24].

Escherichia coli strains genetically engineered to express stress-specific luminescent proteins have also been encapsulated in silica gel coatings [90]. Such an approach not only enables the detection of external chemicals for biosensing but also allows the study of the stress induced by the encapsulation process [91].

11.5.3 Biomedical Applications

11.5.3.1 Immunoassays in Sol-gel Matrices

The immune system is based on the specific recognition of antigens by antibodies. We have seen that antibodies can be trapped within sol-gel matrices in order to build immunoaffinity devices. They can also be used for medical applications in order to make blood tests. Such experiments have been performed using *Leishmania donovani infantum* promastigotes as antigens. These parasitic protozoa were trapped within a silica gel and used to detect the presence of specific antibodies in the serum of humans and dogs. This reaction was followed using the so-called Enzyme Linked ImmunoSorbent Assays (ELISA) which provide a very sensitive method for detecting antibodies. Tests were performed using the standard polystyrene microtitre plates that contain 96 wells in which 96 different samples can be studied simultaneously. Encapsulation is performed directly inside these wells by mixing a suspension of *Leishmania* cells with a pre-hydrolyzed silica sol. Gelation occurs inside the wells within about five minutes. Sera are then poured into the wells and the antigen-antibody association is detected via an enzyme conjugate with an optical probe. Optical density measurements show a clear-cut difference between negative and positive sera, even for highly diluted sera [92, 93].

11.5.3.2 Cell Transplantation

Cell transplantation is a promising technique in order to avoid the usual problems of organ transplantation. However, foreign cells have to fight against immune rejection. Therefore living cells are encapsulated within porous membranes that shield the cells from immune attack. Organic hydrogels and biopolymers are currently developed for such applications but recent work shows that sol-gel silica could also be used for cell transplantation. Several papers have been published recently describing the encapsulation of single cells or cell aggregates within silica capsules or alginate microspheres coated by a silica layer [94, 95].

The islets of Langerhans are involved in the metabolism of glucose and the production of insulin. *In vitro* experiments show that trapped cells still respond to glucose addition. *In vivo* experiments have even been performed via the transplantation of encapsulated islets into a diabetic mouse. The cells continue to function and to mimic a natural organ. Urinary excretion of glucose falls to almost zero while blood sugar and blood insulin concentrations remain normal. The fine porosity of the gel protects transplanted islets against antibody aggression but permits nutrients to reach the cell and by-products to escape. After one month of transplantation, the surgically removed transplant showed no evidence of fibrosis. Such encapsulated cells are under study at Solgene Therapeutics LLC in the USA [96].

Another process, called "biosil", was developed in Italy for the encapsulation of swine hepatocytes and rat liver. In this process, living cells are deposited onto a substrate and then partially covered by a silica film via the gas phase. This method is used for extracorporeal bioartificial organs. They have already functioned as a bridge to liver transplantation in patients with fulminant hepatic failure [97]. Biogel transplants, if viable for extended lengths of time, could emerge as a viable treatment for diseases such as diabetes.

11.6 The Future of Sol-gel Bioencapsulation

Sol-gel glasses offer several advantages compared to organic polymeric matrices, which are nowadays widely used in biotechnology. Hard porous glasses do not swell in water and protect biospecies against external aggression (pH, temperature, solvents, antibodies etc.). The bioactivity of trapped species can even be enhanced via the chemical control of the sol-gel matrix, as shown for lipases, which are now commercially available.

Preserving the viability of cells within sol-gel matrices appears to be a real challenge. Recent studies indicate that encapsulated bacteria may survive the gelation procedure and remain able to continue normal metabolic activity within the gel matrix. The anaerobic formation of metabolites was observed in a gel containing sulfate-reducing bacteria, providing evidence of the metabolic conversion of lactate to acetate [36]. Recent experiments show that trapped *E. coli* bacteria retain their metabolic activity toward glycolysis for several weeks [98]. This might be one of the major advances of sol-gel chemistry that opens up new possibilities for the design of whole-cell-based bioreactors and biosensors [99, 100].

Sol-gel encapsulation offers an easy and generic immobilization procedure that could compete with more usual encapsulation processes. However, many problems have still to be overcome before efficient devices can be obtained. The size of the pores is still difficult to control. Aging phenomena have almost never been studied [101]. Checking the bioactivity of encapsulated species once or even for some days is not enough. Long-term stability and reproducibility have still to be obtained on a routine basis. Progress still has to be made in order to improve the viability of cells within silica gels, but these first results are really promising and sol-gel encapsulation could provide an alternative issue for bio-immobilization.

References

1 L. L. Hench, *Biomaterials* **1998**, *19*, 1419–1423.
2 M. Vallet-Regi, *J. Chem. Soc. Dalton Trans.* **2001**, *2*, 97–108.
3 H. Weetall, *TIBTECH* **1985**, *3*, 276–280.
4 *Immobilization of enzymes and cells*, ed G. F. Bickerstaff, *Methods in Biotechnology*, Human Press Inc., Totowa 1997.
5 C. T. Kresge, M. E. Leonowicz, W. J. Roth, J. C. Vartuli, J. S. Beck, *Nature* **1992**, *359*, 710–712.
6 U. Ciesla, F. Schüth, *Microporous Mesoporous Mater.* **1999**, *27*, 131–149.
7 J. Felipe Diaz, K. J. Balkus, Jr., *J. Mol. Catal. B: Enzymatic* **1996**, *2*, 115–126.
8 M. E. Gimon-Kinsel, V. L. Jimenez, L. Washmon, K. J. Balkus, Jr., *Stud. Surf. Sci. Catal.* **1998**, *117*, 373–380.
9 H. Takahashi, B. Li, T. Sasaki, C. Miyazaki, T. Kajino, S. Inigaki, *Microporous Mesoporous Mater.* **2001**, *44–45*, 755–762.

10 H. H. P. Yiu, P. A. Wright, N. P. Botting, *Microporous Mesoporous Mater.* **2001**, *44–45*, 763–768.

11 H. H. P. Yiu, P. A. Wright, N. P. Botting, *J. Mol. Catal. B: Enzymatic* **2001**, *15*, 81–92.

12 H. Furukawa, T. Watanabe, K. Kuroda, *Chem. Commun.* **2001**, 2002–2003.

13 J. Deree, E. Magner, J. G. Wall, B. K. Hodnett, *Chem. Commun.* **2001**, 465–466.

14 H. Hata, S. Saeki, T. Kimura, Y. Sugahara, K. Kuroda, *Chem. Mater.* **1999**, *11*, 1110–1119.

15 Y.-J. Han, G. D. Stucky, A. Butler, *J. Am. Chem. Soc.* **1999**, *121*, 9897–9898.

16 M. Vallet-Regi, A. Ramila, R. P. del Real, J. Pérez-Pariente, *Chem. Mater.* **2001**, *13*, 308–311.

17 N. T. Whilton, P. J. Vickers, S. Mann, *J. Mater. Chem.* **1997**, *7*, 1623–1629.

18 S. Aisawa, S. Takahashi, W. Ogasawara, Y. Umetsu, E. Narita, *J. Solid State Chem.* **2001**, *162*, 52–62.

19 A. I. Khan, L. Lei, A. J. Norquist, D. O'Hare, *Chem. Commun.* **2001**, 2342–2343.

20 J.-H. Choy, S.-Y. Kwak, J.-S. Park, Y.-J. Jeong, J. Portier, *J. Am. Chem; Soc.* **1999**, *121*, 1399–1400.

21 J.-H. Choy, S.-Y. Kwak, Y.-J. Jeong, J.-S. Park, *Angew. Chem. Int. Ed.* **2000**, *39*, 4042–4045.

22 *Special Issue on Hybrid Materials*, ed D. Loy, *Mater. Res. Soc. Bull.* **2001**, *26*, 364–408.

23 L. M. Ellerby, C. R. Nishida, F. Nishida, S. A. Yamanaka, B. Dunn, J.S. Valentine, J. I. Zink, *Science* **1992**, *255*, 1113–1115.

24 A. Coiffier, T. Coradin, C. Roux, O. M. M. Bouvet, J. Livage, *J. Mater. Chem.* **2001**, *11*, 2039–2044.

25 G. Carturan, R. Dal Monte, M. Muraca, *Encapsulation of Viable Animal Cells for Hybrid Bioartificial Organs*, PCT International Application, EP no. 9602265, May 28, **1996**.

26 I. Gill, A. Ballesteros, *J. Am. Chem. Soc.* **1998**, *120*, 8587–8598.

27 I. Gill, E. Pastor, A. Ballesteros, *J. Am. Chem. Soc.* **1999**, *121*, 9487–9496.

28 Y. Miao, S. N. Tan, *Anal. Chim. Acta* **2001**, *437*, 87–93.

29 A. Pierre, P. Buisson, *J. Mol. Catal. B: Enzymatic* **2001**, *11*, 639–647.

30 T. Keeling-Tucker, M. Rakic, C. Spong, J. D. Brennan, *Chem. Mater.* **2000**, *12*, 3695–3704.

31 K. Nakane, T. Ogihara, N. Ogata, Y. Kurokawa, *J. Appl. Polym. Sci.* **2001**, *81*, 2084–2088.

32 A. Kros, M. Gerritsen, V. S. I. Sprakel, N. A. J. M. Sommerdjik, J. A. Jansen, R. J. M. Nolte, *Sensors and Actuators B* **2001**, *81*, 68–75.

33 I. Brasack, H. Bottcher, U. Hempel, *J. Sol-Gel Sci. Techn.* **2000**, *19*, 479–482.

34 J.-P. Jolivet, *Metal Oxide Chemistry and Synthesis*, Wiley, Chichester, 2000.

35 R. B. Bhatia, C. J. Brinker, A. K. Gupta, A. K. Singh, *Chem. Mater.* **2000**, *12*, 2434–2441.

36 K. S. Finnie, J. R. Bartlett, J. L. Woolfrey, *J. Mater. Chem.* **2000**, *10*, 1099–2003.

37 D. Avnir, S. Braun, O. Lev, M. Ottolenghi, *Chem. Mater.* **1994**, *6*, 1605–1614.

38 J. I. Zink, J. S. Valentine, B. Dunn, *New J. Chem.* **1994**, *18*, 1109–1115.

39 J. Livage, *C. R. Acad. Sci. Paris IIb* **1996**, *322*, 417–427.

40 I. Gill, A. Ballesteros, *TIBTECH* **2000**, *18*, 282–296.

41 H. Böttcher, *J. Prakt. Chem.* **2000**, *342*, 427–438.

42 J. Livage, T. Coradin, C. Roux, *J. Phys.: Condens. Matter* **2001**, *13*, R673–R691.

43 I. Gill, *Chem. Mater.* **2001**, *13*, 3404–3421.

44 C. R. Lloyd, E. M. Eyring, *Langmuir* **2000**, *16*, 9092–9094.

45 A. K. Williams, J. T. Hupp, *J. Am. Chem. Soc.* **1998**, *120*, 4366–4371.

46 U. Künzelmann, H. Böttcher, *Sensors and Actuators B* **1997**, *39*, 222–228.

47 S. Sheltzer, S. Rappoport, D. Avnir, M. Ottolenghi, S. Braun, *Biotechnol. Appl. Biochem.* **1992**, *15*, 227–235.

48 L. Zheng, J.D. Brennan, *Analyst* **1998**, *123*, 1735–1744.

49 T. K. Das, I. Khan, D. L. Rousseau, J. M. Friedman, *J. Am. Chem Soc.* **1998**, *120*, 10268–10269.

50 R. Wilson, A. P. F. Turner, *Biosensors & Bioelectronics* **1992**, *7*, 165–185.

51 S. Braun, S. Shtelzer, S. Rappoport, D. Avnir, M. Ottolenghi, *J. Non-Cryst. Solids* **1992**, *147–148*, 739–743.

52 S. A. Yamanaka, F. Nishida, L. M. Ellerby, C. R. Nishida, B. Dunn, J. S. Valentine, J. I. Zink, *Chem. Mater.* **1992**, *4*, 495–497.

53 S. Sheltzer, S. Braun, *Biotechnol. Appl. Biochem.* **1994**, *19*, 293–305.

54 P. Audebert, C. Demaille, C. Sanchez, *Chem. Mater.* **1993**, *5*, 911–913.

55 P. Audebert, C. Sanchez, *J. Sol-Gel Sci. Techn.* **1994**, *2*, 809–812.

56 I. Willner, E. Katz, *Angew. Chem. Int. Ed.* **2000**, *39*, 1180–1218.

57 J. Wang, D. S. Park, P. V. A. Pamidi, *J. Electroanalyt. Chem.* **1997**, *434*, 185–189.

58 S. Sampath, O. Lev, *Anal. Chem.* **1996**, *68*, 2015–2021.

59 G. Gun, M. Tsionsky, O. Lev, *Anal. Chim. Acta* **1994**, *294*, 261–270.

60 S. Sampath, I. Pankratov, J. Gun, O. Lev, *J. Sol-Gel Sci. Techn.* **1996**, *7*, 123–128.

61 L. Kuselman, B. K. Losefzon, O. Lev, *Anal. Chim. Acta* **1992**, *256*, 65–68.

62 J. Li, L. S. Chia, N. K. Goh, S. N. Tan, *J. Electroanalyt. Chem.* **1999**, *460*, 234–241.

63 R. Nagata, K. Yokoyama, H. Durliat, M. Comtat, S. A. Clark, I. Karube, *Electroanalysis* **1995**, *7*, 1027–1031.

64 K. Ramanathan, B. R. Jönsson, B. Danielsson, *Anal. Chim. Acta* **2001**, *427*, 1–10.

65 M. T. Reetz, *Adv. Mater.* **1997**, *9*, 943–954.

66 R. D. Schmid, R. Verger, *Angew. Chem. Int. Ed.* **1998**, *37*, 1608–1633.

67 R. Wang, U. Narang, P. N. Prassad, F. V. Bright, *Anal. Chem.* **1993**, *65*, 2671–2675.

68 J. D. Jordan, R. A. Dunbar, F. V. Bright, *Anal. Chim. Acta* **1996**, *332*, 83–91.

69 N. Aharonson, M. Altstein, G. Avidan, D. Avnir, A. Bronshtein, A. Lewis, K. Lieberman, M. Ottolenghi, Y. Polevaya, C. Rottman, J. Samuel, S. Shalom, A. Strinkowski, A. Turnaiansky, *Mater. Res. Soc. Symp. Proc.* **1994**, *346*, 519–530.

70 A. Bronshtein, N. Aharonson, A. Turniansky, M. Altstein, *Chem. Mater.* **2000**, *12*, 2050–2058.

71 B. Hock, A. Dankwardt, K. Kramer, A. Marx, *Anal. Chim. Acta* **1995**, *311*, 393–405.

72 A. Turniansky, D. Avnir, A. Bronshtein, N. Aharonson, M. Altstein, *J. Sol-Gel Sci. Techn.* **1996**, *7*, 135–143.

73 A. Bronshtein, N. Aharonson, A. Turniansky, M. Altstein, *Chem. Mater.* **1997**, *9*, 2632–2639.

74 E. H. Lan, B. Dunn, J. I. Zink, *Chem. Mater.* **2000**, *12*, 1874–1878.

75 J. Wang, P. V. A. Pamidi, K. R. Rogers, *Anal. Chem.* **1998**, *70*, 1171–1175.

76 *Immobilized living cell systems*, eds R. G. Willaert, G. V. Baron, L. De Backer, Wiley, Chichester, 1996.

77 R. Armon, J. Starosvetzky, I. Saad, *J. Sol-Gel Sci. Techn.* **2000**, *19*, 289–292.

78 M. R. Peralta-Perez, G. Saucedo-Castaneda, M. Gutierrez-Rojas, A. Campero, *J. Sol-Gel Sci. Techn.* **2001**, *20*, 105–110.

79 G. Carturan, R. Campostrini, S. Dirè, V. Scardi, E. De Alteris, *J. Mol. Cat.* **1989**, *57*, L13-L16.

80 M. Uo, K. Yamashita, M. Suzuki, E. Tamiya, I. Karube, A. Makishima, *J. Ceram. Soc. Jap.* **1992**, *100*, 426–429.

81 E. J. A. Pope, *J. Sol-Gel Sci. Techn.* **1995**, *4*, 225–230.

82 G. M. Gadd, C. White, *TIBTECH* **1993**, *11*, 353–359.

83 M. Al-Saraj, M. S. Abdel-Latif, I. El-Nahal, R. Baraka, *J. Non-Cryst. Solids* **1999**, *248*, 137–140.

84 J. Szilva, G. Kuncova, M. Patzak, P. Dostalek, *J. Sol-Gel Sci. Techn.* **1998**, *13*, 289–294.

85 G. Pressi, R. Dal Toso, R. Dal Monte, G. Carturan, *J. Sol-Gel Sci. Techn.*, in press.

86 R. Campostrini, G. Carturan, R. Caniato, A. Piovan, R. Filippini, G. Innocenti, E. M. Cappelletti, *J. Sol-Gel Sci. Techn.* **1996**, *7*, 87–98.

87 G. Carturan, R. Dal Monte, G. Pressi, P. Verza, *J. Sol-Gel Sci. Techn.* **1998**, *13*, 273–276.

88 S. Fennouh, S. Guyon, C. Jourdat, J. Livage, C. Roux, *C. R. Acad. Sci. Paris IIc*, **1999**, 625–630.

89 S. Fennouh, S. Guyon, C. Jourdat, J. Livage, C. Roux, *J. Sol-Gel Sci. Techn.* **2000**, *19*, 647–649.

90 J. R. Premkumar, O. Lev, R. Rosen, S. Belkin, *Adv. Mater.* **2001**, *13*, 1773–1775.

91 J. R. Premkumar, O. Lev, R. S. Marks, B. Polyak, R. Rosen, S. Belkin, *Talanta* **2001**, *55*, 1029–1038.

92 J.-Y. Barreau, J. M. Da Costa, I. Desportes, J. Livage, L. Monjour, M. Gentilini, *C. R. Acad. Sci. Paris III* **1994**, *317*, 653–657.

93 J. Livage, C. Roux, J. M. Da Costa, I. Desportes, J.-F. Quinson, *J. Sol-Gel Sci. Techn.* **1996**, *7*, 45–52.

94 E. J. A. Pope, K. Braun, C. M. Peterson, *J. Sol-Gel Sci. Techn. 1997*, **8**, 635–639.

95 G. Carturan, G. Dellagiacoma, M. Rossi, R. Dal Monte, M. Muraca, *Sol-Gel Optics IV, SPIE Proc.* **1997**, *3136*, 366–369.

96 K. P. Peterson, C. M. Peterson, E. J. A. Pope, *Proc. Soc. Exp. Bio. Med.* **1998**, *218*, 365–369.

97 R. Dal Monte, *J. Sol-Gel Sci. Techn.* **2000**, *18*, 291–294.

98 N. Nassif, C. Roux, T. Coradin, J. Livage, O. Bouvet, unpublished results.

99 D. A. Stenger, G. W. Gross, E. W. Keefer, K. M. Shaffer, J. D. Andreadis, W. Ma, J. J. Pancrazio, *TIBTECH* **2001**, *19*, 304–309.

100 T. Klaus-Joerger, R. Joerger, E. Olsson, C.-G. Granqvist, *TIBTECH* **2001**, *19*, 15–20.

101 K. K. Flora, J. D. Brennan, *Chem. Mater.* **2001**, *13*, 4170–4179.

Index

p